Advanced Structured Materials

Volume 92

Series editors

Andreas Öchsner, Faculty of Mechanical Engineering, Esslingen University of Applied Sciences, Esslingen am Neckar, Germany
Lucas F. M. da Silva, Department of Mechanical Engineering, University of Porto, Porto, Portugal
Holm Altenbach, Lehrstuhl für Technische Mechanik, Institut für Mechanik, Fakultät für Maschinenbau, Otto-von-Guericke-Universität, Magdeburg, Germany

AF793483

Common engineering materials reach in many applications their limits and new developments are required to fulfil increasing demands on engineering materials. The performance of materials can be increased by combining different materials to achieve better properties than a single constituent or by shaping the material or constituents in a specific structure. The interaction between material and structure may arise on different length scales, such as micro-, meso- or macroscale, and offers possible applications in quite diverse fields.

This book series addresses the fundamental relationship between materials and their structure on the overall properties (e.g. mechanical, thermal, chemical or magnetic etc.) and applications.

The topics of *Advanced Structured Materials* include but are not limited to

- classical fibre-reinforced composites (e.g. class, carbon or Aramid reinforced plastics)
- metal matrix composites (MMCs)
- micro porous composites
- micro channel materials
- multilayered materials
- cellular materials (e.g. metallic or polymer foams, sponges, hollow sphere structures)
- porous materials
- truss structures
- nanocomposite materials
- biomaterials
- nano porous metals
- concrete
- coated materials
- smart materials

Advanced Structures Material is indexed in Google Scholar and Scopus.

More information about this series at http://www.springer.com/series/8611

Andreas Öchsner · Holm Altenbach
Editors

Engineering Design Applications

Springer

Editors
Andreas Öchsner
Faculty of Mechanical Engineering
Esslingen University of Applied Sciences
Esslingen am Neckar
Germany

Holm Altenbach
Lehrstuhl für Technische Mechanik, Institut für Mechanik, Fakultät für Maschinenbau
Otto-von-Guericke-Universität
Magdeburg
Germany

ISSN 1869-8433 ISSN 1869-8441 (electronic)
Advanced Structured Materials
ISBN 978-3-030-07702-0 ISBN 978-3-319-79005-3 (eBook)
https://doi.org/10.1007/978-3-319-79005-3

© Springer International Publishing AG, part of Springer Nature 2019
Softcover re-print of the Hardcover 1st edition 2019
This work is subject to copyright. All rights are reserved by the Publisher, whether the whole or part of the material is concerned, specifically the rights of translation, reprinting, reuse of illustrations, recitation, broadcasting, reproduction on microfilms or in any other physical way, and transmission or information storage and retrieval, electronic adaptation, computer software, or by similar or dissimilar methodology now known or hereafter developed.
The use of general descriptive names, registered names, trademarks, service marks, etc. in this publication does not imply, even in the absence of a specific statement, that such names are exempt from the relevant protective laws and regulations and therefore free for general use.
The publisher, the authors and the editors are safe to assume that the advice and information in this book are believed to be true and accurate at the date of publication. Neither the publisher nor the authors or the editors give a warranty, express or implied, with respect to the material contained herein or for any errors or omissions that may have been made. The publisher remains neutral with regard to jurisdictional claims in published maps and institutional affiliations.

Printed on acid-free paper

This Springer imprint is published by the registered company Springer International Publishing AG part of Springer Nature
The registered company address is: Gewerbestrasse 11, 6330 Cham, Switzerland

Preface

Different engineering disciplines such as mechanical, materials, computer and process engineering provide the foundation for the design and development of improved structures, materials and processes. The modern design cycle is characterized by an interaction of different disciplines and a strong shift to computer-based approaches where only a few experiments are performed for verification purposes. A major driver for this development is the increased demand for cost reduction, which is also connected to environmental demands. In the transportation industry (e.g. automotive or aerospace), this is connected with the demand for higher fuel efficiency, which is related to the operational costs and the lower harm for the environment. One way to fulfil such requirements is lighter structures and/or improved processes for energy conversion. Another emerging area is the interaction of classical engineering with the health and medical sector. This volume gives an overview on recent developments in the mentioned areas of modern engineering design application.

We would like to express our sincere appreciation to the representatives of Springer, in particular to Dr. Christoph Baumann, Senior Editor Engineering, who made this volume possible.

Esslingen am Neckar, Germany — Prof. Dr.-Ing. Andreas Öchsner, D.Sc.
Magdeburg, Germany — Prof. Dr.-Ing. habil. Dr. h. c. mult. Holm Altenbach

Contents

A Remedy for a Family of Dissipative, Unconditionally Stable Explicit Integration Methods

Shuenn-Yih Chang and Chiu-Li Huang

Abstract A family of dissipative structure-dependent integration methods has been developed for structural dynamics. In general, this family of methods can integrate favorable numerical properties together, such as the unconditional stability, explicit formulation, second-order accuracy and controllable numerical damping. Due to the unconditional stability and explicit formulation, it is very computationally efficient for solving general structural dynamics problems. However, an adverse high-frequency overshooting behavior in the steady-state response might be generally experienced if this family of methods is applied to carry out the time integration. The cause of this overshoot is explored. It seems that the incomplete formulation of the difference equation for the displacement increment is responsible for this overshoot. Consequently, this overshooting can be completely eliminated after adding an appropriate load-dependent term into the difference equation for the displacement increment.

Keywords High-frequency overshoot · Steady-state response · Unconditional stability · Explicit formulation · Time integration method · Dynamic analysis

1 Introduction

Structure-dependent integration methods are a new type of time integration methods. It is well known that the coefficients of the two difference equations are generally scalar constants for conventional integration methods while they will become functions of the initial structural properties and the step size for structure-dependent integration methods. The first structure-dependent integration method was developed by Chang in 2002 and it can simultaneously have the unconditional stability

S.-Y. Chang (✉)
National Taipei University of Technology, Taipei, Taiwan, ROC
e-mail: changsy@ntut.edu.tw

C.-L. Huang
Fu Jen Catholic University, New Taipei City, Taiwan, ROC
e-mail: 003951@mail.fju.edu.tw

© Springer International Publishing AG, part of Springer Nature 2019

A. Öchsner and H. Altenbach (eds.), *Engineering Design Applications*,
Advanced Structured Materials 92, https://doi.org/10.1007/978-3-319-79005-3_1

and explicit formulation together [1]. Some other structure-dependent integration methods were proposed to improve numerical properties later [2–8]. Recently, some dissipative structure-dependent integration methods were successfully developed by using an asymptotic equation of motion [9–13]. All of the family methods were enhanced with desirable high-frequency numerical damping. One of them is a free parameter controlled family method [9]. In fact, the parameter p is applied to govern the numerical properties. This family method is referred as Chang dissipative method and will be abbreviated as CDM for brevity herein.

In general, CDM is performed very well in the general dynamic analysis and it is computationally efficient in the step-by-step solution of a nonlinear structural dynamic problem, where the total response is dominated by the low frequency modes while the high frequency responses are of no interest. However, it might experience a new type of overshoot in the steady-state response of a high frequency mode rather than in the early transient response. Notice that an overshoot in the early transient response has been found by Goudreau and Taylor in 1972 [14]. Clearly, the overshoot in this study is different from that found by Goudreau and Taylor since it only occurs under dynamic loading. In this work, the root cause of this overshoot is thoroughly studied and a load-correction scheme is proposed to overcome the overshoot.

2 Chang Dissipative Integration Method

The general formulation of CDM for a single degree of freedom system can be written as:

$$\begin{aligned} & ma_{i+1} + c_0 v_{i+1} + \frac{2p}{p+1} k_{i+1} d_{i+1} - \frac{p-1}{p+1} k_i d_i = \frac{2p}{p+1} f_{i+1} - \frac{p-1}{p+1} f_i \\ & d_{i+1} = \beta_0 d_{i-1} + \beta_1 d_i + \beta_2 (\Delta t) v_i + \beta_3 (\Delta t)^2 a_i \\ & v_{i+1} = v_i + \frac{3p-1}{2(p+1)} (\Delta t) a_i - \frac{p-3}{2(p+1)} (\Delta t) a_{i+1} \end{aligned} \tag{1}$$

where the coefficients β_0 to β_3 are found to be:

$$\begin{aligned} & \beta_0 = -\frac{1}{D}\left[\frac{p-1}{8}\left(\frac{2}{p+1}\right)^3 \Omega_0^2\right] \\ & \beta_1 = 1 + \frac{1}{D}\left[\frac{p-1}{8}\left(\frac{2}{p+1}\right)^3 \Omega_0^2\right] \\ & \beta_2 = \frac{1}{D}\left(1 - \frac{p-3}{p+1}\xi\Omega_0\right) \\ & \beta_3 = \frac{1}{D}\left\{\frac{1}{2} - \frac{1}{2}\left[\left(\frac{2}{p+1}\right)^2 + \frac{p-3}{p+1}\right]\xi\Omega_0\right\} \end{aligned} \tag{2}$$

where ξ is a viscous damping ratio; $\Omega_0 = \omega_0 (\Delta t)$ and $\omega_0 = \sqrt{k_0/m}$ is the natural frequency of the system determined from the initial stiffness k_0.

3 Overshoot Behavior

To illustrate that an unusual overshoot may occur in the steady-state response of a high frequency mode for CDM, the following equation of motion is solved.

$$\ddot{u} + \omega_0^2 u = \varepsilon \omega_0^2 \sin(\bar{\omega} t) \tag{3}$$

where $\bar{\omega}$ is a driving frequency; and ε is a constant for adjusting the response amplitude. The exact solution of this equation for the initial conditions of $u(0) = u_0$ and $\dot{u}(0) = \dot{u}_0$ can be obtained from the fundamental theory of structural dynamics and is found to be:

$$u(t) = u_0 \cos(\omega_0 t) + \left(\frac{\dot{u}_0}{\omega_0} - \frac{\varepsilon\theta}{1-\theta^2} \right) \sin(\omega_0 t) + \frac{\varepsilon}{1-\theta^2} \sin(\bar{\omega} t) \tag{4}$$

where $\theta = \bar{\omega}/\omega_0$ is a frequency ratio; u_0 and $\dot{u}_0$ are the initial displacement and velocity, respectively. There will be no transient response in the total response if the initial conditions are chosen to be $u_0 = 0$ and $\dot{u}_0 = \omega_0 \varepsilon \theta/(1-\theta^2)$.

Three systems are simulated by taking $\omega_0 = 1$, 100 and 1000 rad/s corresponding to S1, S2 and S3. In addition, the driving frequency $\bar{\omega} = 1.5$ rad/s is assumed for each analysis. This leads to $\theta = 1.5$, 1.5×10^{-2} and 1.5×10^{-3} in correspondence to S1, S2 and S3. To demonstrate the unusual overshoot that might be experienced for CDM, two members of CDM are adopted to calculate the responses. The member of $p = 1$ is referred as CDM1 while that of $p = 0.75$ is referred as CDM2. Notice that CDM1 has no numerical damping while CDM2 can have a desired numerical damping. A time step of $\Delta t = 0.1$ s is adopted for time integration. The calculated results are plotted in Fig. 1. This time step is small enough to accurately integrate the steady-state response since the value of $\Delta t/\bar{T}_0$ is as small as $3/40\pi$, where $\bar{T}_0 = 2\pi/\bar{\omega}_0 = 4\pi/3$. It has been shown by Chang that a harmonic loading can be very accurately represented if $\Delta t/\bar{T}_0$ is chosen to be less than 1/12 [15]. In Fig. 1 it is shown that either CDM1 or CDM2 can provide accurate solutions for S1 while an overshoot is found in the results of S2 and S3. In Fig. 1a–c it can be seen that the overshoot becomes more significant as the natural frequency increases; and there is almost no overshoot for a low frequency mode. Similar phenomena are also seen in Fig. 1d–f except for the amount of overshoot. This is because that CDM2 can have a desired numerical damping to suppress or even remove the high frequency mode and thus the results obtained from CDM2 show less significant overshoot for S2 and S3 when compared to CDM1. It is evident that CDM experiences an overshoot in the steady-state response of a high frequency mode.

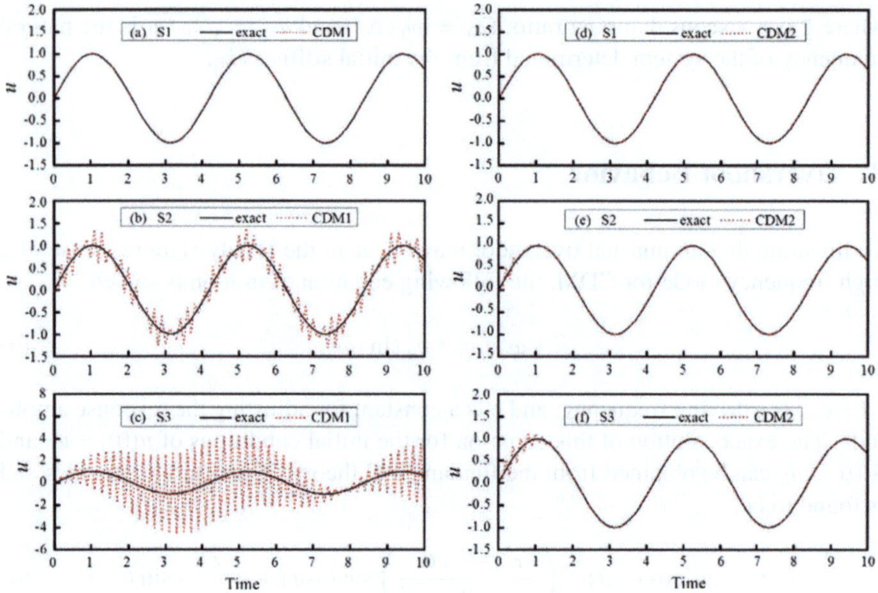

Fig. 1 Forced vibration responses to sine loading for using CDM1 and CDM2

4 Cause of Overshoot

In order to investigate the cause of the overshoot in the steady-state response of a high frequency mode, a local truncation error is constructed from a forced vibration response rather than a free vibration response since the dynamic loading is closely related to the overshoot. A local truncation error is defined as the error committed in each time step by using the differential equation to replace the difference equation. The derivations of the local truncation error for an integration method can be found in the references [1–5] and will not be elaborated herein.

After very complex algebraic manipulations, the local truncation error for CDM for zero viscous damping is found to be:

$$E = \frac{1}{D}\left\{\left[\frac{p(p-2)}{(p+1)^2} + \frac{1}{12}\right](\Delta t)^2 \dddot{u} + \frac{1}{(p+1)^2}\frac{1}{m}(\Delta t)^2 \ddot{f}_i\right\} \tag{5}$$

where

$$D = 1 - \frac{p-3}{p+1}\xi\Omega_0 + \frac{p}{4}\left(\frac{2}{p+1}\right)^3 \Omega_0^2 \tag{6}$$

Clearly, a second-order accuracy can be generally achieved for CDM. It seems that the last error term in Eq. 5 plays the key issue to cause an overshooting phenomenon in the steady-state response of a high frequency mode for CDM since it can be replaced by the other three terms of the second derivative of the equation of motion:

$$\tfrac{1}{m}(\Delta t)^2 \ddot{f}_i = \Omega_0^2 \ddot{u}_i + 2\xi\Omega_0 (\Delta t) \dddot{u}_i + (\Delta t)^2 \ddddot{u}_i \tag{7}$$

After substituting this equation into Eq. 5, it will become:

$$E = \frac{1}{D}\left\{\left[\frac{(p-1)^2}{(p+1)^2} + \tfrac{1}{12}\right](\Delta t)^2 \ddddot{u}_i + \frac{1}{(p+1)^2}\Omega_0^2 \ddot{u}_i\right\} \tag{8}$$

In this equation, it is found that the second term is quadratically proportional to Ω_0. Thus, the steady-state response might be almost unaffected by low frequency modes or a small value of Ω_0 while it might be drastically affected by high frequency modes or a large value of Ω_0. Consequently, this local truncation error seems to explain why an overshoot behavior might occur in the steady-state response of a high frequency mode while there is no such overshoot for a low frequency mode.

5 A Proposed Remedy

Since the overshoot in the steady-state response is an adverse behavior for CDM, it is better to eliminate it to obtain an accurate solution. For this purpose, we want to propose a remedy to remove the adverse overshoot behavior in the steady-state response of a high frequency mode. It seems that a slight modification of the formulation of the CDM can be applied to remove the adverse overshoot behavior. Since this modification should not affect the original numerical properties of CDM, a load-dependent term is simply introduced into the difference equation for displacement increments. This is because any load-dependent term will not affect the eigen-analysis of the amplification matrix and thus numerical properties. As a result, the modified form of CDM can be expressed as:

$$\begin{aligned}
& m a_{i+1} + c_0 v_{i+1} + \frac{2p}{p+1} k_{i+1} d_{i+1} - \frac{p-1}{p+1} k_i d_i = \frac{2p}{p+1} f_{i+1} - \frac{p-1}{p+1} f_i \\
& d_{i+1} = \beta_0 d_{i-1} + \beta_1 d_i + \beta_2 (\Delta t) v_i + \beta_3 (\Delta t)^2 a_i + q_{i+1} \\
& v_{i+1} = v_i + \frac{3p-1}{2(p+1)} (\Delta t) a_i - \frac{p-3}{2(p+1)} (\Delta t) a_{i+1}
\end{aligned} \tag{9}$$

where q_{i+1} is a load-dependent term. This modified form of the CDM can be referred to as the MCDM herein for brevity. Based on this formulation, the local truncation error for MCDM can be also obtained by using the same procedure to derive that for CDM and it is found to be:

$$E = \frac{1}{D}\left\{\left[\frac{(p-1)^2}{(p+1)^2} + \frac{1}{12}\right](\Delta t)^2 \dddot{\ddot{u}}_i + \frac{1}{(p+1)^2}\Omega_0^2 \ddot{u}_i\right\} - q_{i+2} + q_{i+1} \tag{10}$$

Apparently, this equation is very similar to Eq. 8, except that the last two error terms are closely related to the load-dependent term, which is intentionally introduced into the difference equation for displacement increments.

In Eq. 10, the load-dependent term q_{i+1} must be properly determined so that the quadratic error term $\Omega_0^2 \ddot{u}_i / D(p+1)^2$ in Eq. 8 can be completely removed. After a series of algebraic manipulations, a possible choice of q_{i+1} is found to be:

$$\begin{aligned} q_{i+1} = & \frac{1}{D}\frac{1}{(p+1)^2}\frac{1}{m}(\Delta t)^2 (f_{i+1} - f_i) \\ & + \frac{1}{D}\frac{p-1}{(p+1)^3}\frac{1}{m}(\Delta t)^2 (f_{i+1} - 2f_i + f_{i-1}) \end{aligned} \tag{11}$$

Notice that the extra loading term is also structure dependent and the denominator is the same as that for β_0 to β_3. After determining the extra loading term, the local truncation error for CDM is found to be:

$$E = \frac{1}{D}\left[\frac{(p-1)^2}{(p+1)^2} + \frac{1}{12}\right](\Delta t)^2 \dddot{\ddot{u}}_i \tag{12}$$

It is apparent that the adverse error term as shown in Eq. 8 is no longer in this local truncation error. This strongly indicates that the overshoot in the steady-state response of a high frequency mode is eliminated by the load-dependent term.

6 Numerical Confirmations

In order to verify the effectiveness of the proposed scheme, two numerical examples are examined by using MCDM. Notice that MCDM1 and MCDM2 are in correspondence to CDM1 and CDM2. Apparently, the only difference between CDM and MCDM in the formulation is that the latter has a load-dependent term.

6.1 Example 1

The illustrated example for a forced vibration response of a single degree of freedom system is solved again by using MCDM and the calculated results are plotted in Fig. 2. It is noteworthy that the calculated results obtained from MCDM1 and MCDM2 for a time step of $\Delta t = 0.1$ s are overlapped together with the exact solutions in each plot

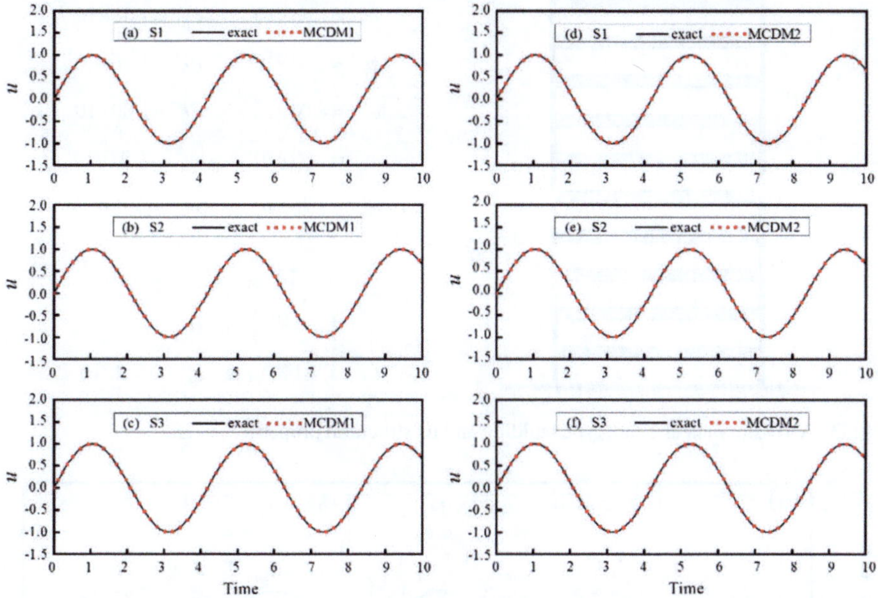

Fig. 2 Forced vibration responses to sine loading for using MCDM1 and MCDM2

of this figure. Thus, it is verified that the proposed scheme can effectively eliminate the adverse overshooting behavior in the steady-state response of a high frequency mode.

6.2 *Example 2*

In order to show that the adverse overshoot might experience in general dynamic analysis, a 10-story shear-beam type building is used to illustrate that an overshoot in the steady-state response of a high frequency mode may occur in practical applications. The building and its structural properties are shown in Fig. 3, where only a dynamic loading of $10k_1 \sin(3t)$ is intentionally imposed upon the bottom story of the building. The lowest natural frequency of the building is found to be 16.52 rad/s, whereas the highest natural frequency is found to be 10^4 rad/s. Both CDM1 and CDM2 as well as MCDM1 and MCDM2 with a time step of $\Delta t = 0.02$ s are applied to calculate the numerical results. The results obtained from CDM1 and MCDM1 are plotted in Fig. 4. For comparison, the solution obtained from the use of the constant average acceleration method (AAM) with $\Delta t = 0.01$ s is considered as a reference solution.

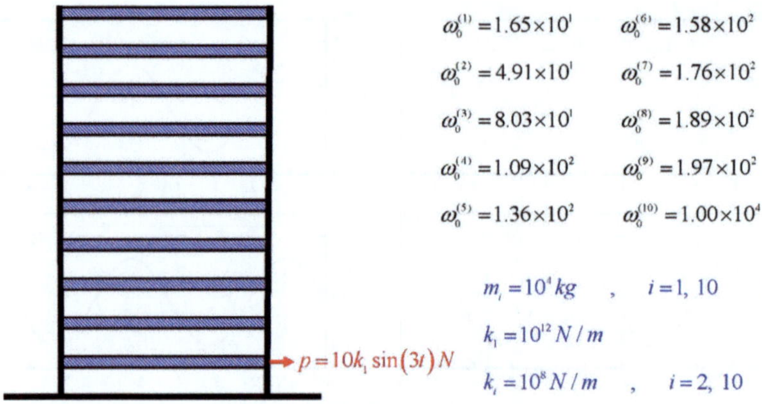

Fig. 3 A 10-story shear-beam type building and its structural properties

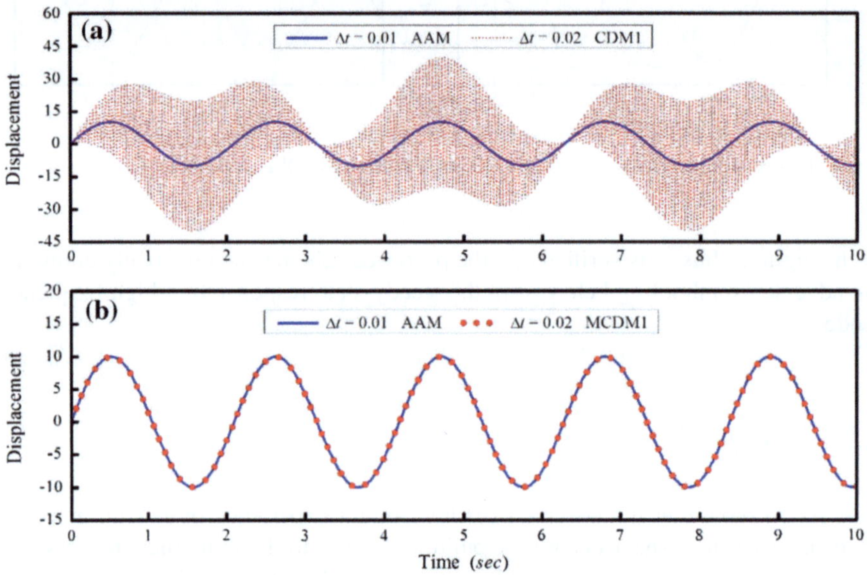

Fig. 4 Forced vibration response to 10-story building at bottom story

As shown in Fig. 4a, the response obtained from CDM1 for the bottom story of the 10-story building exhibits a very significant overshoot behavior. Contrary to this, there is almost no overshoot in Fig. 4b for the responses obtained from MCDM1 since the calculated results are coincided with the reference solutions. Notice that the displacement response at the bottom story is dominated by the steady-state response since the period of this displacement response is exactly equal to $2\pi/3$ s, which is the period of the sine loading. It is evident from this example that the proposed remedy can effectively remove the adverse overshoot in the steady-state response of a high frequency mode.

7 Conclusions

The one-parameter controlled Chang dissipative structure-dependent family method is very computationally efficient since it can simultaneously integrate the unconditional stability and explicit formulation together. In addition, it also has a second-order accuracy and desired numerical damping. However, it might experience an unusual overshoot in the steady-state response of a high frequency mode. This adverse overshooting can be revealed by the local truncation error constructed from a forced vibration response rather than a free vibration response. In addition, it can be used to derive a remedy to eliminate the adverse overshoot in the steady-state response of a high frequency mode. The remedy is to adjust the difference equation for displacement increment by adding an extra load-dependent term. The effectiveness of this remedy is confirmed by numerical examples in addition to the analytical study. As a result, the additional load-dependent term must be included in the general formulation of the one-parameter controlled Chang dissipative family method so that the adverse overshoot can be automatically eliminated.

References

1. Chang, S.Y.: Explicit pseudodynamic algorithm with unconditional stability. J. Eng. Mech. ASCE **128**(9), 935–947 (2002)
2. Chang, S.Y.: Improved explicit method for structural dynamics. J. Eng. Mech. ASCE **133**(7), 748–760 (2007)
3. Chang, S.Y.: An explicit method with improved stability property. Int. J. Numer. Method Eng. **77**(8), 1100–1120 (2009)
4. Chang, S.Y.: A new family of explicit method for linear structural dynamics. Comput. Struct. **88**(11–12), 755–772 (2010)
5. Chang, S.Y.: Development and validation of a new family of computationally efficient methods for dynamic analysis. J. Earthquake Eng. **19**(6), 847–873 (2015)
6. Chen, C., Ricles, J.M.: Development of direct integration algorithms for structural dynamics using discrete control theory. J. Eng. Mech. ASCE **134**(8), 676–683 (2008)
7. Gui, Y., Wang, J.T., Jin, F., Chen, C., Zhou, M.X.: Development of a family of explicit algorithms for structural dynamics with unconditional stability. Nonlinear Dyn. **77**(4), 1157–1170 (2014)
8. Tang, Y., Lou, M.L.: New unconditionally stable explicit integration algorithm for real-time hybrid testing. J. Eng. Mech. ASCE **143**(7), 04017029-1-15 (2017)
9. Chang, S.Y.: Numerical dissipation for explicit, unconditionally stable time integration methods. Earthquakes Struct. Int. J. **7**(2), 157–176 (2014)
10. Chang, S.Y.: A family of non-iterative integration methods with desired numerical dissipation. Int. J. Numer. Meth. Eng. **100**(1), 62–86 (2014)

11. Chang, S.Y.: Dissipative, non-iterative integration algorithms with unconditional stability for mildly nonlinear structural dynamics. Nonlinear Dyn. **79**(2), 1625–1649 (2015)
12. Chang, S.Y., Tran, N.C., Wu, T.H.: A one-parameter controlled dissipative unconditionally stable explicit algorithm for time history analysis. Sci. Iranica **24**(5), 2307–2319 (2017)
13. Kolay, C., Ricles, J.M.: Development of a family of unconditionally stable explicit direct integration algorithms with controllable numerical energy dissipation. Earthq. Eng. Struct. Dyn. **43**, 1361–1380 (2014)
14. Goudreau, G.L., Taylor, R.L.: Evaluation of numerical integration methods in elasto-dynamics. Comput. Methods Appl. Mech. Eng. **2**, 69–97 (1972)
15. Chang, S.Y.: Accurate representation of external force in time history analysis. J. Eng. Mech. ASCE **132**(1), 34–45 (2006)

Development and Characterization of AA5083 Reinforced with SiC and Al_2O_3 Particles by Friction Stir Processing

Eman M. Zayed, N. S. M. El-Tayeb, M. M. Z. Ahmed and Ragaie M. Rashad

Abstract In this work, friction stir process (FSP) was performed to enhance the surface properties of the aluminum alloy 5083 by incorporating SiC of 6 μm average size, Al_2O_3 of 3 μm average size and mixture of SiC/Al_2O_3 particles in the alloy. Tensile tests, Vickers micro hardness measurements, wear tests, and optical and scanning electron microscopies (SEM) examination were conducted on the surface of Aluminum alloy 5083 to analyze the processed zones of composite samples. Results of tensile tests revealed that ultimate tensile strength of 120 MPa was obtained at 600 rpm. The average hardness increased by 30% with increasing the content of ceramic particles and the maximum hardness values obtained for the hybrid surface composite (50% Al_2O_3 and 50% SiC) developed by 2 passes FSP and processing parameters of 600 rpm. The addition of ceramic powder (SiC) as a reinforcement into AA5083 improved the wear resistance of the received AA5083 by 40%. Hybrid composite of 50% SiC and 50% Al_2O_3 exhibited a superior wear resistance compared to the 100% SiC or 100% Al_2O_3 at a normal load of 20 N.

Keywords Aluminum alloy AA5083 · FSP · Surface modification
Metal matrix composites · Hybrid surface composite

E. M. Zayed (✉) · N. S. M. El-Tayeb · M. M. Z. Ahmed
Mechanical Engineering Department, Faculty of Engineering, The British University in Egypt (BUE), El Shorouk, Egypt
e-mail: eman.zayed@bue.edu.eg

N. S. M. El-Tayeb
e-mail: nabil.eltayeb@bue.edu.eg

M. M. Z. Ahmed
e-mail: mohamed.zaky@bue.edu.eg

M. M. Z. Ahmed
Department of Metallurgical and Materials Engineering, Faculty of Petroleum and Mining Engineering, Suez University, Suez 43721, Egypt

R. M. Rashad
Design and Production Department, Faculty of Engineering, Cairo University, Cairo, Egypt
e-mail: RagaieM.Rashad@bue.edu.eg

© Springer International Publishing AG, part of Springer Nature 2019
A. Öchsner and H. Altenbach (eds.), *Engineering Design Applications*,
Advanced Structured Materials 92, https://doi.org/10.1007/978-3-319-79005-3_2

1 Introduction

Aluminum and its alloys have versatile properties; they are lightweight in addition, their high strength to weight ratio and excellent resistance to corrosion make them suitable for use in structural applications like military, aerospace and transportation industries. However, in some applications, the surface needs to have better mechanical properties such as improved strength and better hardness [1]. Therefore, the reinforcement of surface layer of the component by hard ceramic particles is highly desirable to achieve the required properties while the substrate still maintains the original structure with good ductility and thermal conductivity [2].

In the past, attempts to fabricate metal matrix composites by adding particulate material to the molten alloy followed by casting have not been successful [3]. It has been postulated that the major difficulty in this approach is that the interfacial reaction between reinforcement and metal-matrix cannot be avoided in addition to the formation of some detrimental phases. Therefore, if processing of surface composite is carried out at temperatures below melting point of the substrate, the problems mentioned above can be avoided [4].

Friction stir processing has been evolved as a successful solid state processing technique based on the principle of friction stir welding developed by Mishra [5, 6]. FSP leads to a localized severe plastic deformation of the material and thus can be used for homogenization of microstructure and for grain refinement, thereby resulting in improved material properties. Little or no macroscopic shape change may occur despite the local severity of the plastic deformation [7].

In the view of the above it can be concluded that FSP can be employed to enhance the surface properties (viz. micro-hardness, wear resistance, among others) of metals and their alloys. However, the work pertaining to enhancement of surface properties of aluminum/aluminum alloys through friction stir processing (FSP) is rather scant. In this regard the aim of this study is to investigate the effect of incorporating the surface of AA5083 with micro-sized SiC and Al_2O_3 particles and mixture of them through friction stir processing on its mechanical properties and to study the influence of FSP parameters on the microstructure of the developed metal matrix composite (MMC). In addition to the effect of varying the composition of reinforcement material on mechanical properties (i.e. hardness and tensile) and tribological properties (i.e. wear) were studied with respect to the process parameters.

2 Experimental Procedure

2.1 Material Used

In this study, a cold rolled aluminium alloy (AA5083) plate of 10 mm thickness, 200 mm length, and 50 mm width has been used for the purpose of carrying out FSP. A groove was machined at the center of the substrates with a width of 3 mm and a

Table 1 Nominal chemical composition of 5083 Al alloy (wt%) [8]

	Si	Fe	Cu	Mn	Cr	Mg	Ti	Zn	Al
wt%	0.40 max	0.40 max	0.10 max	0.40/1.0	0.0/0.25	4.0/4.90	0.15 max	0.25 max	Balance

depth of 6 mm. The nominal chemical composition of the alloy 5083 is summarized in Table 1.

The surface of AA5083 is incorporated through FSP with micro-sized SiC and Al_2O_3 powder to produce the (MMC).

2.2 *Friction Stir Processing (Equipment)*

Figure 1 shows the FSP setup used for carrying the development of MMCs.

A FSP tool of H13 tool steel with shoulder of 25 mm diameter and square probe of 6 mm width and 6 mm length was used to perform the FSP with a schematic shown in Fig. 2, before the start of the actual FSP. The top surface of the grooves was closed using penless tool as shown in Fig. 3a of 25 mm diameter shoulder to prevent the escape of the packed particles during the process and then the FSP was then performed at rotation rate of 400, 600, 800 rpm, constant traverse speed of 60 mm/min, depth of 6 mm, and tool tilt angle was set at 3° with the conventional tool as manifested in Fig. 3b.

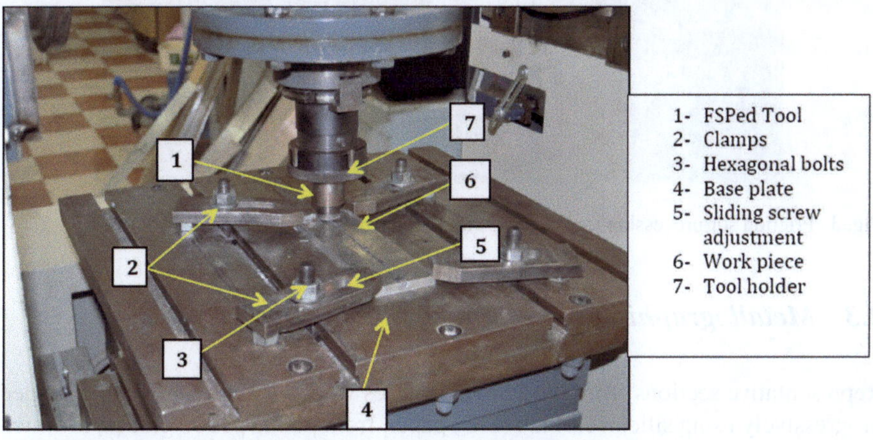

Fig. 1 Clamping arrangement during friction stir processing

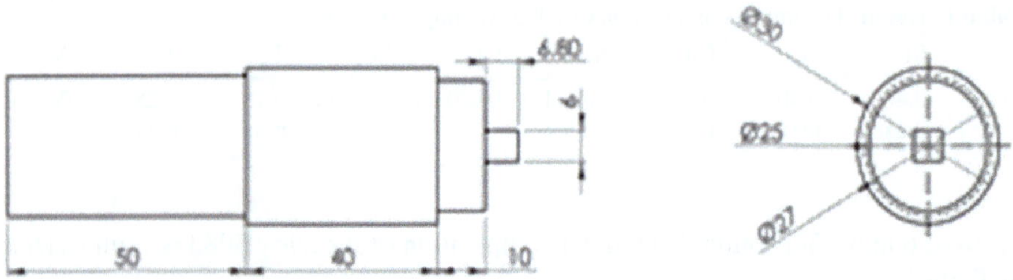

Fig. 2 Schematic illustration of the used FSP tool dimension

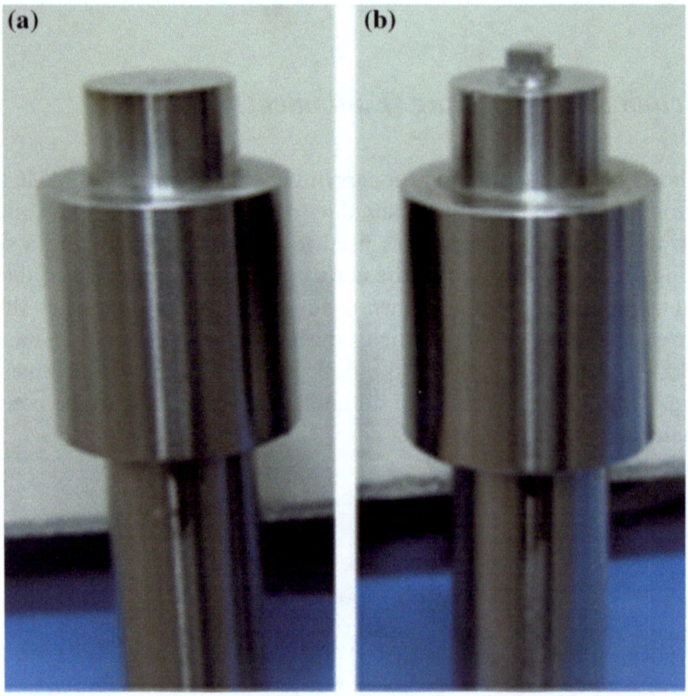

Fig. 3 Friction stir processing tool **a** pinless tool **b** conventional FSP tool

2.3 Metallographic Preparation

Representative sections from the FSPed samples were cut and then were grounded progressively using silicon carbide (SiC) paper, from p60 to p1200 grades, in which each stage progressively removes and replaces the larger surface scratches with smaller ones.

The samples were then polished on 6 and 3 μm diamond cloths for 3–5 min, and finally with an alumina suspension for 5–10 min. Note that after each grinding and polishing step, the samples were washed thoroughly using methanol and air-dried.

2.4 Microstructural Characterization

Optical Microscopy (OM)

The microstructure of the prepared specimens was examined using an optical microscope linked to a high-resolution digital camera. The microstructures were revealed after etching using Keller's reagent.

Scanning Electron Microscope (SEM)

Quanta 250 Field Emission Gun—Scanning Electron Microscope (FEI model: Quanta FEG-250-SEM) was used to examine the reinforcement ceramic particles (SiC and Al_2O_3) and worn surface of the friction stir processed specimens and then they were further analyzed using Energy Dispersive X-Ray Spectroscopy (EDX) within the SEM instrument.

2.5 Mechanical Properties Characterization

To verify the feasibility of using FSP to enhance the mechanical properties of AA 5083, tensile strength and micro-hardness testing at ambient temperature was conducted for FSPed specimens and compared to the As-received AA5083.

Micro-hardness Test

Hardness measurements were carried out on the polished samples. The micro hardness was measured using Vickers hardness tester, the test was carried out at load of 10 HV. The reported results are the average of 10 readings.

Tensile Test

In order to determine the mechanical properties of the materials after FSP, tensile specimens were cut out from the processed zone (PZ) and machined to the specified dimensions according to *ASTM E8* as shown in Fig. 4 of 10 mm and gauge length 50 mm parallel to the processing direction, and tested at room temperature under uniaxial tensile conditions using a universal tensile testing machine. All results were based on the average of three samples from three different processed plates.

2.6 Tribological Characterization

Dry wear behavior of the base metal and surface composite layer (SCL) was studied using a pin on ring tribo-tester. The wear tests were conducted for base material at normal force of 20 N, sliding distance of 800 m and constant track diameter 370 mm. The wear rate is calculated depending on the weights before and after the wear test. The wear behavior of surface composite layer processed after the second pass with

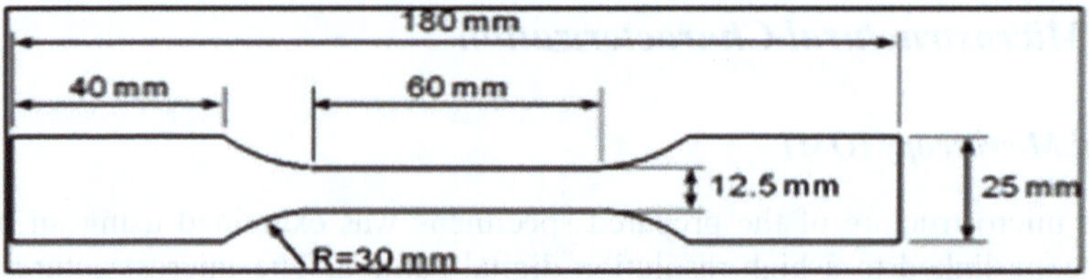

Fig. 4 Schematic diagram showing the dimensions of the tensile specimens used in the uni axial tensile experiment

Fig. 5 Surface appearance of friction stir processed surface at 400 rpm and 100% SiC

additives SiC and Al_2O_3 particles at different FSP conditions were compared with the wear behavior of base material.

3 Results and Discussion

Figures 5 and 6 show the surface appearance of some of the processed surfaces produced at different composition of ceramic particles and rotational speeds. The processed surface showed a considerable amount of flash. A good appearance with a little flash was observed at the lower rotational speed of 400 rpm. The higher flash associated with the higher rotational speed may be attributed to the higher plunging forces applied.

3.1 *Microstructural Evolutions*

3.1.1 Optical Microstructure

From the micrograph images shown in Fig. 7, it was observed that the agglomeration size of ceramic particles is reduced by increasing number of FSP passes which resulted in improvement of reinforcement particles distribution. Furthermore, pro-

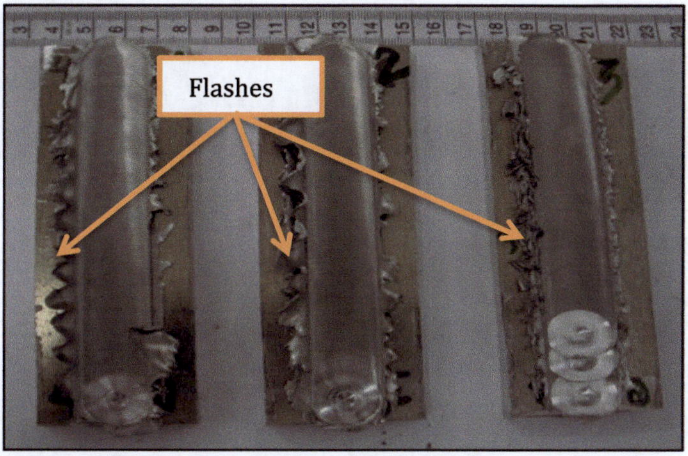

Fig. 6 Surface appearance of friction stir processed surface at 800 rpm and 100% SiC

cessing ceramic particles dispersed more homogenously in the base metal at low rpm, microstructures show that good dispersion areas than with high rpm, since there are voids formed due to low action of homogeneity that resulted from high rotation. This result is in a good agreement with Kwon et al. [9] who studied the FS processed Al 1050 alloy and found the hardness and tensile strength of the FS processed 1050 aluminum alloy were observed to increase significantly with decreased tool rotation speed.

Formation of onion rings is one of the characteristics of FSP, which is found in Fig. 7c.

Figure 7f represents the micro-structure of the hybrid composition of SiC and Al_2O_3 particles at rotational speeds 400 rpm. As shown, clustering of reinforcement particles was observed in Fig. 7f due to the large size of SiC powders, Increasing the number of FSP passes can eliminate this feature.

3.1.2 SEM Microstructure

Figures 8a–f illustrate the SEM micrographs of the nugget cross-sections for processing parameters 400, 600 and 800 rpm for additive composition SiC, Al_2O_3 and mixture of them. Figure 8c shows a uniform distribution of Al_2O_3 powders in the stir zone, Clustering occurs at 800 rpm as shown in Fig. 8f.

Another important note should be mentioned is the alteration of the shape of the nugget with variation of processing parameters. Mishra and Ma [5] represented two types of stirring zone (nugget), basin-shaped nugget and elliptical nugget. For 400 rpm, elliptical-shaped nugget, was observed in Fig. 8e.

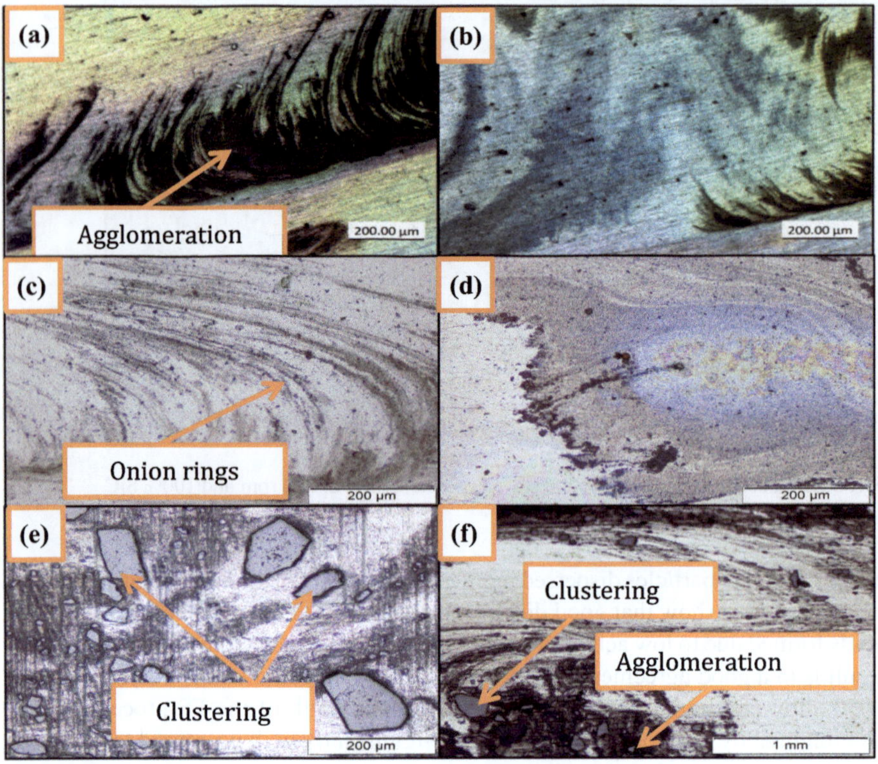

Fig. 7 Optical microscopy images for processed specimens with SiC and Al_2O_3 at different rotational speeds

3.2 Hardness

3.2.1 Effect of Reinforcement Particles Type on Hardness

Micro hardness of friction stir processed specimens was taken using Vickers hardness machine under load of 10 HV. The average hardness values are shown in Table 2 and Fig. 10. It is observed that Vickers hardness values depend on the rotational speeds and this can be interpreted by the frictional heat generated as a result of rubbing of the tool with the matrix. The quantity of frictional heating generated is dependent upon the tool rotation speed. As the tool rotation increases, the amount of frictional heating causing softening and sliding of material during FSP will also increase. Also coarsening of grains due to high heat generated was the main reason for the lowest hardness of those FSPed samples which have been produced at the higher rotation rate compared to the other with lower rotation rate.

It is observed from Fig. 9 that the hybrid surface composites having equal composition (50% SiC and 50% Al_2O_3) exhibit more hardness than the surface composite

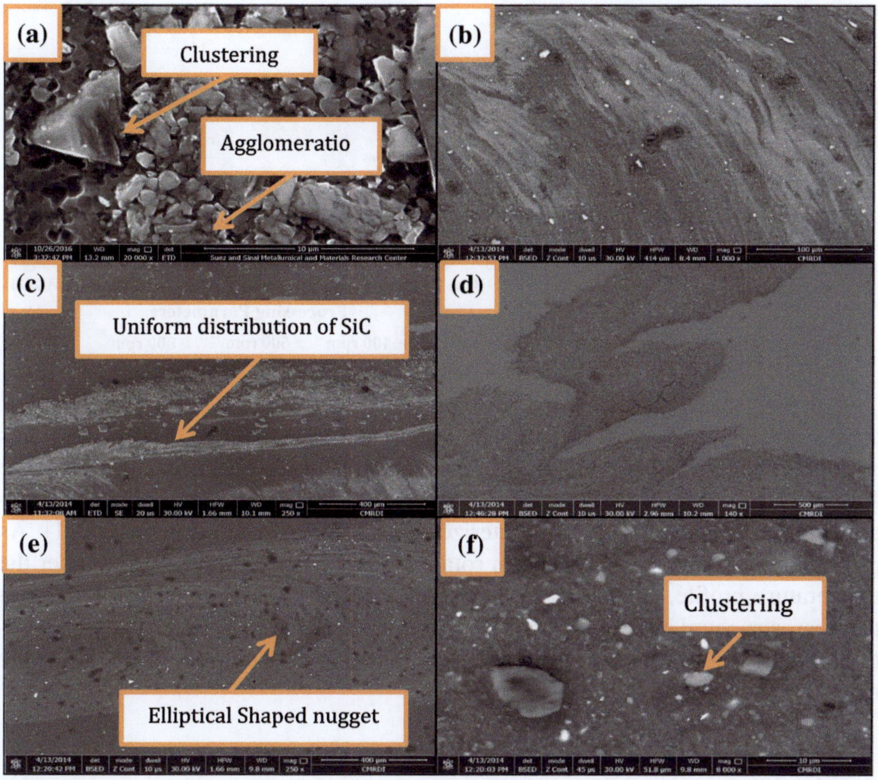

Fig. 8 SEM images for processed specimens with SiC and Al_2O_3 at different rotational speeds

Table 2 Hardness average values

Composition	Rotational speed (rpm)	Hardness avg. (HV)
SiC	400	110.8
	600	106.6
	800	108.4
Al_2O_3	400	101.8
	600	62.34
	800	84.4
SiC and Al_2O_3	400	115.5
	600	117
	800	117

fabricated using other composition of ceramic particles and this result is in a good agreement with Keerthivel et al. [10].

Figures 10 a–c show the micro hardness profile measured across top surface of the FSPed specimens. It is observed that the hardness of processed zone is signifi-

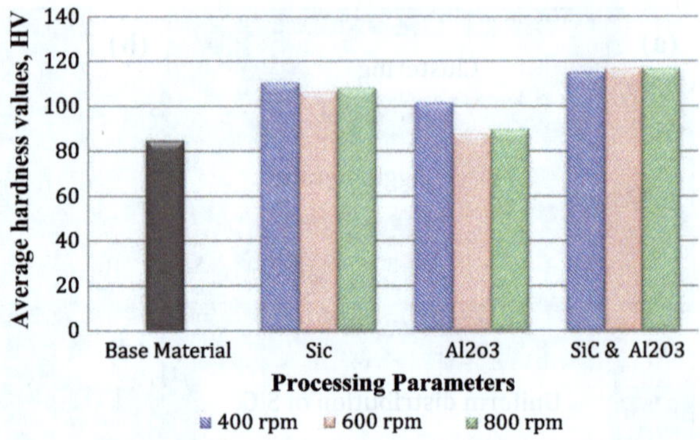

Fig. 9 Effect of the content of ceramic powders and rotational speeds in rpm on the average hardness, HV

cantly higher than that observed for the as-received AA5083 base alloy which has an average hardness of 85 HV. The highest value is observed to be 230 HV for the specimen that underwent 2 passes of FSP at 50% SiC and 50% Al_2O_3. The Vickers hardness value has been characterized as rotational speed dependent [11]. As previously mentioned, the change in rotational and traverse speeds would alter the temperature profile, as well as the exposure time, in the stir zone; this will affect the microstructural characteristics. It can be observed that either increasing the rotational speed and/or changing the type and composition of ceramic powder affect the mechanical properties.

3.3 *Evaluation of Tensile Strength*

A tensile test of FSPed specimens was carried out. Figure 11 a–c show results of the tensile tests for FSPed specimens with different additive compositions at different rotational speeds. Table 3 shows the ultimate tensile strength (UTS) as well as the 0.2% offset yield values for each FSPed specimen. Two processed specimens were tested for each processing condition and the failure loads were averaged.

The effect of the rotational speed on the ultimate tensile strength of the processed zones is shown in Fig. 12. When the processing is carried out at 600 rpm and with an additive composition of 50% SiC and 50% Al_2O_3, ultimate tensile strength reached to the maximum value of 120 MPa. Conversely by further increasing the rotational speed to 800 rpm, tensile strength slightly decreased again. The reason behind decreasing the UTS values is the incorporation of ceramic particles, which increase the brittleness of the FSPed specimens. However the ductility of the material was greatly improved.

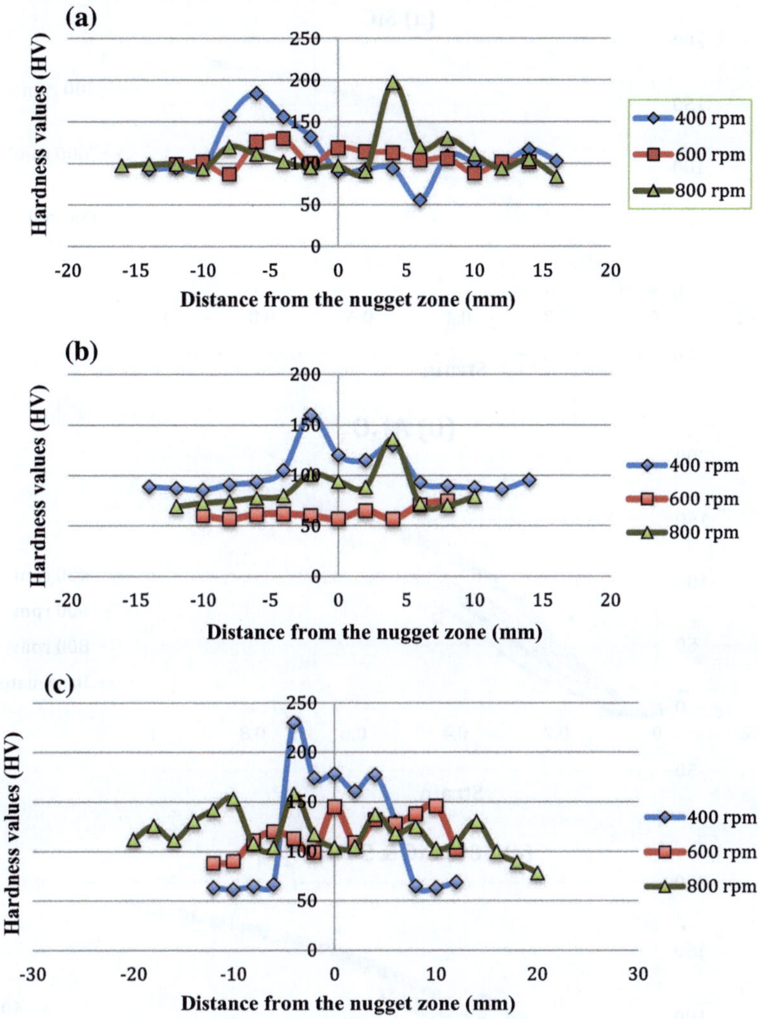

Fig. 10 Distance from the center of FSP zone (mm) versus Hardness (HV) for compositions **a** SiC **b** Al_2O_3 **c** SiC and Al_2O_3

3.4 Wear Characteristics

Figures 13 a–c show the variation of wear rate in (g/m) of the base materials and FSPed specimens at normal force 20 N. Figures 13 a–c reveal that the weight loss of the as received material is higher than the SCL samples. It is known [12] that the reinforced SiC ceramic particles resist the plastic liquidity of the material by increasing the resistance and hardness of the material at high temperatures. Additionally, the SiC particles keep the material debris on the surface, and some of this

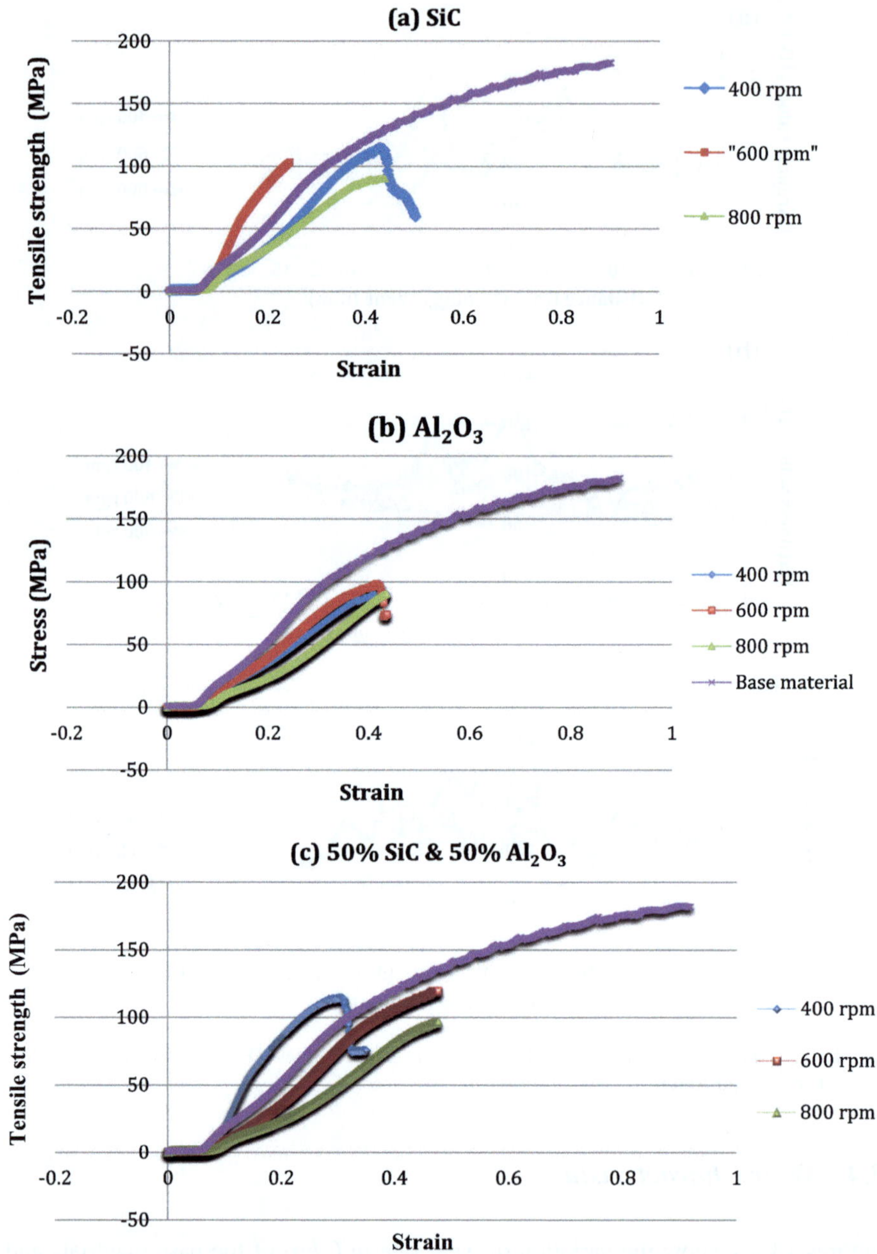

Fig. 11 Effect of rotational speeds on tensile strength of FSPed specimens with additive composition **a** 100% SiC, **b** 100% AL_2O_3 and **c** 50% SiC and 50% Al_2O_3 at different rotation speed 400, 600 and 800 rpm compared with the base material

Table 3 Average yield strength and ultimate tensile strength for as received AA5083 and FSPed samples at different processing parameters, tested at room temperature

Composition	Rotational speed	0.2% offset yield strength (MPa)	Ultimate tensile strength (MPa)
100% SiC	400 rpm	60	110.5
	600 rpm	100	100.5
	800 rpm	80	90
100% Al_2O_3	400 rpm	80	80.5
	600 rpm	85	100
	800 rpm	80	90
50% SiC & 50% Al_2O_3	400 rpm	100	110.5
	600 rpm	90	120
	800 rpm	80	90.5
Base material	As received AA 5083	70	180.5

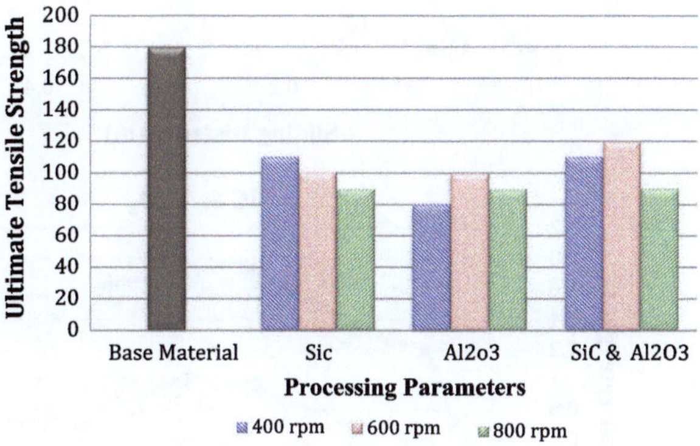

Fig. 12 Effect of rotational speed on UTS for FSP samples at 400, 600 and 800 rpm and different additive composition

debris accumulates around the particles during wear by remaining stable at the points where they already exist. This is how the wear resistance of the composite material increased [13].

The hard 50% SiC and 50% Al_2O_3 reinforcement particles of 800 rpm exhibited the lowest wear rate among the tested specimens. This is primarily due to the fact that the hard dispersoids, present on the surface of the composites, act as protrusions, which protect the matrix from severe contact with the counter surfaces and thus resulting in less wear in composites as compared to that in the case of alloy for all loads.

The effect of the rotational speed on the wear rate in (g/m) of the processed zones at different rotational speeds is shown in Fig. 14. Specimens were tested at 400 rpm. At an additive composition 100% SiC, exhibited the least wear rate among the tested specimens. However, by further increasing the rotational speed to 600 rpm, the wear

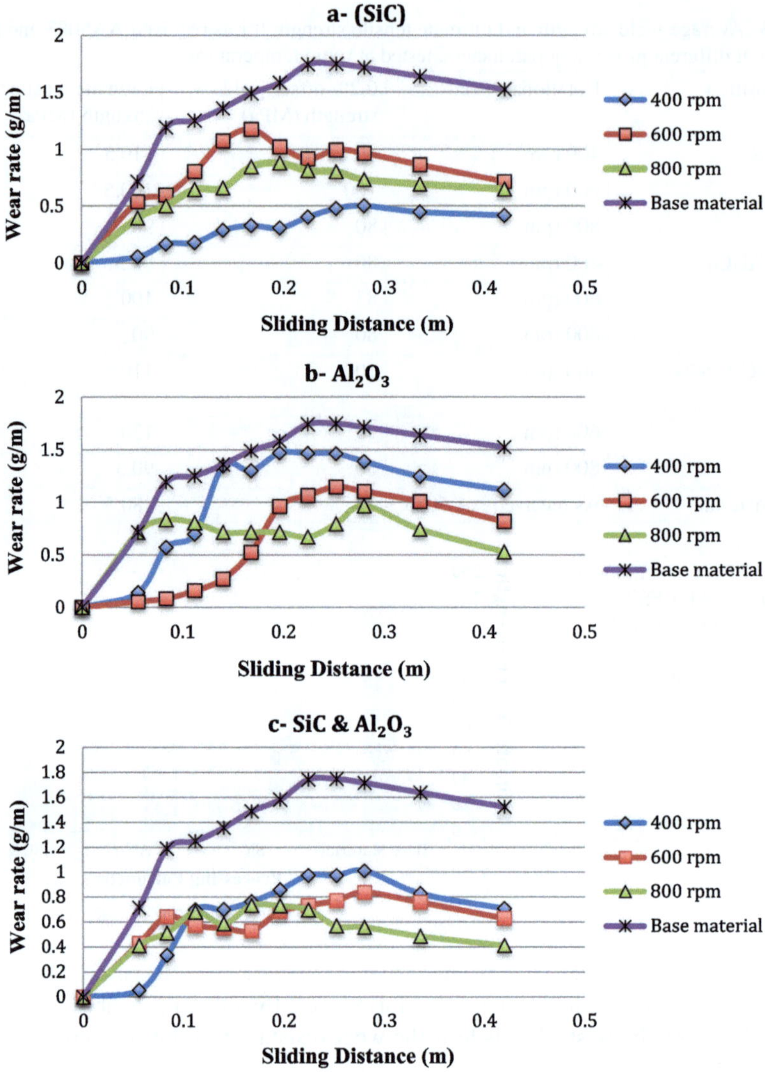

Fig. 13 Wear rate versus Sliding distance of FSPed specimens show the effect of rotational speeds and additive composition on wear rate (g/m) at normal force of 20 N

rate increased again due to the incorporation of ceramic particles that increase the wear resistance of the FSPed specimens.

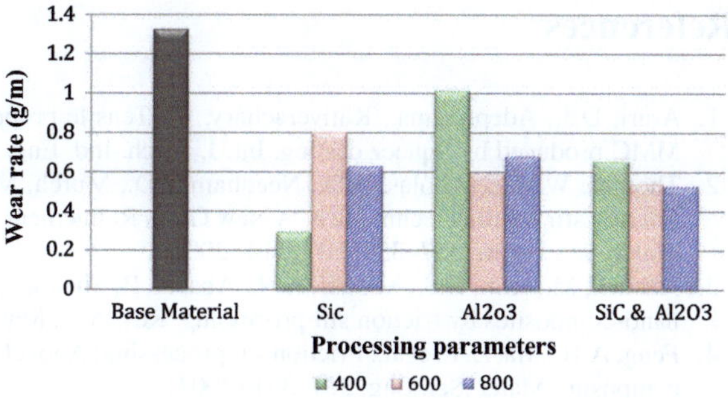

Fig. 14 Effect of the content of ceramic powders on wear rate in (g/m) at different rotational speeds 400, 600 and 800 rpm

4 Conclusion

Friction Stir Processing is a new thermo mechanical processing technique, which can be used to induce a local modification in microstructure and mechanical properties of a material. In the present study, FSP was successfully used for incorporation of SiC and Al_2O_3 powders in AA 5083. Results were compared to previous studies, in order to obtain a better understanding of the processing conditions. The main findings are summarized below:

- The distribution of reinforcement ceramic particles in the stir zone became more uniform and homogenous by increasing the number of processing passes due to the repeated stirring action of the tool.
- Decreasing the rotational speed is more effective in distribution of reinforcement particles.
- The average hardness increased by 30% by increasing the content of ceramic particles.
- The maximum hardness value for the produced aluminum MMCs is 116 HV and was obtained from the hybrid surface composite having equal composition of (50% Al_2O_3 and 50% SiC) developed by 2 passes of FSP and processing parameters of 800 rpm and 60 mm/min.
- From tensile experiments, the ultimate tensile strength reached the maximum value of about 120 MPa although by further increasing the rotational speed to 800 rpm, tensile strength slightly decreased again. However, the ductility of the material was greatly improved after FSP, as voids were eliminated from the microstructure.
- The wear properties of the Al5083 alloy were improved by 40% with the addition of SiC/Al_2O_3 particles.
- Hybrid composition of 50% SiC and 50% Al_2O_3 showed a superior wear resistance compared to the 100% SiC or 100% Al_2O_3 at a normal load of 20 N.

Acknowledgements The Department of Metallurgical and Materials Engineering at Suez University is acknowledged for carrying out the FSP experiments at the FSW laboratory.

References

1. Aruri, D.J., Adepukuma., Kativerachary, B.: Tensile properties of AA 6061-T6/SiC surface MMC produced by Squeez casting. Int. J. Mech. Ind. Eng. (IJMIE) **1**, 26–30 (2011)
2. Thomas, W.M., Nicholas, E.D., Needham, J.D., Murch, M.G., Templesmith, P., Daws, C.J.: Friction stir welding technique b "A New Grain Refinement Technique in Commercial Alloys". Mater. Sci. Forum **357–359**, 507–514 (2001)
3. Azizieh, M., Kim, H.S., Kokabi, A.H., Abachi, P., Shahraki, B.K.: Fabrication of AZ31/Al2O3 nano-composites by friction stir processing. Rev. Adv. Mater. Sci. **28**, 85–89 (2011)
4. Feng, A.H., Ma, Z.Y., et al.: Friction stir processing: Anovel technique for fabrication of surface composite. Mater. Sci. Eng. 307–331 (2003)
5. Mishra, R.S., Ma, Z.Y., et al. Friction stir welding and processing. (2006)
6. Mishra, R. S., Kim H. S., Kokabi, A.H., Abachi, P., Shahraki, B.K., Azizieh, M.: Fabrication of AZ31/Al2O3 nano-composites by friction stir processing. Rev. Adv. Mater. Sci. **28**, 85–89 (2000)
7. Gan, W-Y., Zheng, Z., Zhang, H., Tao, P.: Evolution of microstructure and hardness of aluminum after friction stir processing. Trans. Nonferr. Metals Soc. China **24**, 975–981 (2014)
8. Dolatkhah, A., Golbabaei, P., Besharati Givi, M.K., Molaiekiya, F.: Investigating effects of process parameters on microstructural and mechanical properties of Al5083/SiC metal matrix composite fabricated via friction stir processing. Mater. Des. **37**, 458–464 (2012)
9. Kwon, M.J., Peel.: The friction-stir welding of dissimilar aluminium alloys. Doctor of Philosophy Thesis (2005)
10. Keerthivel., Adem, K., Ilyas, U., Eren, C.: Surface modification of aluminium by friction stir processing. J. Mater. Process. Tech. **211**, 313–317 (2011)
11. Mazaheri, Y., Karimzadeh, F., Enayati, M.H.: Tribological behavior of A356/Al2O3 surface nanocomposite prepared by friction stir processing. Metall. Mater. Trans. A **45A**, 2250–2259 (2014)
12. Shafiei-Zarghani, A., Kashani-Bozorg, S.F., Zarei-, Hanzaki A.: Wear assessment of Al/Al2O3 nano-composite surface layer produced using friction stir processing. Wear **270**, 403–412 (2011)
13. Shafiee, M., Celik, O.N., Sert, A.: Wear behaviour of SiC-reinforced surface composite Al7075-T651 aluminioum alloy produced using friction stir processing. Indian J. Eng. Mater. Sci. **21**, February (2013)

Influence of the Change in Relative Dielectric Constant on Partial Discharge in Insulators

Ivica Kuzmanić and Igor Vujović

Abstract This paper covers several topics relating to a single problem. The primary concern is partial discharge (PD). The second concern is the influence of environmental factors on PD, which is calculated by modifying the relative dielectric constant. The third concern is the manner of presentation of the obtained data. Finally, the basic concern is predicting what will happen to the PD when a parameter changes a material's characteristics, what the margin of tolerance is and consequently, what will happen to a dielectric-dependent system under PD. The study of partial discharge in dielectrics is important for the reliability of electrical systems of vital importance aboard vehicles. This paper examines the effect of relative permittivity change to PD in dielectric materials. The topic was previously researched, but only for cables. PD modeling is carried out in Simulink using an equivalent electrical circuit. Simulation is carried out using the Matlab programming language using an m-script and Simulink model with parameter changes executed by script loop. Another topic of this paper is how to present the voluminous data obtained (in this case 3.5 GB), dealing with large data handling and visualization. 3D graphs with multi-curve representation in the same figure are used for visualization of the obtained results.

1 Introduction

This paper investigates dielectric breakdown in electrical insulation and its correlation with the change in the dielectric constant due to various environmental factors. Each topic has been the subject of separate research, but they have never been combined in a single paper. The impact of various environmental factors on relative permittivity is researched in, for example [1, 2]. As a separate problem, partial discharge (PD) has been researched in studies such as [3, 4]. Electrical breakdown is within the scope of interest of many studies. The stretch dependence, for example,

I. Kuzmanić (✉) · I. Vujović
Faculty of Maritime Studies, University of Split, Ruđera Boškovića 37, 21000 Split, Croatia
e-mail: ikuzman@pfst.hr

I. Vujović
e-mail: ivujovic@pfst.hr

© Springer International Publishing AG, part of Springer Nature 2019
A. Öchsner and H. Altenbach (eds.), *Engineering Design Applications*,
Advanced Structured Materials 92, https://doi.org/10.1007/978-3-319-79005-3_3

of the electrical breakdown strength and dielectric constant for elastomer substances was investigated in [4]. The dielectric properties are analyzed for specific materials, e.g. in [5–7]. However, influence of the dielectric constant change was not studied in depth. PD simulation models are presented in [8–12]. All the developed models have some differences in the equivalent circuit, which represent the void. Hence, it is interesting to compare these models and the obtained results. The impact of change in the dielectric constant was neither computed nor simulated in any of the currently available literature.

Since the results obtained in this research contain a large amount of data that is difficult to interpret as is, visualization of the dataset is required to arrive at conclusions. The visualization of copious data is the best way to draw conclusions from large data sets. When results depend on several factors, surface visualization and 3D graphs are the only available options. We combine these fields in one compact problem in an attempt to determine the impact of the change in the relative dielectric constant to PD. This type of research is rare due to a number of unknown dependencies influencing the usefulness of research. We tried to visualize a large amount of data to present them in a form convenient to making conclusions.

Finally, this issue also has a wider utility, i.e. identifying how to apply the findings in actual products. It can be viewed as low level analysis for higher level applications, such as design of autonomous vehicles (such as in [13]), which must account and compensate for the influence of climate change in everyday operation [14].

This paper is organized as follows. First, PD models will be discussed; then follow the presentation and discussion of results; finally, conclusions is given in the final section.

2 Partial Discharge Model

There are several models presented in [3, 10, 15] which can be used to study the PD phenomenon and develop simulations. PD creates a void in the insulator. In general, the void can be represented with an equivalent circuit, consisting of three capacitors in two branches. The cylindrical void inside the insulator and the remaining part of the insulation are placed in the serial branch of the equivalent circuit. The parallel branch represents the lumped capacitance of the rest of the solid insulation. Although all referenced models have this part in common, different additional resistors and capacitors are used. Capacitance of the void can be expressed as (see [3]):

$$C_{void} = \varepsilon_0 \varepsilon_r \cdot \frac{a - 2b}{c} \cdot b = f(\varepsilon_r) \tag{1}$$

$$C_{remaining\_insulation} = \varepsilon_0 \varepsilon_r r^2 \pi / (c - h) = f(\varepsilon_r) \tag{2}$$

$$C_{lumped} = \varepsilon_0 \frac{r^2 \pi}{h} \neq f(\varepsilon_r) \tag{3}$$

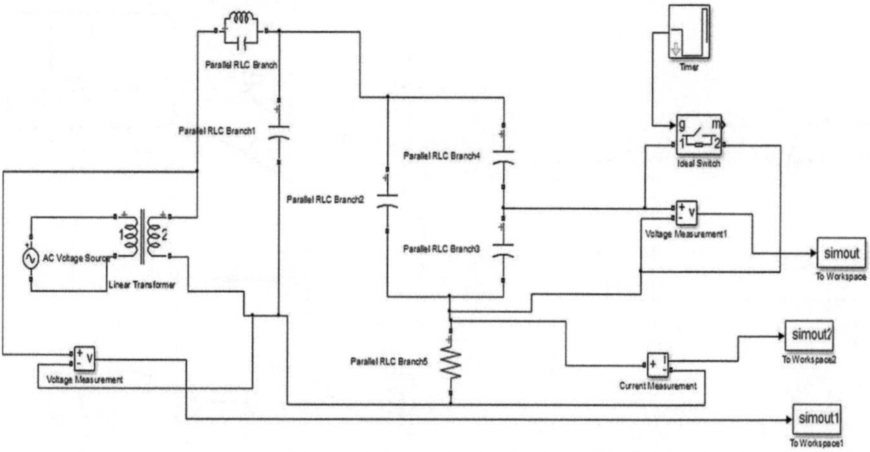

Fig. 1 Simulation model 1 (based on [3])

where:

- *a* denotes the length of the sample insulator,
- *b* denotes the breadth of the sample,
- *c* denotes thickness of the sample,
- *r* denotes the radius of the cylindrical void, and
- *h* denotes the height of the void.

The usual size is such that all parameters can be expressed in mm units. A discharge is calculated by the voltage drop across the void and remaining capacitance in the serial branch. Smaller voids are known to take longer time to damage the insulation, extending the overall process.

The model presented in [3] is modified to simulate the effect of the relative dielectric constant change. Figure 1 illustrates a Simulink model, automatically called up by the Matlab script with change in parameters. Changes in parameters are generated by loop-programming. The obtained time sequences are exported to Matlab workspace by simout blocks, stored in the multidimensional matrix structure.

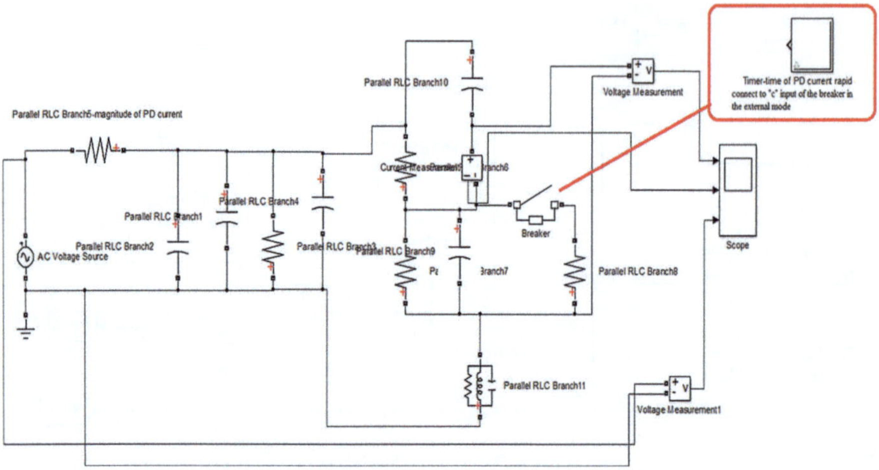

Fig. 2 Simulation model 2 (based on [10])

Model 2, based on [10], is of interest for power grids, including ship's power grids. Power grids are usually characterized by mid to high level voltage. The model consists of: major parameters, radial distances, resistances, and capacitances. Major parameters are: the radius of the conductor (r_c), the distance between the conductor core and the cable outer-sheath (r_s), insulation thickness (D_{ins}), correction factor (K_{ef}), and the relative dielectric constant. The last major parameter is calculated by adding the void location to r_c. The correction factor is obtained by (according to [10]):

$$K_{ef} = \frac{3\varepsilon_r}{1 + 2\varepsilon_r} \tag{4}$$

Radial distances, r_a, and r_b, are distances between the conductor core and the conductor sheath, and the distance between the conductor core and insulator sheath. Additional information about the void is provided by the radius of the void, r_v.

There are several possible variations of model 2. As illustrated in Fig. 2, the breaker can be externally or internally initialized. The red rectangular shows alternative triggering of the breaker. The block in red can be linked to the breaker's third terminal, which is not visible in the configuration shown in Fig. 2.

The third simulation model is based on [15]. An extended PD equivalent circuit with four capacitors (of the defect, of the healthy series with the defect, parallel capacitance in specimen, and capacitance of local charges accumulation) and two resistors is introduced. This critical part is depicted in Fig. 3a. The simulation model is presented in Fig. 3b.

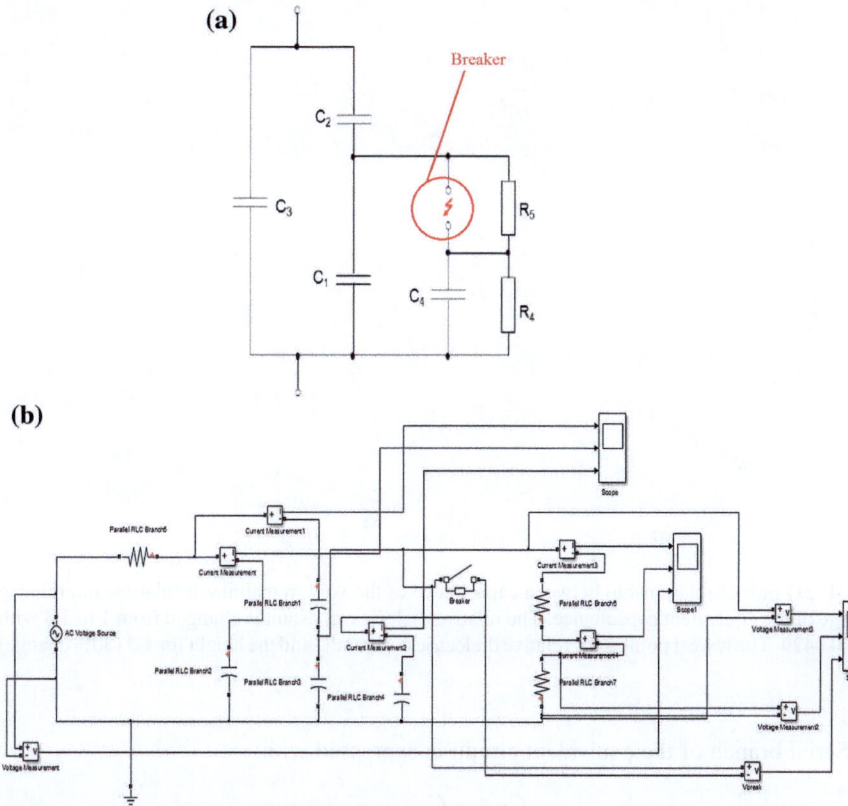

Fig. 3 **a** Critical part of the circuit—void equivalent circuit, **b** simulation model 3 [15]

3 Results

From (1) to (3), it is obvious that the relative change in the dielectric constant causes relative, linear change of the void and equivalent capacitance. Namely, dependence on the relative dielectric constant in (1) and (2) can be formulated as:

$$C_{void} = \varepsilon_r \cdot k_1 \quad (5)$$

$$C_{remaining\_insulation} = \varepsilon_r \cdot k_2 \quad (6)$$

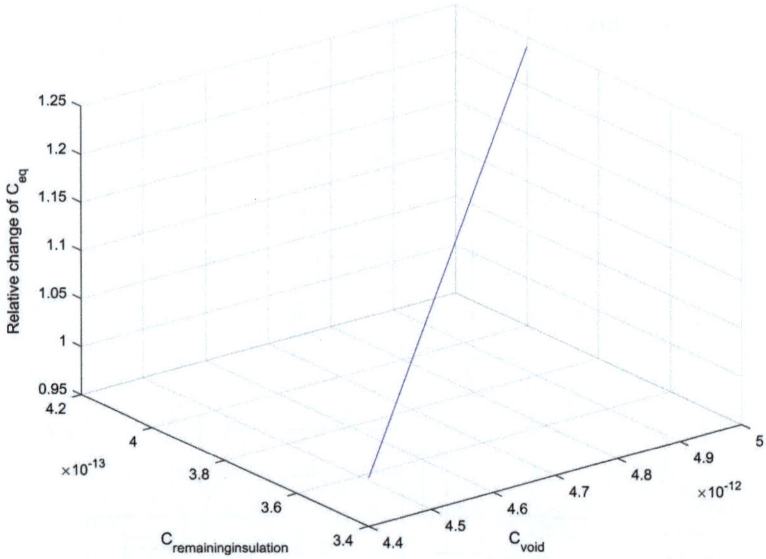

Fig. 4 3D plot of relationship between capacitance of the void, remaining insulation and relative change of the equivalent capacitance. The relative dielectric constant is changed from 1 to 1.3 with step 0.0429. The lowest point is for relative dielectric constant 1, and the height for 1.3 (30% change)

Serial branch of the equivalent circuit is expressed with:

$$C_{serial} = \frac{C_{Void} \cdot C_{remaining\_insulation}}{C_{Void} + C_{remaining\_insulation}} \tag{7}$$

$$C_{eq} = C_{serial} + C_{lumped} \tag{8}$$

Total equivalent capacitance changes linearly with the change of the relative dielectric constant:

$$\frac{C'_{serial}}{C_{serial}} = \frac{\frac{C'_{Void} \cdot C'_{remaining\_insulation}}{C'_{Void} + C'_{remaining\_insulation}}}{\frac{C_{Void} \cdot C_{remaining\_insulation}}{C_{Void} + C_{remaining\_insulation}}} = \frac{\frac{1.1 C_{Void} \cdot 1.1 C_{remaining\_insulation}}{1.1 (C_{Void} + C_{remaining\_insulation})}}{\frac{C_{Void} \cdot C_{remaining\_insulation}}{C_{Void} + C_{remaining\_insulation}}} = \frac{1.1^2}{1.1} = 1.1 \tag{9}$$

$$\frac{C'_{eq}}{C_{eq}} = \frac{C'_{serial} + C'_{lumped}}{C_{serial} + C_{lumped}} = \frac{1.1 C_{serial} + C_{lumped}}{C_{serial} + C_{lumped}} \tag{10}$$

Actual dependence on real-life values is linear, although values can be changed to produce non-linear effects (Fig. 4).

The simulation is performed using the Matlab script language. The program calls up the Simulink model in every "for-loop" with different parameter values. Input variables are varied as follows:

- relative dielectric constant from 1 to 10 in 0.1 steps,
- radius of the cylindrical void in millimeters from 0.1 to 1 in 0.1 steps,
- height of the cylindrical void in millimeters from 0.5 to 10 in 0.1 steps.

Other parameters used in this program are: $a = 3$ mm, and $b = 14.72$ mm. As previously mentioned, a denotes the length of the sample insulator, and b the breadth of the insulator sample.

Output variables from the Simulink model are exported to the workspace and stored in a high-dimensional matrix. Alternative matrix sparsity allows the use of structures to save memory.

The influence of change in ε_r on the PD current in model 1 is given in Fig. 5. Time is expressed in sample order, not in seconds. Dielectric constant index is the position of the dielectric constant value in the vector (size 1×100). Current is expressed in amperes. A wide solid area is just appearance, because there are too much points to observe separate data points.

Figure 6 shows examples of results for simulation model 2. The first serial RLC branch has a direct influence on the magnitude of the PD current, though not on the occurrence thereof. Figure 6a shows results for this block with value R = 1 Ω, and Fig. 6b for R = 1 kΩ.

Figure 7 shows results for the simulation model 3.

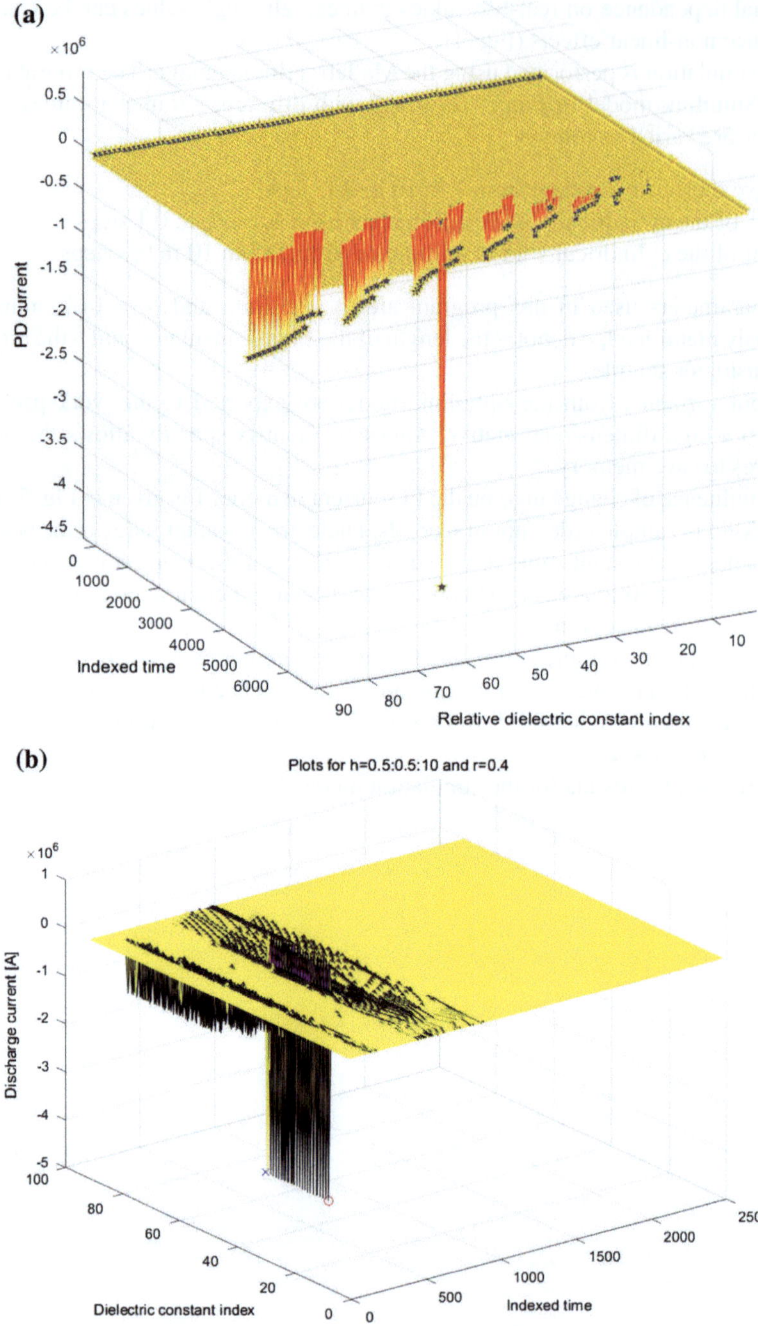

Fig. 5 Model 1—some results: **a** 3D plot of PD dependence on relative dielectric constant, **b** 3D plot of PD current dependence vs relative dielectric constant with change of geometric properties of the insulator (blue x is mark for h = 0.5, red circle is mark for h = 10 mm), **c** 3D simulation results (red circle is mark for h = 10 mm, and blue x is mark for h = 0.5 mm)

(c)

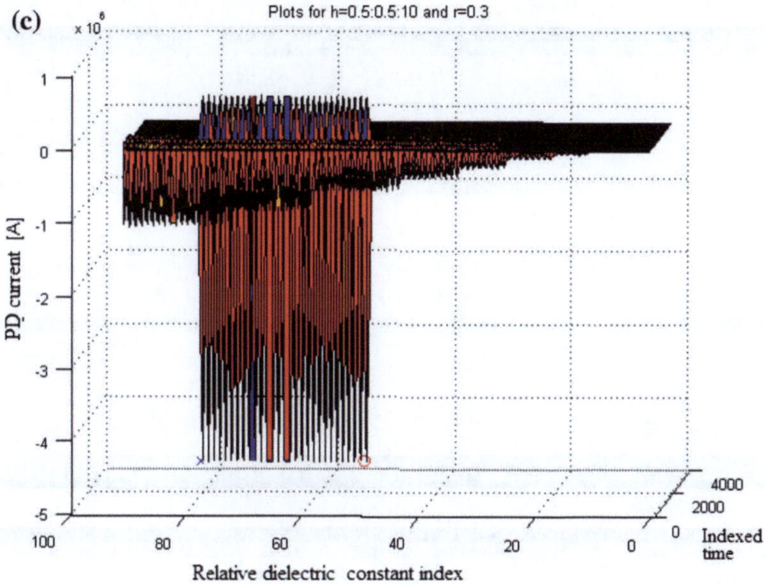

Fig. 5 (continued)

(a)

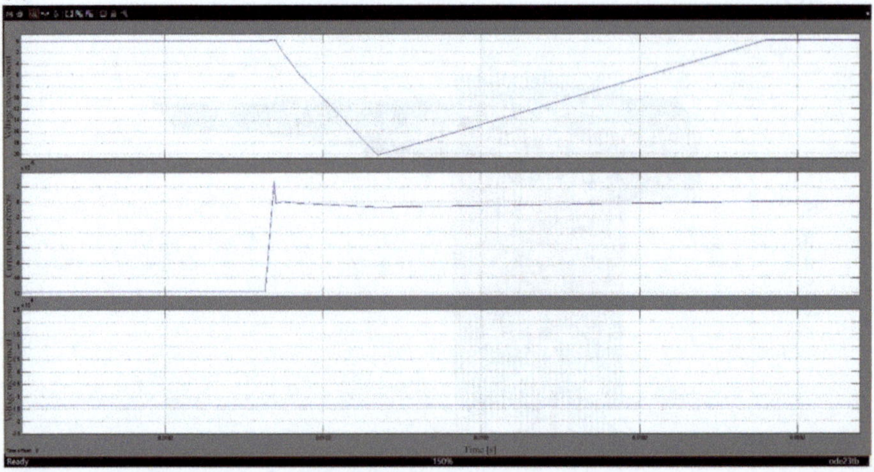

(b)

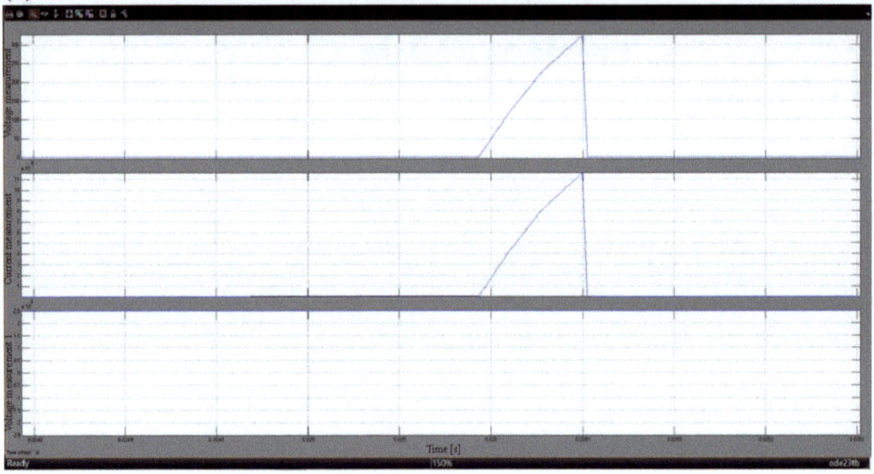

Fig. 6 Model 2—some results, zoomed parts: **a** for serial resistance 1 Ω, **b** for serial resistance 1 kΩ (no observable graph due to time zoom in the lowest subplot—no impulse at the zoomed time), **c** wider zoom, **d** unzoomed result for R = 1 kΩ (the braker's initial state = 0)

(c)

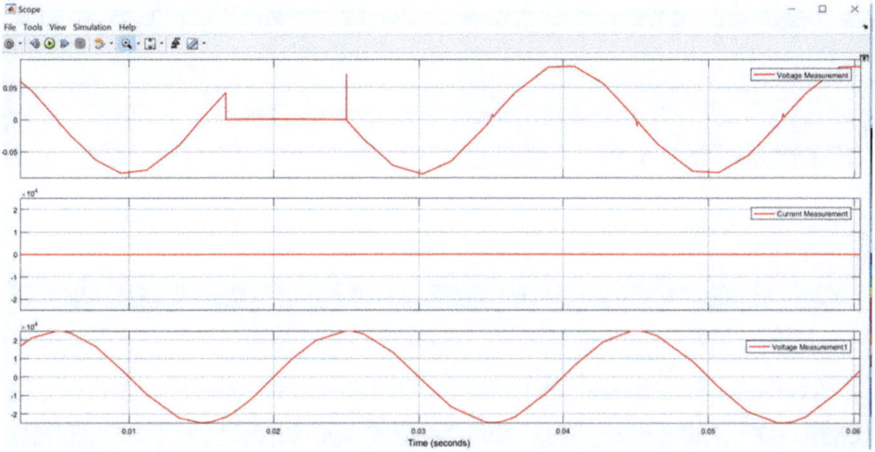

(d)

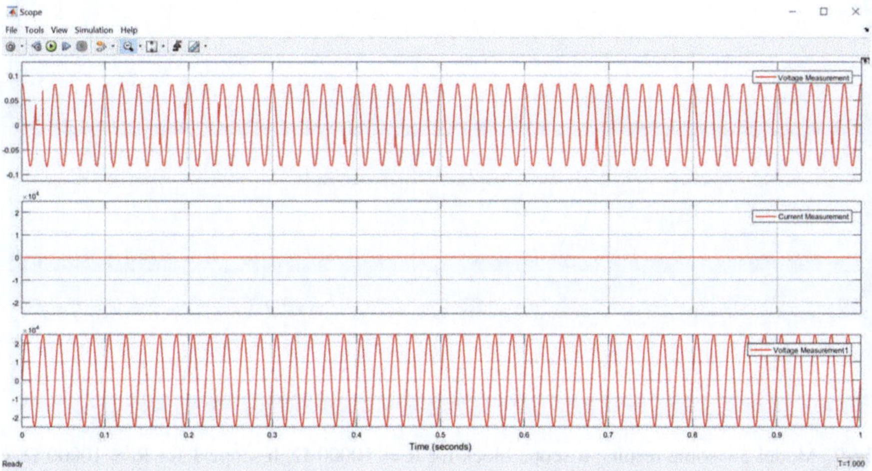

Fig. 6 (continued)

(a)

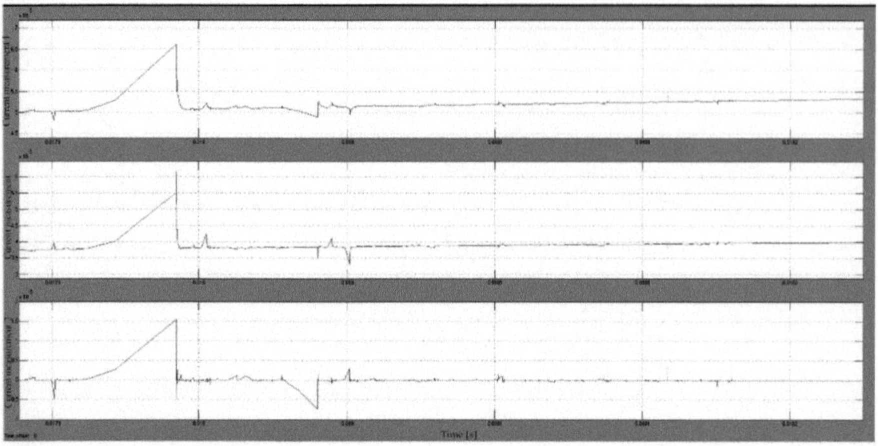

(b)

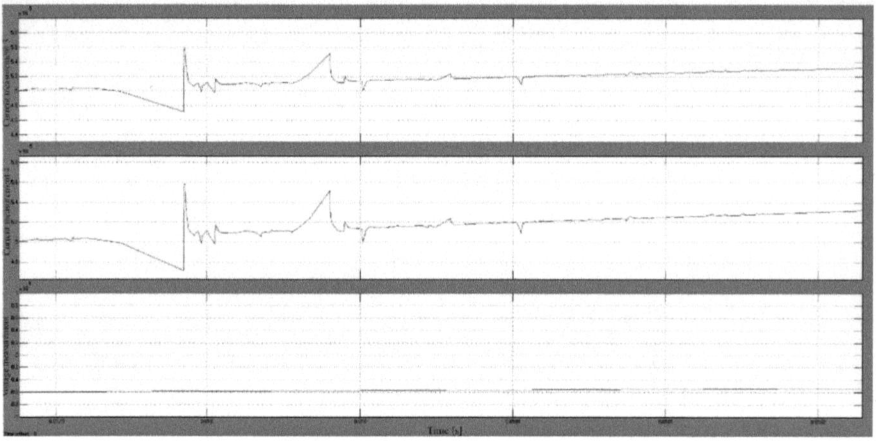

Fig. 7 Model 3—some results: **a** scope block for R = 10000 Ω, **b** scope 1 for R = 10000 Ω, **c** scope 2 for R = 10000 Ω, **d** scope block for R = 1 Ω, **e** scope 1 for R = 1 Ω, **f** scope 2 for R = 1 Ω

(c)

(d)

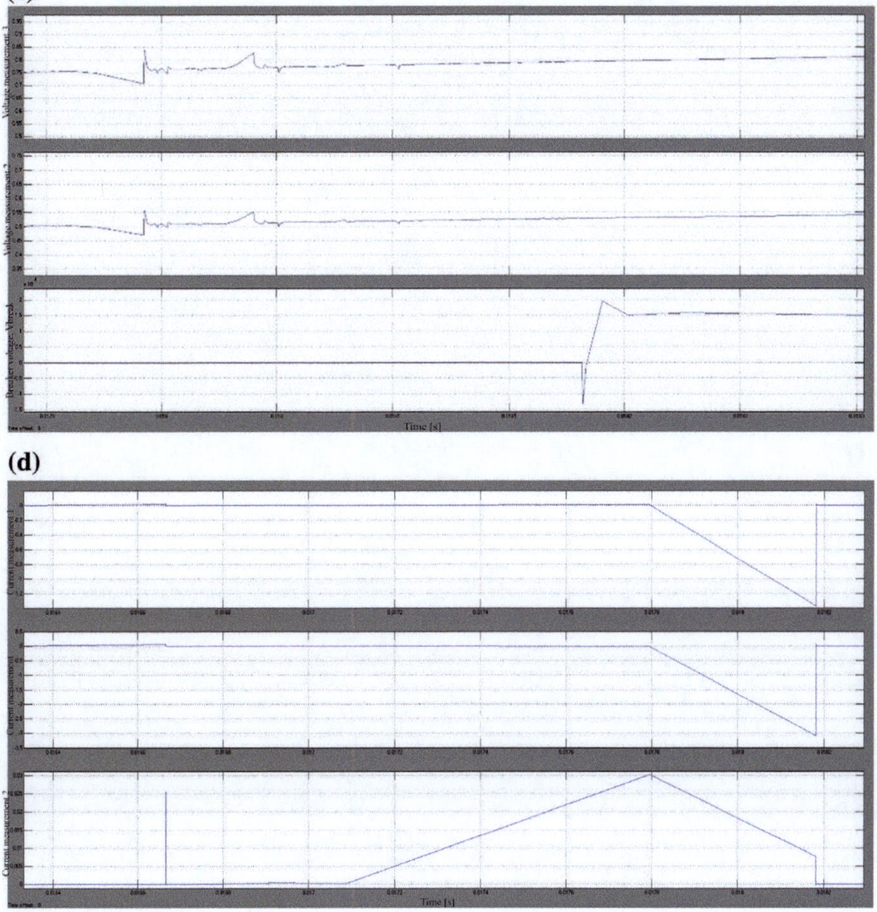

Fig. 7 (continued)

(e)

(f)

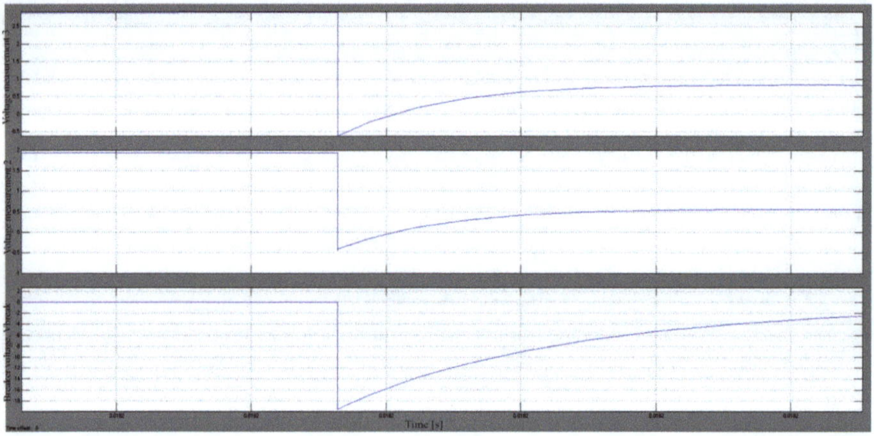

Fig. 7 (continued)

4 Conclusions

Slowly advanced degradation of the insulator results in the creation of a void, and PD occurrence. The speed of insulator degradation and the progress of the PD depend on the size of the void, the applied voltage, type of material used and other factors.

This paper compares three PD simulation models and presents some of the results obtained. The change in the relative dielectric constant, regardless of causation, was shown to effect change in the PD current and capacitance of the void.

The focus of the paper is not on large data as such, but on 3.5 GB of data obtained by simulation, difficult to comprehend without the application of visualization techniques. An attempt was made to visualize this large quantity of data with 3D graphs.

References

1. Kuzmanić, I., Vujović, I., Šoda, J.: Multiobjective optimization of dielectric material's selection in marine environment. In: Öchsner, A., Altenbach, H. (eds.) Properties and Characterization of Modern Materials. Springer, Singapore (2017)
2. Vujović, I., Kulenović, Z., Kuzmanić, I.: New algorithm for optimal dielectric material selection in marine environment. Brodogradnja **66**, 39–48 (2015)
3. Kavitha, D., Alex, N., Nambiar, T.N.P.: Classification and study on factors affecting partial discharge in cable insulation. J. Electr. Syst. **9**, 346–354 (2013)
4. Tröls, A., Kogler, A., Baumgartner, R., Kaltseis, R., Keplinger, C., Schwödiauer, R., Graz, I., Bauer, S.: Stretch dependence of the electrical breakdown strength and dielectric constant of dielectric elastomers. Smart Mater. Struct. (2013). https://doi.org/10.1088/0964-1726/22/10/104012
5. Yuliastuti, E.: Analysis of dielectric properties comparison between mineral oil and synthetic ester oil. MSc Thesis, Faculty of Electrical Engineering, Mathematics, and Computer Science, Delft University of Technology (2010)
6. Shakun, A.: Soft elastomeric material with improved dielectric permittivity. MSc Thesis, Tempere University of Technology (2014)
7. Taslak, E., Kocatepe, C., Arıkan, O., Kumru, C.F.: Electrical analysis of corn oil as an alternative to mineral oil in power transformers. Int. J. Electr. Comput. Energ. Electron. Commun. Eng. **9**, 833–837 (2015)
8. Sharma, P., Bhanddakkar, A.: Simulation model of partial discharge in power equipment. Int. J. Electr. Electron. Res. **3**, 149–155 (2015)
9. Kartalović, N., Kovačević, D., Milosavljević, S.: An advanced model of partial discharge in electrical insulation. Facta universitatis (Niš). Electron. Energ. **24**, 41–55 (2011)
10. Ehineni, K.O.: Partial discharge modeling by Simulink. MSc Thesis, Eastern Mediterranean University, Gazimagusta, North Cyprus (2014)
11. Sima, W., Jiang, C., Lewin, P., Yang, Q., Yuan, T.: Modeling of the partial discharge process in a liquid dielectric: effect of applied voltage, gap distance, and electrode type. Energ. **6**, 934–952 (2013)
12. Sharma, P., Bhanddakkar, A.: A review on partial discharge behavior in insulators. Int. J. Eng. Res. Rev. **3**, 121–127 (2015)

13. Vukšić, M., Josipović, S., Čorić, A., Kraljević, A.: Underwater ROV as inspection and development platform. Trans. Marit. Sci. ToMS (2017). https://doi.org/10.7225/toms.v06.n01.005
14. Mudronja, L., Matić, P., Katalinić, M.: Data-based modelling of significant wave height in the Adriatic sea. Trans. Marit. Sci. ToMS (2017). https://doi.org/10.7225/toms.v06.n01.001
15. Arief, Y.Z., Izzati, W.A., Adzis, Z.: Modeling of partial discharge mechanisms in solid dielectic material. Int. J. Eng. Innov. Technol. **1**, 315–320 (2012)

Dynamic and Noise Characteristics of Lead Screw in Vehicle Power Seat Adjuster

Sung-Yuk Kim and Key-Sun Kim

Abstract This study investigates the effects of the dynamic characteristics of the lead screw on the operating noise of the seat adjuster through experiments and numerical analysis. The dynamic characteristics of the lead screw were determined through a modal test in three boundary conditions, namely free-free, clamped-pinned, and the change in position of the gearbox on the lead screw. The operating noise was measured using an experimental device that minimized elements other than the lead screw. Lastly, the modal test results of the leadscrew were applied to the mathematical model, and numerical analysis was carried out. It was found that the cause of resonance noise and vibration instability of the lead screw was the incomplete restraint of the front bracket.

Keywords Seat adjuster · Lead screw · Noise · Resonance
Dynamic characteristics

1 Introduction

The vehicle seat is always in contact with the driver and its functions are very important. In addition, as drivers are spending increasingly more time behind the wheel, there is an increasing demand for seat comfort and quality [1]. As a result, power seats, which have been used primarily in high-end automobiles in the past, are becoming increasingly popular and have even begun to be installed in compact cars. This has led to considerable technological advances, such as position control systems, temperature management systems, hot wires, posture revising systems, and 12-way power

S.-Y. Kim
Department of Mechanical Engineering, Graduate School, Kongju National University, 1223-24, Cheonan Daero, Seobuk-Gu, Cheonan-Si, Chungnam 31080, Republic of Korea
e-mail: sykim5641@naver.com

K.-S. Kim (✉)
Division of Mechanical & Automotive Engineering, Kongju National University, 1223-24, Cheonan Daero, Seobuk-Gu, Cheonan-Si, Chungnam 31080, Republic of Korea
e-mail: keysun@kongju.ac.kr

© Springer International Publishing AG, part of Springer Nature 2019
A. Öchsner and H. Altenbach (eds.), *Engineering Design Applications*,
Advanced Structured Materials 92, https://doi.org/10.1007/978-3-319-79005-3_4

seats. However, most of these seat control systems are powered by electric motors, which causes consumer claims to increase annually due to various vibrations during operation and noises that have not occurred in conventional manual seats. In the case of a seat adjuster, it is a linear motion system that adjusts the seat position back and forth [2]. It has the largest size, wide operating range, and various noise sources, so continuous research is under way to improve operating noise [3]. In 1995, Cerrato et al. [4] conducted a sound quality assessment of the operating noise of the seat adjuster. In 1997, Pickering et al. [5] investigated the cause of squeal noise between the lead screw and the drive block respectively. In 2010, Kang et al. [6] constructed a lead screw-nut friction noise model using a continuum model for squeak noise in a seat screw system. In 2016, Kim et al. [7] identified the dynamic characteristics of slide rails and used experiments to confirm noise amplification due to structural resonance occurring in the 200–300 Hz region. However, the effect of the noise due to the dynamic characteristics of the lead screw was not confirmed.

Therefore, in this study, we investigated the effects of the dynamic characteristics of the lead screw on the operating noise of the seat adjuster through experiments and numerical analysis which were not clarified in previous research by Kim [7]. The dynamic characteristics of the lead screw were determined by a modal test using an impact hammer in an experimental device true to actual seat conditions and the operating noise was measured in the same experimental device. And, by constructing the lead screw-gearbox model using the energy method, the dynamic response of the lead screw according to the movement of the seat adjuster was analyzed through numerical analysis. As a result, we were able to ascertain the relationship between the dynamic characteristics of the lead screw and the operating noise generated by the seat adjuster.

2 Operating Mechanism of Seat Adjuster

The driving mechanism of the power seat adjuster causes the DC motor to rotate the worm screw and the worm wheel through the shaft. The worm screw and worm wheel are meshed inside the gearbox and the gearbox is coupled to the upper rail. The worm wheel is hollow and threaded in this hole. The thread of the worm wheel is engaged with the lead screw, and when the worm wheel rotates, the gearbox and the upper rail move together in a straight line. The forces acting on the lead screw are thrust, the tangential force, and the radial force, all of which are caused by the torque of the motor. Moreover, the weight of the seat and the force exerted by the person's load are transmitted to the seat rail. That is, the lead screw is only affected by the force transmitted by the torque of the motor. The DC motor used in this study consisted of 2 brushes and 10 slots. The reduction gear used a 2-start worm screw and 12-tooth worm wheel. Figure 1 shows the driving mechanism of the seat adjuster.

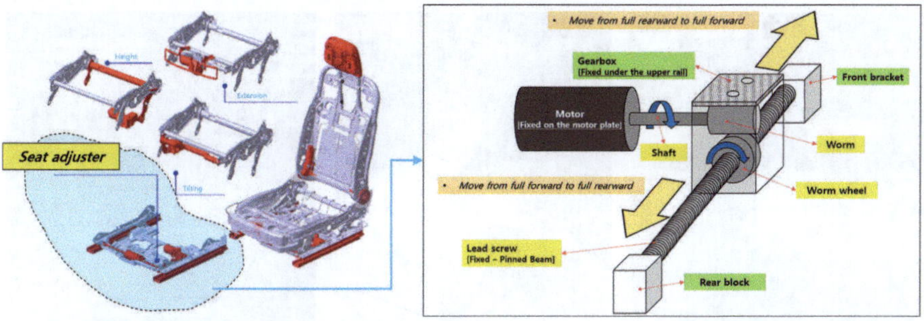

Fig. 1 Operating mechanism of power seat adjuster

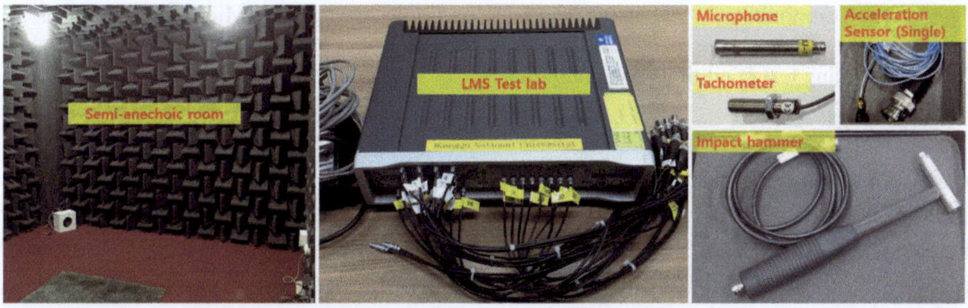

Fig. 2 Test room, measurement/analysis system and sensors

3 Experimental Conditions and Methods

3.1 Environmental Conditions and Equipment

All of the experiments performed in this study were conducted in a semi-anechoic room. The environmental conditions of the semi-anechoic room were about 20 ± 5 °C in temperature, about 40 ± 10% in humidity, and about 25 dB in background noise. Measurement and analysis were carried out using the LMS test lab to perform noise measurements and modal tests. The sensors used for operating noise and modal testing are PCB piezotronics, INC. 1/2 inch microphones, a single axis accelerometer and an impact hammer. Figure 2 shows the semi-anechoic room, instrumentation, and each sensor.

3.2 Operating Test of Seat Adjuster

The components required for the operating test of the seat adjuster were the DC motor, reduction gear & gearbox, flexible shaft, and lead screw, all of which were part of the commercial seat. In order to minimize the noise and vibration caused by

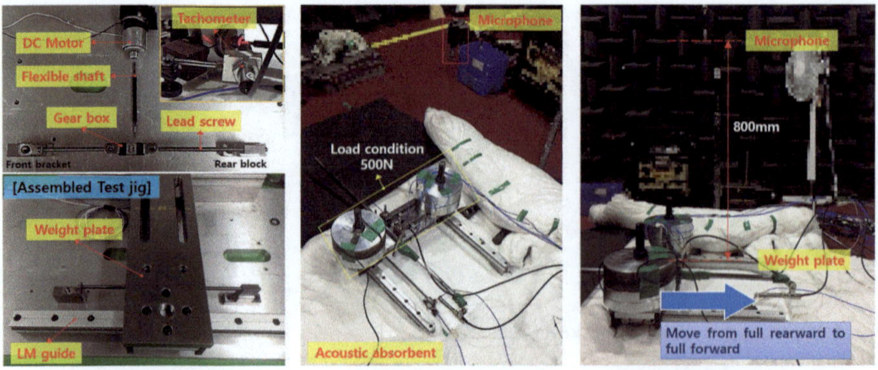

(a) Components (b) Load condition (c) Microphone position

Fig. 3 Operating test conditions and components

other seat components and frames, the experimental device as shown in Fig. 3a was used. The lead screw is mounted on each rail of the seat, but only one is used for the experimental device. In order to reproduce the seat rail, the LM guide was used to move back and forth while minimizing the influence of friction. The load was 500 N as shown in Fig. 3b, and the position of the microphone was 800 mm in the vertical direction on the weight plate, taking into consideration the location of the human ear. The operating noise measured the movement from full rearward to full forward. Figure 3c shows the operation of the experimental device and the location of the microphone.

3.3 Modal Test of Lead Screw

To analyze the dynamic characteristics of the lead screw, modal tests were carried out in three conditions using an impact hammer. The test conditions were as follows. The first was the free-free boundary condition as shown in Fig. 4, the second was the clamped-pinned boundary condition in Fig. 5, and the third condition was the lead screw being installed on the experimental device, and the gearbox was moved at 75 mm intervals, which is shown in Fig. 6. The first and second are the most basic experiments, and the basic data obtained from them was used for the numerical analysis. The third test is aimed at deriving the natural frequency of the lead screw which changes continuously with the position of the gearbox. In particular, the third test was conducted on the assumption that resonance noise would occur if the changing natural frequency of the lead screw was close to or coincident with the harmonic components of the motor.

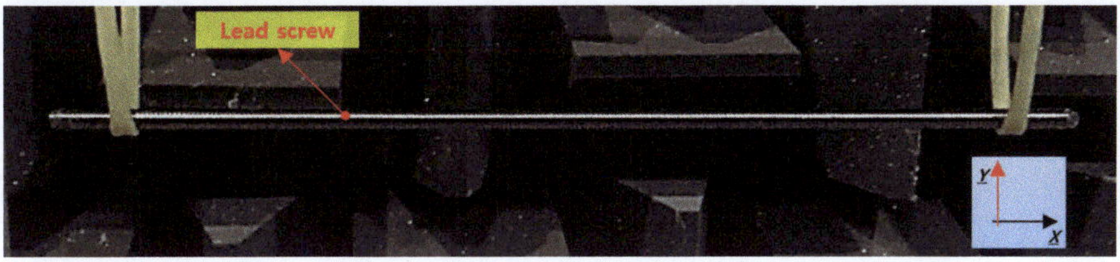

Fig. 4 Free-free boundary condition for modal test of lead screw

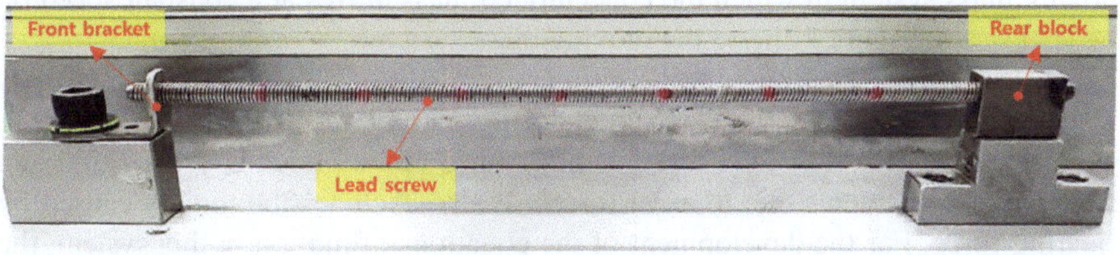

Fig. 5 Clamped-pinned boundary condition for modal test of lead screw

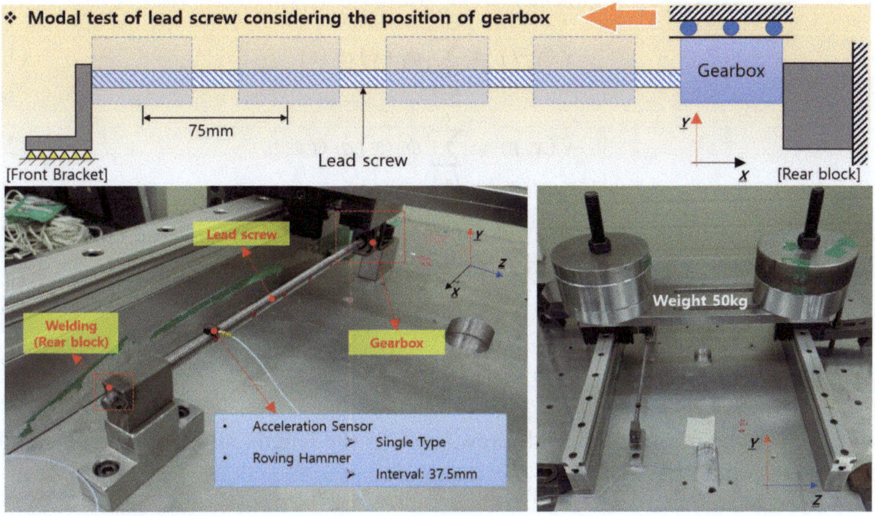

Fig. 6 Moving gearbox boundary condition for modal test of lead screw

4 Derivation of Equations of Motion

The thrust F_T, tangential force F_P, and radial force F_S are applied to the lead screw when the seat adjuster is actuated. They are all generated from the torque of the motor, and the constraint of the lead screw is clamped-pinned. The load generated by the radial force acts on the lead screw as a vertical load form of the gearbox which performs a uniform motion of a straight line. Therefore, it can be expressed as a

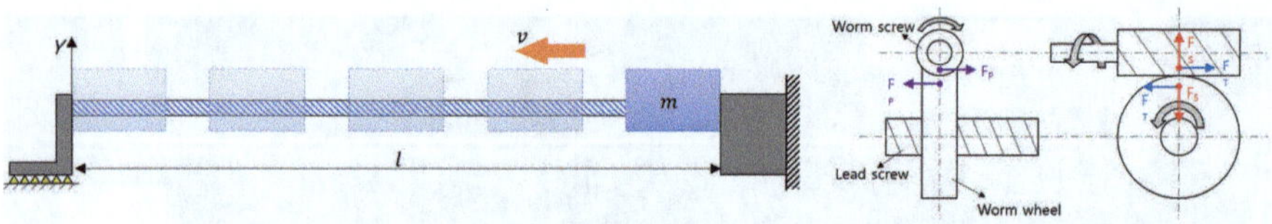

Fig. 7 A mathematical model of simple beam under a moving mass with constant speed

moving mass system in which the concentrated mass moves at a constant speed on the lead screw. Figure 7 shows the moving mass system.

The lead screw is assumed to be a Euler-Bernoulli beam, and the moving mass is always in contact with the lead screw during travel, except for all effects due to friction and gravity. The lead screw is a clamped pinned boundary condition and excludes displacements due to axial forces. It also excludes the torsion caused by shearing force. For the development of the equations of motion in this system, the lateral displacement and velocity of lead screw are expressed by modal expansion form as shown in Eqs. (1) and (2), respectively, using the assumed mode method.

$$y(x,t) = \sum_{i=1}^{n} \phi_i(x)\, q_i(t) \tag{1}$$

$$\dot{y}(x,t) = \sum_{i=1}^{n} \phi_i(x)\, \dot{q}_i(t) \tag{2}$$

where $\phi_i(x)$ is the mode shape function and $q_i(t)$ is the generalized coordinates. The kinetic energy T_B of the lead screw is expressed as Eq. (3).

$$T_B = \frac{1}{2} \int_0^l \frac{m_b}{l} \left[\sum_{i=1}^{n} \phi_i(x)\, \dot{q}_i(t) \right]^2 dx \tag{3}$$

where m_b is the mass of the lead screw, and l is the length, similarly, the potential energy is given in Eq. (4).

$$U_B = \frac{1}{2} \int_0^l EI \left[\sum_{i=1}^{n} \phi_i''(x)\, q_i(t) \right]^2 dx \tag{4}$$

where *EI* is the flexural rigidity of the lead screw.

Next, if the horizontal displacement of the moving mass m_m is x_m, the horizontal velocity v_m is expressed as Eq. (5).

$$v_m = \int_0^t v\,dt \tag{5}$$

The moving mass m_m is in contact with the lead screw, where the lateral displacement $\boldsymbol{y_m}$ is

$$y_m = y\,(x_m, t) = \sum_{i=1}^{n} \phi_i\,(x_m)\,q_i\,(t) \tag{6}$$

$$\dot{y}\,(x_m, t) = \sum_{i=1}^{n} \phi_i\,(x_m)\,\dot{q}_i\,(t) \tag{7}$$

The kinetic energy T_m of the moving mass is given in Eq. (8).

$$T_m = \frac{1}{2}\left[v\frac{\partial}{\partial x}\left(\sum_{i=1}^{n} \phi_i\,(x_m)\,q_i\,(t)\right) + \sum_{i=1}^{n} \phi_i\,(x_m)\,\dot{q}_i\,(t)\right]^2 \tag{8}$$

Therefore, using the assumed mode method, the equation of motion of the moving mass system is derived from the Lagrange equation and is shown in Eq. (9).

$$\begin{aligned} \frac{\mathrm{d}}{\mathrm{dt}}\left(\frac{\partial T}{\partial \dot{q}_i}\right) - \left(\frac{\partial T}{\partial q_i}\right) + \left(\frac{\partial U}{\partial q_i}\right) = \quad & \ddot{q}\left[\frac{m_b}{l}l\int_0^1 \{\phi_i(x)^2\}\,dx + m_m\{\phi_i\,(x_m)\}^2\right] \\ & +q\left[EI\int_0^1 \{\phi_i''(x)^2\}\,dx - m\vartheta^2\{\phi_i'\,(x_m)\}^2\right] = 0 \end{aligned} \tag{9}$$

The mode shape coefficient σ_n and the mode shape function $\phi_i\,(x)$ of the lead screw are given as Eqs. (10) and (11).

$$\sigma_n = \frac{\cosh \beta_n l - \cos \beta_n l}{\sinh \beta_n l - \sin \beta_n l} \tag{10}$$

$$\phi_i\,(x) = \cosh \beta_n x - \cos \beta_n x - \sigma_n\,(\sinh \beta_n x - \sin \beta_n x) \tag{11}$$

Since the above equation of motion is a second-order ordinary differential equation, it is transformed into a first-order ordinary differential equation by substituting variables and is expressed as a matrix.

$$[\bar{M}]\{\dot{X}\} + [\bar{K}]\{X\} = 0 \tag{12}$$

5 Results and Discussions

The operating noise of the seat adjuster generally occurs in the region below 800 Hz, and the measurement range of the modal test is set to be more than twice as high as

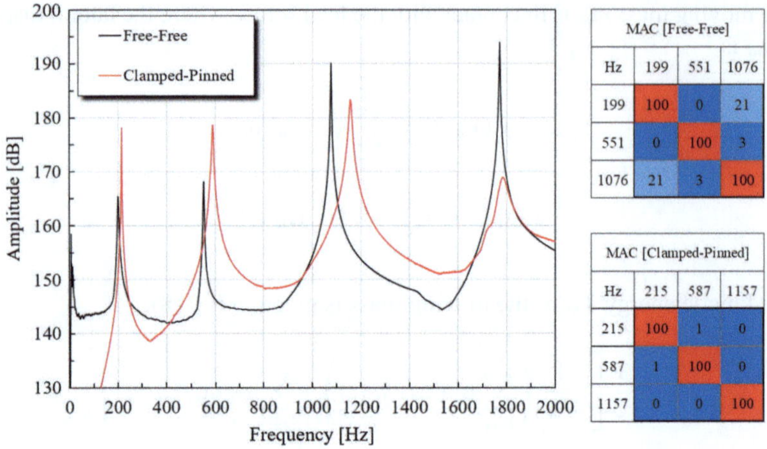

Fig. 8 FRF curve and MAC under free-free and clamped-pinned condition

Table 1 Natural frequencies of lead screw under free-free and clamped-pinned condition

List	Free-free	Clamped-pinned
Mode 1 [Hz]	199.0	214.6
Mode 2 [Hz]	550.6	586.7
$(\beta_1 l)^2$	22.4	24.1
$(\beta_2 l)^2$	61.7	65.8

800 Hz. In addition, we confirmed the orthogonality through the Modal Assurance Criterion (MAC) for each mode.

In the free-free boundary condition, the first and second natural frequencies of the lead screw were 199 and 551 Hz, respectively. In the clamped-pinned boundary condition, it was 214.6 and 586.7 Hz. Especially, in the second test, it was lower than the natural frequency occurring in the general Clamped-Pinned boundary condition. This is because the front bracket is threadedly meshed with the lead screw, and the contact area is also small. On the other hand, the rear block is welded to the lead screw, so it is completely fixed. In conclusion, because the stiffness of the front bracket is relatively smaller than rear block, it is concluded that natural frequencies are lower than the generally used clamped-pinned boundary condition. Figure 8 shows FRF and MAC derived from two tests. Table 1 shows the natural frequencies and weighted natural frequencies.

Next is the result of a modal test according to the position of the gearbox. Each natural frequency was derived by the FRF sum and MAC as shown in Fig. 9.

First, the natural frequency was higher in the rear position than the forward position of the gearbox, because the stiffness of the front bracket was relatively low compared to the rear block. Secondly, as the position of the gearbox moves to the center, the natural frequency gradually increases. It was analyzed that the gearbox acts as a constraint, which increases the natural frequency at the center position where the

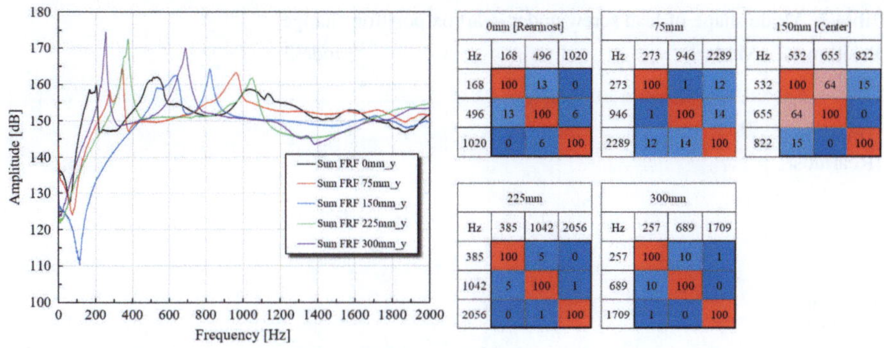

Fig. 9 FRF curve and MAC under gearbox position change

Table 2 Natural frequencies of lead screw under gearbox position change

Unit: Hz	0 mm [Rearmost]	75 mm	150 mm [Center]	225 mm	300 mm [Forefront]
Mode 1_1	167.5	272.5	532.3	384.9	257.1
Mode 1_2			654.6		
Mode 2	495.6	945.7	822.2	1042.2	689.1

length of the lead screw is divided by half. And when the position of the gearbox is located at the center, the orthogonality for the two eigenmodes, which was 535 and 655 Hz, was 64%, which were a very similar mode. This was caused by the 1st mode appearing sequentially from both sides based on the gearbox. Tables 2 and 3 show the natural frequencies and eigenmodes derived from each position.

Based on the above test results, the resonance noise was analyzed. The graph on the left in Fig. 10 shows the harmonic components of the motor and the first natural frequency of the leadscrew. The right shows that the operating noise data is converted to the spectrogram. The operating noise was amplified between 75 and 225 mm, which was found to follow the 1st natural frequency line of the leadscrew. The noise was the largest at about 510 Hz, which corresponds to the 10th order of the motor. In the case of the 10th order, it was determined that the motor noise was generated by the commutator. In conclusion, we found that the commutator noise of the seat adjuster was significantly affected by the resonance noise due to the lead screw.

Finally, numerical analysis was performed to analyze the dynamic behavior of the lead screw. Table 4 shows the system parameters used for the numerical analysis. Figure 11 shows the results of the dynamic behavior of the lead screw. In the 1st mode, the vibration of the lead screw linearly increased as the position of the gearbox moved forward. In the 2nd mode, the vibration of the lead screw was smaller than that of the 1st mode. However, it also tended to increase as the position of the gearbox moved forward. The divergent of these vibrations appeared regardless of the moving speed of gearbox and radial force of the worm screw. As a result, it was analyzed that the

Table 3 Mode shape of lead screw under gearbox position change

Position of gearbox	Mode 1	Mode 2
0 mm [Rearmost]		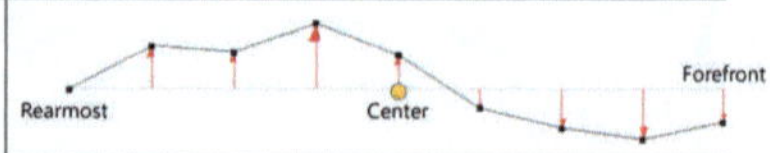
75 mm		
150 mm [Center]		–
	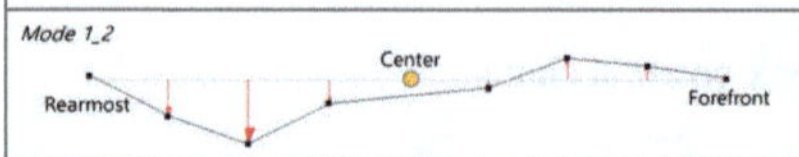	–
225 mm	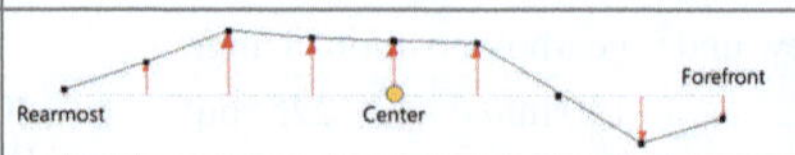	
300 mm [Forefront]	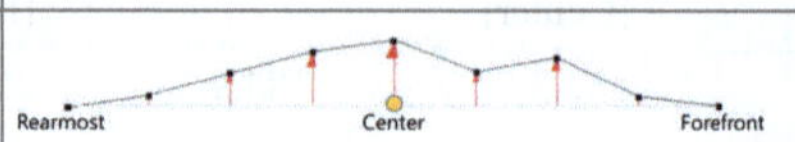	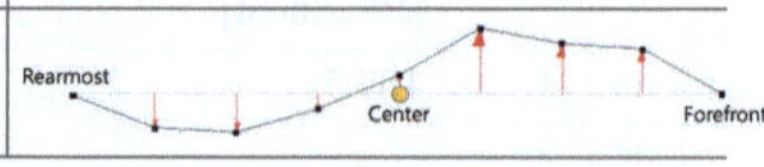

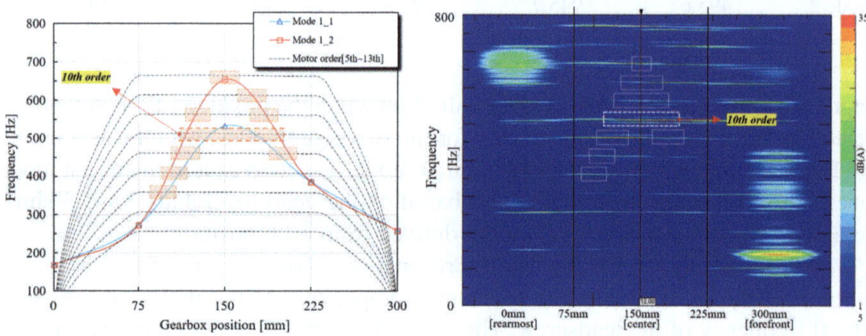

Fig. 10 Resonance noise between motor and lead screw

low stiffness of the front bracket described above has an imperfect constraint, which causes the instability of the oscillation as the gearbox moves forward.

6 Conclusion

1. The resonance noise of the lead screw occurred in the range of 300–700 Hz of the operating noise. It was found that the resonance noise occurred in the region where the harmonic components of the motor were overlapped with the curve in which the natural frequencies of the lead screw changed continuously as the

Table 4 Nominal values of system parameter

Parameter	Symbol	Value
Young's modulus of lead screw	E	210 GPa
Length of lead screw	L	0.3 m
Lead of lead screw	l	0.003 m
Pitch diameter of lead screw	d_{p1}	0.0015 m
Pitch diameter of worm screw	d_{p2}	0.008527 m
Pressure angle of worm screw	a_s	20°
Torque of worm screw	T	0.077 Nm
Radial force of worm screw	F_s	6.2 N
Motor speed	RPM	3000 rpm

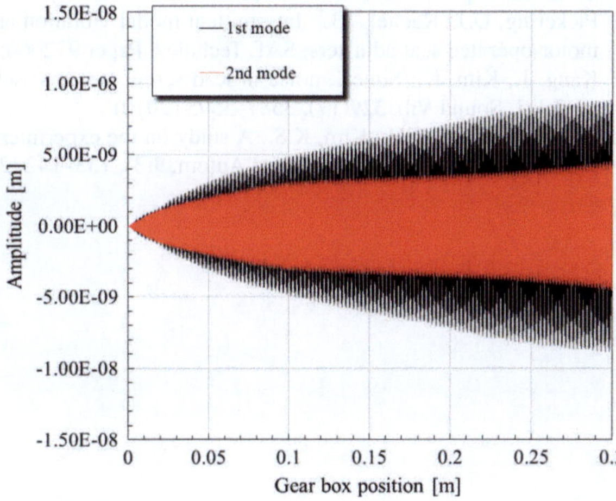

Fig. 11 Dynamic behavior of the lead screw at the position of the gear box

position of gearbox. In particular, the commutator noise of the motor, which occurred at about 510 Hz, is the noise that is constantly present among operating noise of the actual seat adjuster. That is, the amplification of the commutator noise generated by the seat adjuster is closely related to the dynamic characteristics of the lead screw.

2. To analyze the dynamic behavior of the lead screw, numerical analysis was performed using the data derived from the modal test. As a result, the instability caused by the divergence of oscillation was found in the 1st and 2nd modes. This is thought to cause unstable behavior for the seat adjuster system because it continually changes the force applied to the lead screw.
3. The resonance noise and instability of the leadscrew were caused by incomplete constraint of the front bracket. Therefore, additional research is needed on the effects of the constraint of the lead screw.

Acknowledgements This research was supported by Basic Science Research Program through the National Research Foundation of Korea(NRF) funded by the Ministry of Education(No. 2017R1D1A1B03033881).

References

1. Kim, S.Y., Jeon, O.H., Kim, K.S.: A study on the abnormal noise characteristics for moving device of automotive power seat. Int. Inf. Inst. Inf. **20**(5B), 3667–3674 (2017)
2. Kang, J., Seo, K., Kim, K.: Experimental investigation of friction noise in lead screw system under mode-coupling. J. Mech. Sci. Technol. **29**(12), 5183–5188 (2015)
3. Laux, P.C., Kleiss, J.A., Huff, J.E.: Power seat adjuster noise metric development & correlation to subjective response data. SAE Technical Paper 2005-01-2474:1–6 (2005)
4. Cerrato, G., Crewe, A., Terech, J.: Sound quality assessment of powered seat adjusters. SAE Technical Paper 951288:435–440 (1995)
5. Pickering, D.J., Rachel, T.L.: Intermittent modal vibration and squeal sounds found in electric motor-operated seat adjusters. SAE Technical Paper 972060:1517–1526 (1997)
6. Kang, J., Kim, K.: Squeak noise in lead screw systems: self-excited vibration of continuous model. J. Sound Vib. **329**(17), 3587–3595 (2010)
7. Kim, S.Y., Jeon, O.H., Kim, K.S.: A study on the experimental analysis of noise from vehicle power seat slide rail. Int. J. Control Autom. **9**(3), 133–142 (2016)

ICT-Based Integrated Operation Management System Design and Plant Growth Monitoring System Development for Plant Factories

Y. C. Kim and M. T. Cho

Abstract Plant factory aim to achieve excellent productivity by automatically controlling plant growth in an enclosed environment. Because plant factories are very complex systems with various sensors and controls, an integrated operating system that controls the sophisticated and complex operation of these devices is very important. In this paper, by designing the plant analysis system of plant based on ICT based on the core technologies such as overcoming the limit of the place, the sun instead of light, automation, nutrient supply system, temperature, etc. and solar power generation, it increased the freshness of agricultural products through production and created a new market by merging and combining with IT and BT industry. It also improved convenience of farming work and improved comfort of working space by automatic control and robot development. Then, the method of measuring the development status of plants using two cameras was studied. Using the plant measurement algorithm, you can build a system that can get cheap and reliable data with just a camera and a simple system. From the construction of such a system, we have developed a monitoring system that can efficiently measure the size of plants growing in plant factories. By using this system, it is possible to save the time required to maintain and manage the size measurement data of the plant and to save the expenses accordingly, and it is possible to obtain the effect of improving the work efficiency of the plant manager, we confirmed the performance of the proposed system.

Keywords Plant factory · Integrated operating system · Plant growth · IT · ICT

Y. C. Kim
Division of Mechanical and Automotive Engineering, Kongju National University, 1223-24, Cheonan Daero, Seobuk-gu, Cheonan-si, Chungnam 31080, Republic of Korea
e-mail: yckim59@kongju.ac.kr

M. T. Cho (✉)
Department of Electrical and Electronic Engineering, Daewon University College, 316 Daehak Road, Jechen-si, Jecheon, Chungbuk Province 27135, Republic of Korea
e-mail: mtcho@daewon.ac.kr

© Springer International Publishing AG, part of Springer Nature 2019

A. Öchsner and H. Altenbach (eds.), *Engineering Design Applications*,
Advanced Structured Materials 92, https://doi.org/10.1007/978-3-319-79005-3_5

1 Introduction

Plant factory can capture greenhouse gases in cities by collecting carbon dioxide that is soaring in urban air and supplying them to plant factories. In addition, plant factories can save energy by using LED, which is a typical technology of greenhouse gas saving, as a light source. On the other hand, in order to manage the plants that grow continuously in the plant, the process of confirming the size of the plant has been essential. However, in order to measure the size of a plant in a plant, the manager had to manually measure the size of each plant and manage the data. Therefore, there is a demand for a system that can effectively monitor the growth of plants within a plant factory that cultivates and manages plants in large quantities [1–4]. And plant factory is a plant production system of artificial ecosystem, which controls the growth environment artificially and produces agricultural products of uniform quality according to the production plan like the manufacturing plant. These plant factories are not activated well, but the problem is that the operating costs, high production costs and production technologies, are not profitable except for high construction costs. There is a need to overcome this and this can be solved through standardization of facilities/production/operation.

In order to solve the problem, a monitoring system capable of efficiently measuring the size of a plant growing in a plant has been developed. By using this system, it is possible to save the time required to maintain and manage the size measurement data of the plant and to save the expenses accordingly, and it is possible to obtain the effect of improving the work efficiency of the plant manager, we confirmed the performance of the proposed system.

2 ICT and Plant Factories

The plant is completely surrounded by artificial environment, but man is monitored, but there is no hand, computer and robot are controlled. Also, unfamiliar utilities are growing plants. Instead of the sun, the LED is lit for 24 h, halving the harvest time. Oxygen and carbon dioxide are constantly discharged, reducing the amount of dioxins emitted by other industries. Plant plants artificially control the environment to control temperature, humidity, and carbon emissions, and create an environment where plants can grow optimally. Plant movement and monitoring are both computerized. One of the advantages of such a system is its many possibilities. It can be cultivated anywhere, like a ship or a ship. But there are challenges to overcome right now. In the absence of constant climatic conditions, such as the Middle East, re-usable energy is needed, such as solar energy, to provide a stable and sustainable energy supply, but it is still expensive. The most problematic is that plant factories require a lot of energy and electricity to cool down. For the success of plant factories, energy-saving technologies must be preceded.

2.1 Plant Factory Service Model

Figure 1 shows the linkage of unit services by layering IT service modules as service models of plant factories. In the development of the service model, the design is top-down, and the system development needs to be developed from the bottom-up, that is, the lower technology. This is a community model in which producers and consumers of plants coexist with each other. It is an integrated total and history management, ERP (purchasing, distribution and accounting), growth management (plant growth monitoring and control) Information database) and Knowledge DB (Growth Environment Model, Growth Manual).

2.2 System Architecture

The basic framework for the development of plant factory systems is highly efficient and large-scale management when constructed based on EDA. EDA is an Event Driven Architecture that consists of a system that responds to an event when an event occurs. It is composed of a unit system and a loosely coupled system. Event can be transferred to work with new system or new system flexibly. Figure 2 shows the structure of Publish/Subscribes in EDA.

Figure 3 is a system architecture diagram based on the Event Driven Architecture (EDA) architecture. The integrated management system consists of three structures, and three unit systems of growth management system, growth management system and growth analysis system are loosely interlocked in the form of Pub-Sub.

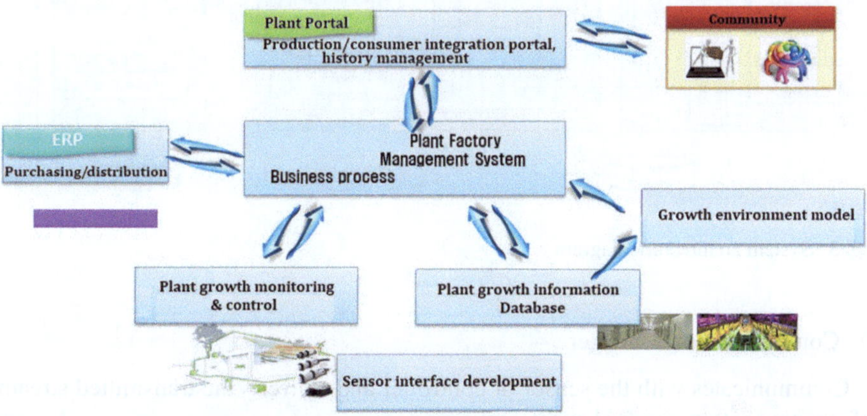

Fig. 1 Plant factory service community model

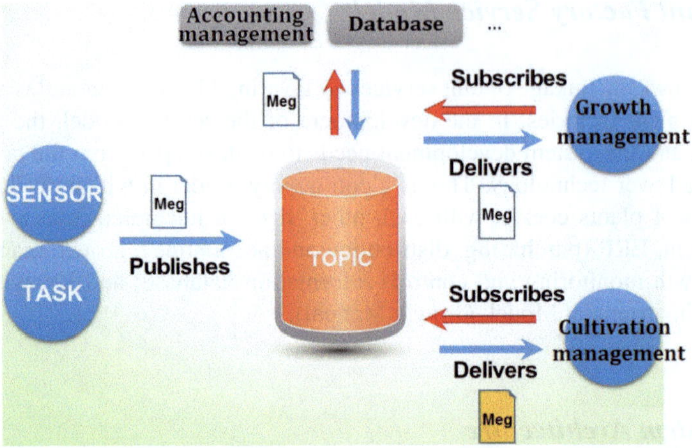

Fig. 2 Structure of publish/subscribes in EDA

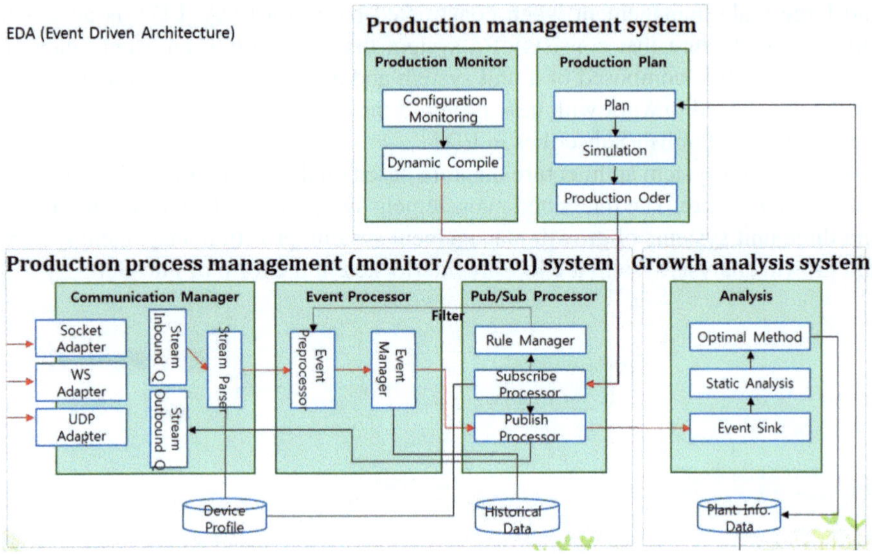

Fig. 3 System architecture diagram

① Communication Manager

Communicates with the sensor or controller and converts the transmitted stream information into a standard event conversion.

② Event Processor

Analysis of received events according to growing conditions.

③ Publish/Subscribe Processor

EDA architecture interoperates with other systems, registers event types and specific results that require transmission, and distributes events to requesters based on the results of event analysis.

④ Production Plan

Production planning/growth method (growth manual) and simulation of production according to growth method.

⑤ Production Monitor

Monitoring screen, method configuration and production, environmental monitoring and control.

⑥ Experiment Analysis

Analyze the growth results by production lot and prepare optimized growth manual according to the result.

2.3 Plant Factory Management Process

Plant management processes register farm information, crop information and growing manuals with basic information on the farm, and seasonal crop production plans are established according to the sales plan and predict harvest. In addition, schedule and environmental conditions for cultivated crops and work are set according to cultivation manuals. When production is started according to the production plan, it automatically monitors the growing environment and growth condition according to the growing manual and performs the farming according to the work schedule. The environmental condition analyzed based on the information collected during the cultivation period is reflected in the existing cultivation manual, and a new cultivation manual is created, registered and reflected in the future production plan.

The registration information for basic information on plant cultivation is as follows.

① Plant factory or cultivation farm is registered, and plant factory is subdivided into field and section. It registers the manager, location and cultivation method by operating unit, and can reflect climate information with location information.
② Method is information about recipe manual, recipe, and it manages task by schedule and environment condition at this time. At this time. Environmental conditions are registered at the time of cultivation by sensing and environment registration of the operator.

The next step is to register the production plan. The management of lot unit is used for optimal cultivation environment analysis and history management of shipment crops. Assign a field or section based on the current cultivation status at the time of production planning. Next, select the crop and cultivation manual and confirm the work schedule. Based on this, the work schedule and yield are predicted before the plantation is prepared. First of all, Asset, that is, registered equipment and sensor are assigned to Section for management and control of cultivation environment. When registering an asset and starting cultivation, it monitors and controls information about each section and informs the task according to work schedule.

2.4 Plant Factory Monitoring and Control

Figure 4 shows the scenario of plant factory production management that can be used as an analysis data by controlling the environment according to the growth manual of the environmental sensor monitoring information and creating the agrochemical journal. When the cultivation starts in the plan, the environmental information is monitored through the sensor installed at the cultivation management stage and the environment such as environment and nutrient solution is controlled according to the cultivation method selected at the cultivation planning stage. In addition, breeding status and cultivation daily log are prepared.

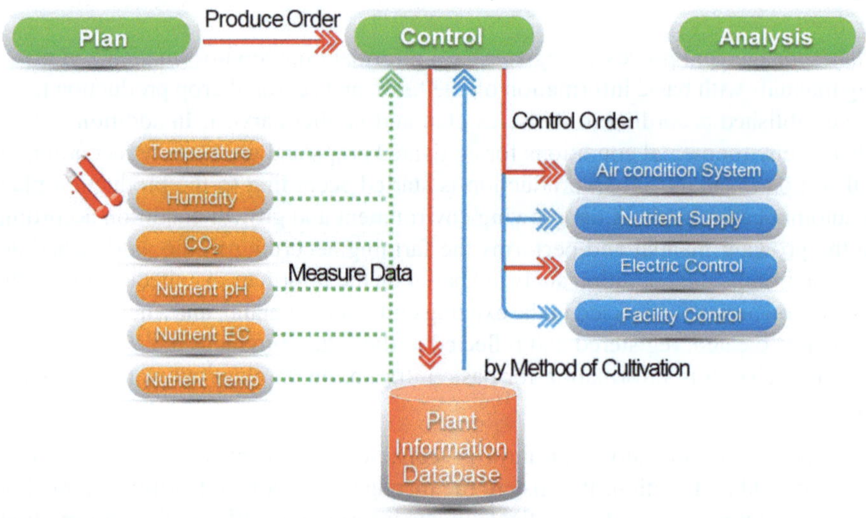

Fig. 4 Plant factory production management scenario

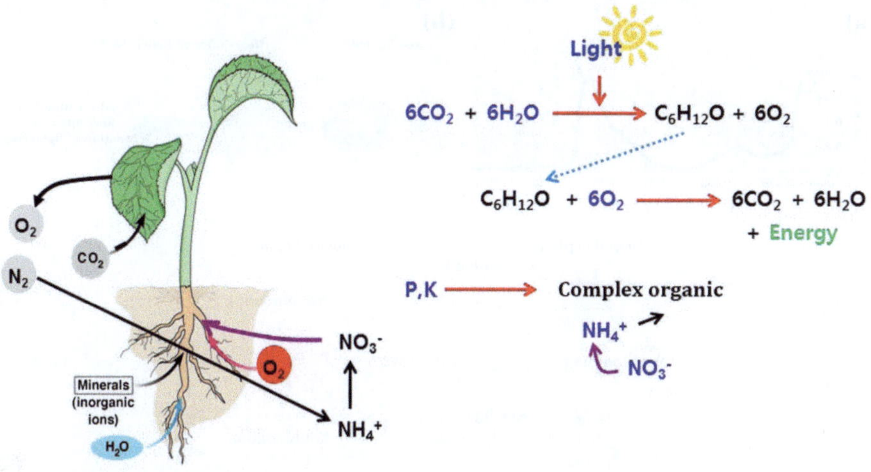

Fig. 5 Metabolic process of plants

3 Plant Growth Measurement

3.1 Growth of Plants

In a natural ecosystem, photosynthesis is the most important chemistry for securing energy in the form of chemicals using light. As shown in Fig. 5, plants acquire energy sources by photosynthesis, absorb respiration and various organic matter, and grow through biosynthesis. Cellular respiration is a metabolic process in which living organisms decompose organic compounds to obtain energy, which is a complex chemical action. What is shown here is a factor for the growth of the plant and shows how it will produce a high quality product when the conditions are met. Of course, this is not all of the elements, but it should be managed when forming an artificial eco system.

3.2 Biorthythm of Plants

A plant's biorhythm is a periodic system called a bio-clock, and people are able to grow up well if they manage the same way as humans, just as humans become harmed by the sunrise and biorhythm activity and sleeping. Figure 6 shows the absorbance of red light and near infrared light.

The photo period regulates flowering according to the relative length of the night and day, while the circadian rhythm is involved in the flowering, motion, and bloom in a 24-h cycle of biorhythm, the rhythm of the circadian rhythm does not appear but is controlled by a time measuring device (biological clock) inside the living body.

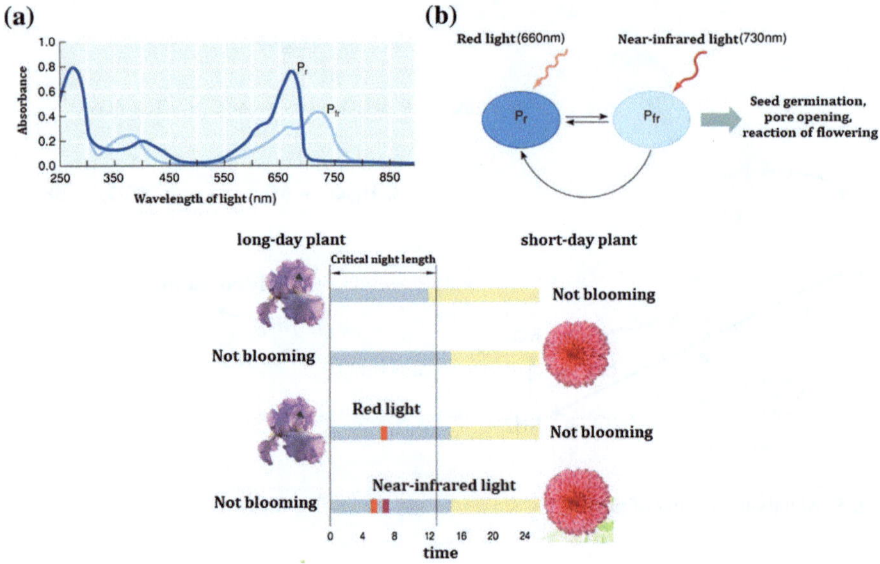

Fig. 6 Absorbance according to red light and near infrared light

4 Vision-Based Growth Measurement

In order to predict the crop production for the planned production, database of cultivation environment information, image and production volume was created, and real—time image leaf area analysis program and middleware were developed.

The photosynthetic efficiency was measured by measuring the amount of carbon dioxide consumed by the combination of red (630 [nm]) and blue (450 [nm]) 1: 1.

Also, photosynthetic requirement of plants and physiological metabolism of plants were measured for simultaneous cultivation of plants with similar photosynthetic efficiency. This was done by measuring the plant height, root width, and leaf area in the image information and by regression analysis by variable selection and multivariate analysis by best subset regression.

4.1 Image Processing

The shape of the growth is measured in 1–3 dimensional form, and the plant is evaluated on the average or at a certain point of time by changing the growth pattern during the day. Four methods are used for image processing.

① Point Processing

And processes the pixel based on the value possessed by the pixel.

② Area Processing

Image processing using mask and convolution.

③ Topological Processing.

Processing for changing the arrangement of pixels by arbitrary geometric transformation.

④ Frame Processing

Generate pixel values based on operations involving two or more different images.

4.2 Machine Learning

It is used for image processing or field recognition through machine analysis. It is classified into 4 kinds.

① Supervised Learning

A function is deduced from training data. In general, an attribute of the input object is included as a vector, and the desired result is displayed for each vector.

② Unsupervised Learning

Inferred from unlabeled data, autonomous learning derives patterns from data, such as density estimation of statistics.

③ SVM

And the distance between the support vectors in the second plane.

④ K-means

Reflectivity data The distance between the normal leaf and the stained leaf of the cluster is calculated.

5 Conclusion

Conventional plant factory integrated operation system is controlled and simple monitoring through devices installed in plant factories. We also used equipment implemented by industry developers' experience in monitoring and controlling the plant factory environment. In particular, the automatic control method for the growth environment of the integrated plant management system of the existing plant factory has a static structure to which the algorithm that does not consider the characteristics of crops is applied, so that the whole program must be modified in order to apply the growth algorithm of other crops is.

Therefore, the following results were obtained by the development of integrated operation management system of ICT-based plant factory presented in this paper.

① Increase freshness of agricultural products through order production and planned production.
② Creating a new market through convergence with IT and BT industry.
③ Enhancement of convenience of agricultural work and improvement of work space comfort by automatic control and robot development.
④ Provide citizens with the opportunity to experience and learn the entire process of plant growth.
⑤ Give life to the citizens who are tired of the city life like the oasis in the city center.
⑥ Prevention of environmental pollution through efficient recycling of resources.
⑦ Produce and supply various agricultural products stably throughout the year without being influenced by climate.

These artificial ecosystem plant production systems are able to produce uniform quality agricultural products according to the production plan by artificial control of the growing environment and production automation. However, the high production cost, high operation cost, lack of exclusive breed, If the problem is solved in the plant factories in terms of stable profitability, it will contribute to the improvement of the competitiveness of farmers and the increase of profits.

References

1. Analog Devices.: 12-Bit R/D Converter with Reference Oscillator. Analog Devices (2003)
2. Analog Devices: Using the ADSP-2100 family volume 1. Analog Devices Rev **1**, 51–66 (1990)
3. Texas Instruments.: TMS320F240 DSP Solution for Obtaining Resolver Angular Position and Speed. Application Report SPRA605 (2000)
4. Chung, S., Kim, H.M., Kim, S.D.: Formulation of stable Bacillus subtilis AH18 against temperature fluctuation with highly heat-resistant endospores and micropore inoranic carriers. Appl. Microbiol. Biotechnol. **76**, 217–224 (2007)

Cable Railway Simulation: A Two-Span Oscillator Model

Markus Wenin, Michael Irschara, Stephan Obexer, Maria Letizia Bertotti and Giovanni Modanese

Abstract In the dynamic behaviour of a cable railway oscillations of cables and cars play an important role. We present a simple model to describe and investigate oscillations of a cable, spanned over a support and charged with an arbitrary number of point loads with arbitrary masses. We construct a time-dependent propagator, which contains the full intrinsic information of the mechanical system and represents a linear map between the initial state, $t = 0$ (initial condition of a set of linear differential equations) and the state at a time t. We consider undamped and damped oscillations, where damping is introduced by a phenomenological way (Onsager's lineare ansätze). A numerical example is given.

Keywords Cable railway · Cable oscillations · Initial state independent description

M. Wenin (✉)
CPE Computational Physics and Engineering, Weingartnerstr. 28, 39011 Lana, BZ, Italy
e-mail: markus.wenin@cphysics.com

M. Irschara
I&M Ingenieure, Kapuzinerplatz 9, 39031 Bruneck, BZ, Italy
e-mail: michael.irschara@im-ing.it

S. Obexer
Ropes GmbH, Staatstr. 15, 39030 Vintl, BZ, Italy
e-mail: s.obexer@ropes.cc

M. L. Bertotti · G. Modanese
Faculty of Science and Technology, Free University of Bozen/Bolzano, Piazza Università 1, 39100 Bolzano, BZ, Italy
e-mail: marialetizia.bertotti@unibz.it

G. Modanese
e-mail: giovanni.modanese@unibz.it

© Springer International Publishing AG, part of Springer Nature 2019
A. Öchsner and H. Altenbach (eds.), *Engineering Design Applications*,
Advanced Structured Materials 92, https://doi.org/10.1007/978-3-319-79005-3_6

1 Introduction

Apart from a quasi static approach, which gives the fundamental numerical quantities needed for the design of a cable ropeway [1, 2], dynamical investigations and especially oscillations gain more and more importance [3–7]. Many analytical and semi-analytical, partially older, works exist about oscillations of cables [8–11]. In practice, vibrations of cables are induced by wind, and in particular when a cable car passes a support [12, 13].

In this contribution we develop a simple model to study in-plane oscillations of a cable spanned over a support, such that two cable spans form an interacting oscillatory system. The treatment of damping is discussed. Our focus is to bring the problem into an ideal form for a solution using higher level programming languages such as *Mathematica* or *Matlab* [14, 15]. Furthermore, we develop a description where the dynamics of the system can be described independent of its initial state by introduction of a linear map (propagator). Such an approach is widely used in quantum mechanics [16] and has the advantage to allow a discussion of the intrinsic properties of the mechanical system.

As usual, the presented model is invalid for higher oscillation frequencies of the cable, corresponding to shorter wavelengths, because we cannot neglect the bending stiffness of a real rope [17]. The paper is organized as follows: Sec. II describes the two-span model and introduces the basic quantities. The computation of the equilibrium is outlined. In Sec. III we discuss the dynamics of the cable and the loads, where we consider the undamped and the damped case respectively. We use the Lagrangian formalism to derive the governing equations. Sec. IV is devoted to numerical studies and Sec. V to a summary. The appendix contains some mathematical material needed to handle the constraints in the Lagrangian of the system.

2 Two-Span Model System

We consider a simple model to study the effects of energy transfer between two spans. The inelastic cable is fixed at the end points and frictionless movable on the support. It is characterized by a linear mass density ρ_L; g denotes the earth's acceleration. The empty cable configuration (the catenary) is given, when we fix the cable tension F_0 at one end (usually the valley side) of the line [1]. In order to obtain a system suitable for numerical studies, we discretize the cable into $N_{1,2}$ rigid parts of length $\Delta l_{1,2} = L_{1,2}/N_{1,2}$, where $L_{1,2}$ are the lengths of spans 1,2 respectively. Therefore the condition of constant total cable length is fulfilled. The mass of such an empty chain element is $m_i^{(1,2)} = \rho_L \Delta l_{1,2}$ and is centered in the middle, whereas the mass vector for a charged cable, written for the first span, is $\{m_1^{(1)}, m_2^{(1)}, \dots m_{k-1}^{(1)}, m_k^{(1)} + M^Z, m_{k+1}^{(1)}, \dots m_{N_1}^{(1)}\}$, where M^Z is the mass of the load at position k. In this way a quite simple generalization to an arbitrary number of different point loads is possible. We use the angles $\phi_j^{(1,2)}$ of the chain elements with respect

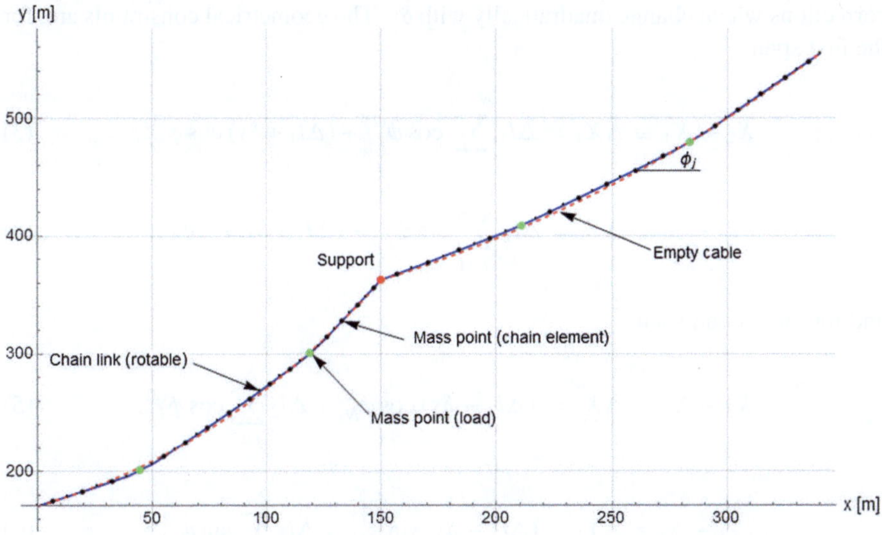

Fig. 1 Sketch of the model considered in this work. The cable is replaced by a chain with N_1 and N_2 chain elements for two spans respectively. For $N_{1,2} \to \infty$ the solution for the empty cable approaches the catenary equation $y = a \cosh\left(\frac{x - x_m}{a}\right) - c$, with appropriate constants a, x_m, c for both spans

to the x- axis and the cable displacement at the support position δs as generalized coordinates (see Fig. 1). The coordinates of the endpoints and the support are denoted as $\{X_j, Y_j\}$, $j = 1, 2, 3$. We start with the potential energy of the cable, consisting of the two parts,

$$U^{(1)} = g \sum_{i=1}^{N_1-1} m_i^{(1)} \left\{ Y_1 + \Delta l_1 \sum_{j=1}^{i-1} \sin\phi_j^{(1)} + \frac{1}{2} \Delta l_1 \sin\phi_i^{(1)} \right\}$$
$$+ g m_{N_1}^{(1)} \left\{ Y_1 + \Delta l_1 \sum_{j=1}^{N_1-1} \sin\phi_j^{(1)} + \frac{1}{2} (\Delta l_1 + \delta s) \sin\phi_{N_1}^{(1)} \right\}, \qquad (1)$$

for the first span, and

$$U^{(2)} = g m_1^{(2)} \left\{ Y_2 + \frac{1}{2} (\Delta l_2 - \delta s) \sin\phi_1^{(2)} \right\}$$
$$+ g \sum_{i=2}^{N_2} m_i^{(2)} \left\{ Y_2 + (\Delta l_2 - \delta s) \sin\phi_1^{(2)} + \Delta l_2 \sum_{j=2}^{i-1} \sin\phi_j^{(2)} + \frac{1}{2} \Delta l_2 \sin\phi_i^{(2)} \right\}, \qquad (2)$$

for the second, respectively. One should remark that in this approach we make a small error because we treat $m_{N_1}^{(1)}$ and $m_1^{(2)}$ as constant, which means that we neglect the small contribution $\pm\rho_L \delta s$. However δs is a small quantity and this term leads to corrections which change quadratically with δs. The geometrical constraints are, for the first span

$$X_2 - X_1 \equiv \Delta X_1 = \Delta l_1 \sum_{j=1}^{N_1-1} \cos\phi_j^{(1)} + (\Delta l_1 + \delta s)\cos\phi_{N_1}^{(1)}, \tag{3}$$

$$Y_2 - Y_1 \equiv \Delta Y_1 = \Delta l_1 \sum_{j=1}^{N_1-1} \sin\phi_j^{(1)} + (\Delta l_1 + \delta s)\sin\phi_{N_1}^{(1)}, \tag{4}$$

and for the second span

$$X_3 - X_2 \equiv \Delta X_2 = (\Delta l_2 - \delta s)\cos\phi_{N_2}^{(2)} + \Delta l_2 \sum_{j=2}^{N_2} \cos\phi_j^{(2)}, \tag{5}$$

$$Y_3 - Y_2 \equiv \Delta Y_2 = (\Delta l_2 - \delta s)\sin\phi_{N_2}^{(2)} + \Delta l_2 \sum_{j=2}^{N_2} \sin\phi_j^{(2)}. \tag{6}$$

The numerical local minimization is a straightforward and quite stable procedure, if we use the values for the empty cable as initial guess (calculated from continuum cable model) and $\delta s = 0$. For the equilibrium values of the generalized coordinates we use the symbols $\psi_j^{(1)}, \delta l_{eq}, \psi_j^{(2)}$,

$$\left\{\psi_j^{(1)}, \delta l_{eq}, \psi_j^{(2)}\right\} = \min\left\{U^{(1)} + U^{(2)}\right\}, \tag{7}$$

subject to the constr. Eqs. 3–6.

3 Dynamics in the Linear Regime

A. Undamped System

To study the dynamics of the two-span system we use the Euler-Lagrange equations, and we develop the theory as in [18]. To do this we need to expand the potential energy at the minimum, particularly to incorporate the constraints in a correct way. As we have four constraints we can eliminate the four variables, $\phi_1^{(1)}, \phi_{N_1}^{(1)}, \phi_1^{(2)}$ and $\phi_{N_2}^{(2)}$. The corresponding computations are outlined in the appendix. As a result we obtain a set of coupled harmonic oscillator equations, which are well suited for a numerical treatment.

1. ***Kinetic energy***

The coordinates of the i-th mass point within the first span are

$$x_i^{(1)} = X_1 + \Delta l_1 \sum_{j=1}^{i-1} \cos\phi_j^{(1)} + \frac{1}{2}\Delta l_1 \cos\phi_i^{(1)}, \tag{8}$$

$$y_i^{(1)} = Y_1 + \Delta l_1 \sum_{j=1}^{i-1} \sin\phi_j^{(1)} + \frac{1}{2}\Delta l_1 \sin\phi_i^{(1)}, \quad i = 1 \ldots N_1 - 1. \tag{9}$$

The midpoint of the last chain element has the coordinates

$$x_{N_1}^{(1)} = X_1 + \Delta l_1 \sum_{j=1}^{N_1-1} \cos\phi_j^{(1)} + \frac{1}{2}(\Delta l_1 + \delta l_{eq} + \delta l) \cos\phi_{N_1}^{(1)}, \tag{10}$$

$$y_{N_1}^{(1)} = Y_1 + \Delta l_1 \sum_{j=1}^{N_1-1} \sin\phi_j^{(1)} + \frac{1}{2}(\Delta l_1 + \delta l_{eq} + \delta l) \sin\phi_{N_1}^{(1)}. \tag{11}$$

Here $\phi_j^{(1)}$ and δl are time-dependent variables, $\phi_1^{(1)}$ and $\phi_{N_1}^{(1)}$ are eliminated by the constraints using Eqs. 27 and 28. Similar expressions hold for the second span, where $\delta l_{eq} + \delta l$ enters in $x_1^{(2)}$, $y_1^{(2)}$ with an opposite sign. The kinetic energy T is given as usual by

$$T = T^{(1)} + T^{(2)} = \frac{1}{2}\sum_{i=1}^{N_1} m_i^{(1)} \left[\left(\dot{x}_i^{(1)}\right)^2 + \left(\dot{y}_i^{(1)}\right)^2\right] + \frac{1}{2}\sum_{i=1}^{N_2} m_i^{(2)} \left[\left(\dot{x}_i^{(2)}\right)^2 + \left(\dot{y}_i^{(2)}\right)^2\right]. \tag{12}$$

2. ***Equations of motion***

If we denote with $\mathbf{q} = \left\{\delta l, \delta\phi_2^{(1)}, \ldots \delta\phi_{N_1-1}^{(1)}, \delta\phi_2^{(2)}, \ldots \delta\phi_{N_2-1}^{(2)}\right\}$ the vector of the generalized coordinates, measured from the equilibrium position, we can write the 'stiffness' matrix elements as

$$K_{ik} = \left(\frac{\partial^2(U^{(1)} + U^{(2)})}{\partial q_i \partial q_k}\right)_{q_i=0}. \tag{13}$$

The mass matrix elements M_{ik} are defined through the quadratic form ($N_{tot} = N_1 + N_2 - 3$)

$$T = \frac{1}{2}\sum_{i,k=1}^{N_{tot}} M_{ik}\dot{q}_i\dot{q}_k, \quad M_{ik} = \left(\frac{\partial^2(T^{(1)} + T^{(2)})}{\partial \dot{q}_i \partial \dot{q}_k}\right)_{q_i=0}. \tag{14}$$

The matrix elements K_{ik} and M_{ik} are quick to compute analytically by a Computer Algebra System as *Mathematica* (with symmetric matrix and ability to parallelize). The Lagrangian

$$L = T - U^{(1)} - U^{(2)} = \frac{1}{2} \sum_{i,k=1}^{N_{tot}} \{M_{ik}\dot{q}_i\dot{q}_k - K_{ik}q_iq_k\}, \tag{15}$$

gives as Euler-Lagrange equations the coupled system of harmonic oscillators

$$\sum_{k=1}^{N_{tot}} \{M_{ik}\ddot{q}_k - K_{ik}q_k\} = 0, \quad i = 1 \ldots N_{tot}. \tag{16}$$

With the matrices $K = (K_{ik})$ and $M = (M_{ik})$ the eigenvalue equation for the eigenvectors, labeled by n, $\mathbf{u}_n$ and eigenvalues Ω_n^2 reads ($\hat{1}_{N_{tot}}$ is the $N_{tot} \times N_{tot}$ identity matrix)

$$\left(M^{-1}K - \Omega_n^2\hat{1}_{N_{tot}}\right)\mathbf{u}_n = 0, \quad n = 1, \ldots N_{tot}. \tag{17}$$

We remark that the eigenvectors in general are not orthogonal (but we assume always normalized to 1). Using the eigenvalues and eigenvectors we can write the solution with initial conditions $\mathbf{q}(0)$, $\dot{\mathbf{q}}(0)$ as

$$\mathbf{q}(t) = \sum_n \left\{c_n e^{i\Omega_n t} + c_n^* e^{-i\Omega_n t}\right\}\mathbf{u}_n, \tag{18}$$

where, with $U_i = \mathbf{q}(0) \cdot \mathbf{u}_i$, $V_i = \dot{\mathbf{q}}(0) \cdot \mathbf{u}_i$, $A_{mn} = \mathbf{u}_m \cdot \mathbf{u}_n$, $B_{mn} = i\Omega_n A_{mn}$, the coefficients in vector notation are given as

$$\mathbf{c} = \frac{1}{2}\left(A^{-1}\mathbf{U} + B^{-1}\mathbf{V}\right). \tag{19}$$

For the special case $\mathbf{U} = 0$, i.e. the system is in equilibrium position at $t = 0$, we obtain for the generalized coordinates $\mathbf{q}(t) = i \sum_{n=1}^{N_{tot}} \left(c_n - c_n^*\right) \sin(\Omega_n t)\,\mathbf{u}_n$.

B. The Damped System

To introduce damping in the system, we have to add in Eq. 16 an additional term,

$$\sum_{k=1}^{N_{tot}} \{M_{ik}\ddot{q}_k - K_{ik}q_k + \alpha_{ik}\dot{q}_k\} = 0, \tag{20}$$

where the symmetric matrix (α_{ik}) in our case is a diagonal matrix, $\alpha_{11} = \mu_1 \geq 0$, $\alpha_{kk} = \mu_2 \geq 0$, $k = 2, \ldots N_{tot}$, and all other elements are zero. The parameter μ_1 is related to the dynamic friction resistance of the cable on the support and one can assume proportionality to the static normal force on the support. Here we assume, for simplicity, a constant value. The parameter μ_2 can be identify as the aerodynamic drag and the inner friction resistance of the cable, as a result of small relative displacements of the single cable wires during the motion. For a numerical computation of the eigenvalues it is convenient to introduce a new quantity $\mathbf{Q}(t) = \{\mathbf{q}(t), \mathbf{p}(t)\}$ to transform the system of ODE's from second to first order, by setting $\dot{\mathbf{q}}(t) = \mathbf{p}(t)$. One obtains ($\widehat{O}_{N_{tot}}$ is the $N_{tot} \times N_{tot}$ zero matrix)

$$\begin{pmatrix} \dot{\mathbf{q}} \\ \dot{\mathbf{p}} \end{pmatrix} = \begin{pmatrix} \widehat{O}_{N_{tot}} & \hat{1}_{N_{tot}} \\ -M^{-1}K & -M^{-1}\alpha \end{pmatrix} \begin{pmatrix} \mathbf{q} \\ \mathbf{p} \end{pmatrix}, \tag{21}$$

or

$$\dot{\mathbf{Q}}(t) = H\mathbf{Q}(t), \tag{22}$$

where H is the time-independent coefficient matrix from above. We define the new eigenvalue problem with complex eigenvalues using the ansatz $\mathbf{Q} = e^{i\Omega_n t}\mathbf{v}_n$,

$$H\mathbf{v}_n = i\Omega_n \mathbf{v}_n, \quad n = 1 \ldots 2N_{tot}. \tag{23}$$

The prefactor i ensures the usual definition of the eigenvalues: Re(Ω_n) describes oscillations and Im(Ω_n) damping, respectively. Using matrix exponentials, the solution of Eq. 22 is given by

$$\mathbf{Q}(t) = S(t)\mathbf{Q}(0), \quad S(t) = e^{Ht}, \quad S(0) = \hat{1}_{2N_{tot}}, \tag{24}$$

showing linearity in the initial condition. The numerical computation of the matrix exponential using the eigen system of H or a power series is a standard procedure [19]. Conducted in the spirit of quantum mechanics, this strategy is particularly useful when one is interested on the numerical integration of Eq. 22 for many initial conditions (for instance a distribution of initial conditions around a mean value to simulate uncertainty in a measurement process), because one evaluates the propagator S(t) one time only. It allows one to study the system's properties without consideration of the initial state, because all intrinsic information is captured by H. Obviously Eq. 24 is general and also valid for vanishing damping. For the case with damping, the limit $S(t \to \infty) = \widehat{O}_{2N_{tot}}$ holds; any initial state is projected onto the equilibrium state.

4 Numerical Demonstration

To demonstrate the presented theoretical concepts, we consider a concrete numerical example. We assume a cable with $\rho_L = 1$ kg/m, the lengths of the empty cable spans are $L_1 = 247.5$ m, $L_2 = 272.1$ m, the tension at valley side F_0

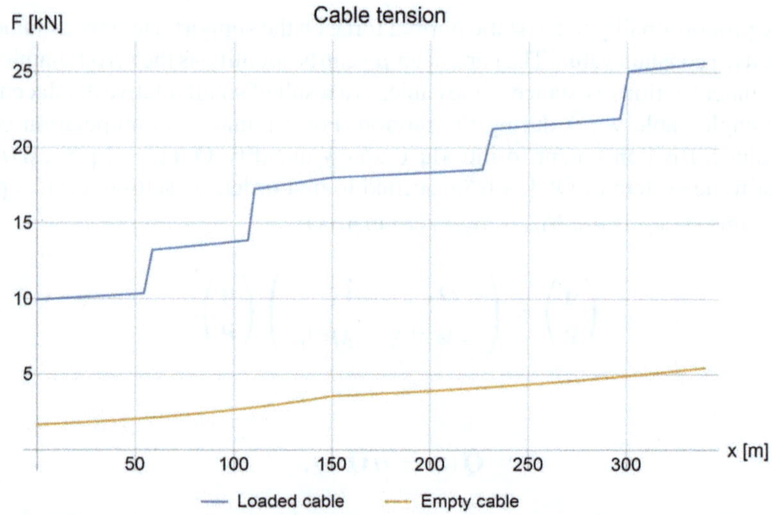

Fig. 2 Cable tension as a function of position for the empty and the loaded cable in equilibrium. When the equilibrium cable angles are known (computed by solving Eq. 7) the equations for the balance of forces are sufficient to calculate the cable tension as a function of position. In the case of the loaded cable the tension is discontinuous at the positions of the mass points and continuous at the support position

= 1.68 kN. The mass points are localized at the arc lengths (from the left) $s = \{74.2, 160.8, 349.5, 451.9\}$ m and all have masses of 400 kg. With a discretization of $N_1 = N_2 = 40$ elements one obtains $\Delta l_1 = 6.2$ m, $\Delta l_2 = 6.8$ m, and $\delta l_{eq} = -0.02$ m. To give an example of the dynamics we use the initial condition $\mathbf{q}(0) = \{0, \ldots 0\}$, $\dot{\mathbf{q}}(0) = \{0.7$ m/s, $0, \ldots 0\}$. For the damping constants we set (estimations) $\mu_1 = 1000$ kg/s and $\mu_2 = 500$ kg/s.

Figure 2 shows the cable tension of the empty and the loaded cables respectively, obtained after the computation of the equilibrium state. It is an important quantity for practical engineering, to ensure safety against breaking [2]. Figures 3–4 show the matrices (K_{ik}) and (M_{ik}) for the loaded cable respectively, where the sub-blocks and the presence of the mass points are visible (patterns). These are computed symbolically according to Eqs. 13–14, and evaluated thereafter numerically. In Fig. 5 we show the eigenfrequencies for the loaded cable without damping. It is interesting to compare the result with the eigenfrequencies of a fixed, weightless (but not massless) string, where the eigenvalue problem is solvable analytically. For small values of n the frequencies are comparable, whereas for higher values the x-dependent cable tension and the presence of the support shows an effect. Figure 6 is a matrixplot of H, showing the structure of Eq. 21. It is a non-symmetric matrix (even in the sub-blocks) with complex eigenvalues as shown in Fig. 7, which is a plot of the eigenfrequencies in the complex plane. Due to the constant values for $\mu_{1,2}$ the damping has more relative influence on the motion of the empty cable than on the loaded cable. Figure 8 shows the cable displacement on the support $\delta l(t)$ for the loaded cable with and without

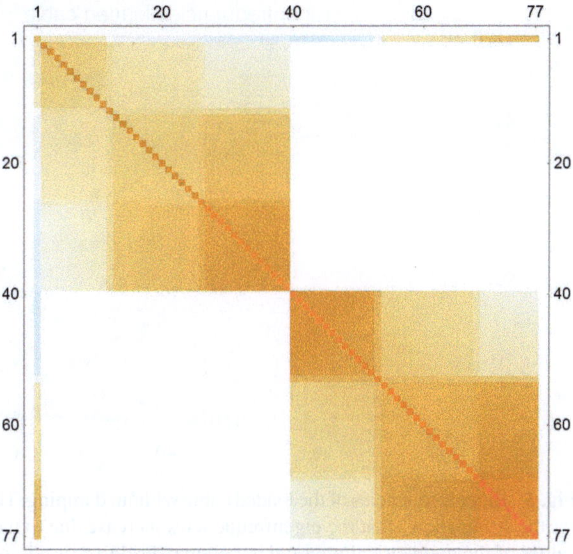

Fig. 3 Block structure of the symmetric (K_{ik}) matrix. The bands in the first row/column arise from the constraints

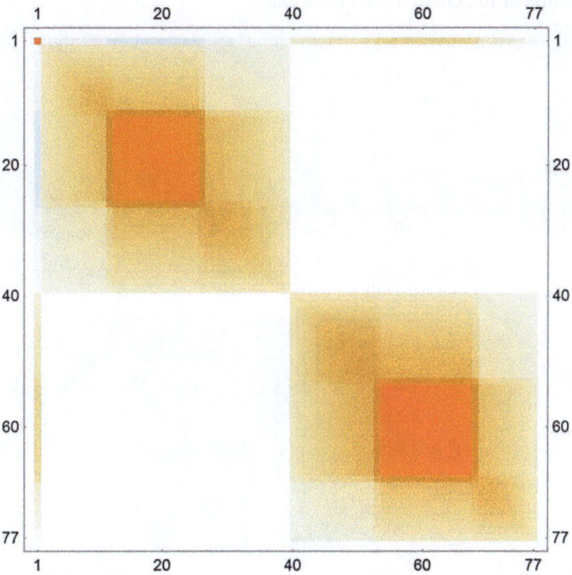

Fig. 4 Block structure of the symmetric (M_{ik}) mass matrix. The bands in the first row/column arise from the constraints

damping. In the latter case the presence of higher order terms in the Fourier series, Eq. 18, is hard to see (the amplitude for $n = 3$ is about 1% of the amplitude for $n = 1$ and all other amplitudes are negligible). The quasi-period of the oscillations is about 9 s.

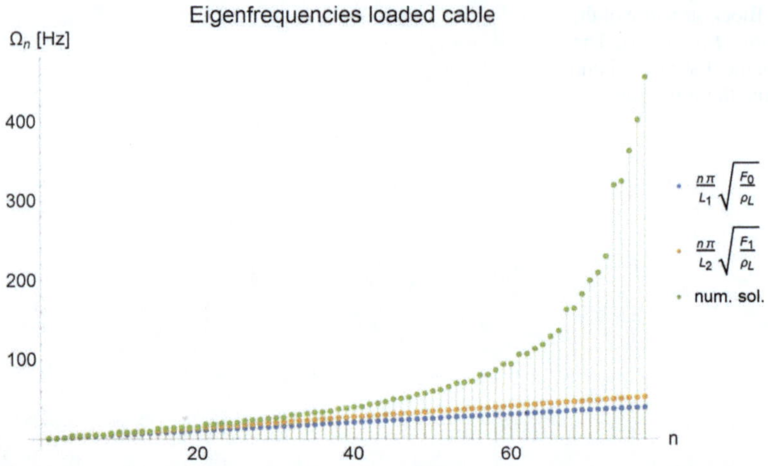

Fig. 5 Eigenfrequencies of the loaded cable without damping. The comparison with the eigenvalues for the string shows that the eigenfrequencies increase due to the higher cable tension. The chosen values of the geometrical/physical parameters lead to a quasi-continuum in the spectrum for lower frequencies. The higher frequencies however are not of interest because of damping and of the limits of the chain model for a cable

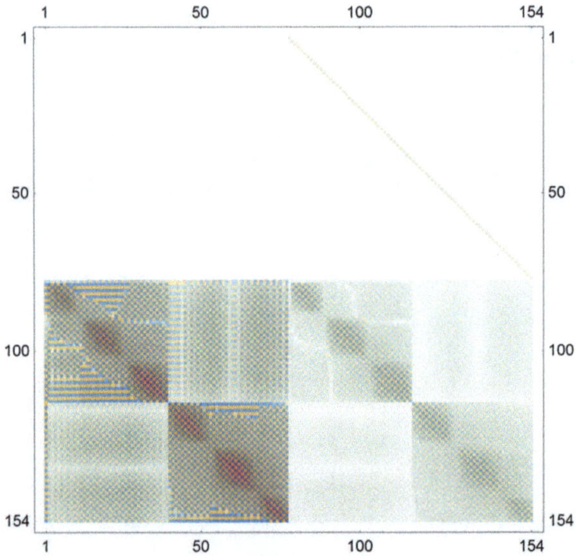

Fig. 6 Block structure of the H matrix, containing the four blocks with sub-blocks according to Eq. 21

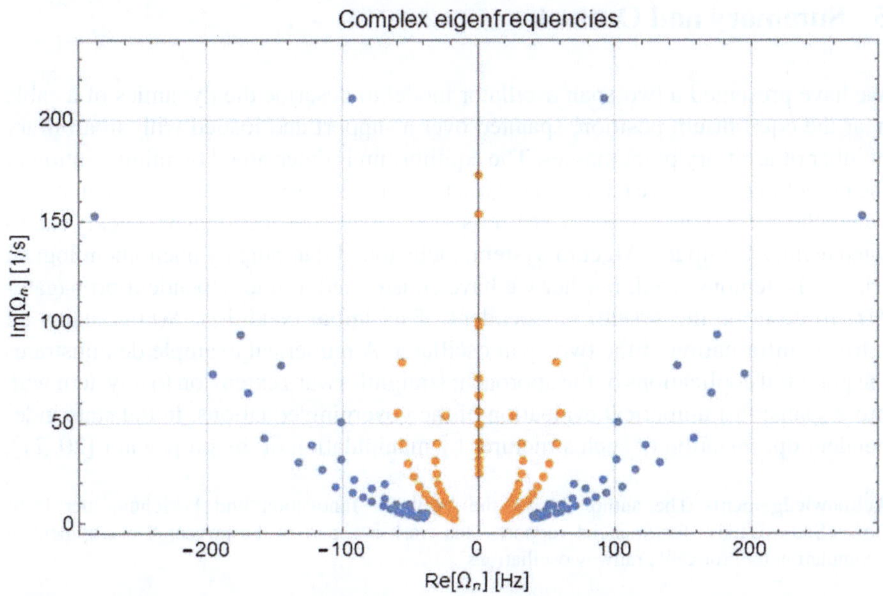

Fig. 7 Eigenfrequencies in the complex plane, where orange points correspond to the empty and blue points to the loaded cable respectively. These are symmetric along the imaginary axis to guarantee a real solution by superposition and have strictly positive imaginary part to ensure decay of oscillations. For overdamped systems some eigenvalues are located at the imaginary axis. One can see a decrease of the maximum frequency (≈300 Hz) compared with the undamped case (greater than 400 Hz, see Fig. 8)

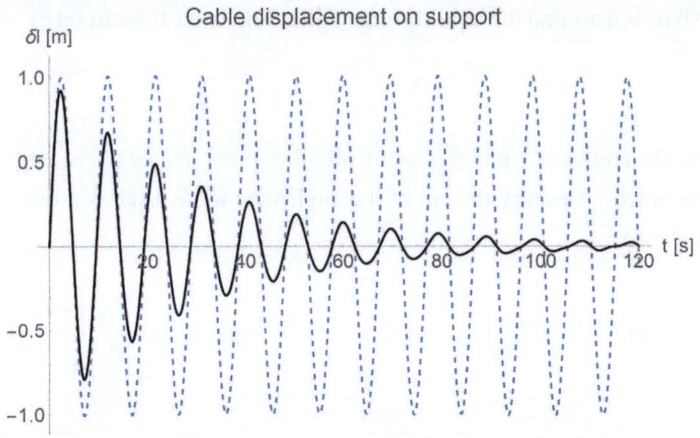

Fig. 8 Cable displacement on the support for the loaded, undamped (blue dashed line) and the damped two-span oscillator system (black solid line)

5 Summary and Outlook

We have presented a two-span oscillator model to describe the dynamics of a cable near the equilibrium position, spanned over a support and loaded with an arbitrary number of arbitrary point masses. The equilibrium is determined by minimization of the potential energy. We have derived a set of coupled harmonic oscillator equations, where the coefficient-matrices ('stiffness'—and mass matrices) can be quickly calculated using a Computer Algebra System; inclusion of damping by phenomenological friction is demonstrated. Further we have constructed a time-dependent propagator $S(t)$ to describe the system independent of its initial condition. $S(t)$ contains all intrinsic information of the two-span oscillator. A numerical example demonstrates the potential applications of the approach: straightforward extension to a system with more spans; fast numerical evaluation of the governing equations; initial state independent optimization of such structures by manipulation of the propagator [20, 21].

Acknowledgements The authors thank the 'Amt für Innovation und Forschung der Prov. Bozen/Südtirol/Italy' for financial support. This work is a part of the project 'Development of a Simulation tool for cable railway oscillations'.

Appendix: Treatment of Constraints

In this appendix we give the explicit expressions for the angles $\phi_1^{(1)}, \phi_{N_1}^{(1)}, \phi_1^{(2)}$ and $\phi_{N_2}^{(2)}$, needed to derive the Lagrangian L. The angles are correct up to second order of a series expansion around the equilibrium angles. For δs we have to set $\delta s = \delta l_{eq} + \delta l$ in Eqs. 3–6.

1. ***Span 1***

We express the angles $\phi_1^{(1)}$ and $\phi_{N_1}^{(1)}$ using all other inner angles $\phi_2^{(1)} \ldots \phi_{N_1-1}^{(1)}$. As we need in the potential energy the sin of the angles we write a series around $\psi_1^{(1)}$ and $\psi_{N_1}^{(1)}$ respectively,

$$\sin\phi_1^{(1)} = \sin\left(\psi_1^{(1)} + \delta\phi_1^{(1)}\right) = \sin\psi_1^{(1)} + \cos\psi_1^{(1)}\delta\phi_1^{(1)} - \sin\psi_1^{(1)}\frac{\left(\delta\phi_1^{(1)}\right)^2}{2} + \cdots, \tag{25}$$

$$\sin\phi_{N_1}^{(1)} = \sin\left(\psi_{N_1}^{(1)} + \delta\phi_{N_1}^{(1)}\right) = \sin\psi_{N_1}^{(1)} + \cos\psi_1^{(1)}\delta\phi_{N_1}^{(1)} - \sin\psi_{N_1}^{(1)}\frac{\left(\delta\phi_{N_1}^{(1)}\right)^2}{2} + \cdots. \tag{26}$$

It is advantageous to split $\delta\phi^{(1)}_{1,N_1} = \delta\varphi^{(1)}_{1,N_1} + \delta\gamma^{(1)}_{1,N_1}$ further into two parts, where the first parts $\delta\varphi^{(1)}_{1,N_1}$ solves Eqs. 3 and 4 for the linear expansion only. We obtain

$$\delta\varphi^{(1)}_1 = \frac{R^{(1)}_c \cos\psi^{(1)}_{N_1} - R^{(1)}_s \sin\psi^{(1)}_{N_1}}{\Delta l_1 \sin\left(\psi^{(1)}_1 - \psi^{(1)}_{N_1}\right)}, \tag{27}$$

$$\delta\varphi^{(1)}_{N_1} = \frac{R^{(1)}_s \sin\psi^{(1)}_1 - R^{(1)}_c \cos\psi^{(1)}_1}{\left(\Delta l_1 + \delta l_{eq} + \delta l\right) \sin\left(\psi^{(1)}_1 - \psi^{(1)}_{N_1}\right)}. \tag{28}$$

Here we have defined the auxiliary quantities

$$R^{(1)}_s = \Delta Y_1 - \Delta l_1 \sum_{j=2}^{N_1-1} \sin\phi^{(1)}_j - \Delta l_1 \sin\psi^{(1)}_1 - \left(\Delta l_1 + \delta l_{eq} + \delta l\right) \sin\psi^{(1)}_{N_1}, \tag{29}$$

and

$$R^{(1)}_c = \Delta l_1 \cos\psi^{(1)}_1 + \left(\Delta l_1 + \delta l_{eq} + \delta l\right) \cos\psi^{(1)}_{N_1} - \Delta X_1 + \Delta l_1 \sum_{j=2}^{N_1-1} \cos\phi^{(1)}_j. \tag{30}$$

The terms $\delta\gamma^{(1)}_{1,N_1}$ take into account the next order,

$$\delta\gamma^{(1)}_1 = -\frac{\Delta l_1 \cos\left(\psi^{(1)}_1 - \psi^{(1)}_{N_1}\right)\left(\delta\varphi^{(1)}_1\right)^2 + \left(\Delta l_1 + \delta l_{eq} + \delta l\right)\left(\delta\varphi^{(1)}_{N_1}\right)^2}{2\Delta l_1 \sin\left(\psi^{(1)}_1 - \psi^{(1)}_{N_1}\right)}, \tag{31}$$

$$\delta\gamma^{(1)}_{N_1} = \frac{\Delta l_1 \left(\delta\varphi^{(1)}_1\right)^2 + \left(\Delta l_1 + \delta l_{eq} + \delta l\right)\cos\left(\psi^{(1)}_1 - \psi^{(1)}_{N_1}\right)\left(\delta\varphi^{(1)}_{N_1}\right)^2}{2\left(\Delta l_1 + \delta l_{eq} + \delta l\right)\sin\left(\psi^{(1)}_1 - \psi^{(1)}_{N_1}\right)}. \tag{32}$$

So the correct expansion up to second order is given by (and similar for $\sin\phi^{(1)}_{N_1}$)

$$\sin\phi^{(1)}_1 = \sin\psi^{(1)}_1 + \cos\psi^{(1)}_1 \left(\delta\varphi^{(1)}_1 + \delta\gamma^{(1)}_1\right) - \sin\psi^{(1)}_1 \frac{\left(\delta\varphi^{(1)}_1\right)^2}{2} + \ldots. \tag{33}$$

The term $\left(\delta\gamma^{(1)}_{1,N_1}\right)^2$ leads to terms of the order $\left(\delta\varphi^{(1)}_{1,N_1}\right)^4$ and therefore can be neglect.

2. ***Span 2***

The computations are similar as for the first span, so we list the results for completeness.

$$\delta\varphi_1^{(2)} = \frac{R_c^{(2)} \cos\psi_{N_2}^{(2)} - R_s^{(2)} \sin\psi_{N_2}^{(2)}}{\left(\Delta l_2 - \delta l_{eq} - \delta l\right) \sin\left(\psi_1^{(2)} - \psi_{N_2}^{(2)}\right)}, \tag{34}$$

$$\delta\varphi_{N_2}^{(2)} = \frac{R_s^{(2)} \sin\psi_1^{(2)} - R_c^{(2)} \cos\psi_1^{(2)}}{\Delta l_2 \sin\left(\psi_1^{(2)} - \psi_{N_2}^{(2)}\right)}. \tag{35}$$

The auxiliary quantities read

$$R_s^{(2)} = \Delta Y_2 - \Delta l_2 \sum_{j=2}^{N_2-1} \sin\phi_j^{(2)} - \Delta l_2 \sin\psi_{N_2}^{(2)} - \left(\Delta l_2 - \delta l_{eq} - \delta l\right) \sin\psi_1^{(2)}, \tag{36}$$

and

$$R_c^{(2)} = \Delta l_2 \cos\psi_{N_2}^{(2)} + \left(\Delta l_2 - \delta l_{eq} - \delta l\right) \cos\psi_1^{(2)} - \Delta X_2 + \Delta l_2 \sum_{j=2}^{N_2-1} \cos\phi_j^{(2)}, \tag{37}$$

$$\delta\gamma_1^{(2)} = -\frac{\left(\Delta l_2 - \delta l_{eq} - \delta l\right) \cos\left(\psi_1^{(2)} - \psi_{N_2}^{(2)}\right) \left(\delta\varphi_1^{(2)}\right)^2 + \Delta l_2 \left(\delta\varphi_{N_2}^{(2)}\right)^2}{2\left(\Delta l_2 + \delta l_{eq} + \delta l\right) \sin\left(\psi_1^{(2)} - \psi_{N_2}^{(2)}\right)}, \tag{38}$$

$$\delta\gamma_{N_2}^{(2)} = \frac{\left(\Delta l_2 - \delta l_{eq} - \delta l\right) \left(\delta\varphi_1^{(2)}\right)^2 + \Delta l_2 \cos\left(\psi_1^{(2)} - \psi_{N_2}^{(2)}\right) \left(\delta\varphi_{N_2}^{(2)}\right)^2}{2\Delta l_2 \sin\left(\psi_1^{(2)} - \psi_{N_2}^{(2)}\right)}. \tag{39}$$

The correct expansion up to second order, written for the first angle,

$$\sin\phi_1^{(2)} = \sin\psi_1^{(2)} + \cos\psi_1^{(2)} \left(\delta\varphi_1^{(2)} + \delta\gamma_1^{(2)}\right) - \sin\psi_1^{(2)} \frac{\left(\delta\varphi_1^{(2)}\right)^2}{2} + \cdots. \tag{40}$$

References

1. Czitary, E.: Seilschwebebahnen (2. Auflage). Springer, Wien, (1962)
2. CEN-Norm: Sicherheitsanforderungen für Seilbahnen für den Personenverkehr. Amtsblatt der EU C51, (2009)
3. Beha, R.: Bewegungsverhalten und Kraftwirkungen des Zugseiles und der Fahrzeuge von Zweiseilbahnen zur Berechnung der Dynamik des Gesamtsystems. Dissertation, Institut für Fördertechnik der Universität Stuttgart (1994)
4. Liedl, S.: Bewegungen und Kräfte des Seilsystems und der Fahrzeuge von Seilschwebebahnen im Fahrbetrieb. Dissertation, Herbert Utz Verlag (1999)
5. Kopanakis, G.: Internationale Seilbahn-Rundschau **1**(22) (2010)
6. Bryja, D., Knawa, M.: Computational model of an inclined aerial ropeway and numerical method for analyzing nonlinear cable-car interaction. Comput. Struct. **98**, 1895–1905 (2011)
7. Starossek, U.: Cable dynamics-a review. Struct. Eng. Int. **4**(3), 171–176 (1994)

8. Carrier, G.F.: On the non-linear vibration problem of the elastic string. Q. Appl. Math. **3**(2), 157–165 (1945)
9. Irvine, H.M., Caughey, T.K.: The linear theory of free vibrations of a suspended cable. Proc. R. Soc. Lond. A **341**, 299–315 (1974)
10. Zhou, X., Yan, S., Chu, F.: J. Sound Vib. **133**, 031001 (2011)
11. Yi, Z., Wang, Z., Zhou, Y., Stanciulescu, I.: Modeling and vibratory characteristics of a mass-carrying cable system with multiple pulley supports in span range. Appl. Math. Model. **49**, 59–68 (2017)
12. Engel, E., Löscher, R.: Arbeiten des Institutes für Eisenbahnwesen, Verkehrswirtschaft und Seilbahnen. http://www.eiba.tuwien.ac.at/fileadmin/mediapooleisenbahn/Diverse/Institutshefte/ih31.pdf (2003)
13. Volmer, M.: Stochastische Schwingungen an ausgedehnten Seilfeldern und ihre Anwendung zur Spurweitenberechnung an Seilbahnen. Dissertation ETH Nr. 13379, (1999)
14. Mathematica homepage. https://www.wolfram.com/mathematica/. Accessed 16 Mar 2017
15. Matlab homepage. https://de.mathworks.com/products/matlab.html. Accessed 16 Mar 2017
16. Schmutzer, E.: Grundlagen der Theoretischen Physik Bd. 2. BI Wissenschaftsverlag, (1989)
17. Kang, H.J., Zhao, Y.Y., Zhu, H.P.: Linear and nonlinear dynamics of suspended cable considering bending stiffness. J. Vib. Control **21**(8), 1487–1505 (2013)
18. Landau, L.D., Lifschitz, E.M.: Lehrbuch der Theoretischen Physik I, Klassische Mechanik (14. Auflage). Verlag Harry Deutsch, (2011)
19. Courant, R., Hilbert, D.: Methoden der mathematischen Physik (4. Auflage). Springer, (1993)
20. Thaler, H., Wenin, M., Brunner, J., Reiterer, D., Bertotti, M. L., Modanese, G., Oberhuber, E.: Advanced Structured Materials, pp. 113–124. Springer, (2016)
21. Lee, S.Y., Mote, C.D.: J. Dyn. Syst. Meas. Control **118**, 1666–1674 (1996)

Beams with Quadratic Function's Variation of Height

Matjaž Skrinar and Denis Imamović

Abstract This paper extends the utilisation of the standard beam finite element model with two nodes by presenting derivations of a closed-form stiffness matrix and a load vector for slender beams where the height variation is given as a general quadratic function. All the required stiffness matrix coefficients were obtained from two cantilever substructures: clamped at both the right and left-ends, respectively. Each substructure was subjected to a vertical upward force and an anticlockwise bending moment at the free end and transverse displacement and rotation at the free-end due to both acting loads were evaluated. From the expressions obtained, the nodal force and moment were afterwards vice-versly expressed as functions of nodal displacement and rotation. Further, both supports' reactions due to the applied loads were evaluated from the basic equilibrium, thus completing the coefficients of two rows of the stiffness matrix. The derived matrix and vector thus define an 'exact' finite element for this kind of height variation and all the obtained terms are written entirely in closed-symbolic forms. Numerical examples demonstrated that the elaborated solutions may be effectively implemented for structural analyses, as the presented expressions produced excellent results that were confirmed independently by more thorough 2D models.

Keywords Beam finite element · Stiffness matrix · Load vector · Quadratic function's variation of height

1 Introduction

A beam is not only simple but also the most common type of structural component, subject to transverse loading. Although it is generally a straight bar of an arbitrary

M. Skrinar (✉) · D. Imamović
Faculty of Civil Engineering, Transportation Engineering and Architecture,
University of Maribor, Maribor, Slovenia
e-mail: matjaz.skrinar@um.si

D. Imamović
e-mail: denis.imamovic@um.si

© Springer International Publishing AG, part of Springer Nature 2019
A. Öchsner and H. Altenbach (eds.), *Engineering Design Applications*,
Advanced Structured Materials 92, https://doi.org/10.1007/978-3-319-79005-3_7

cross-section, the majority of research is focused onto beams with constant cross section.

The stiffness matrix and the corresponding load vector, as well as their derivations, can for the prismatic beams be found in numerous references (e.g. [1]).

Several papers (usually in association with cracked beams) consider beams with nonuniform cross section. Cracked stepped beams are considered in the works by Biondi and Caddemi [2], as well as by Skrinar [3], where beams with linear variation of heights are also considered. All these solutions can also be implemented for non-cracked situations.

This paper presents the derivation of a closed-form stiffness matrix and a load vector for slender beams where the height variation is given as a general quadratic function, thus extending the utilisation of two nodes beam finite element model.

2 Mathematical Model Formulation

The discussed finite element considers slender elastic homogeneous Euler-Bernoulli beams subjected to small deflections. The element of total length L is assumed to have a uniform modulus of elasticity E and width b. The quadratic variation of the element's height is given in general form as:

$$h(x) = 2 \cdot \left(A_1 + A_2 \cdot x + A_3 \cdot x^2\right) \tag{1}$$

The finite element has four degrees of freedom altogether: transverse displacement Y_1 and rotation Φ_1 at the left-end (node 1), as well as transverse displacement Y_2 and rotation Φ_2 at the right-end (node 2), seen in Fig. 1. Upward displacements and anticlockwise rotations are taken as positive. The finite element is thus fully compatible with standard beam finite elements.

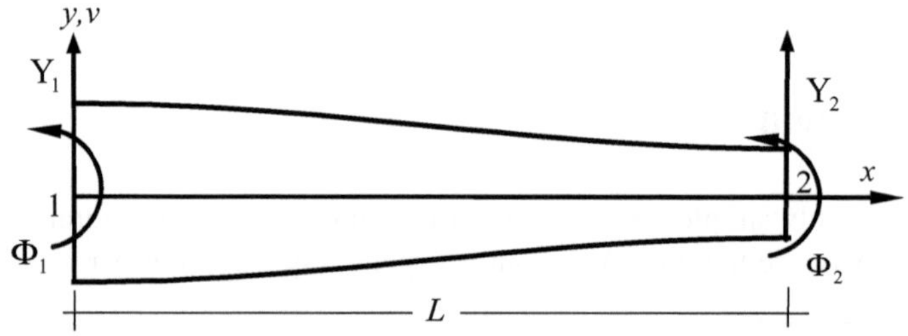

Fig. 1 Beam finite element with nodes and degrees of freedom

3 Derivation of Stiffness Matrix

The four columns of the stiffness matrix are obtained from two separate derivations. Since the finite element has four degrees of freedom altogether, this consequently means that in order to obtain a statically determinate structure, two degrees of freedom must be simultaneously removed. In order to complete this, all the required stiffness matrix coefficients are obtained from two cantilever substructures: clamped at both the right and left-ends, respectively.

3.1 Cantilever Clamped at the Right-End

The derivation of the first two columns of the stiffness matrix considers a cantilever clamped at the right-end. The structure is subjected to a vertical upward force F_{n_1} and an anticlockwise bending moment M_{n_1} at the free end. Since this is a statically determinate structure, the distribution of bending moments due to the applied loads can be obtained from basic equilibriums.

Transverse displacement Y_1 and rotation Φ_1 at the left-end due to the acting loads are obtained by implementing the principle of virtual work. Therefore, a virtual vertical upward force $\delta x_1 = 1$ and virtual anticlockwise bending moment $\delta x_2 = 1$ are applied individually at the considered point. The required nodal transverse displacement Y_1 is obtained afterwards by integrating the diagrams of bending moments for the structure subjected to the simultaneously applied force F_{n_1} and bending moment M_{n_1}, with the diagrams for the structure subjected to the virtual force δx_1. Consequently, the total displacement is a sum of displacements from both loads:

$$Y_1 = Y_1\left(F_{n1}\right) + Y_1\left(M_{n1}\right) = F_{n1} \cdot \left(\int\limits_{x=0}^{L} \frac{x^2}{EI(x)} \cdot dx \right) + M_{n1} \cdot \left(\int\limits_{x=0}^{L} \frac{-x}{EI(x)} \cdot dx \right) \quad (2)$$

Similarly, the nodal rotation Φ_1 is obtained by integrating the diagrams for the structure subjected to the applied loads, with the moments' diagram due to the applied virtual bending moment $\delta x_2 = 1$. The entire rotation also consists of contributions from both loads:

$$\Phi_1 = \Phi_1\left(F_{n1}\right) + \Phi_1\left(M_{n1}\right) = F_{n1} \cdot \left(\int\limits_{x=0}^{L} \frac{-x}{EI\,(x)} \cdot dx \right) + M_{n1} \cdot \left(\int\limits_{x=0}^{L} \frac{1}{EI\,(x)} \cdot dx \right) \quad (3)$$

By introducing the following abbreviations:

$$
\begin{aligned}
k &= \sqrt{4 \cdot A_1 \cdot A_3 - A_2^2} \\
dat &= \arctan\left(\frac{A_2}{k}\right) - \arctan\left(\frac{A_2 + 2 \cdot A_3 \cdot L}{k}\right) \\
a_{23} &= A_2 + A_3 \cdot L \\
a_{123} &= (A_1 + L \cdot a_{23})^2 \\
a_{2233} &= A_2^2 + 10 \cdot A_2 \cdot A_3 \cdot L + 8 \cdot A_3^2 \cdot L^2 \\
a_{23L} &= 3 \cdot A_2 + 8 \cdot A_3 \cdot L \\
a_{42} &= A_2 + 2 \cdot A_3 \cdot L \\
a_{45} &= A_1 + L \cdot a23 \\
a_{46} &= 3 \cdot A_2^2 + 10 \cdot A_2 \cdot A_3 \cdot L + 6 \cdot A_3^2 \cdot L^2 \\
a_{48} &= -6 \cdot A_2 \cdot A_3 + 4 \cdot A_3^2 \cdot L \\
a_{232} &= 11 \cdot A_2 + 10 \cdot A_3 \cdot L \\
i_1 &= \frac{4}{3} \cdot k^5 \cdot b \cdot E \cdot a_{123} \\
a_{213} &= A_2^2 + 2 \cdot A_1 \cdot A_3 \\
a_{50} &= A_2^3 - 4 \cdot A_2^2 \cdot A_3 \cdot L - 10 \cdot A_2 \cdot A_3^2 \cdot L^2 - 5 \cdot A_3^3 \cdot L^3 \\
a_{51} &= A_1 \cdot L \cdot \left(9 \cdot A_2^2 + 10 \cdot A_2 \cdot A_3 \cdot L + 8 \cdot A_3^2 \cdot L^2\right) \\
a_{52} &= 6 \cdot A_1^2 \cdot A_2 + A_2^2 \cdot L^2 \cdot (2 \cdot A_2 + A_3 \cdot L) \\
b_2 &= A_2^2 + 3 \cdot A_2 \cdot A_3 \cdot L - 2 \cdot A_3^2 \cdot L^2 \\
b_6 &= 8 \cdot A_2^2 + A_2 \cdot A_3 \cdot L - 6 \cdot A_3^2 \cdot L^2 \\
b_{31} &= 4 \cdot A_1^2 \cdot A_3 + 2 \cdot A_1 \cdot b_2 + A_2 \cdot L \cdot a_{46} \\
b_{32} &= \frac{A_2^2 \cdot L \cdot a_{23}^2}{A_1} + 2 \cdot A_1 \cdot A_3 \cdot a_{23L} + A_3 \cdot L \cdot a_{2233} \\
b_{33} &= 20 \cdot A_1 \cdot A_3^2 + \frac{A_2^3 \cdot L \cdot a_{23}^2}{A_1^2} - 2 \cdot A_3 \cdot b_6 + \frac{2 \cdot A_2 \cdot a_{50}}{A_1} \\
b_{34} &= \left(\frac{A_2^3 \cdot L \cdot a_{23}}{A_1^2} - \frac{A_2 \cdot A_3 \cdot L \cdot a_{232}}{A_1} + a_{48}\right) \cdot a_{45} \\
b_{35} &= \left(A_2^2 + 6 \cdot A_2 \cdot A_3 \cdot L + 2 \cdot A_3 \cdot \left(A_1 + 3 \cdot A_3 \cdot L^2\right)\right) \cdot dat \\
b_{36} &= -4 \cdot A_3 + \frac{A_2^3 \cdot L}{A_1^2} - \frac{2 \cdot A_2 \cdot (A_2 + 5 \cdot A_3 \cdot L)}{A_1}
\end{aligned} \tag{4}
$$

the first node's displacement and rotation due to applied force F_{n_1} and bending moment M_{n_1} can be expressed as:

$$Y_1\left(F_{n_1}\right) = -F_{n1} \cdot \frac{k \cdot L \cdot b_{31} + 4 \cdot a_{213} \cdot a_{123} \cdot dat}{i_1} = F_{n1} \cdot Y_1\left(\delta x_1\right)$$

$$\Phi_1\left(F_{n_1}\right) = -F_{n1} \cdot \frac{k \cdot L \cdot b_{32} + 12 \cdot A_2 \cdot A_3 \cdot a_{123} \cdot dat}{i_1} = F_{n1} \cdot \Phi_1\left(\delta x_1\right)$$

$$Y_1\left(M_{n_1}\right) = -M_{n_1} \cdot \frac{k \cdot L \cdot b_{32} + 12 \cdot A_2 \cdot A_3 \cdot a_{123} \cdot dat}{i_1} = M_{n_1} \cdot Y_1\left(\delta x_2\right)$$

$$\Phi_1\left(M_{n_1}\right) = M_{n_1} \cdot \frac{k \cdot L \cdot b_{33} - 24 \cdot A_3^2 \cdot a_{123} \cdot dat}{i_1} = M_{n_1} \cdot \Phi_1\left(\delta x_2\right) \qquad (5)$$

3.1.1 Derivation of the First Two Columns and Rows of the Stiffness Matrix

From the Eqs. (2) and (3), the nodal force F_{n_1} and moment M_{n_1} are now vice-versly expressed as functions of nodal displacement Y_1 and rotation Φ_1:

$$F_{n_1} = \frac{\Phi_1\left(\delta x_2\right) \cdot Y_1 - Y_1\left(\delta x_2\right) \cdot \Phi_1}{\Phi_1\left(\delta x_2\right) \cdot Y_1\left(\delta x_1\right) - \Phi_1\left(\delta x_1\right) \cdot Y_1\left(\delta x_2\right)} \qquad (6)$$

$$M_{n_1} = \frac{-\Phi_1\left(\delta x_1\right) \cdot Y_1 + Y_1\left(\delta x_1\right) \cdot \Phi_1}{\Phi_1\left(\delta x_2\right) \cdot Y_1\left(\delta x_1\right) - \Phi_1\left(\delta x_1\right) \cdot Y_1\left(\delta x_2\right)} \qquad (7)$$

The corresponding right support's vertical reaction F_{n_2} (positive upwards) due to the applied loads can now be evaluated from the basic equilibrium of forces and moments as:

$$F_{n_2} = -F_{n_1} \qquad (8)$$

whilst the corresponding reaction bending moment M_{n_2} (positive anticlockwise) at the same location due to applied loads can now be evaluated as:

$$M_{n_2} = F_{n_1} \cdot L - M_{n_1} \qquad (9)$$

By introducing Eqs. 6–7 into Eqs. 8–9, the reactions are also expressed by nodal displacement Y_1 and rotation Φ_1. Consequently, since the stiffness matrix represents the relationship between nodal unit forces (and moments) and nodal displacements (and rotations), Eqs. 6–9 can be utilised to obtain the first two columns (and also the first two rows) of the stiffness matrix's considered finite element.

Therefore, by taking $Y_1 = 1$ and $\Phi_1 = 0$ in Eqs. 6–9, the coefficients of the first row and column of the stiffness matrix can be obtained, whilst the coefficients of the second row and the column are obtained by using $Y_1 = 0$ and $\Phi_1 = 1$. The coefficients of the first row of the stiffness matrix of beam finite element with quadratic height variation (FEQVH) are thus:

$$\frac{\{\Phi_1(\delta x_2),\, -Y_1(\delta x_2),\, -Y_1(\delta x_2),\, \Phi_1(\delta x_2)\cdot L + Y_1(\delta x_2)\}}{\Phi_1(\delta x_2)\cdot Y_1(\delta x_1) - \mathrm{Y}_1(\delta x_2)^2} \tag{10}$$

Furthermore, the coefficients of the second row and the column are:

$$\frac{\{-Y_1(\delta x_2),\, Y_1(\delta x_1),\, Y_1(\delta x_2),\, -Y_1(\delta x_1) - Y_1(\delta x_2)\cdot L\}}{\Phi_1(\delta x_2)\cdot Y_1(\delta x_1) - Y_1(\delta x_2)^2} \tag{11}$$

3.2 Cantilever Clamped at the Left-End

The process of the free-end's vertical displacement and rotation calculation is now repeated for a cantilever, clamped at the left-end. This allows for the remaining third and fourth rows of the stiffness matrix to be derived (due to its symmetry, only three coefficients are actually still unknown).

The structure is subject to a vertical upward-force F_{n_2} and an anticlockwise bending moment M_{n_2} at the free right-end, and the vertical displacement Y_2 and the rotation Φ_2 of the same end are first expressed as functions of both applied loads. A virtual vertically-upward force $\delta x_3 = 1$ and a virtual anticlockwise bending moment $\delta x_4 = 1$ are thus separately applied at the free-end, in order to obtain the transverse displacement and rotation of this point. The transverse displacement Y_2 due to both applied loads is thus obtained by integrating the bending moment diagram due to the applied transverse force δx_3 with the diagrams of moments due to loads F_{n_2} and M_{n_2}:

$$Y_2 = Y_2(F_{n2}) + Y_2(M_{n2}) = F_{n2}\cdot\left(\int_{x=0}^{L}\frac{(L-x)^2}{EI(x)}\cdot dx\right) + M_{n2}\cdot\left(\int_{x=0}^{L}\frac{L-x}{EI(x)}\cdot dx\right) \tag{12}$$

In order to obtain the rotation, the cantilever is further subjected to a virtual bending moment δx_4 at the right-end, and the rotation is obtained by integrating the diagram due to the applied bending moment δx_4 with the diagrams of bending moments due to loads:

$$\Phi_2 = \Phi_2(F_{n2}) + \Phi_2(M_{n2}) = F_{n2}\cdot\left(\int_{x=0}^{L}\frac{L-x}{EI(x)}\cdot dx\right) + M_{n2}\cdot\left(\int_{x=0}^{L}\frac{1}{EI(x)}\cdot dx\right) \tag{13}$$

The second's node displacement and rotation due to applied force F_{n_2} and bending moment M_{n_2} can be expressed as:

$$Y_2\left(F_{n_2}\right) = F_{n_2} \cdot \frac{(k \cdot L \cdot b_{36} - 4 \cdot b_{35}) \cdot a_{123}}{i_1} = F_{n_2} \cdot Y_2\left(\delta x_3\right)$$
$$\Phi_2\left(F_{n_2}\right) = F_{n_2} \cdot \frac{k \cdot L \cdot b_{34} - 12 \cdot A_3 \cdot a_{42} \cdot a_{123} \cdot dat}{i_1} = F_{n_2} \cdot \Phi_2\left(\delta x_3\right)$$
$$Y_2\left(M_{n_2}\right) = M_{n_2} \cdot \frac{k \cdot L \cdot b_{34} - 12 \cdot A_3 \cdot a_{42} \cdot a_{123} \cdot dat}{i_1} = M_{n_2} \cdot Y_2\left(\delta x_4\right)$$
$$\Phi_2\left(M_{n_2}\right) = M_{n_2} \cdot \frac{k \cdot L \cdot b_{33} - 24 \cdot A_3^2 \cdot a_{123} \cdot dat}{i_1} = M_{n_2} \cdot \Phi_2\left(\delta x_4\right) \quad (14)$$

From expressions (12) and (13) the nodal force F_{n_2} and the moment M_{n_2} are vice versa expressed as functions of nodal displacement Y_2 and rotation Φ_2:

$$F_{n_2} = \frac{\Phi_2\left(\delta x_4\right) \cdot Y_2 - Y_2\left(\delta x_4\right) \cdot \Phi_2}{\Phi_2\left(\delta x_4\right) \cdot Y_2\left(\delta x_3\right) - \Phi_2\left(\delta x_3\right) \cdot Y_2\left(\delta x_4\right)} \quad (15)$$
$$M_{n_2} = \frac{-\Phi_2\left(\delta x_3\right) \cdot Y_2 + Y_2\left(\delta x_3\right) \cdot \Phi_2}{\Phi_2\left(\delta x_4\right) \cdot Y_2\left(\delta x_3\right) - \Phi_2\left(\delta x_3\right) \cdot Y_2\left(\delta x_4\right)} \quad (16)$$

Furthermore, the reactions at the left-support, that can actually serve for verification purposes only, are:

$$F_{n_1} = -F_{n_2} \quad (17)$$
$$M_{n_1} = -M_{n_2} - F_{n2} \cdot L \quad (18)$$

By taking $Y_2 = 1$ in Eqs. 15–18, the remaining coefficients of the third row and the column of the stiffness matrix are obtained (to obtain the missing coefficients of the stiffness matrix Eqs. 15–16 are the only ones actually needed due to the symmetry of the matrix), whilst the remaining coefficients of the fourth row and the column are obtained by inserting $\Phi_2 = 1$.

Finally, the coefficients of the third row of the stiffness matrix are thus:

$$\frac{\left\{-\Phi_2\left(\delta x_4\right), Y_2\left(\delta x_4\right) - \Phi_2\left(\delta x_4\right) \cdot L, \Phi_2\left(\delta x_4\right), -Y_2\left(\delta x_4\right)\right\}}{\Phi_2\left(\delta x_4\right) \cdot Y_2\left(\delta x_3\right) - Y_2\left(\delta x_4\right)^2} \quad (19)$$

while the coefficients of the fourth row and the column are:

$$\frac{\{Y_2(\delta x_4), -Y_2(\delta x_3) + Y_2(\delta x_4) \cdot L, -Y_2(\delta x_4), Y_2(\delta x_3)\}}{\Phi_2(\delta x_4) \cdot Y_2(\delta x_3) - Y_2(\delta x_4)^2} \quad (20)$$

4 Derivation of the Load Vector Due to the Uniform Load *q* Over the Whole Finite Element

Similarly to the derivation of the stiffness matrix, the load vector's coefficients are also formally derived from separate derivations on two structures.

A cantilever clamped at the right-end and subjected to a uniformly-distributed load q (positive upwards) is considered during the derivation of the load vector's first two coefficients.

The nodal transverse displacement $Y_{1,q}$ of the free-end is obtained by integrating the diagram of the bending moments for the structure subjected to the applied distributed load q, with the diagram for the structure subjected to the virtual vertically-upward force δx_1. It can be expressed as:

$$Y_{1,q} = \frac{q}{2} \cdot \frac{k \cdot L \cdot (a_{51} + a_{52}) + 12 \cdot A_1 \cdot A_2 \cdot a_{123} \cdot dat}{i_1} \tag{21}$$

Similarly, the nodal rotation $\Phi_{1,q}$ of the free-end is obtained by integrating the moment diagram for the structure subjected to the applied load q, with the diagram due to the applied virtual anticlockwise bending moment $\delta x_2 = 1$. The nodal rotation is thus:

$$\Phi_{1,q} = \frac{q}{2} \cdot \frac{k \cdot L \cdot b_{31} + 4 \cdot a_{213} \cdot a_{123} \cdot dat}{i_1} \tag{22}$$

The originally distributed load is, in the finite element model, replaced by concentrated nodal forces ($F_{n_1,q}$ and $F_{n_2,q}$) and bending moments ($M_{n_1,q}$ and $M_{n_2,q}$), and the substituted loads must provide identical nodal displacements and rotations to that of the genuine one. Since the transverse nodal displacement and rotation due to the force F_{n_1}, and the moment M_{n_1} are already known, Eqs. 6–7, respectively, the substituted nodal force and moment are obtained from a system of two linear equations:

$$Y_{1,q} = Y_1\left(F_{n_{1,q}}\right) + Y_1\left(M_{n_{1,q}}\right) = Y_1(\delta x_1) \cdot F_{n_{1,q}} + Y_1(\delta x_2) \cdot M_{n_{1,q}} \tag{23}$$

$$\Phi_{1,q} = \Phi_1\left(F_{n_{1,q}}\right) + \Phi_1\left(M_{n_{1,q}}\right) = \Phi_1(\delta x_1) \cdot F_{n_{1,q}} + \Phi_1(\delta x_2) \cdot M_{n_{1,q}} \tag{24}$$

that yield the first two coefficients of the load vector as:

$$\begin{aligned} F_{n_{1,q}} &= \frac{\Phi_1(\delta x_2) \cdot Y_{1,q} - Y_1(\delta x_2) \cdot \Phi_{1,q}}{\Phi_1(\delta x_2) \cdot Y_1(\delta x_1) - \Phi_1(\delta x_1) \cdot Y_1(\delta x_2)} \\ M_{n_{1,q}} &= \frac{-\Phi_1(\delta x_1) \cdot Y_{1,q} + Y_1(\delta x_1) \cdot \Phi_{1,q}}{\Phi_1(\delta x_2) \cdot Y_1(\delta x_1) - \Phi_1(\delta x_1) \cdot Y_1(\delta x_2)} \end{aligned} \tag{25}$$

The remaining two terms of the load vector can be derived at in a similar manner from a cantilever clamped at the left-end. However, these coefficients can also be determined from a simple static equilibrium of a cantilever clamped at the right-end:

$$F_{n2,q} = L \cdot q - F_{n_{1,q}} \tag{26}$$

$$M_{n2,q} = F_{n_{1,q}} \cdot L - M_{n1,q} - \frac{L^2 \cdot q}{2} \tag{27}$$

5 Numerical Examples

The applicability of derived at expressions for the stiffness matrix and load vector is demonstrated in three examples with different boundary conditions and applied loads. The relevance of the discussed expressions is confirmed by comparing the obtained results for transverse displacements and the reactions at the supports to the results from alternative 2 D finite element models.

5.1 Example 1—Cantilever Beam with a Concentrated Load at the Free End

In the first example, a 8 m long multi-cracked cantilever clamped at the left end and loaded with a vertical concentrated load of 1 kN at the right end was analysed, as shown in Fig. 2.

For the considered cantilever the height h followed a parabolic distribution given by a polynomial of second degree:

$$h(x) = 0.4 - 0.0625 \cdot x + 0.00390625 \cdot x^2$$

The Young's modulus of the material was $E = 30$ GPa, with Poisson's ratio 0.3, whilst the constant width b of the cantilever was 0.25 m.

The computational model consisted of a single beam finite-element with just two nodes, thus representing the smallest possible finite-element computational model. The model had four degrees of freedom. However, by considering the known zero displacement and rotation at the left end of the element the discrete unknowns (vertical displacement and rotation at the free end) were evaluated from a simple system of two linear equations:

$$\begin{bmatrix} 184864.859381 & -432670.006400 \\ -432670.006400 & 1611064.370722 \end{bmatrix} \cdot \begin{Bmatrix} u_2 \\ \Phi_2 \end{Bmatrix} = \begin{Bmatrix} -1000 \\ 0 \end{Bmatrix}$$

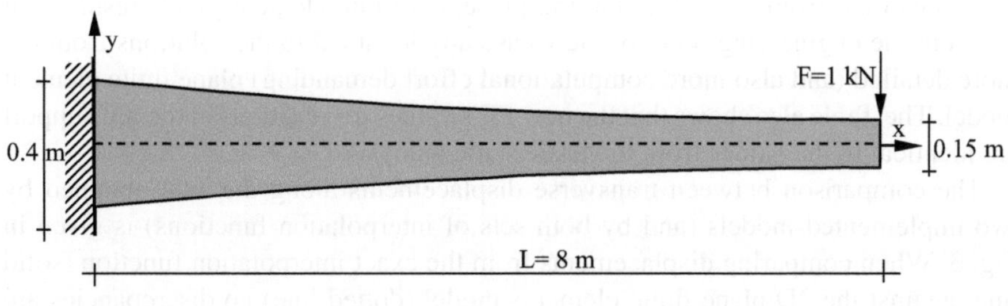

Fig. 2 First example structural setup

The values obtained (vertical displacement and rotation of the free end) further allowed for the computation of nodal vertical reaction force and bending moment of the clamped-end over the element's stiffness matrix. They additionally allowed for the calculation of transverse displacements along the longitudinal axis of the structure. The distribution of transverse displacements within the structure was obtained by introducing the nodal displacements' values into the interpolation functions. Two sets of interpolation functions were used: standard interpolation functions belonging to the beams with constant flexural rigidity, as well as interpolation functions that belong to the quadratic function's variation of height. By implementing standard interpolation functions the obtained approximate transverse displacement function was:

$$v_a(x) = -1.93760 \times 10^{-4} \cdot x^2 - 4.22368 \times 10^{-6} \cdot x^3 \quad 0 \leq x \leq 8\,\text{m}$$

while the implementation of the exact interpolation functions yielded considerable more complex solution:

$$\begin{aligned} v_e(x) = {} & -1.96832 \times 10^{-2} + 6.4 \times 10^{-4} \cdot x + \frac{1.36533 \times 10^{-3} - 1.70667 \times 10^{-4} \cdot x}{0.2 + \left(1.95313 \times 10^{-3} \cdot x - 3.125 \times 10^{-2}\right) \cdot x} \\ & - 1.41011 \times 10^{-2} \cdot \arctan\left(41.31182 \cdot \left(1.953125 \times 10^{-3} \cdot x - 3.125 \times 10^{-2}\right)\right) \\ & - 4.68375 \times 10^{-17} \cdot x \cdot \arctan\left(41.31182 \cdot \left(1.953125 \times 10^{-3} \cdot x - 3.125 \times 10^{-2}\right)\right) \end{aligned}$$

The values obtained from the simplified model were further compared to the results obtained from the more detailed computational model of the considered structure. These results were obtained through a 2D shell element computational model by implementing a SAP2000 v15.0.0 finite element program. The transverse displacements and reactions were obtained from a computational model consisting of 5760 2D shell elements with 6253 joints. In each node, two degrees of freedom were taken into account—vertical and horizontal displacements. Vertical and horizontal displacements of discrete nodal points were obtained by solving approximately 18,720 linear equations.

The results for several significant discrete parameters (transverse displacements, rotation and reactions) from both applied computational models were compared and are summarised in Table 1.

It is evident from the Table that the presented finite element yields results that are from the engineering point of view basically identical to the solutions from the more detailed (and also more computational effort demanding) plane finite element model. The Table also shows that the new FE's values for reactions in the left support are identical to the values from the basic static analysis.

The comparison between transverse displacements along the axis obtained by two implemented models (and by both sets of interpolation functions) is given in Fig. 3. When comparing displacements from the exact interpolation function (solid line) against the 2D plane finite elements model (dotted line) no discrepancies are noticeable. However, when comparing displacements from the standard interpolation

Table 1 Comparisons of the results for the cantilever from two different computational approaches

Parameter/Model	1 FEQVH FE	SAP2000 5760 2D FE
Vertical reaction at the left support	1000 N	1000 N
Bending moment at the left support	8000 Nm	8000 Nm
Displacement [mm] at the free end (x = 8 m)	−14.5632	−14.5659
Rotation [rad] at the free end	-3.9111×10^{-3}	-3.9147×10^{-3}

Fig. 3 Comparison of transverse displacements of cantilever from two applied models

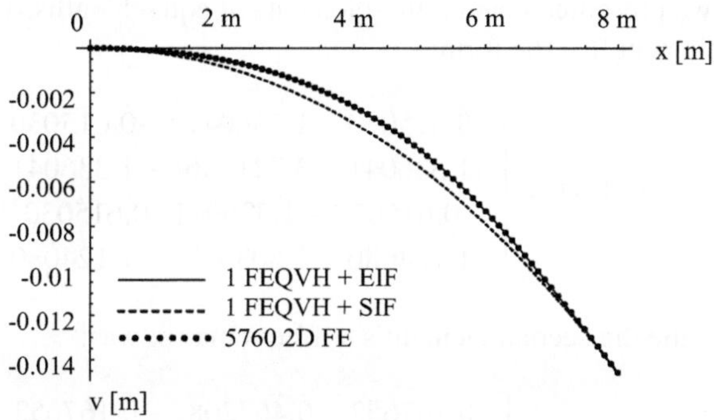

function (dashed line) against the 2D plane finite elements model some discrepancies are noticeable.

Nevertheless, when the structure was additionally modelled by two FEQVH finite elements, the displacements from the standard interpolation function also exhibited excellent agreement with the 2D plane finite elements model displacements.

5.2 Example 2—Simply Supported Beam with a Concentrated Load at the Mid-Span

In the second example, a 8 m long multi-cracked simply supported beam loaded with a vertical concentrated load of 100 kN at the mid-span was analysed, as shown in Fig. 4.

For this structure the height h followed a parabolic distribution in the following form of a second degree polynomial

$$h(x) = 0.4 - 0.00390625 \cdot x^2$$

The Young's modulus of the material was $E = 30$ GPa with Poisson's ratio 0.3, whilst the width b of the beam was 0.25 m. The computational model consisted of

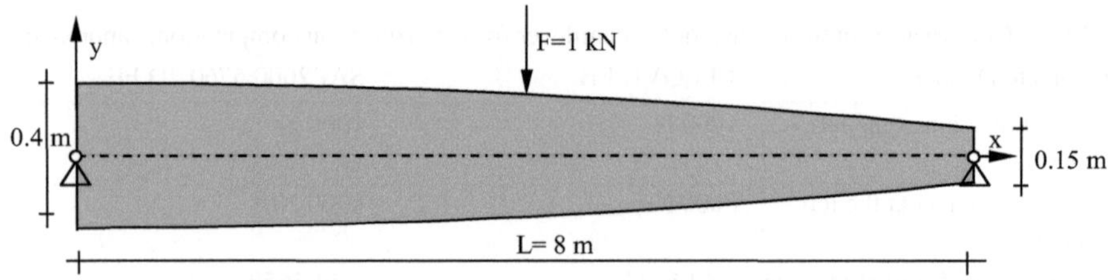

Fig. 4 Second example structural setup

two presented beam finite-elements of equal lengths. The stiffness matrix of the first element had the form:

$$[k_1] = \begin{bmatrix} 0.615030 & 1.336041 & -0.615030 & 1.124080 \\ 1.336041 & 3.741086 & -1.336041 & 1.603077 \\ -0.615030 & -1.336041 & 0.615030 & -1.124080 \\ 1.124080 & 1.603077 & -1.124080 & 2.893242 \end{bmatrix} \times 10^7$$

while the second element's stiffness matrix was:

$$[k_2] = \begin{bmatrix} 0.167652 & 0.467208 & -0.167652 & 0.203399 \\ 0.467208 & 1.487836 & -0.467208 & 0.380997 \\ -0.167652 & -0.467208 & 0.167652 & -0.203399 \\ 0.203399 & 0.380997 & -0.203399 & 0.432599 \end{bmatrix} \times 10^7$$

By considering known displacements at both ends of the element the discrete unknowns (transverse displacement and rotation at the mid-span as well as rotations at both ends) were evaluated from a system of four linear equations:

$$\begin{bmatrix} 3.741086 & -1.336041 & 1.603077 & 0 \\ -1.336041 & 0.782682 & -0.656871 & 0.203399 \\ 1.603077 & -0.656871 & 4.381078 & 0.380997 \\ 0 & 0.203399 & 0.380997 & 0.432599 \end{bmatrix} \cdot \begin{Bmatrix} \Phi_1 \\ Y_2 \\ \Phi_2 \\ \Phi_3 \end{Bmatrix} = \begin{Bmatrix} 0 \\ -100000 \\ 0 \\ 0 \end{Bmatrix}$$

By implementing the obtained discrete values of nodal displacement and rotations, the corresponding nodal vertical reaction forces, as well as the distribution of transverse displacements along the longitudinal axis of the structure were further evaluated. Just like in the first example standard interpolation functions as well as interpolation functions belonging to the quadratic function's variation of height were implemented.

In addition, this example was further analysed by implementing a computational model with 4800 2D shell plane finite elements. The comparison of transverse dis-

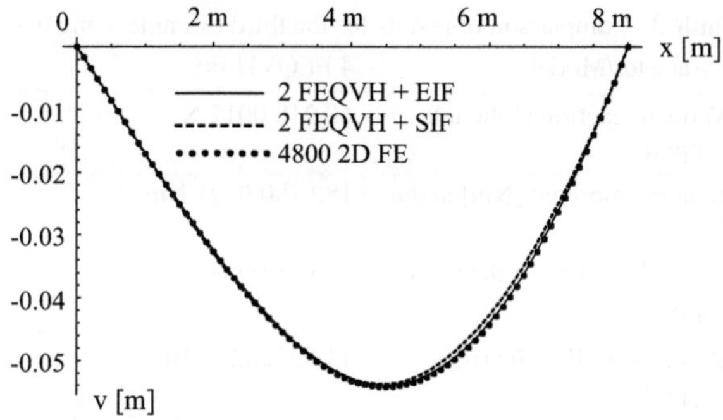

Fig. 5 Comparison of transverse displacements of simply supported beam from two applied models

Table 2 Comparisons of the results for the simply supported beam from two different computational approaches

Parameter/Model	1 FEQVH FE	SAP2000 4800 2D FE
Vertical reaction at the left support (A)	50,000 N	50,000 N
Rotation [rad] at the left support (A)	-1.709520×10^{-2}	-1.736995×10^{-2}
Vertical displacement [mm] at the mid-span	-52.843385	-53.0031
Rotation [rad] at the mid-span	-4.145957×10^{-3}	$-3.9094167 \times 10^{-3}$
Vertical reaction at the right support (B)	50,000 N	50,000 N
Rotation [rad] at the right support (B)	2.849726×10^{-2}	2.877525×10^{-2}

placements along the axis obtained by FEQVH and the 2D plane finite elements model is given in Fig. 5.

When comparing displacements from the exact interpolation function (solid line) against the 2D plane finite elements model (dotted line) no discrepancies are noticeable (just like in the first example). On the other hand, rather small discrepancies are again noticeable when comparing displacements from the standard interpolation function (dashed line) against the 2D plane finite elements model.

Furthermore, the discrete values of some representative quantities are summarised in Table 2.

The Table shows that the FEQVH's values for vertical reactions in the supports are identical to the values from the basic static analysis. Furthermore, it is also evident from the Table that the matching of the results for displacements and rotations is very good.

Table 3 Comparison of results for the third example using two different computational models

Parameter/Model	4 FEQVH FE	SAP2000 6000 2D FE
Vertical reaction at the left support	68,218.0012 N	68,262.45 N
Bending moment [Nm] at the left support	182,180.0121 Nm	182,624.4 Nm
Vertical reaction at the right support	31,781.9988 N	31,737.6 N
Rotation [rad] at the right support	11.935282×10^{-3}	12.0214×10^{-3}
Displacement at mid-span	−34.43406 mm	−34.571 mm

5.3 Example 3—Propped Cantilever with a Uniform Load

In the third example, a multi-cracked propped cantilever was analysed, seen in Fig. 6. It was loaded by a vertical uniform load of 10 kN/m over the whole length of 10 m.

For this structure the height h was varied from 0.2 m at the mid-span to 0.4 m at both ends following the trigonometric distribution:

$$h(x) = 0.3 + 0.1 \cdot \cos(0.2 \cdot \pi \cdot x)$$

The Young's modulus of the material was E = 30 GPa with a Poisson's ratio 0.3, whilst the width b of the cantilever was 0.25 m. Due to the trigonometric distribution of the element's height the computational model required four presented beam finite-elements with five nodes altogether to achieve sufficient description of the height's variation.

In addition, this example was also analysed by implementing a computational model with 6000 2D shell plane finite elements. Figure 7 thus represents the distribution of transverse displacements from both models.

Furthermore, the discrete values of some representative quantities are summarised in Table 3.

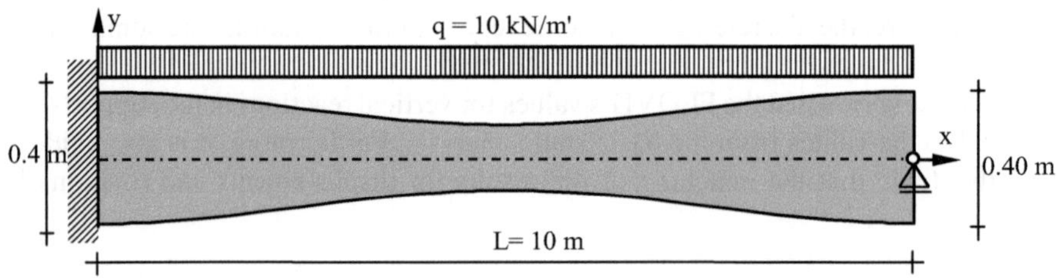

Fig. 6 Third example structural setup

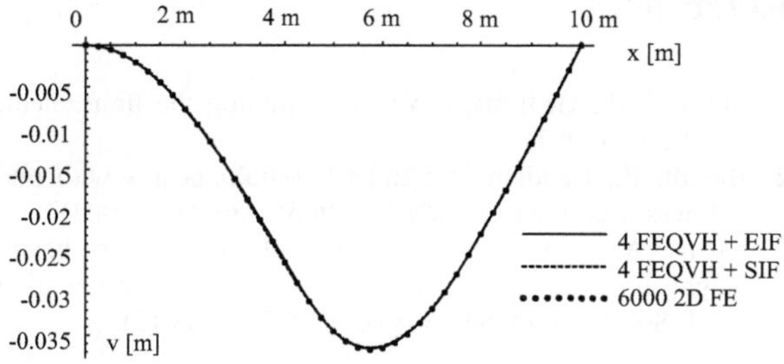

Fig. 7 Comparison of transverse displacements of propped cantilever from two applied models

It is evident from both Fig. 7 and Table 3 that there is an obvious agreement from both models.

6 Conclusions

The stiffness matrix, as well as the load vector due to a uniform continuous load were derived at for a beam finite element with quadratic function's variation of height (FEQVH). Despite the derivation's straightforwardness, all the obtained terms are written entirely in closed-symbolic forms. The finite element is fully compatible with standard beam finite elements and it can be easily incorporated into existing software for FEM structural analysis.

Three numerical examples demonstrated that elaborated solutions can be effectively implemented for structural analyses as the presented expressions produced excellent results that were confirmed independently by more thorough 2D models. The derived at expressions thus offer—although rather short and compact—a reliable and efficient computational model.

Beside implementation in classical structural analysis for structure's response evaluation the presented finite element could find implementation in model updating to take into account any potential non-uniform height variation (as a result of erection imperfection).

Acknowledgements The first author acknowledges the financial support from the Slovenian Research Agency (research core funding No. P2-0129 (A) "Development, modelling and optimization of structures and processes in civil engineering and traffic").

References

1. Smith, I.M., Griffiths, D.V.: Programming the finite element method. John Wiley & Sons, Chichester (1997)
2. Biondi, B., Caddemi, S.: Euler-Bernoulli beams with multiple singularities in the flexural stiffness. Eur. J. Mech. A/Solids **26**(5), 789–809 (2007)
3. Skrinar, M.: Computational analysis of multi-stepped beams and beams with linearly-varying heights implementing closed-form finite element formulation for multi-cracked beam elements. Int. J. Solids Struct. **50**(14–15), 2527–2541 (2013)

Design and Manufacture of a Forearm Prosthesis by Plastic 3D Impression for a Patient with Transradial Amputation Applied for Strum of a Guitar

Beatriz Romero-Ángeles, Daniel Hernández-Campos, Guillermo Urriolagoitia-Sosa, Christopher René Torres-San Miguel, Rafael Rodríguez-Martínez, Jacobo Martínez-Reyes, Rosa Alicia Hernández-Vázquez and Guillermo Urriolagoitia-Calderón

Abstract At the present, the development of prostheses has been extremely wide and focused on restoring the patient to their daily activities. Nevertheless, not much has been done to restore the capability of playing musical instruments, perhaps due to the complexity and sensibility needed to perform certain movements. For example, the fingering movements on a guitar or a piano, are random and require of different amounts of force to be applied to the instrument. Furthermore, the playing of an instrument is accomplished by muscular memory. Nowadays, in order to emulate such movements, the application of robotically programmed prostheses has improved. Nevertheless, this is not a good option for musicians, as this solution do not offer the sensation of actually playing the instrument. One of the options to achieve the mentioned aim, would be the development of mechanical prostheses

B. Romero-Ángeles · D. Hernández-Campos · G. Urriolagoitia-Sosa (✉)
C. R. Torres-San Miguel · R. Rodríguez-Martínez · J. Martínez-Reyes
R. A. Hernández-Vázquez · G. Urriolagoitia-Calderón
Sección de Estudios de Posgrado e Investigación, Unidad Profesional Adolfo López Mateos "Zacatenco", Escuela Superior de Ingeniería Mecánica y Eléctrica, Instituto Politécnico Nacional, Ciudad de México, Mexico
e-mail: guiurri@hotmail.com

B. Romero-Ángeles
e-mail: romerobeatriz97@hotmail.com

C. R. Torres-San Miguel
e-mail: ctorress@ipn.mx

R. Rodríguez-Martínez
e-mail: rafarm68@hotmail.com

J. Martínez-Reyes
e-mail: jacmartinez@esimez.mx

R. A. Hernández-Vázquez
e-mail: alyzia.hv@esimez.mx

G. Urriolagoitia-Calderón
e-mail: urrio332@hotmail.com

© Springer International Publishing AG, part of Springer Nature 2019

A. Öchsner and H. Altenbach (eds.), *Engineering Design Applications*, Advanced Structured Materials 92, https://doi.org/10.1007/978-3-319-79005-3_8

that apply myoelectric technology, however these devises do not possess sufficient sensibility to perform the movements needed to play an instrument. In this research, a mechanical design to be manufactured using 3D impression of a forearm prosthesis for a transradial amputated patient that is able to be used to play the guitar, is presented. The prosthesis is capable to provide the service for basic rehabilitation or train a recently amputated patient recently amputated. The proposed prosthesis was manufactured by the use of a plastic 3D impression machine using ABS-P400 strings. The prosthesis consists of two parts; the first one is the main socket that is attached to the forearm while the second part, that supports the nib guitar. The mechanical design of the prosthesis was based on data obtained by videometry of a musician developing the basic strum on a guitar. With these data was possible to define the basic strum technique, the movements that are involved (displacements, speed, acceleration, forces, etc.) and the joints involved in the strum of a guitar. The obtained results are very encouraging, since the prosthesis can be applied to restore the musician ability to perform the basic tasks to execute on his instrument, as well as, rehabilitation activities. This first proposal could be optimized to produce a final product with higher quality and with a wider area of applicability.

Keywords Biomechanics · Videometry · Mechanical design · 3D impression

1 Introduction

In a general manner, there is a belief that to be a musician involves very simple activities and that it does not required a major effort. Nevertheless, it is necessary to have years of practice and study, and sometimes big economic investments go into a musicians training. Sometimes, injuries have to de overcome to continue with musical activities. Although, seemingly the musician's work does not seem to need a lot of physical activities, actually it is not so. When playing an instrument, most of the time the player has to adopt uncomfortable or unnatural postures for the body, which lead to great repetitive stresses in the limb position that could cause injuries (musculoskeletal disorder *HSE*) [1]. Something worse than any injury, is the amputation of a member, which can easily end a musician´s career. But beyond the physical impossibility, it has to be considered that the severity of the injury could require the amputation of a limb as well as the psychological consequences that it brings.

A person who has suffered an amputation, faces a very difficult duel, which is divided into different phases. These stages are denial, anger, negotiation, depression and finally acceptance. Once acceptance is achieved, the patient can learn to live with this situation [2], this is possible by making several decisions. There are very few documented cases about amputee musicians, one of them was Paul Wittgenstein (see Fig. 1) who was a very virtuous pianist and who unfortunately lost his right arm in the First World War [3]. Another well-known case was Rick Allen (see Fig. 1), drummer member of Def Leppard's band, who in 1984, suffered an automobile accident and

Fig. 1 Musicians with amputation, *Paul Wittstengard*, *Rick Allen* and *Tony Iommi*

lost his left arm [4]. Another one is the case Tony Iommi guitarist for the band Black Sabbath, who uses a prosthetic limb finger to play his instrument (see Fig. 1) [5].

For another point of view, nowadays there are a great number of robots that perform diverse activities and among them it is playing music. Robots can play musical pieces, either in concert with other robots or with people, or as soloists, performing in a really excellent manner [6–8], which, in a sense is logical, as they are programmed precise machines. Nevertheless, Robotics is not the right approach for prosthesis development for musicians, since music is about feeling, practice and a personal touch. Additionally, the great development of prostheses has allowed artificial limbs to be controlled by the patients thought, but without any of them been able to provide sensitivity to the wearer.

2 Prothesis

In developing a prosthesis for a person who lacks a limb, it is intended to emulate as much as possible the natural movements of the limb which will be replaced so the patient to emulate the natural movements of the limb that will be replaced as far as possible, so the patient can continue performing their daily activities as normal as possible. It is for this reason, that the prostheses are required to make the movements as natural as possible [9]. To achieve the desire of functionality of the prostheses, it is indispensable to know the anatomy and functioning of the member it is required to replace. In order to study the body member in depth, in some cases it is possible to make use of technologies such as motion capture. This together with the help of a software providing position data, would help to determine acting forces, movements, trajectories and turns that are to be made by the person [10–12].

A prosthesis can be used by any amputee, as long as it meets certain factors, such as size and geometric shape of the area to be replaced, the limb function and the medical condition of the patient, among others [13]. Currently there are different types of prosthesis, in operation and mode of operation, although basically, prostheses are classified into two different types passive and active [14–17]. Each of the different types of prostheses that have particular characteristics, which makes it more or less effective, depending on the case. In the musical area, it is not yet possible to play an instrument through a prosthesis as would be done with the natural limbs of the body. This is a very difficult task as the execution of an instrument makes use of the so-called *musical memory* [18], which is an ability to, through trial and practice, to automatically interpret a musical theme by an instrument. The development of this type of memory requires four stages of the memorization process: impression, perception, comprehension and retention [19]. Musical memory involves seven different types of memories to be developed, among which a very important one is muscle memory [20]. When performing an instrument, there are often finger movements that are sometimes so fast that they could not be performed by a prosthesis. For example, the strumming on a guitar consists basically in provoking the vibration of the strings (one or more at the same time), usually with the fingers to generate a musical harmony [21]. Six types of movements are described in the strum; grained, dry, *chorlitazo*, double, blow and double *chorlitazo* [22]. However, there are some texts in which they consider the types of strumming of different form and only consider 4 types; thumb and finger, thumb hooked, thumb stationary and all fingers [23].

3 General Anatomical Description of the Wrist

In the strumming of the guitar strings, the movement of the wrist has to be considered in addition to the action of the fingers. The wrist is the distal articular complex of the upper limb, which allows the hand to assume a desired position and function as the end effector of the upper limb. This system has two degrees of freedom which are performed only by the wrist, with pronosupination, that is the rotation of the forearm on along the longitudinal axis, a third degree of freedom is added to the wrist, with which the hand can be oriented in any desired or required position [24]. The wrist articular complex, has a central nucleus called carpi, which is composed of eight small bones. Two joints are formed by this complex; the radiocarpal joint and the mediocarpal joint [24]. The wrist has basically two movements, flexion extension movements [25] and movements of abduction [26]. The range of abduction, or radial deflection, has an amplitude of 15° at its maximum position from the neutral axis (see Fig. 2).

The range of adduction or ulnar deviation may vary depending on the axis from which the movement is taken. Only for this movement the angle goes from 30° to 55°, since different neutral axes are taken into consideration. The range of motion in flexion is 85° at its maximum position and reaches 100° when a force or an external support is applied to the movement, which is already a forced position called passive flexion. In extension, the range of motion is also 70°, reaching 95° by supporting the hand and applying force on a surface, which is called passive extension [27, 28].

There is a third degree of freedom, which is the movement of pronosupination. This movement is performed by the rotation of the arm on its longitudinal axis. In this movement, the joints involved are the distal ulnar radius and proximal ulnar radius.

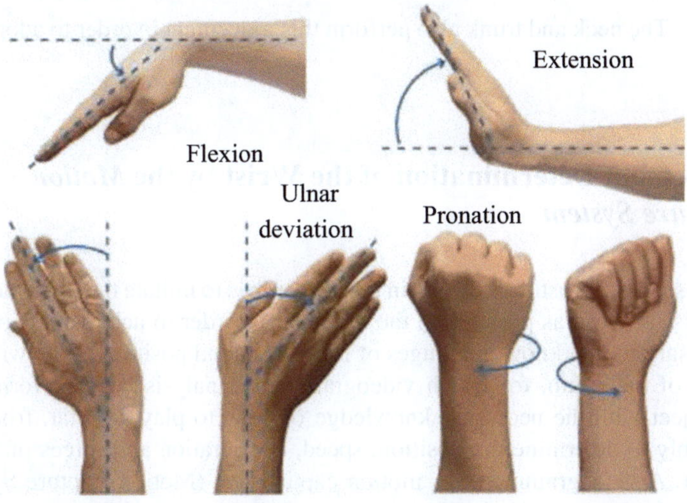

Fig. 2 Movements range of the wrist

By this movement, in conjunction with the above-shown movements of the wrist joint complex, the hand can be positioned, almost in any desired direction. Pronosupination also has a range of motion, which is measured from the neutral axis, which as in the wrist. Although under the analysis of the movement of pronosupination, it is only possible to measure if the forearm is flexed 90° and attached to the body, since, if measured in an extended position, in line to the axis of the arm, the movement gains more range as it is affected by arm rotation [28, 29].

Then, if we consider three positions of the forearm when performing this movement. The anatomical position, which is the aforementioned neutral. So, the supination position is that in which the palm of the hand is positioned upwards, and reaches a maximum range of motion of 90° in the horizontal plane at the end last position is that of pronation, in which the palm of the hand is positioned downwards and has a maximum amplitude of movement of 85° equally on the horizontal plane [28, 29]. By combining the movements of pronosupination, flexo-extension and abduction-adduction. Then we can consider the three positions of the forearm, it is possible to generate another movement, called circumduction, which is a twist in the wrist, a movement made when performing the strumming movement on a guitar. This movement is performed simultaneously on the two axes of the wrist and the axis of pronosupination. The movement performed can be described graphically with an irregular conical shape, because the amplitude of the movements is not always symmetrical. The cone formed, encompasses the sagittal plane and the coronal plane, with its maximum amplitude in the sagittal plane. In the circumference, the amplitude of movements is greater in pronation than in supination, which means that the cone is not completely in pronation [28, 29]. Finally, the circumference movement is not exclusive to the wrist, in general the upper and lower limbs perform this type of movement. The neck and trunk also perform this movement in order to adopt certain postures.

4 Movement Determination of the Wrist by the *Motion Capture System*

The purpose of the prosthesis design in this research is to imitate the movement made by the wrist as close as possible to the real one. In order to achieve this objective, it is necessary to first know the ranges of movement and position of the wrist in the execution of the strum, for which videogrammetry analysis were performed with a test subject with the necessary knowledge on how to play a guitar, from which it is possibly to determine the position, speed, acceleration and forces of the wrist movement. A videogrammetry or motion capture test (Motion Capture System or MoCap System) was carried on to capture and digitize the movement in order to analyze, step by step, the characteristics of the movements made by the wrist in the strumming of a guitar (see Fig. 3).

To obtain position data, in order to analyze the movement that is generated by the wrist when executing a strum, a method of optical capture of movement was used. The working area used is delimited by a quadrilateral structure (4 m^2). There are 16 cameras, positioned in groups of 4 per side in the upper part of the structure (see Fig. 4).

17W Prime cameras, Optitrack® brand, have 360 frames per second capture and 70° field of view horizontally. Passive markers are applied, mounted on a special suit

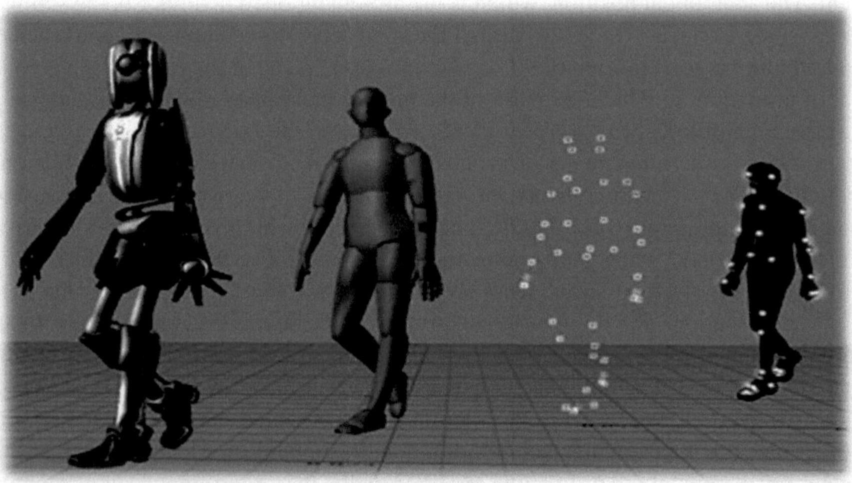

Fig. 3 Motion capture to develop a digital skeleton

Fig. 4 Location of the 16 cameras

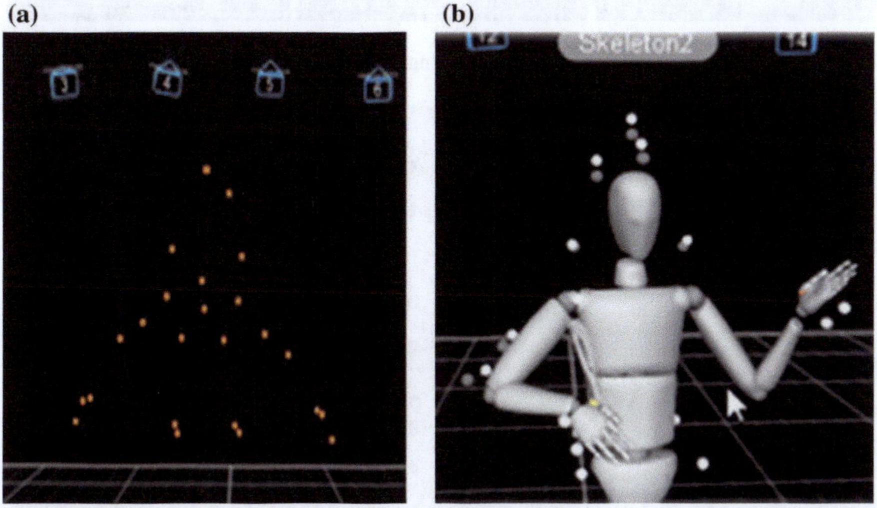

Fig. 5 Representative images of Videometry software. **a** Dot detection process. **b** Skeleton construction

and whose distribution in the body is obtained by means of a software. This equipment captures the movement of the right upper limb and a calibration is performed through a series of parameters established by the software (axes and planes). A digital skeleton is generated to capture the movement of the person. The markers are displayed on the screen as a dotted set (see Fig. 5a) and a 3D virtual skeleton is generated (see Fig. 5b).

The movement that is made when strumming a guitar, consists of four steps. It starts with the arm in a neutral position, with the pick guitar between the thumb and forefinger and supported on the sixth string (see Fig. 6a). Next, the arm is moved down to the first string (see Fig. 6b) and continues with an upward movement from first string to sixth string (see Fig. 6c). It continues the cycle by performing a downward movement on all strings (see Fig. 6b).

When capturing the movement data and selecting the most useful one, the strumming movement is performed in a time interval of 0.8 s and only 3 of the 25 markers

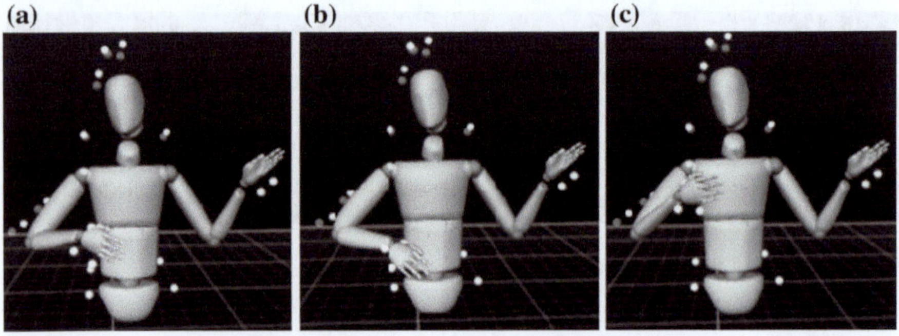

Fig. 6 Strum cycle. **a** Initial position. **b** Descending. **c** Ascending

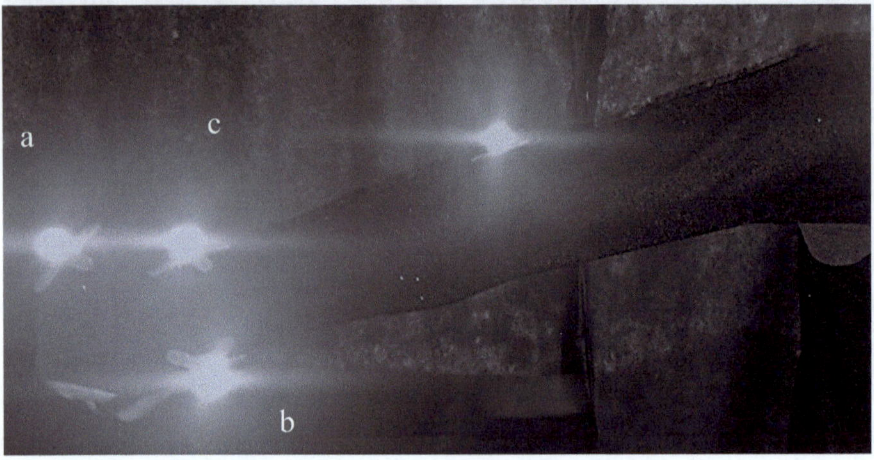

Fig. 7 Location of the markers

that make up the skeleton will be analyzed. The data of the markers to be used are as follows (see Fig. 7):

1. *RHand_1*.—Marker placed at the level of the ridge of the proximal phalanx of the little finger of the right hand (a).
2. *RHand_2*.—Marker on the radial head of the right limb (b).
3. *RHand_3*.—Marker on the ulnar head of the right limb (c).

By the movement it is possible to delimit and determine displacements, forces, speeds, accelerations and angles in the extremity. Figure 8 shows the positions of the markers in the aforementioned steps on the coronal plane (XY plane). Figures 9 and 10 show the movements from the perspective of the sagittal plane and the transverse plane, respectively.

The traveled distance by the hand when performing the strumming movement (see Table 1) the middle point (see Table 2) [30] is established. The trajectories traversed

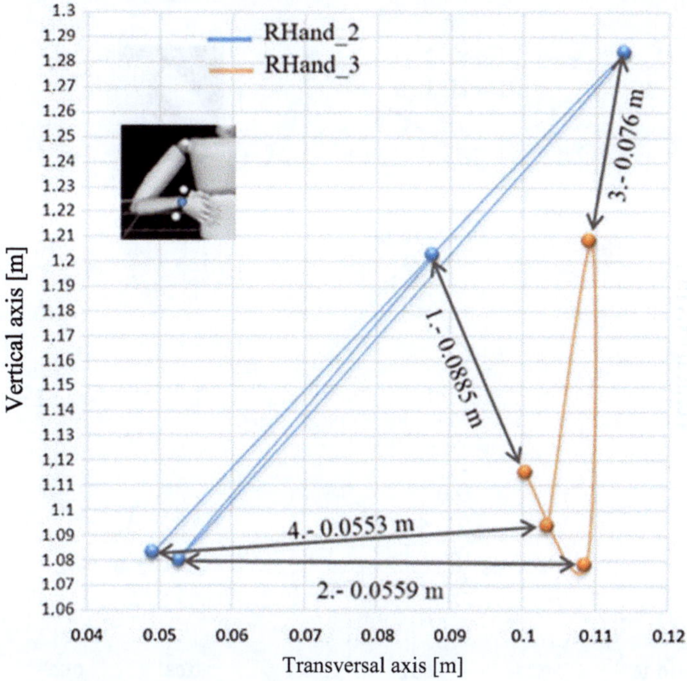

Fig. 8 Markers view from the coronal plane

Table 1 Distance between markers depending on the position

Position	Distance between markers *RHand_2* and *RHand_3*		
	Coronal plane (m)	Sagittal plane (m)	Transversal plane (m)
1	0.0885	0.091095	0.027835
2	0.0559	0.07464991	0.093219
3	0.076	0.091359	0.05106
4	0.0553	0.074858	0.0918

by the wrist are obtained by performing the strumming. The trajectory is divided into three movements: first the neutral position to the lower maximum point, the second is from the lowest point to the upper maximum point, and the third one is the movement when it is at the maximum upper position and descends to the contact with the sixth string.

In Fig. 11 it can be seen the trajectory followed by the midpoint of the wrist determined between the markers *RHand_2* and *RHand_3* on the coronal plane. Table 3 shows the distances traveled in each movement [30].

At this point it is possible to determine the force, speed and acceleration produced between each stride step (Table 4) [31, 32]. The mass used to obtain the forces is the

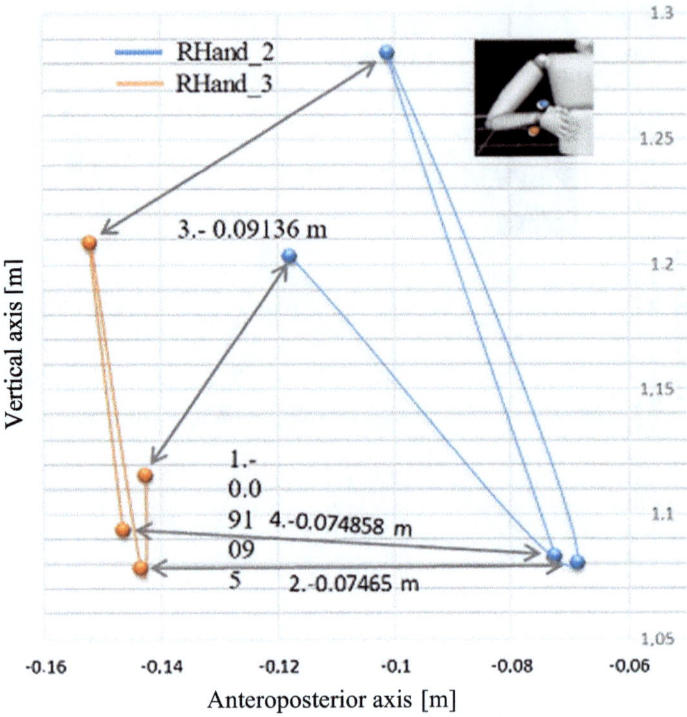

Fig. 9 Markers view from the sagittal plane

Table 2 Middle points location

Position	Middle point x (m)	Middle point y (m)
1	0.0940255	1.1589
2	0.0807375	1.078944
3	0.111608	1.246112
4	0.09986	1.15889

Table 3 Distances between movements

Movement	Distance (m)
1	0.08
2	0.17
3	0.096

sum of the masses of the forearm and hand, which is 2.3 and 2.52% of the total body weight, respectively [33], with the patient having a bodyweight of 70 kg.

It is also necessary to obtain certain angles from the movement of the wrist, to be applied into the design of the prosthesis. These angles are those for abduction and adduction, angles of extension and flexion-extension. The abduction and adduction angles will be obtained for two dimensions on the coronal plane. To determine the

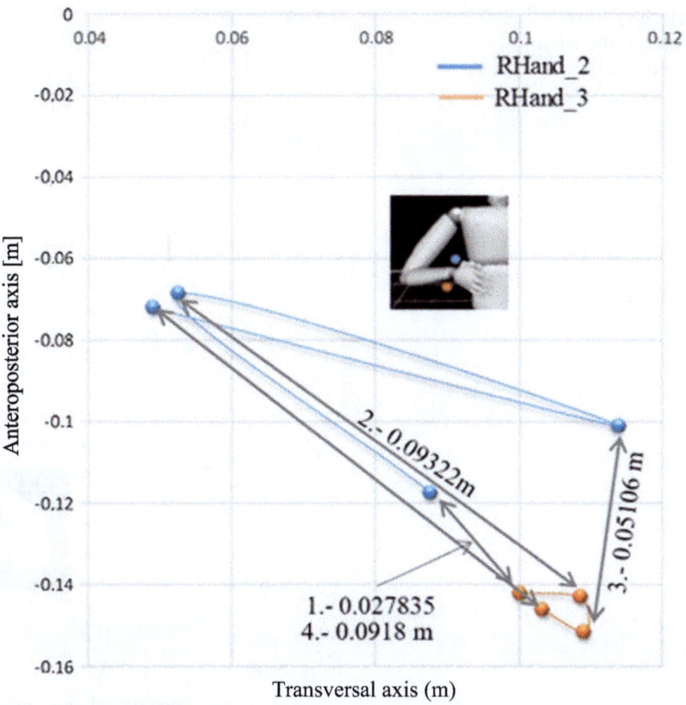

Fig. 10 Markers view from the transversal plane

Table 4 Magnitudes of the wrist on the coronal plane

Movement	Speed (m/s)	Acceleration (m/s^2)	Force (N)
1	0.0405	2.026	2.262
2	0.511	1.521	2.463
3	0.606	2.275	3.663

maximum angles reached when strumming is performed, the positions of the markers *RHand_1* and *RHand_3* (see Fig. 12) will be used. Angles of motion were determined graphically (see Figs. 13, 14 and 15) for various positions and the final result is shown in Table 5. The line segment formed by points AB, markers *RHand_2* and *RHand_3* (sagittal plane) (see Fig. 16). In Fig. 17, the maximum angle of rotation is shown when reaching the maximum position from the sagittal plane. In Fig. 18, the top maximum position is shown from the sagittal plane.

Table 5 Degrees of abduction and adduction

	Abduction	Adduction
Real	32.55°	4.36°

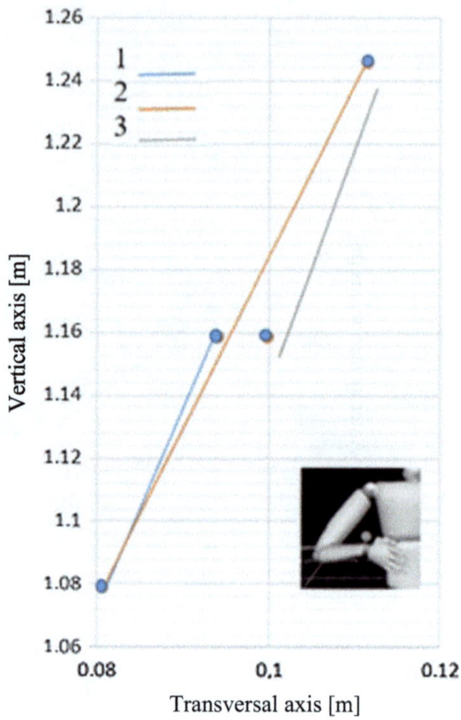

Fig. 11 Trajectory traversed by the midpoint of the wrist in the coronal plane

5 Prosthesis Development [34]

With the data obtained in the previous section it is possible to develop a prosthesis that fulfills the function of the lost organ. A prosthesis driven by body movement, is made up of several components, that is the socket, the suspension system and/or the harnesses, the wrist device and the terminal device [34]. As there was no patient available for this investigation, the prothesis was designed in accordance with data based on medical literature concerning [35]. When performing transradial amputations, the surgeon attempts to maintain the greatest possible length of the stump, so that a prosthesis can be fitted. The ideal length for a stump below the elbow is 18 cm in length [36]. For the development of the prosthesis, the following steps were performed [37], which highlights the application of a 3D copier to manufacture the prosthesis:

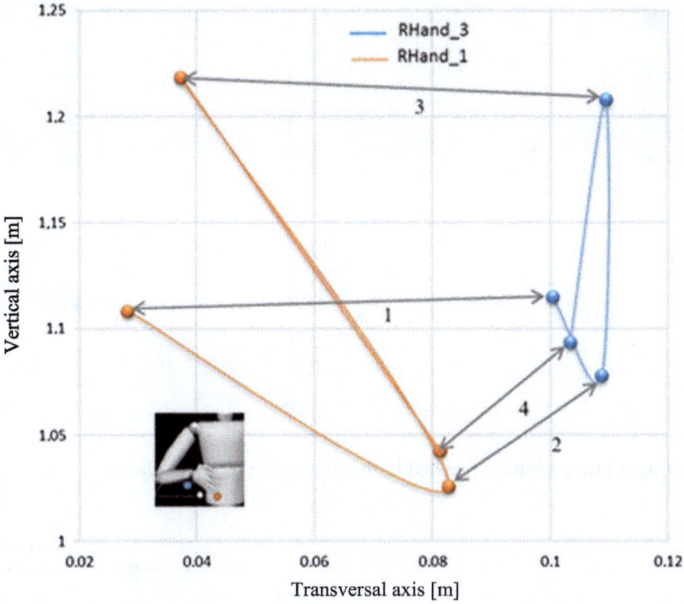

Fig. 12 Movement in four steps from the markers *RHand_1* and *RHand_3*

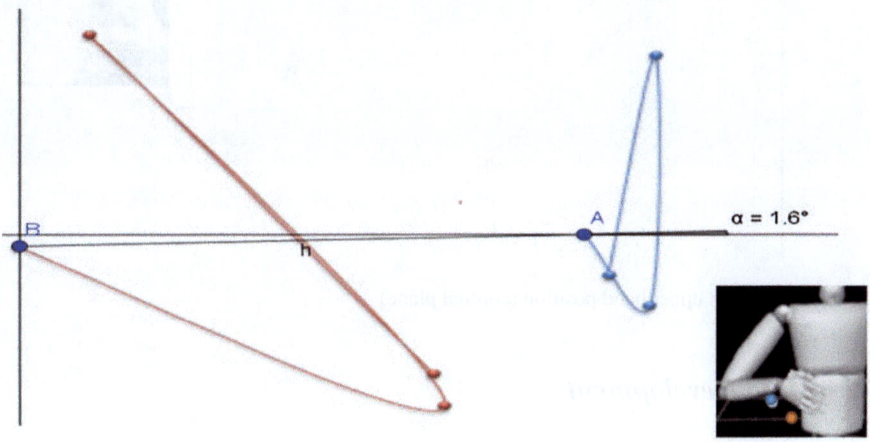

Fig. 13 Neutral hand location (coronal plane)

5.1 Case of Evaluation

Initially, the patient's situation is evaluated, and all the most relevant characteristics are known and a decision on the type of prosthesis to be use is derived from the activities performed by the patient.

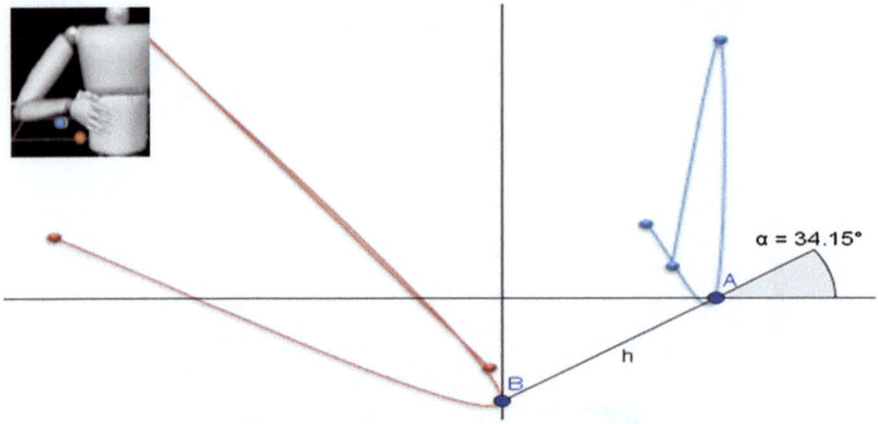

Fig. 14 Maximum lower position reached in the movement (coronal plane)

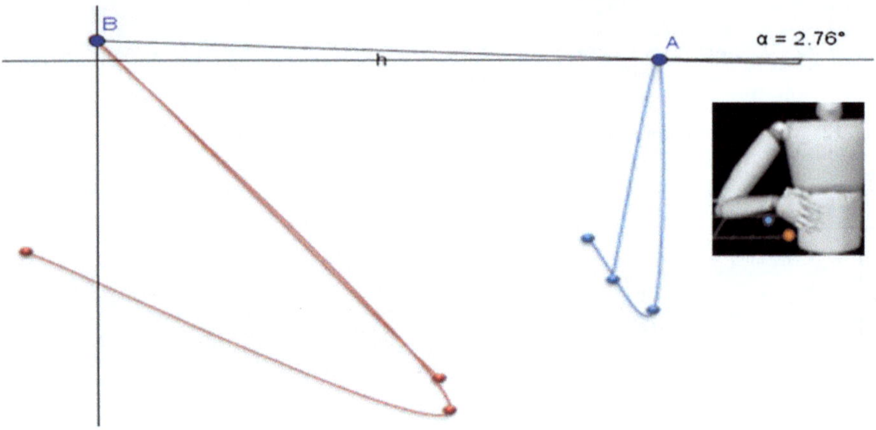

Fig. 15 Maximum upper hand position (coronal plane)

5.2 *Mold Development*

The shape of the antebellum stump is taken, and the desired gypsum mold is fabricated to subsequently develop the socket.

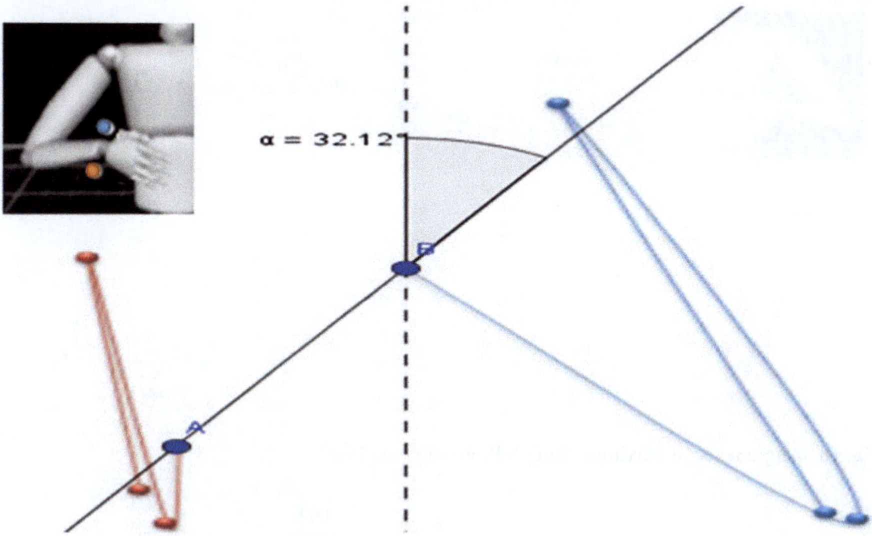

Fig. 16 Angle between markers at neutral position-longitudinal axis (sagittal plane)

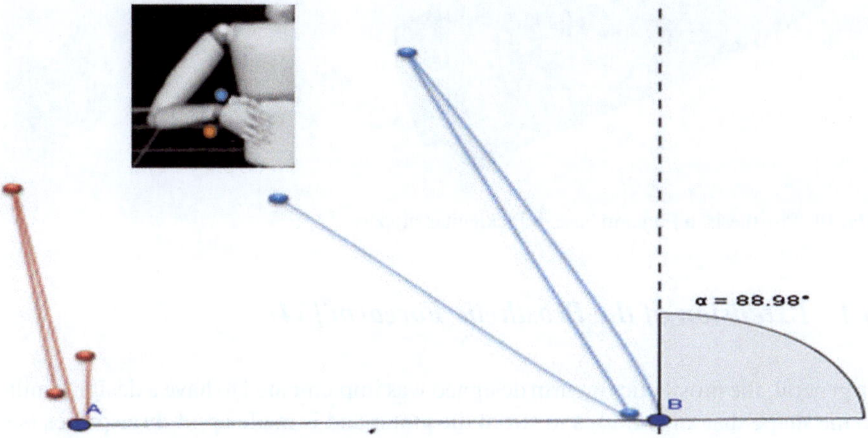

Fig. 17 Maximum turning angle (sagittal plane)

6 Socket Design and Development

Probably the socket is the most important parts of a limb prosthesis. Its importance lies in that it is the basis of the prosthesis and the connection with the stump (the body) [38].

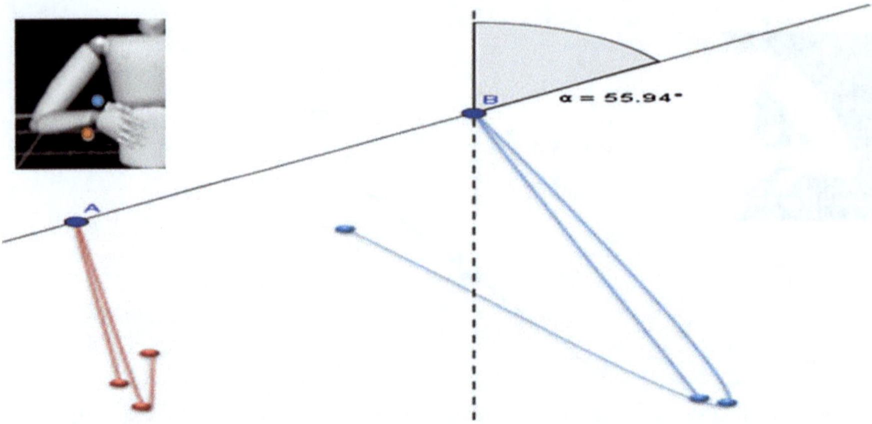

Fig. 18 Angle taken at maximum top position (sagittal plane)

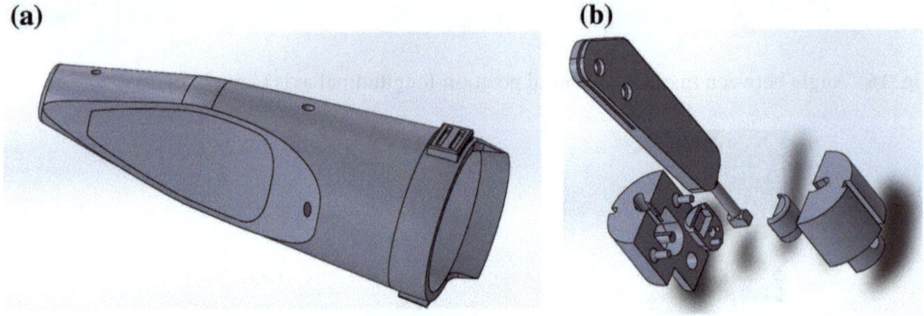

Fig. 19 Prosthesis. a Forearm base. b Pick guitar support

6.1 Extension of the Prosthetic Forearm [34]

In general, the prosthetic forearm designed was implemented to have a design similar to the shape that an arm uses to strum the guitar and is made up of three pieces (see Fig. 19). The three pieces are: the arm, the guitar pick holder, and the socket. The socket for the prosthesis is included in the same piece of the extension of the forearm.

6.2 Terminal Device Design

For the design of the terminal device, it was required to perform certain motion tests to capture the movements and positions taken by the wrist when performing the strumming on the strings of a guitar. This device will basically consist of a moving part, which will emulate the movement of the wrist. The design of the terminal device

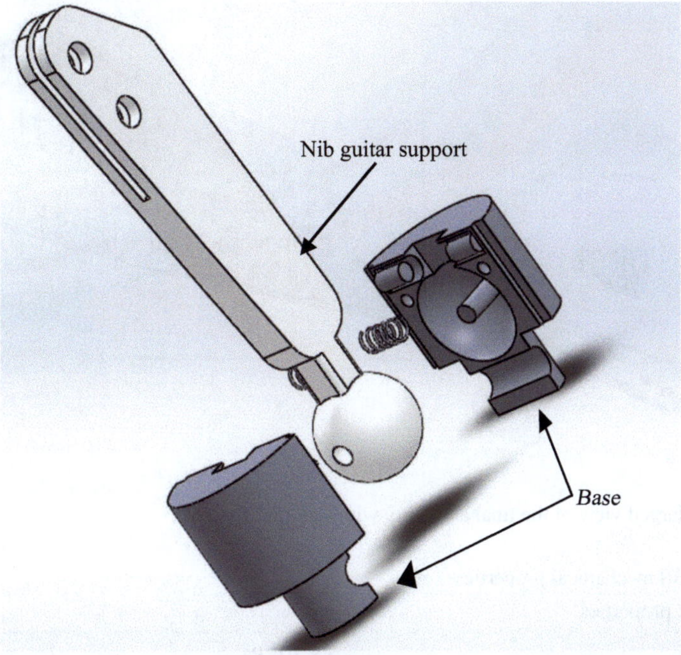

Fig. 20 Enlarged view of the terminal device

is formed with two parts namely the base and guitar pick holder. As well, it has two springs, which will assist the movement, limiting and stabilizing the pick guitar holder in the center of the bases (see Fig. 20). The final assembly of the prosthesis is shown in Fig. 21.

7 Numerical Evaluation

Through the analyses developed in previous sections it is possible to perform numerical analyses of the developed prosthesis. By means of the Finite Element Method and a commercial computational program it is possible to determine the stresses and strains to which the proposed prosthesis system is subjected. In this section, the numerical analysis developed is presented, in order to establish the effects of loading into the prosthesis [20]. The 3D model of the prosthesis produced for a patient weighing 70 kg, was modelling using *ANSYS Workbench V.11*. The numerical analyzes were performed in a structural manner with quasistatic considerations and applying 3D modeling elements. The elements used were solid high order elements with 20 nodes per element and considering 6 degrees of freedom (UX, UY, UZ Rot X, Rot Y and Rot Z). The mesh was performed in a free manner, this is because the model had an

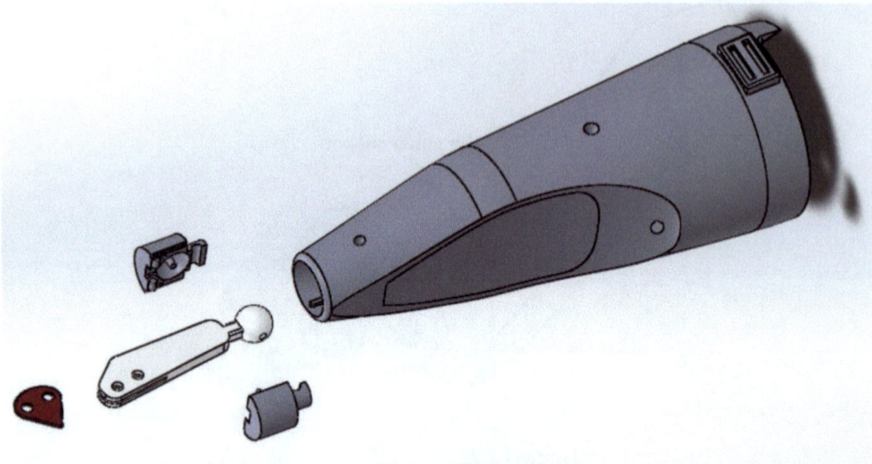

Fig. 21 Enlarged view of the final assembly with terminal device

Table 6 ABS mechanical properties

Mechanical properties	ABS
Yield stress	3×10^7 Pa
Elastic modulus	2×10^9 Pa
Poisson ratio	0.394
Density	1020 kg/m^3

irregular shape (Fig. 11) [20]. The analysis was developed on the complete device and it has a great deal of attention the union of the pen holder with the prosthesis. The magnitude of the force with which it was worked was of 3.663 N, which is the maximum value previously calculated by the videometric. Also, the prosthesis was built using a 3D printer, mainly using ABS filaments. The mechanical properties of the ABS used for the analysis of this work are shown in Table 6.

Figure 22 shows the device in assembly with the guitar pick at the support with its discretized generated mesh which was automatically controlled with the finest parameter.

Figures 23 and 24 shown the *von Mises* stresses and zoom up in the main areas of interest, respectively. The results of the general displacement of the guitar pick support are presented in Fig. 25. Figure 26 shows the effects of loading on the base of the pick guitar support.

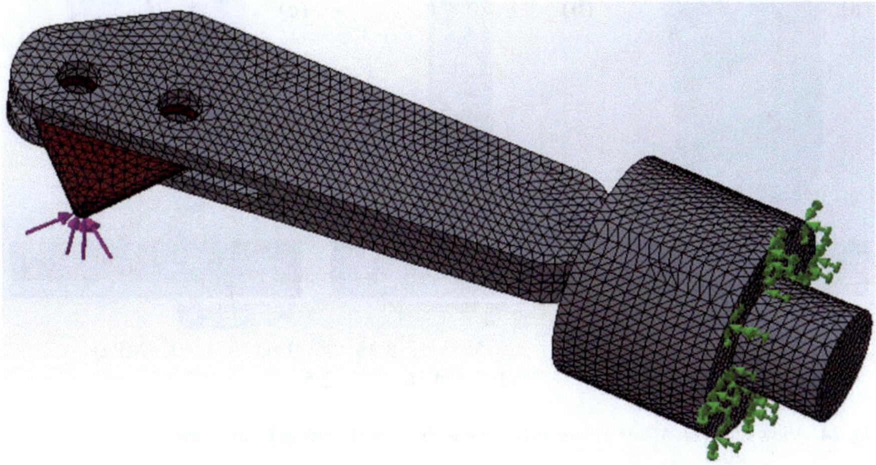

Fig. 22 Assembly and meshing of the complete device and pick guitar

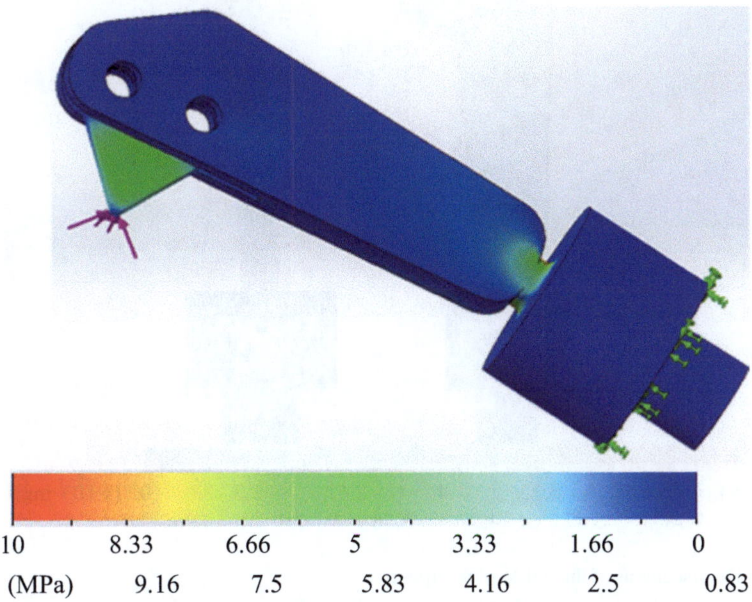

Fig. 23 *von Mises* stress on guitar pick support

8 Prosthesis Development by 3D Printing

The fabrication of the prosthesis was done using 3D printing technology. The material with which it was printed was mentioned in the numerical analysis (ABS polymer) of

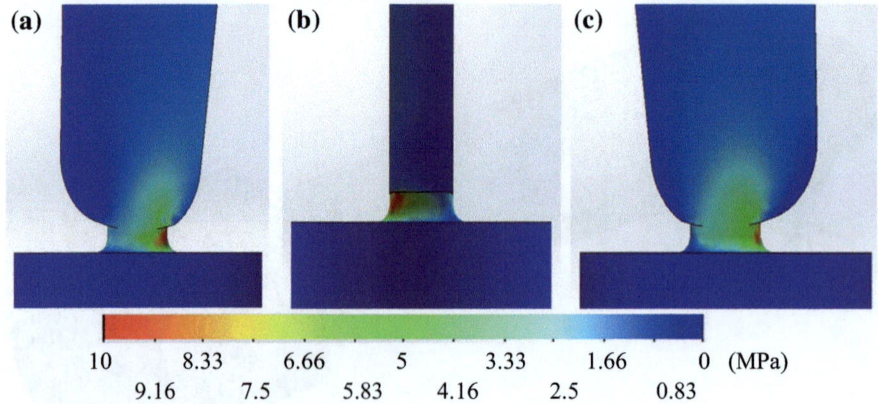

Fig. 24 Views for *von Mises Stress*. **a** Left view. **b** Frontal view. **c** Right view

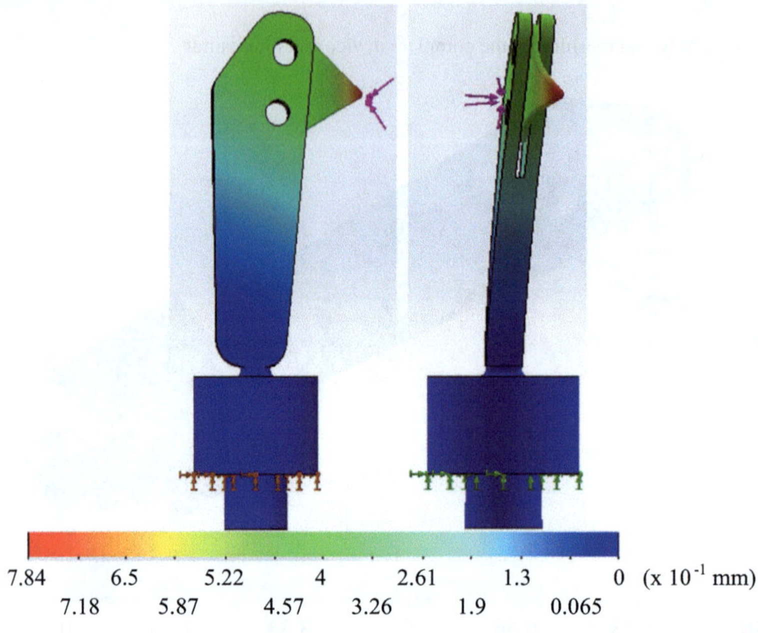

Fig. 25 Displacement of the guitar pick support

which the mechanical properties for this material were already mentioned during the analysis. The pick guitar used for the prosthesis is from the *Jim Dunlop*®, brand Jazz III model, made by Nylon. The guitar pick model is specified, since a great variety of polymers are used to manufacture nibs (guitar picks), to mention a few, such

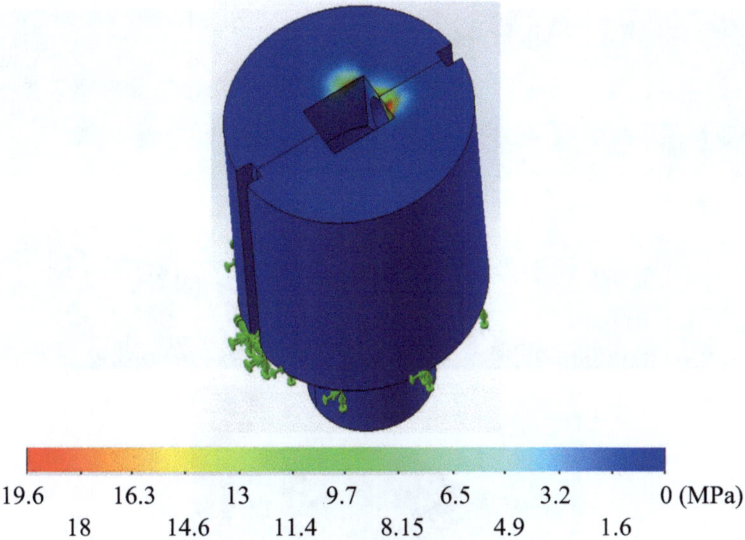

Fig. 26 *von Mises* stress from the base of the pen holder

as *Tortex*®, *Derlin*®, ABS, Steel, Aluminum, among others. Those made of some metallic material, are not usually used. In the next figures the 3D printed prosthesis and its components are physically shown (Figs. 27 and 28).

9 Discussions and Conclusions

The loss of a limb or its motor functions, produces important changes in the lives of those who suffer it. In the case of people whose modus vivendi depends on that limb, this translates this into a total and radical change. Musicians are a clear example in which such as loss causes a drastic life change, since they cannot play their instrument properly. There are two possible consequences: ending their musical career and a way of life. In addition, when these patients are to be rehabilitated, prostheses must fulfill not only an aesthetic function which could pass to second term, but a functional one. That is why it is necessary to develop prostheses that allow users to perform a specific function of each limb. The prosthesis proposed in the present study provides a better functionality and ease of movement. It is designed for the right forearm, specifically for a guitarist with a transradial amputation with which it is possible to perform the strumming movement on the strings of a guitar. It consists of a mobile terminal device for the prosthetic forearm, with movements similar to those made by the wrist when strumming a guitar. The terminal device was designed taking into account the motion captured tests performed, which allowed to know the force with which the wrist moves when performing the strumming. In this way, the prosthesis is only a

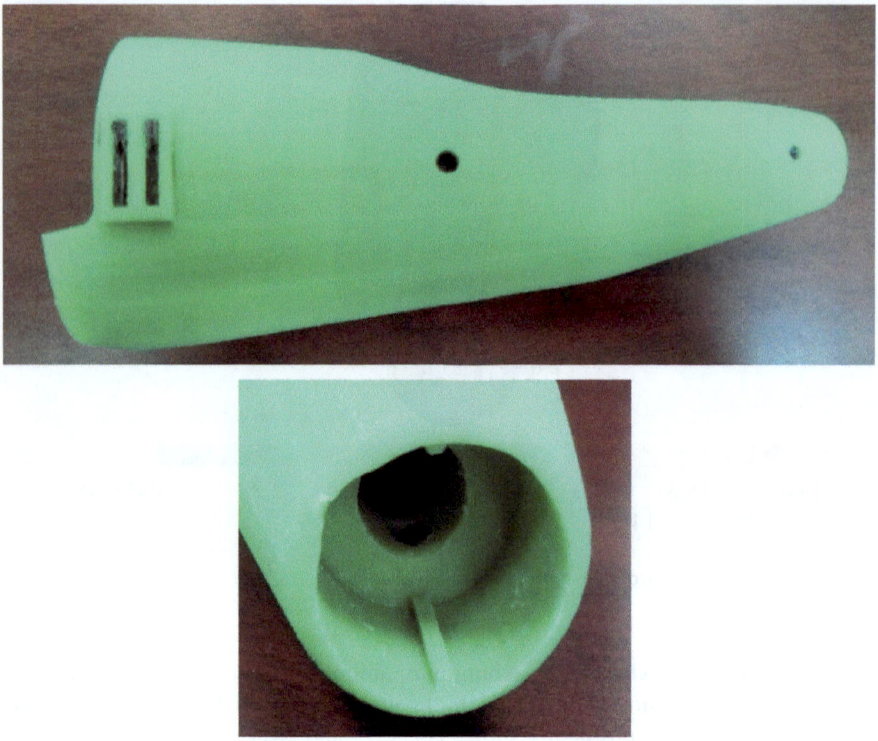

Fig. 27 Prosthesis socket

mechanical device, driven by the force of the movement of the stump of the person who uses it. As well as using the videometric and the interpretation of the graphs obtained from the markers, the forces acting on the arm movement were obtained to perform the strumming of the guitar. With this data, the resistance of the device could be analyzed when making contact with the strings. It should be mentioned that although a guitar string is a flexible element, it was taken as if it was a rigid surface, at the moment of contact of the pen with the rope. There are other variables that can be taken into account such as vibration, tension variants and other factors including in the string and even though they were not taken into account, the design obtained gives the wearer the satisfaction of playing a guitar through movements performed.

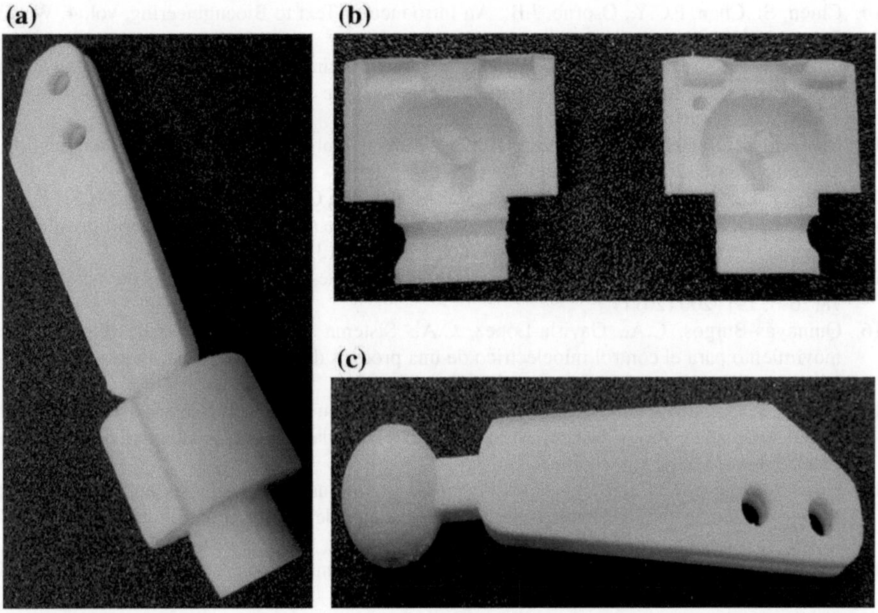

Fig. 28 Guitar pick support. **a** Assembly. **b** Base. **c** Support

Acknowledgements The authors gratefully acknowledge the financial support from the Mexican government by Consejo Nacional de Ciencia y Tecnología (CONACyT) and the Instituto Politécnico Nacional.

References

1. Zetterberg, C., Backlund, H., Karlsson, J., Werner, H., Olsson, L.: Musculoskeletal problems among male and female music students. Med. Probl. Perform. Artist. **13**(4), 160–166 (1998)
2. Santiago, C.H.: Duelo por la Pérdida de un Miembro (pp. 12–21), Tesina de Diplomado, Asociación Mexicana de Tanatología, A.C. (2014)
3. Reich, H.: Rediscovered Score Pianist's Last Legacy. Chicago Tribune, Saturday, 1 July 2017
4. Wilkening, M.: The Day Def Leppard Drummer Rick Allen Lost His Arm in Car Crash, p. 1. Ultimate Classic Rock (2014)
5. Childers, C.: Black Sabbath's Tony Iommi Loses Tips of Fingers—Moments that Nearly Destroyed Rock, p. 1. Loudwire (2000)
6. Shibuya, K.: Violin playing robot and Kansei. Musical Robot. Interact. Multimodal Syst. **74**, 179–193 (2011)
7. Maes, L., Raes, G.W., Rogers, T.: The man and machine robot orchestra at Logos. Comput. Music J. **35**(4), 28–48 (2011)
8. Solis, J., Takanishi, A., Hashimoto, K.: Development of an anthropomorphic saxophone-playing robot. In: Brain, Body and Machine, Proceedings of an International Symposium on the Occasion of the 25th Anniversary of the McGill University Centre for Intelligent Machines, pp. 175–186 (2010)
9. Wills, D., Prosthesis. Stanford University Press (1995)

10. Chien, S., Chen, P.C.Y., Osorno, J.B.: An Introductory Text to Bioengineering, vol. 4. World Scientific (2008)
11. de Armas-Ramírez, N., Sánchez, A.P., de la Celda-Brovkina, A.: Habilidades investigativas en biomecánica; Propuesta de talleres para la preparación de entrenadores deportivos. Arrancada **16**(29), 64–77 (2016)
12. Marasco, P.D., Hebert, J.S., Orzell, B.M.: Artificial limbs for upper extremity amputation. Technol. Adv. Surg. Trauma Crit. Care 609–619 (2015)
13. Murphy, D.: Levels of amputation. Fundam. Amputation Care Prosthet. **15**, 20–21 (2013)
14. Brito, J., Quinde, M., Cusco, D., Calle, J.: Study of the state of the art of hand prostheses. Ingenius: Revista de Ciencia y Tecnología **9**, 57–64 (2013)
15. Arzola, N., Loaiza, J.L.: Evolución y tendencias en el desarrollo de prótesis de mano. Dyna **78**(169), 191–200 (2011)
16. Quinayás-Burgos, C.A., Gaviria-López, C.A.: Sistema de identificación de intención de movimiento para el control mioeléctrico de una prótesis de mano robótica. Ingeniería y Universidad **19**(1), 27–50 (2015)
17. Burgos, C.Q., Albán, O.A.V.: Diseño y Construcción de una Prótesis Robótica de Mano Funcional Adaptada a Varios Agarres, M.Sc. thesis, Universidad del Cauca, Facultad de Ingeniería Electrónica y Telecomunicaciones, pp. 27–28 (2010)
18. Cuartero-Soler, M., Payri, B.: Tipos de memoria, aptitudes y estrategias en el proceso de memorización de obras pianísticas. Revista Electrónica de Música en la Educación **26**, 32–54 (2010)
19. Estarriaga, L.I.L, de Landa, C.N.: El desarrollo de la memoria musical. Revista Arista Digital **24**, 2–6 (2012)
20. Collazos, J.G.: La memoria musical en alumnos de piano en conservatorios españoles; Revisión y propuesta. Dedica; Revista de Educação e Humanidades **10**, 13–27 (2016)
21. Castillo, R.R.: Técnicas de ejecución de la guitarra flamenca; Historia, desarrollo, aporte y mecanismos. Revista Káñina; Artes y letras **39**(2), 215–234 (2015)
22. Ögmundsdóttir, E.Ö.: Flamenco; Una introducción desde su origen hasta nuestros días. El grito del pueblo 55–58 (2014)
23. Chicón, I.B.: La Guitarra Flamenca como Recurso Didáctico en el Aula de Música No. 10, pp. 2–5 (2007)
24. Drake, R., Vogl, A.W., Mitchell, A.W.: Gray's Anatomy for Students, vol. 3, pp. 130–131. Elsevier Health Sciences (2014)
25. Palastanga, N., Field, D., Soames, R.: Anatomía y Movimiento Humano; Estructura y Funcionamiento, pp. 161–168. Editorial Paidotribo (2007)
26. Medina-Gonzalez, C., Benet-Rodríguez, M., Marco-Martínez, F.: El complejo articular de la muñeca; Aspectos anatofisiológicos y biomecánicos, características, clasificación y tratamiento de la fractura distal del radio. MediSur **14**(4), 430–446 (2016)
27. Martín-Alfaro, R., Navarro-Navarro, R., Ruiz-Caballero, J.A., Jiménez-Díaz, J.F., Brito-Ojeda, E.: Biomecánica de la articulación de la muñeca. XIX Jornadas Canarias de Traumatología y Cirugía Ortopédica 103–106 (2005)
28. Kapandji, A.I.: Fisiología Articular; Tomo 1; Hombro, Codo, Pronosupinación, Muñeca, pp. 104–109 and 148–153. Mano, Editorial Médica Panamericana (2006)
29. Thompson, C.W., Floyd, R.T.: Manual de Kinesiología Estructural, pp. 74–75. Paidotribo (1999)
30. Sullivan, M.: Trigonometría y Geometría Analítica, pp. 13–16. Pearson Educación (1997)
31. Lleó, A., Lleó, L.: Gran Manual de Magnitudes Físicas y sus Unidades, pp. 110–114. Ediciones Díaz de Santos, S. A. (2011)
32. Mott, R.L.: Diseño de Elementos de Máquinas, pp. 26–27. Pearson Educación (2006)
33. Charney, P., Malone, A.: ADA Pocket Guide to Nutrition Assessment, pp. 160–161. American Dietetic Associati (2009)
34. Hernandez-Campos, D.: Diseño de una Prótesis de Antebrazo Derecho para Asistir la Ejecución de Guitarra por Rasgueo, M.Sc. thesis, Instituto Politécnico Nacional, SEPI ESIME Zacatenco (2012)

35. Zambudio, R.: Prótesis, Ortesis y Ayudas Técnicas, pp. 137–152. Elsevier, España (2009)
36. Osorio, L., Módulo de amputados. El Portal de la Salud 15–17 (2012)
37. Camelo-Ulloa, K.J.: Construcción de un Encaje o Socket para Prótesis de Miembro Inferior con Amputación Transfemoral. B.Sc., Universidad La Salle, pp. 23–32 (2007)
38. Lake, C.: The evolution of upper limb prosthetic socket design. J. Prosthet. Orthot. **20**(3), 85–92 (2008)

Numerical Analysis of the Knee Articulation Leg. The Angular Position Parameters and Forces Acting on the Joint Were Obtained and Applied into the Corresponding, During Different Stages of the Gait Process

Guillermo Urriolagoitia-Sosa, Beatriz Romero-Ángeles, Daniel Méndez-Romero, Rafael Rodríguez-Martínez, Christopher René Torres-San Miguel, Jacobo Martínez-Reyes, Rosa Alicia Hernández-Vázquez and Guillermo Urriolagoitia-Calderón

Abstract The knee is a diarthrodial joint or a wide mobility joint (or a wide mobility joint), which involves an extremely complex mechanical analysis. The knee possesses a great stability in complete extension to support the corporal weight, and has the necessary mobility to perform diverse daily activities (jump, gait, trot, run, among others) and efficiently orients the foot in relation to the irregularities of the ground. At present, a general understanding is possessed of the forces acting onto the bone structural components of the knee joint during daily activities. On the other hand, by applying Finite Element algorithms it is possible to numerically simulate the

---

G. Urriolagoitia-Sosa (✉) · B. Romero-Ángeles · D. Méndez-Romero · R. Rodríguez-Martínez
C. R. Torres-San Miguel · J. Martínez-Reyes · R. A. Hernández-Vázquez ·
G. Urriolagoitia-Calderón
Instituto Politécnico Nacional, Escuela Superior de Ingeniería Mecánica y Eléctrica, Sección de Estudios de Posgrado e Investigacion, Unidad Profesional Adolfo López Mateos "Zacatenco", Ciudad de México, Mexico
e-mail: guiurri@hotmail.com

B. Romero-Ángeles
e-mail: romerobeatriz97@hotmail.com

R. Rodríguez-Martínez
e-mail: fararm68@hotmail.com

C. R. Torres-San Miguel
e-mail: ctorress@ipn.mx

J. Martínez-Reyes
e-mail: jacmartinez@esimez.mx

R. A. Hernández-Vázquez
e-mail: alyzia.hv@esimez.mx

G. Urriolagoitia-Calderón
e-mail: urrio332@hotmail.com

© Springer International Publishing AG, part of Springer Nature 2019
A. Öchsner and H. Altenbach (eds.), *Engineering Design Applications*,
Advanced Structured Materials 92, https://doi.org/10.1007/978-3-319-79005-3_9

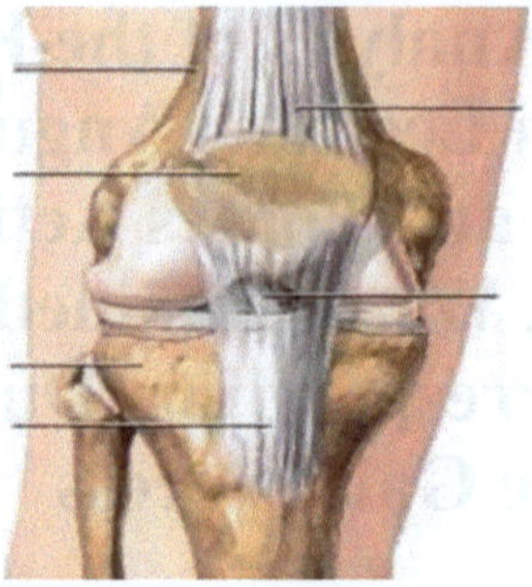

Fig. 1 General knee anatomy

anatomic systems that constitute the human body. This algorithm has turned out to be an important tool to determine research behavior of the bone is not an organism, to determinate the behavior of bones from a mechanical point of view. Additionally, they are applied as a foundation for prosthesis design and numerical model generation to solve problems related to clinical conditions. For example, the degenerative osteoarthritis of the knee is a chronic degenerative disease that is active in persons between the ages of 50 to 60 years old and involves severe wear of the joint. This condition can be accelerated by multiple circumstances, the main one big effect of articulated overload (obesity, knee deformation, meniscus injuries, among others). In this paper we propose that a static structural analysis of the knee joint, which involves three phases of the human gait: normal support, contact (foot-ground) and balancing of the leg. The angular position parameters and the forces acting on the joint were obtained, and applied into the corresponding numerical analyses for each gait phase. The numerical analyses are based on the kinetic and kinematic studies of the knee, to determine the orientation and range of mobility of the joint. The numerical model of the knee joint was developed from a Computed Axial Tomography scan which would assure bio-fidelity in the numerical evaluation. Obtaining as results: the von Misses equivalent stress, the maximum and minimum principal stresses and the total displacement. It was determined by this research, that the knee capacity to support the loads in each step of the gait process and it helped to establish a data base for the design and development of joint knee prosthesis.

Keywords Biomechanics · Finite element method · Mechanical design

1 Introduction

The knee is the biggest and most complex joint of the whole human body, it qualifies as a diarthrodial joint. In the knee there are united 3 bones [the low end of the femur, the top end of the tibia and the patella (kneecap or kneepan)] (see Fig. 1) [1].

The muscles and bones that affect the movement of the knee articulation are a mechanical system, which includes a muscular torque. Such a muscular moment consists of two forces; the muscle force and the bone force acting as a lever. Nevertheless, the muscular force in the human body is extremely difficult to be exactly measured, therefore the muscular force variation showed can be obtained by measuring the muscular length changes, including the total length and its respective period, although the variations or periods are uncertain regarding the force variations [2]. In this respect, any biological knee component that is damaged will result in an abnormal movement or pain. As time goes by, osteoarthritis will arise due to the constant use and degeneration of the cartilage and meniscus, and the normal activities of the patients would be affected. The osteoarthritis affects the quality of life an important manner, specially of ageing and elderly persons, even putting their life risk Additionally, older patients not only suffer from the pain, but they also need to be cared and supported by the family and society [3]. The understanding of the acting forces inside bones and tissues of the knee joint during daily activities is of vital importance to protect this biological system of possible injuries and fractures. Also, this knowledge is applied to the design of protheses and the generation of numerical models with the aim to comprehend the biomechanical behavior of the knee. Knee osteoarthritis is a chronic degenerative disease that is commonly present in persons between the ages of 50 and 60 years old, and it involves the joint wear and can hasten multiple circumstances to overload the joint. A recent report from the *World Health Organization* (*WHO*), indicates that such a disease as osteoarthritis is close to be the fourth most important reason of disability in women and the eighth in men, with a high annual financial cost which are attributed to this pathology that is near to 60 trillion dollars [4].

The numerical modeling of the biological systems that constitute the human body by the application of the Finite Element Method (FEM), has turned nowadays into an important research tool to determine the behavior that presents a bone or series of bones as a mechanical structure. Bendjaballah et al. [5] present a numerical model with an application in a Finite Element Method software, which considers the femur, tibia, patella, cartilage and the modeling of the four principal ligaments (as springs) in a non-linear manner, to investigate and obtain the knee biomechanical properties. In the same year, Heegard et al. [6] developed a 3D model to analyze the patellar biomechanics during the knee passive flexion. Likewise, Beynnon et al. [7] developed an analytical model of the knee sagittal plane to study how the ligaments are responsible to control the cinematic of the joint, Xu-Shu et al. [8] determined the pressure distribution at the knee contact surfaces under the double leg action standing at 0° and 90° knee flexion. The contact pressure results and their pressure distribution were obtained by a numerical simulation applying Finite Element algorithms (see Fig. 2).

In this research a static structural analysis of the knee joint is proposed, which involves three phases of the human gait: normal support, contact (foot-ground) and balancing of the leg. The angular position parameters and the acting forces are obtained, which are applied into the corresponding numerical analyses for each gait phase. The numerical analyses are based on the kinetic and kinematic studies of the

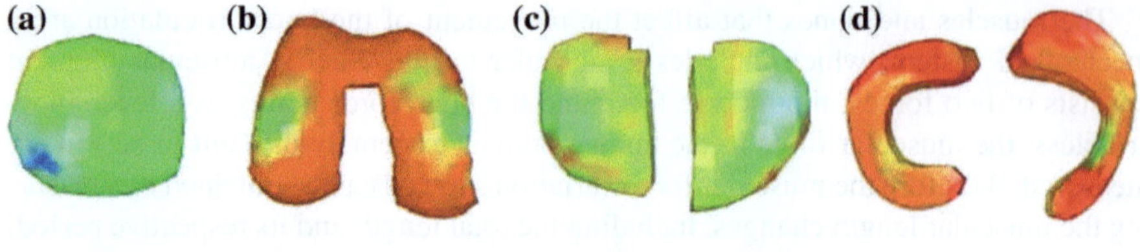

Fig. 2 Maximum compressive stress produced by a 200 N loading during two legs standing [8]. **a** Patellar cartilage. **b** Femur cartilage. **c** Tibia cartilage. **d** Meniscus

knee, to determine the orientation and range of mobility of the joint. The numerical model of the knee joint was developed from a Computed Axial Tomography Scan which would assure bio-fidelity in the numerical evaluation. Obtaining as results: the equivalent von Misses stress, the maximum and minimum principal stresses and the total displacement of the knee. It can be determined by this research. The knee capacity can be determined by this research, helping to establish a database for the design and development of the joint knee prosthesis.

2 Gait Cycle and Loading Considerations

The gait cycle is the sequence of events that takes place between two consecutive repetitions which causes the movement of the body from one point to the next one. Therefore, it is adopted as a cycle beginning at the instant in which one of the feet contacts the ground (habitually by the heel). If it is taken as the origin of the contact made by the right foot, the cycle would end in the following support of the same foot. The left foot would experience the same series of events [9]. In a general manner, during a complete gait cycle, each leg will experience a support phase (the foot is on the ground) and an oscillation phase (the foot is situated at the air). The support phase begins with the heels initial contact and finishes with the takeoff of the foot. The oscillation phase goes through the instant that the foot takes off until the following contact takes place [10].

In relation to the duration of the gait cycle, the support phase directly depends (in normal conditions) on the spontaneous speed adopted as the subject walks and is about 60% of the entire gait cycle [9], so the oscillation phase represents the remaining 40% (see Fig. 3).

Nevertheless, the period length of each one of these phases will strongly depend on the speed. If the velocity is increased the oscillation proportion against the support phase will also be amplified. If the speed is reduced, the oscillation period length will be zero at the double leg support stages [11].

In normal conditions, in the monopodial gait support of the person, the body weight acts in a manner eccentric to the knee, producing an irregular movement. The inertia forces due to the acceleration of the body masses have limited influences into the produced movement. This irregular adduction moment is equivalent to the

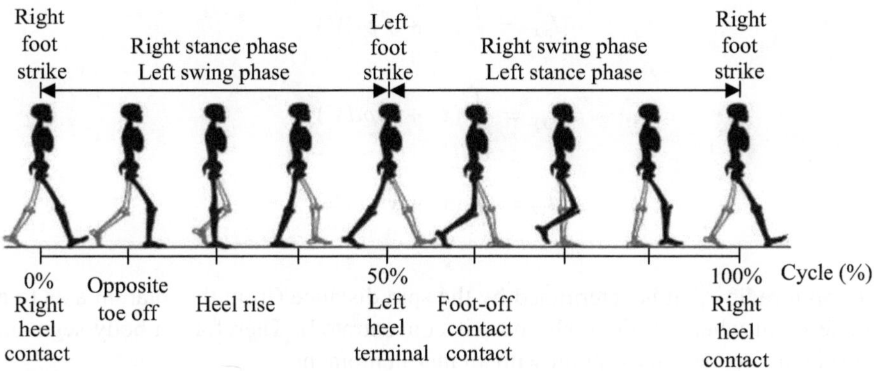

Fig. 3 Gait cycle

applicable weight multiplied by the distance related to the load center of the knee, and this must be counter by the adduction moment that has been produced by the active-passive tension of the knee lateral elements [12]. The resultant force represents the acting load from the femur into the tibia that produces a vector which passes through the loading center of the knee, which necessarily has the knee articulation in equilibrium. Nevertheless, it is important to establish that the load distribution on the biological surfaces are extremely sensible to the location point of the load center [13], which is located at the joint center in the frontal plane.

3 Anthropomorphic Inertia Moment Determination [14]

The body resistance to rotation can be determined by the body or body segments of the inertia moment *I*:

$$I \int_{m} r^2 dm \tag{1}$$

where *m* is the body mass and *r* are the distance from the rotation axis to the body. The incremental mass can be described as ρ *dV*, and for a body with a constant density ρ, the inertia moment can be determined by the body volume integration:

$$I = \rho \int_{v} r^2 dV \tag{2}$$

The last general expression can be expressed in terms of rotation by:

$$I_{xx} = \int_{v} (y^2 + z^2 \rho dV)$$

$$I_{yy} = \int_{v} (x^2 + z^2 \rho dV)$$

$$I_{zz} = \int_{v} (x^2 + y^2 \rho dV) \quad (3)$$

The spin radio r_g, it is determined by the spin distance (from the rotation axis to a single point where all the body mass is concentrated). Therefore, a body segment must be treated as a mass point with an inertia moment.

$$I = mr_g^2 \quad (4)$$

The inertia moment with respect to an I parallel axis, must be in relation with the inertia moment with respect of the body mass center I_{cm}, which can be described as:

$$I = I_{cm} + md^2 \quad (5)$$

where d is the perpendicular distance between the two-parallel axis.

4 Static Analysis of the Loads Acting into the Knee [15, 16]

At the support phase (where the foot is at maximum contact with the ground), the gravity axis must be set across the joint articulation axis. With the knee at full extension, the distance that produces a moment must be zero (see Fig. 4a). Therefore, there is no muscular force to maintain equilibrium at this point. The compressive force at the articulation is equal to half of the body weight. On the other hand, when the knee bent the gravity line falls behind the articulation axis (see Fig. 4b). Additionally, to establish an optimal free body diagram of the knee joint must have the next considerations (see Fig. 4c):

(a) The magnitude, direction and the application axis of the forces reacting against the ground (G).
(b) The direction and application axis of the patellar ligament force.
(c) The application point of the articular reaction force.

The action force line of the muscle is intersected by the action line reaction force of the foot (see Fig. 5). On the other hand, the last line (reaction force) should cross this point of intersection, in order that the three forces acting on the point are concurrent. As before and considering the equilibrium of the forces it makes it necessary that the reaction force should be traced using this point, which serves as reference to be applied as the force on the articular surface (see Fig. 4).

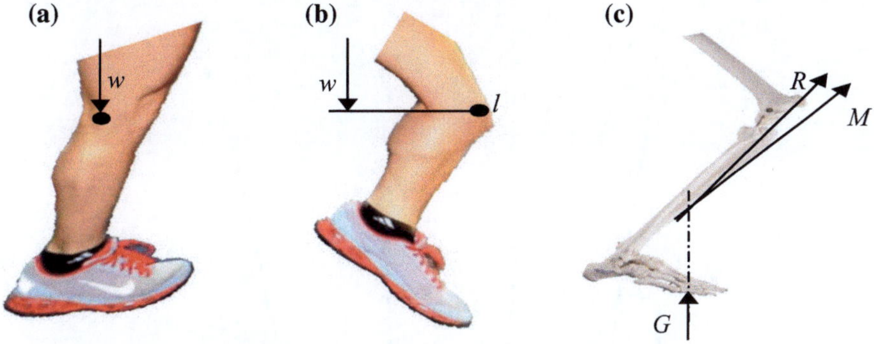

Fig. 4 General free body diagram of the knee joint. a Full support phase (full foot contact). b Right foot strike. c Skeleton right foot strike

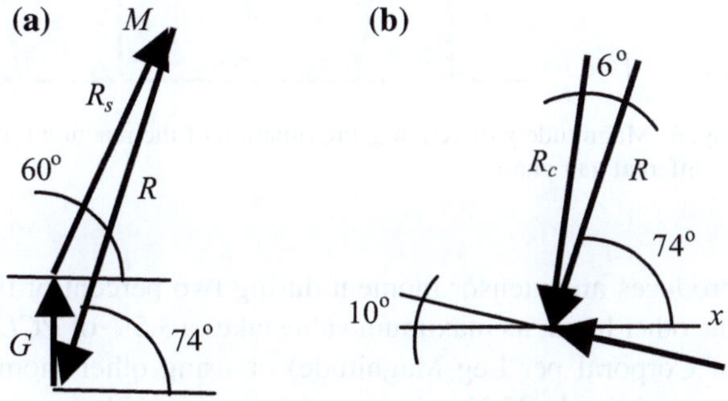

Fig. 5 Muscular forces and angles configuration. a General forces configuration. b General forces reactions configuration

Finally, a vectoral force can be traced, using the vectors *G*, *M* & *R* (see Fig. 5). Once the force triangle is solved, can be found the values of the three vectors can be found, using any solution method. It should be remembered, that the resultant muscular forces or reactions.

According to the antecedents, it can be shown that the articular compression force (*Rc*) is perpendicular at the tibial tray, and it is also parallel to the tibial mechanical axis. Another main consideration is that the sliding force (*Rs*) is equal to 20 degrees of flexion, and that both the thigh and leg exist at an angle of 10° with respect to the vertical axis.

5 Knee Kinematics Analyses

During the supporting phase of the gait cycle, to align the resultant force with the knee, generate four cases of moment (extend, flexion and combined extend-flexion) are generated, which are present during swinging phases. At the time to start of the gait cycle, the reaction force is placed forward from the tibia (see Fig. 6). So, it

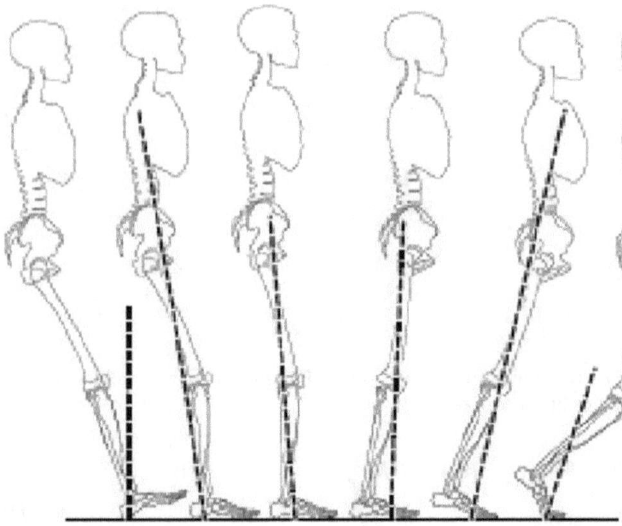

Fig. 6 Magnitude y direction approximation of the reaction forces with respect of the knee joint at different gait phases

produces an extensor moment during two percent of the initiated of gait cycle. On the other hand, its maximum value takes a 3.5% of *PCLM* in anatomic units (Percent of Corporal per Leg Magnitude) or using other moment units, that magnitude is approximately 25 Nm, in accordance with [17].

At the loading member with constant increments, the start of the gait cycle, in the initial support phase, it is quickly aligned, but now the patient, backward at the body center. At 7% of the gait cycle, the reaction force vector passes through the mechanical center of the knee. Following the gait cycle, the mechanical axis is of importance, due to arising having the maximum value in the gait cycle. Which has a moment value of 51 Nm when this phase reaches the final position at the flexion stage. Now, at the beginning of the monopodial support phase of the gait cycle, the opposite phenomenon, arises, namely that mass. It occurs that the mass of the body is inverted. This action implies that the mass of the body with respect to the supported member is inverted, which produces that the resultant reaction force that is lower than the one at the initial gait cycle. In this case, the flexure moment generated diminishes at the knee.

On the other hand, approximately at the half phase of the support phase, a neutral medial alignment (null flexion moment) is presented. Finally, the body continues its movement to the support foot and the resultant reaction vector force takes a position at the anterior plane, which makes a rising extensor moment to appear. The maximum magnitude of the extensor moment reaches 30 Nm [18].

6 Static Determination of the Acting Forces and Angles in Each Case of Study

The static determination of the acting forces and angles was performed taking research considerations developed by González-Jemio et al. [19], Le Veau [16] and Méndez-Romero [20]. Free body diagrams and formulations are also established by basic static theory [21]. Furthermore, for the static analyses a weight of a 70 kg was taken as reference. The free body diagram for each case is shown in Fig. 7. Table 1 the static development to determine the forces and angles acting on the knee joint can be seen.

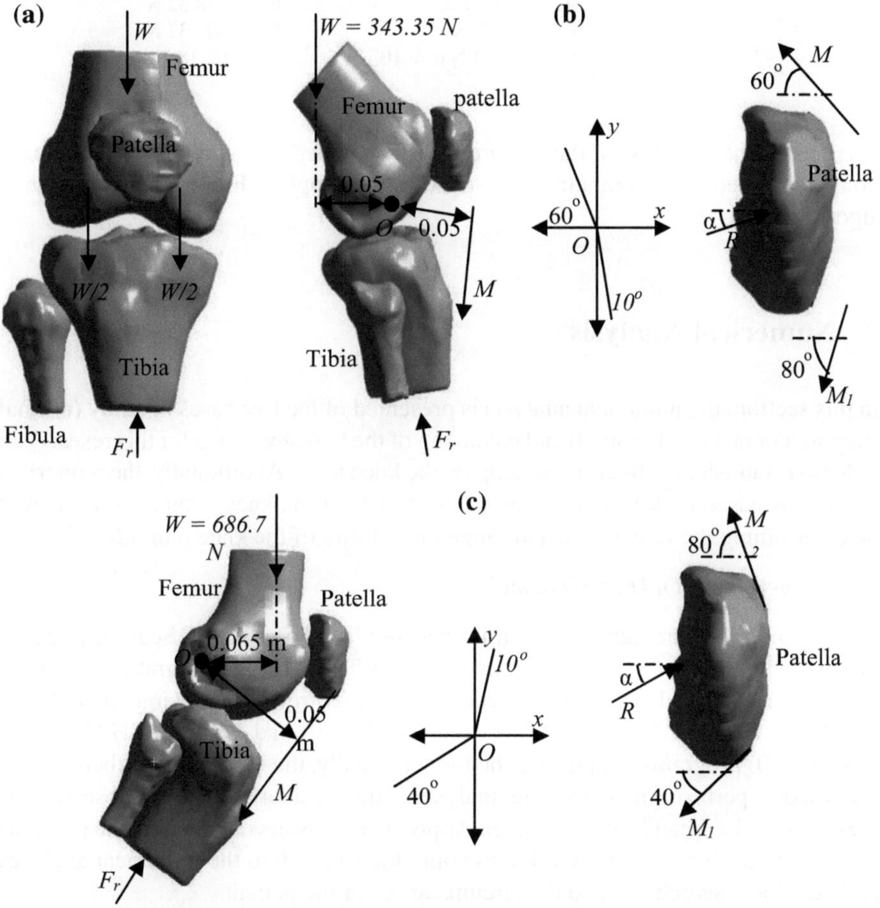

Fig. 7 Free body diagrams of the knee joint at different gait phases. **a** Normal support. **b** Contact (foot-ground). **c** Balancing of the leg

Table 1 Results from the static analyses

Normal support	Contact (foot-ground)	Balancing of the leg
Weight distribution (W)		
W = mg W = (70)(9.81) 6.86.7 N W = W/2 = 343.35 N	W = mg W = (70)(9.81) 6.86.7 N W = W/2 = 343.35 N	W = mg W = (70)(9.81) 6.86.7 N W = W/2 = 343.35 N
Moment (M) with respect to reference point O		
M = 0	M = 343.35 Nm	M = 892.71 Nm
Forces angle		
F_r = 343.35 N α = 0°	F_{rx} = 59.96 N F_{ry} = 685.45 N F_r = 688.07 N	F_{rx} = 683.85 N F_{ry} = 1260.52 N F_r = 1434.07 N
	R_x = 231.29 N R_y = 40.78 N R = 234.84 N α = 10°	R_x = 838.32 N R_y = 305.32 N R = 892.18 N α = 20°

The obtained results in Table 1 are directly applied into the numerical simulations developed by a computational software that applied Finite Element Method algorithms.

7 Numerical Analysis

In this section, the numerical analysis is presented of the tree cases of study (normal support, contact (foot ground) and balancing of the leg) developed for this research in order to establish the effects of loading on the knee joint. Additionally, the numerical analysis is based on kinetic and kinematic studies of the knee, which were focused on determining the orientation and range of mobility of the knee joint [20].

(A) *Generation of a knee 3D model*

The 3D model of the human knee joint was produced from a CT Scan (Computed Tomography Scan) of a patient weighing of 70 kg. The SCAN IP program for imaging processing was used, allowing to generate .stl files from the previous model (see Fig. 8). A COPY CAD 10.120 was used (see Fig. 9) and POWER SHAPE 2010 (see Fig. 10) programs to optimize the model. Finally, the ANSYS Workbench V.11 was used to perform the numerical analyses. The considerations from observations presented by Le Veau [16] were taken. Applying the results obtained by the previous basic static analysis it was possible to apply the forces into the numerical analyses and kept the research close to the circumstances of the patient.

(B) *Numerical considerations to perform the analyses for all cases of study*

The numerical analyses were performed in a structural manner, with quasistatic considerations and using 3D modeling elements. The elements used were solid of

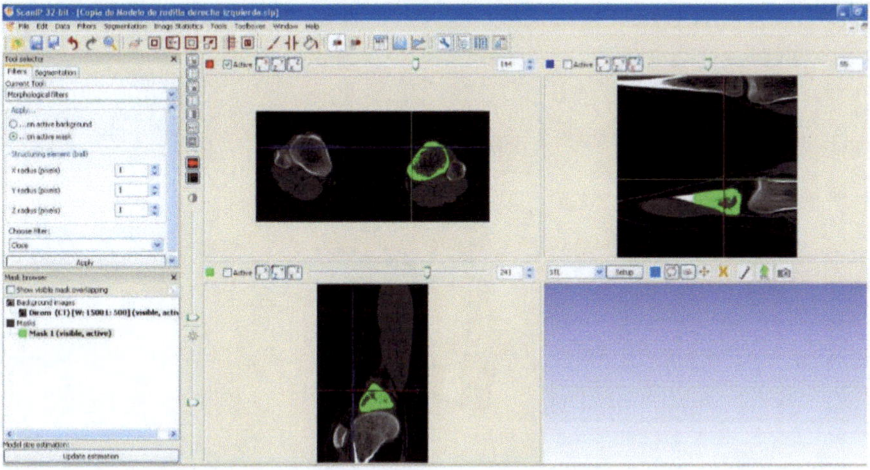

Fig. 8 SCAN IP program for imaging processing

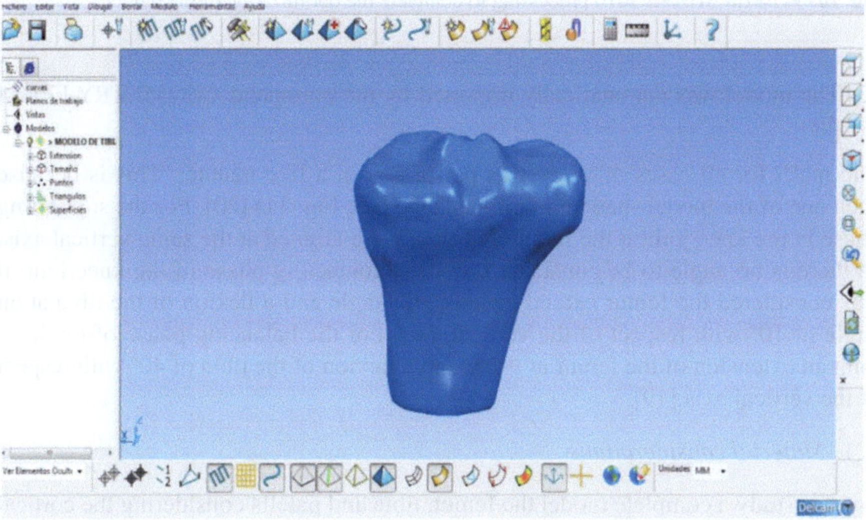

Fig. 9 Application of COPY CAD 10.120 software to optimize the model

high order elements with 20 nodes per element and considering 6 degrees of freedom (UX, UY, UZ Rot X, Rot Y and Rot Z). The next considerations were also applied as well:

1. The file generated to produce the knee model by the POWER SHAPE computer software was exported as a file with the PARASOLID extension.
2. The model was produced in mm and the working units were set up as metric with its equivalences.

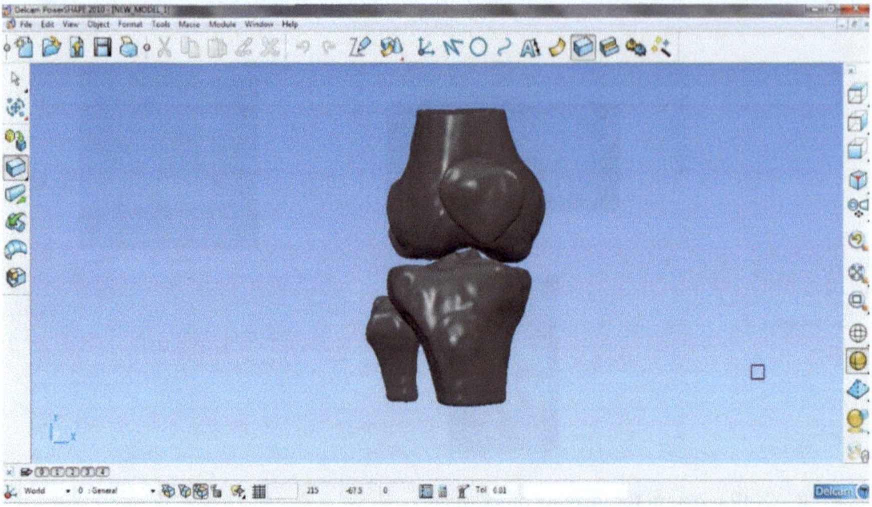

Fig. 10 POWER SHAPE 2010 processing to complete the optimization of the model

3. The model was automatically imported by the command GEOMETRY-FROM file.

The mesh for all cases of study was performed in a free manner. This is because each one of the models had an irregular shape (see Fig. 11) [20]. For the supporting phase in the knee joint it the femur and tibia were aligned at the same vertical axis, so there is no angle to be considered. For the contacting phase of the knee joint it was considered the femur extends with a 30° angle and a flexion of the tibia at an angle of 10º with respect of the vertical axis. For the balancing phase of the knee joint an extension of the femur at −10° and a flexion of the tibia of 40º with respect of the vertical axis [19].

(C) *Material considerations*

For this study a complete model the femur, tibia and patella considering the cortical bone with orthotropic properties was built to fulfill the objective of analyzing the knee joint mechanical behavior under normal conditions. The bone material properties are presented in Table 2.

For all the analyses presented in this research, the movement of the femur was restricted in the *xy* plane, to simulate the effects of the muscles. All degrees of freedom in the tibia were restricted at the inferior part. Finally, a fixed support at the inferior part of the femur and internal part of the patella was applied, to simulate an articular cartilage existence.

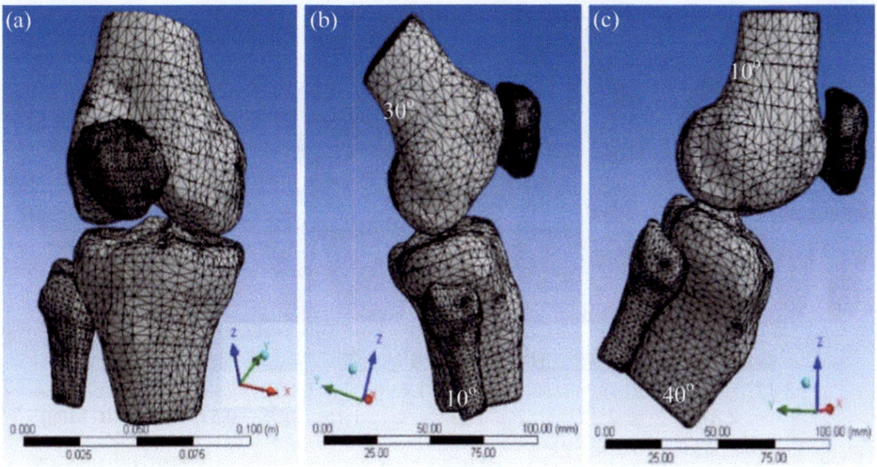

Fig. 11 Developed of the mesh in the right knee model in a free manner. **a** Support phase. **b** Contact phase. **c** Balancing phase

Table 2 Material properties [22]

	Cortical bone	
	E_x	21.9
Elasticity modulus (GPa)	E_y	14.6
	E_z	11.6
	μ_{xy}	0.2
Poisson ratio	μ_{yz}	0.3
	μ_{xz}	0.1
	τ_{xy}	6.99
Shear modulus (GPa)	τ_{yz}	5.29
	τ_{xz}	6.29
Density (kg/m^3)	Total	18,000

8 Results

All the results were performed in a period of 0–1 *s* and the results can be observed as a function of the equivalent *von Mises* stresses, as well of the maximum and minimum principal stresses and total displacements.

(A) *Results for the bipodal support phase*

In this case study, where the knee joint position is total extended, the main objective is to know the compressive and tensile resistance of the biological system. In Fig. 12 the equivalent stress field for the *von Mises* stress can be observed for the knee

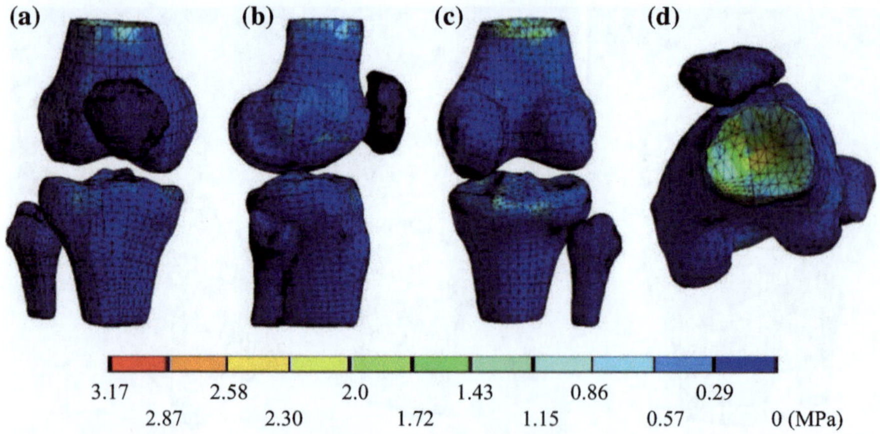

Fig. 12 *von Misses* equivalent stress for the knee joint at the bipodal support. **a** Frontal view. **b** Lateral view. **c** Rear view. **d** Top view

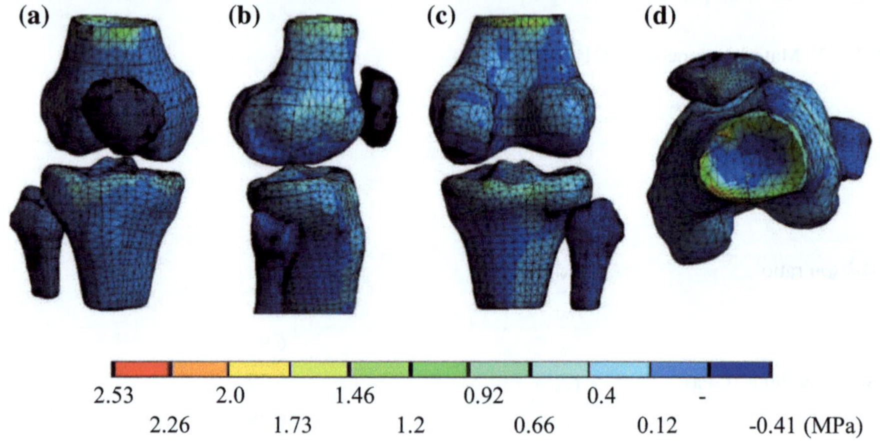

Fig. 13 Maximum principal stress for the knee joint at the bipodal support. **a** Frontal view. **b** Lateral view. **c** Rear view. **d** Top view

system at different planes and it is possible to determine the maximum effects of the loading system applied to the knee joint. Additionally, Figs. 13 and 14 present the maximum and minimum principal stresses fields, respectively. Finally, Fig. 15 shows the total displacement field of the knee joint. To optimize the observation of the knee configuration several views are presented for each one of the figures.

(B) *Results for the contact phase*

In Figs. 16, 17, 18 and 19 the force effects caused at the knee joint due to the contact phase are presented, showing the *von Mises* stress, maximum and minimum principal

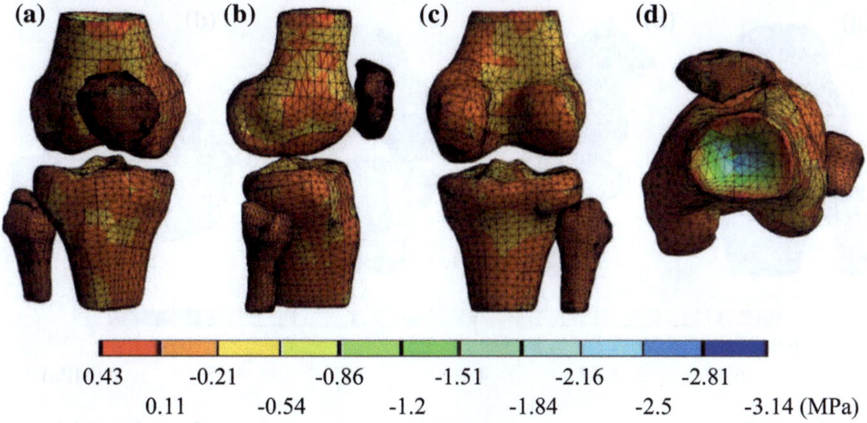

Fig. 14 Minimum principal stress for the knee joint at the bipodal support. **a** Frontal view. **b** Lateral view. **c** Rear view. **d** Top view

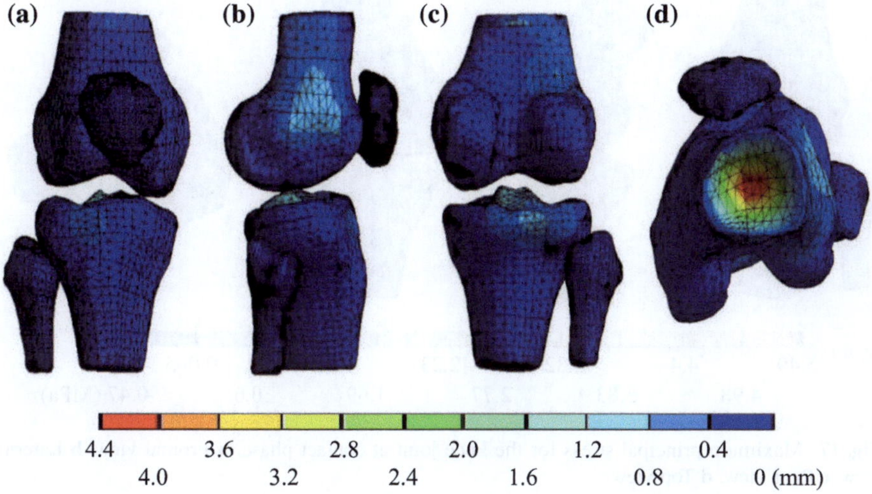

Fig. 15 Total displacement for the knee joint at the bipodal support. **a** Frontal view. **b** Lateral view. **c** Rear view. **d** Top view

stresses and the total displacement, respectively. Different views of the same bones are considered to reach a better understanding of the bone structures behavior.

(C) *Results for the balancing phase*

During the gait cycle, the balancing phase is the most critical, as there is a critical instant in which the knee is supporting the entire weight of the person. Figures 20, 21, 22 and 23 the force effects caused at the knee joint are presented, for the equivalent *von*

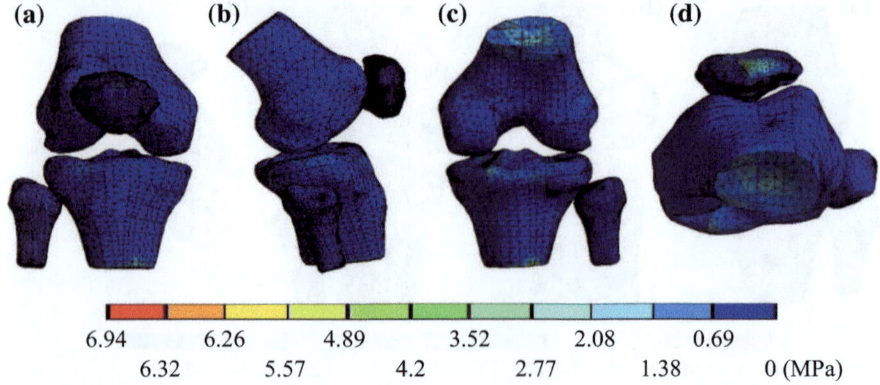

Fig. 16 *von Misses* equivalent stress for the knee joint at the contact phase. **a** Frontal view. **b** Lateral view. **c** Rear view. **d** Top view

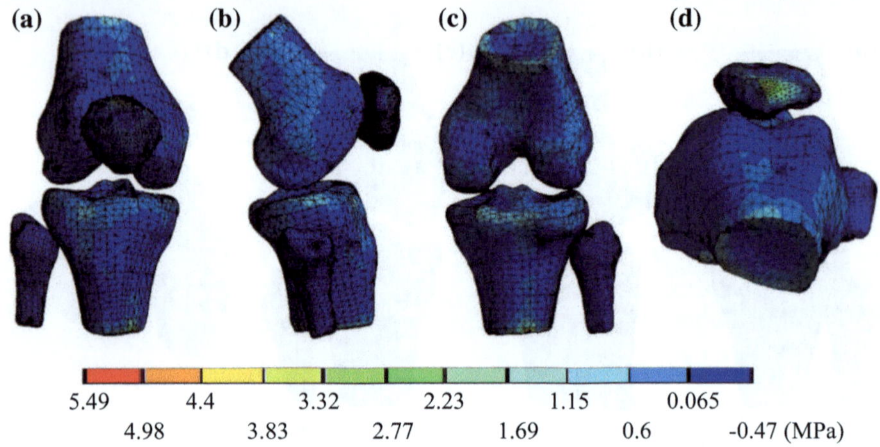

Fig. 17 Maximum principal stress for the knee joint at contact phase. **a** Frontal view. **b** Lateral view. **c** Rear view. **d** Top view

Mises stress, maximum and minimum principal stresses and the total displacement, respectively. Several views are also shown to achieve a better understanding.

9 Discussion and Conclusions

In this study a computational model was generated based on computer tomography, which allows to obtain a greater biofidelity and has proven to be more accurate in the results obtained after an analysis by Finite Element Method (FEM). For this case study, the mechanical properties of a cortical bone were considered. As a result, at

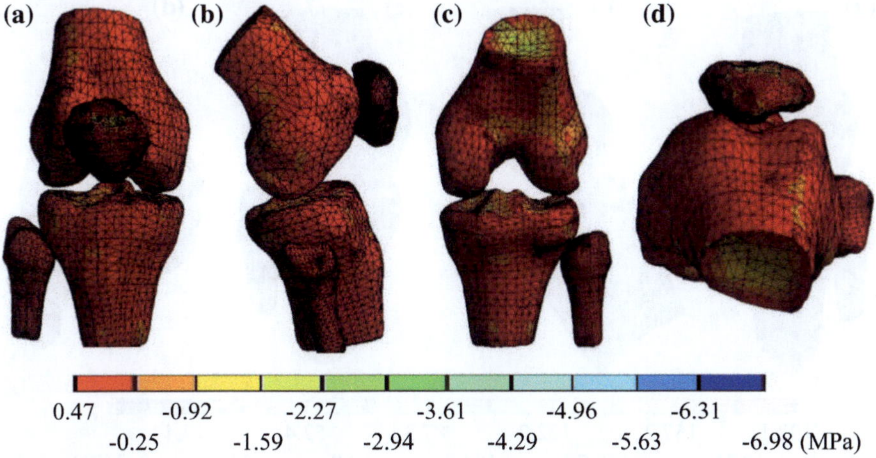

Fig. 18 Minimum principal stress for the knee joint at the contact phase. **a** Frontal view. **b** Lateral view. **c** Rear view. **d** Top view

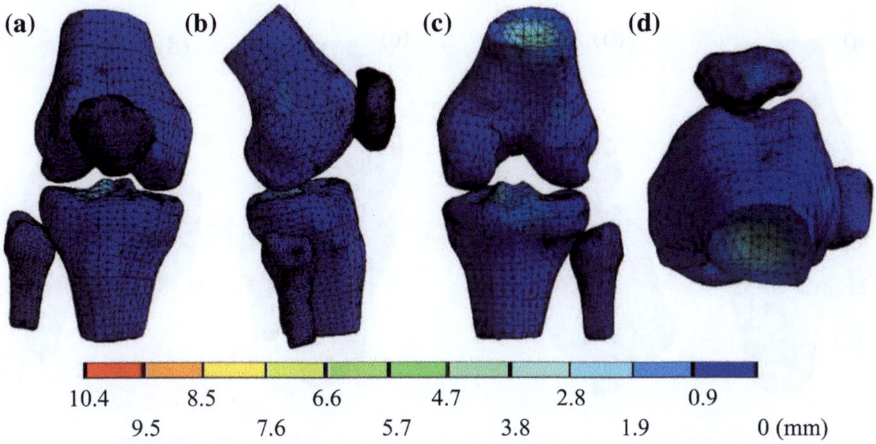

Fig. 19 Total displacement for the knee joint at the contact phase. **a** Frontal view. **b** Lateral view. **c** Rear view. **d** Top view

the bipodal phase there is a higher stress concentration on the tibial plates, since the weight is applied completely on the lower osseous element. Therefore, the tibia becomes the most punished osseous element at this phase, as it has a load parallel to the axis which allows the patella to be free of stress. Unlike the bipodal phase where the load was distributed over the tibial plates at the contact phase, there is a tensile force along the patellar tendon as well as a flexion movement on the femur. The tibia is subjected to extension and the load is not distributed on the tibial plateau. These stress effects are reflected in each of the study cases as stress concentrations and

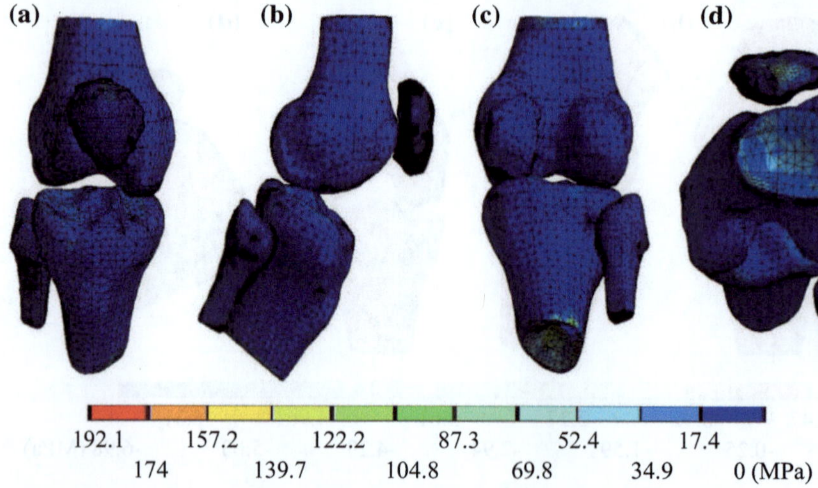

Fig. 20 *von Misses* equivalent stress for the knee joint at the balancing phase. **a** Frontal view. **b** Lateral view. **c** Rear view. **d** Top view

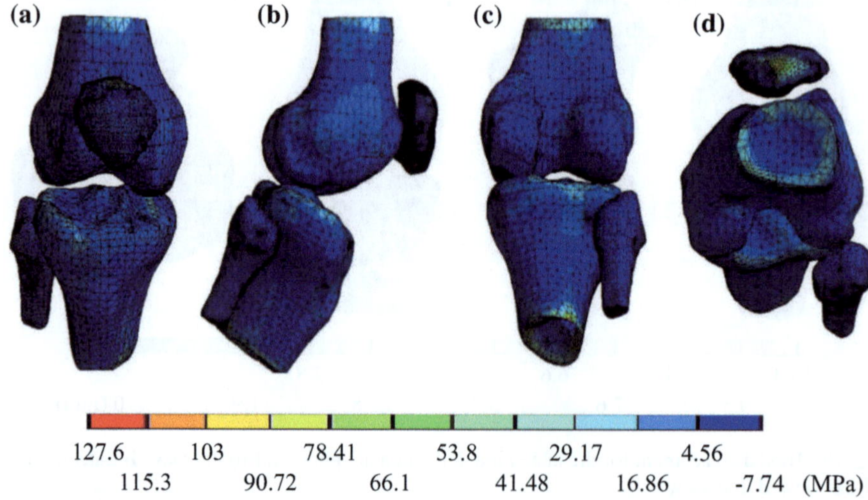

Fig. 21 Maximum principal stress for the knee joint at balancing phase. **a** Frontal view. **b** Lateral view. **c** Rear view. **d** Top view

the most sensible zones of the knee joint can be determine. Additionally, it can be stated that the tibial plates are no longer as punished with high stress concentrations as they were at the bipodal station. Finally, in the swinging phase, having one of the two lower limbs in the air is considered to be placed in a monopodial form, that is 100% of the load acts on a single joint and behaves as in the contact phase, since it has a proper movement since it has a movement of the same cycle. The elements of

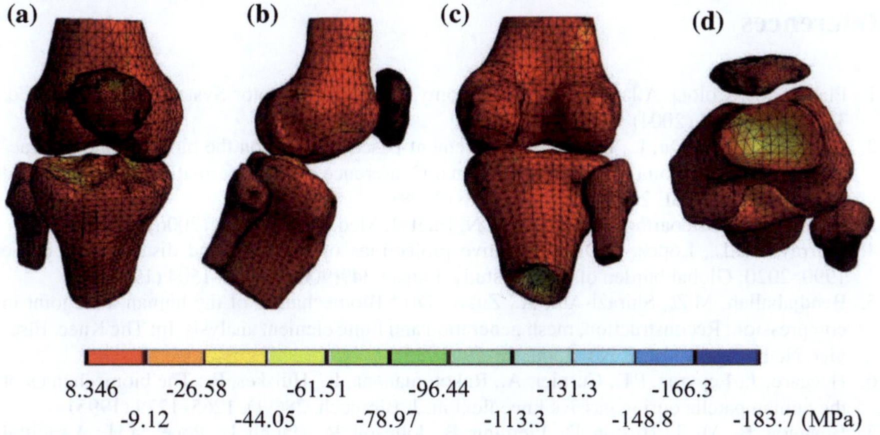

Fig. 22 Minimum principal stress for the knee joint at the balancing phase. **a** Frontal view. **b** Lateral view. **c** Rear view. **d** Top view

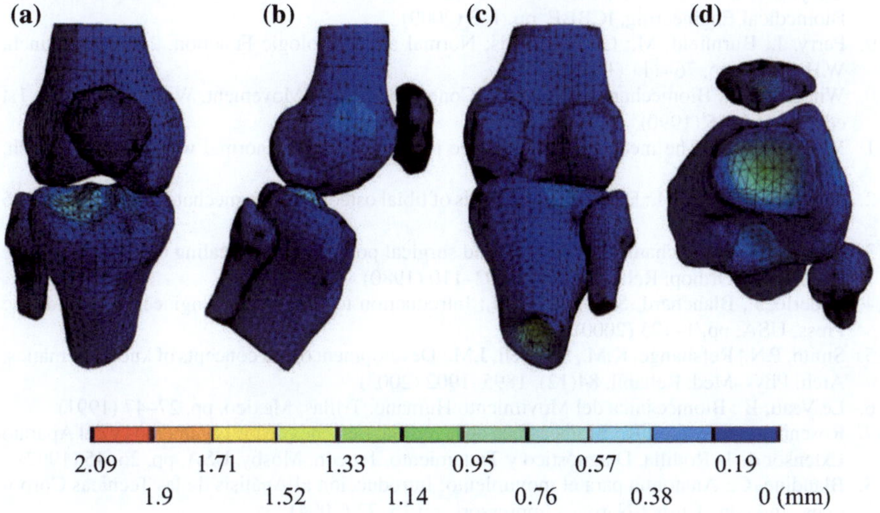

Fig. 23 Total displacement for the knee joint at the balancing phase. **a** Frontal view. **b** Lateral view. **c** Rear view. **d** Top view

the knee are flexed, and the load is not distributed over the whole area of the tibial plates. That is why the areas most punished or where there are a greater number of stress concentrators are on the tibia and femur.

Acknowledgements The authors gratefully acknowledge the financial support from the Mexican government by Consejo Nacional de Ciencia y Tecnología (CONACyT) and the Instituto Politécnico Nacional.

References

1. Platzer, W.: Colour Atlas of Human Anatomy, Vol. 1; Locomotor System, 5th edition, Ed. Thieme, pp. 194 (2004)
2. Yu, L., Tang, X., Du, L., Rao, C., Measurement-based research on the biomechanics characteristics of a knee joint. In: 2nd International Conference on Bioinformatics and Biomedical Engineering; ICBBE 2008, pp. 1233–1236 (2008)
3. Felson, D.T.: Osteoarthritis of the knee. N. Engl. J. Med. **354**, 84–848 (2006)
4. Murray, C.J.L., López, A.D.: Alternative projections of mortality and disability by cause 1990–2020; Global burden of disease study. Lancet **349**(9064), 1498–1504 (1997)
5. Bendjaballah, M.Z., Shirazi-Adl, A., Zukor, D.J.: Biomechanics of the human knee joint in compression; Reconstruction, mesh generation and finite element analysis. In: The Knee. Elsevier, Netherlands, Vol. 2, No. 2, pp. 69–79 (1995)
6. Heegard, J., Leyvraz, P.F., Curnier, A., Rakotomanana, L., Huiskes, R.: The biomechanics of the human patella during passive knee flexion. J. Biomech. **28**(11), 1265–1279 (1995)
7. Beynnon, B., Yu, J., Huston, D., Fleming, B., Johnson, R., Haugh, L., Pope, M.H.: A sagittal plane model of the knee and cruciate ligaments with application of a sensitivity analysis. J. Biomech. Eng. **118**(2), 227–239 (1996)
8. Zhang, X. S., Gou, Y. and Chen, W.: 3D finite element method modelling and contact pressure analysis of the total knee joint in flexion. In: International Conference on Bioinformatics and Biomedical Engineering, ICBBE, pp. 1–3 (2009)
9. Perry, J., Burnfield, M.: Gait Analysis; Normal and Pathologic Function, 2nd edn., Bunch, W.H. (eds.), pp. 76–111 (1992)
10. Winter, D. A.: Biomechanics and Motor Control of Human Movement. Wiley, New York, 1st edn., pp. 61–65 (1990)
11. Morrison, J.B.: The mechanics of the knee joint in relation to normal walking. J. Biomech. **3**(1), 51–61 (1970)
12. Fernández-Fairén, M.: Biomechanical basis of tibial osteotomy. Biomechanical **iv**(7), 122–126 (1996)
13. Maquet, P.: Biomechanics of the knee and surgical possibilities of healing osteoarthritis knee joints. Clin. Orthop. Relat. Res. **146**, 102–110 (1980)
14. Enderle, J., Blanchard, S., Bronzino, J.: Introduction to Biomedical Engineering. Academic Press, USA, pp. 1–125 (2000)
15. Smith, P.N., Refshange, K.M., Scarvell, J.M.: Development of the concepts of knee kinematics. Arch. Phys. Med. Rehabil. **84**(12), 1895–1902 (2003)
16. Le Veau, B.: Biomecánica del Movimiento Humano. Trillas, Mexico, pp. 27–47 (1991)
17. Rosenberg, A., Micos, R.: Biomecánica de la rodilla, Lesiones de los Ligamentos y del Aparato Extensor de la Rodilla. Diagnóstico y Tratamiento, 1st edn. Mosby, USA, pp. 26–45 (1992)
18. Blandine, C.: Anatomía para el movimiento; Introducción al Análisis de las Técnicas Corporales, 2nd edn., Curbet-Narques Impressors, pp 15–22 (2004)
19. González-Jemio, F., Mustafá-Milán, O., Antezana-Arzabe, A.: Alteraciones biomecánicas articulares en la obesidad. Gaceta Médica Boliviana **34**(1), 5256 (2011)
20. Méndez-Romero, D.: Análisis Numérico de las Fuerzas y Efectos que se Involucran en el Movimiento de la Rodilla, M. Sc. Thesis, SEPI ESIME Zacatenco, Instituto Politécnico Nacional (2012)
21. Hibbeler, R. C.: Engineering Mechanics; Statics. Pretince Hall, USA (2010)
22. Jorge-Rellán, J., Guerrero-Rosales, M.A., Vigil-Castiello, F.J., Álvarez-Iglesias, B., Maestro-Fernández, A., Femández-Lombardía, J., Paz-Aparicio, A., Rodríguez-López, L., García-González, P., Fernández-Lombardía, M.I.: Mechanical behaviour simulation of the posterolateral fascicle fixation by means of a biocompatible screw for anterior cruciate ligament repairs. Trauma Fund MAPFRE **23**(2), 77–83 (2012)

Automatic Modeling of Surfaces with Identical Slopes

Victorina A. Romanova, Marina Rynkovskaya and Vyacheslav Ivanov

Abstract The visualization problem of the formation process of surfaces with the same slope (generator angle) is considered. Among them, there are surfaces with different guides, such as a parabola, a circular cosine wave, a spherical evolvent. The possibility of helicoidal surface formation is also considered, since their generators have the same angle of inclination in all positions to the horizontal plane. The principles and techniques of surface formation in the AutoCAD environment are described using the functional language AutoLisp. The tasks of visualizing the formation of surfaces with the same slope with the guide parabola and the general cosine, as well as Monge surfaces and different types of helicoids are solved by means of authors' programs coded in AutoLisp language, tools and the color palette of the AutoCAD program. The generator in all presented cases is a straight line with the same inclination angle α to the guide line (surfaces with parabola, circle cosine and evolvent) or to the horizontal plane xy (helicoids).

Keywords Autolisp · Modeling · Surface with the same slope · Helicoids
Monge surface

1 Introduction

At present, architects are not satisfied with simple geometric images for new structures, as there is a demand to use more complex geometric shapes, which may be

V. A. Romanova · M. Rynkovskaya (✉) · V. Ivanov
Department of Architecture and Civil Engineering, Peoples' Friendship University of Russia (RUDN University), 6 Miklukho-Maklaya St, 117198 Moscow, Russian Federation
e-mail: marine_step@mail.ru

V. A. Romanova
e-mail: v.a.r-victoryna@mail.ru

V. Ivanov
e-mail: i.v.ivn@mail.ru

© Springer International Publishing AG, part of Springer Nature 2019
A. Öchsner and H. Altenbach (eds.), *Engineering Design Applications*,
Advanced Structured Materials 92, https://doi.org/10.1007/978-3-319-79005-3_10

created by different applied software packages such as MathCAD, ArchiCAD, AutoCAD, 3ds Max, Kompas and others.

The MathCAD package is a powerful mathematical modeling tool, based on the object-oriented language C++, and it allows the input of the initial mathematical functions on the display screen and specifies the result type. To construct the surface, it is necessary to specify the projection functions of the surface equation in the orthogonal Cartesian coordinate system xyz. For complex surface types, these equations are usually complicated and it is difficult to form such surfaces. Vector equations of surfaces with the use of special vectors are more convenient. These equations are more compact and reflect the way and nature of surface formation. In Ivanov and Romanova [1], a method is presented using vectorial equations to construct surfaces in the MathCad system, where the formation of surfaces occurs almost instantaneously.

Software packages such as 3ds Max, ArchiCAD, Compass and others, allow the user to create diverse scenes using primitives as building blocks. The result of working in any 3D editor is a static image, calculated by the program. But in these systems the result can be an animated movie.

One of the leading CAD software packages is AutoCAD, as it creates high-quality images of objects, including 3D graphics, contains the built-in functional language AutoLisp, which allows to supplement the software product with custom functions that extends the capabilities of the AutoCAD system, for instance, there are tools to show the formation of surfaces and to create mini-videos.

Nowadays, architects and engineers often choose the surfaces with the same slope, Monge surfaces [2–4] and different types of helicoids [5] when designing projects of various types of buildings and structures. Marine and mechanical engineers are also interested in different types of surfaces [6]. Mathematians were interested in various types of surfaces and their analysis long ago [7], while visualization of the surfaces has become possible after popularization of personal computers. There are several works on surfaces modeling research [1, 8–10], and the modeling of the complex surfaces has become an actual part of investigations.

2 Methods and Materials

Principles for the formation of analytic surfaces are described in the encyclopedia by Krivoshapko and Ivanov [11]. Methods for the formation of analytic surfaces are presented in the monograph by Ivanov and Romanova [1]. The main way to create these surfaces is the kinematic one which is based on the movement of the generating line along the guiding line, the construction of an array of generating lines, and the formation of the surface through the Loft function of the AutoLisp language.

The approach to form the surfaces in the AutoCAD environment consists of two stages:

Fig. 1 Initial position

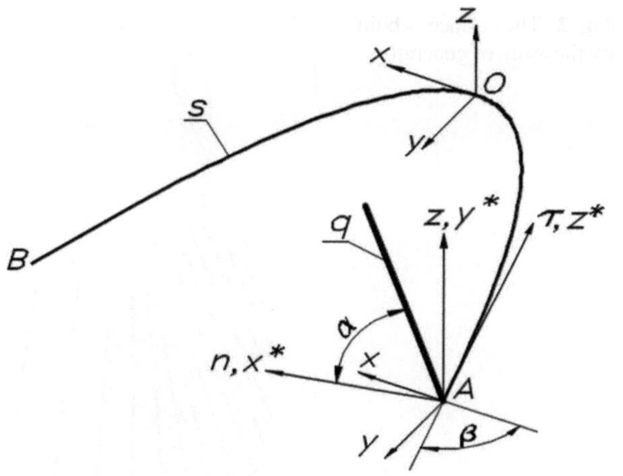

- to form a block of surface compartments, each of which is located in the layer fixed behind it;
- to insert a block with the “frozen” layers in the drawing and “thaw” layers sequentially.

As a result, the gradual appearance of compartments on the screen of the monitor provides the effect of the surface formation process.

We consider the formation of a surface with the same slope, when the guide line is a parabola

$$y = k \cdot x^2. \tag{1}$$

The initial position of the surface elements is shown in Fig. 1, where

s is a guide line,
q is a generator,
n is a normal to the guide line s,
α is the inclination angle of the generator q to the normal n,
β is an angle between the x axis and the tangent to parabola in the point A.

The generatrix line, when moving along the guiding line s, creates an array of generators which are necessary for constructing the surface (Fig. 2). In the program, an array of generators is built in a cycle in which the generator is drawn from each point of the guide, while the function *Ssget* creates an array of these Ssr lines. It should be noted that the generators form an angle α with the normal to the guide. In this connection, the coordinate system should be oriented so that the x-axis coincides with the normal to the line s. To do this, the program calculates the angle β, which is between the tangent line at the given point of the guide and the x-axis

$$\beta = \arctan 2kx. \tag{2}$$

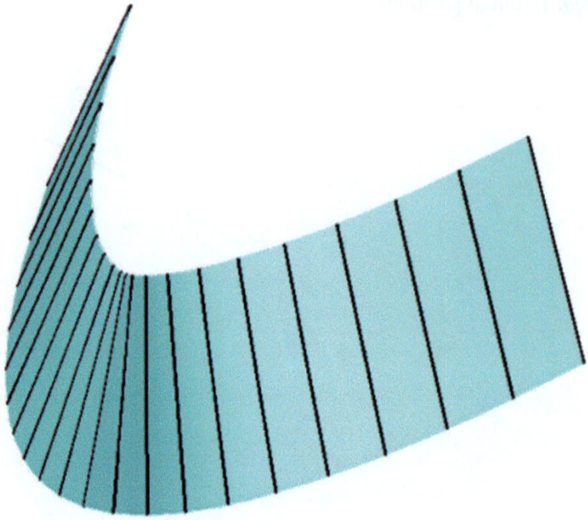

Fig. 2 The surface is built by the array of generators

The coordinate axes move to a given point, first rotating by an angle β around the z axis, and then by an angle 90° around the x axis and by 90° around the y axis. After these transformations of the coordinate system, the generator is drawn from the origin at an angle α to the normal n, which coincides with the x-axis.

The surface is constructed by the function *Loft*, which takes the elements of the array *Ssr* as sections. Figure 1 shows the surface and the array of generators along which it is constructed.

To form the surface according to the second method, "defrosting the layers", a number of layers are created programmatically in the drawing (using the *Slvent* user program), the number of which corresponds to the planned number of surface compartments.

The *Slope-2* program-function creates an array of generators in each layer and forms a surface section along the forming lines. This process is performed in a cycle with the parameter $i(i \leq n)$, where n is the number of compartments.

A fragment of the program, where Para-pt is a user-defined function for calculating the coordinates of the parabola points, USK1 is a user-defined function for converting a coordinate system:

```
(while (<= x xmax)
(setq nsloy (strcat "vent" (itoa i)) ); to create a new layer name
(command "layer" "s" nsloy "")
(setq pt2 (para-pt))
(setq y11 (* 2.0 k x))
(setq beta (atan y11))
(setq betg (/ (* beta 180.0) pi))
(command "ucs" "na" "r" "ucs1")
(command "ucs" "z" 15)
```

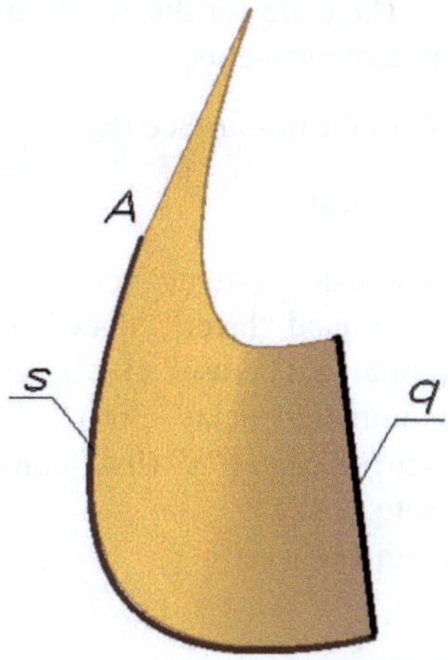

Fig. 3 The surface formation

```
(USK1)
(command "pline" '(0 0) (polar '(0 0) alfar lobr) ""); drawing of the generator
(setq en2 (entlast))
(command "change" en2 """" "p" "c" "4" "lw" 0.3 """")
(setq ssr1 (cons en2 ssr1));
(setq x (+ x dx))
(setq i (1+ i))
(command "delay" 500)
(command "loft" "mo" "su"); surface compartments formation
(foreach ent ssr1
(command ent))
(command """" """")
)
```

A block is created from the surface compartments and a referral to the user's function *Slope*, which provides the surface formation, is entered from the command line of the display. This function creates a series of layers in the drawing for the surface compartments by means of the auxiliary function *Slvent*, performs the "freezing" of the layers, inserts the compartment block in the "frozen" layers and sequentially "thaws" the layers with the *Form-surface* function. All these actions provide a visualization of the surface formation process. The function program *Form-surface* provides visualization of the surface formation process. The surface is formed by following the moving generator (Fig. 3).

The code for the *Form-surface* function is shown below, where k is the number of compartments.

```
(defun form-surface (k)

  (repeat k

(setq nsloyi (strcat "vent" (itoa i)) )
(command "layer" "thaw" nsloyi "")
(command "erase" s1 ""); removal of the previous compartment by the current one
(command "delay" 300)
(setq s2 (ssget "x" (list ( cons 8 (substr nsloyi 1 6)) )))
(setq s1 s2)
(setq i (+ i 1))
)
)
```

The problem of the surface formation can also be solved if the generator is given parametrically.

We consider here the formation of a surface with the same slope with a guiding line in the form of a circular cosine with the following parametric equations

$$x(\varphi) = r_0\,[1 + \mu \cdot [1 + \cos(p\varphi)]] \cdot \cos\varphi$$
$$y(\varphi) = r_0\,[1 + \mu \cdot [1 + \cos(p\varphi)]] \cdot \sin\varphi.$$

In this case, the angle β is determined through the derivatives of the functions $x(\varphi)$ and $y(\varphi)$:

$$\beta = \arctan\frac{dy}{dx} \tag{3}$$

Figure 4 shows the initial position of the surface elements such as the guide curve s and the generator q, and the angles α and β are also shown. The superposition of the x axis with the normal to the line s is performed by the user function *USK1*.

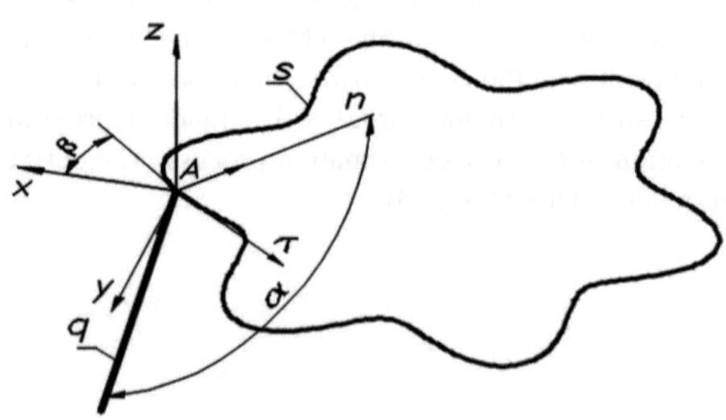

Fig. 4 The initial position of the surface elements

Fig. 5 The formation of the surface

Fig. 6 Types of surfaces in the plan of the general cosine

Further, the surface forming process follows the algorithm described above ("defrosting" the layers with surface compartments) (Fig. 5).

Figure 6 shows two types of surfaces on the plane of the general cosine.

2.1 *Monge Surfaces Formation*

Both surfaces considered above refer to normal surfaces. The Monge surfaces also belong to such surfaces and can be formed by rolling the plane Ψ with the generator of the straight line q along a circular cone. Figure 7 shows the initial position of the cone, tangent plane Ψ and generator q.

The trajectories of the points of the generator are the spherical evolvents corresponding to the circles of the cone which are touched by the points. Each evolvent lies on its sphere (Fig. 8).

Figure 9 shows the initial position of the cone tangent to the plane Ψ and forming the line q.

Fig. 7 Initial position of surfaces elements

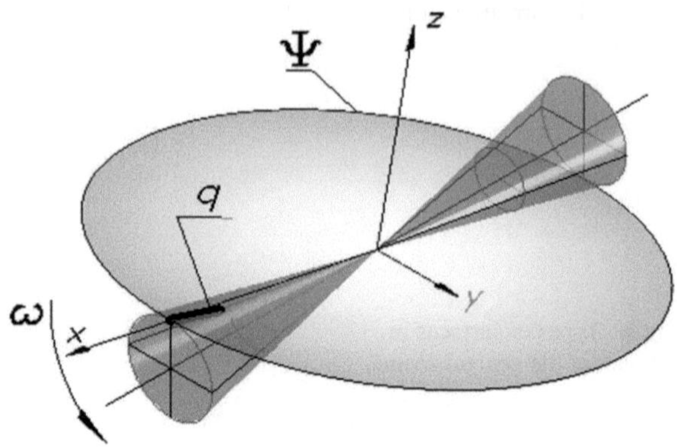

Fig. 8 The trajectories of the extreme points of the generator *q* (spherical evolvents S_1, S_2)

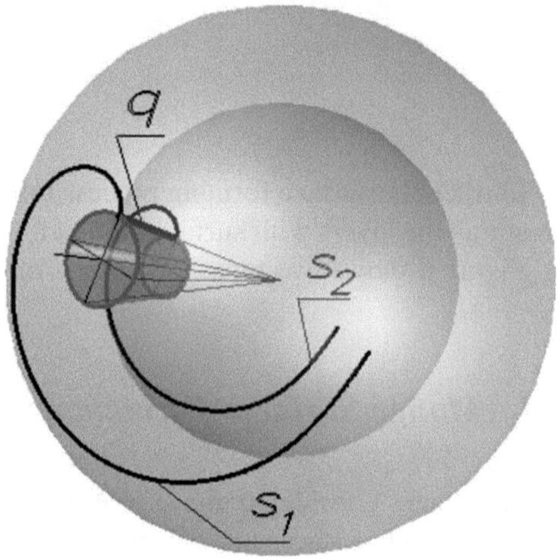

Fig. 9 Flat Ψ rolling along a cone

For each value of the angle of the plane Ψ rotation, the coordinates of the points belonging to the spherical evolvents of two cone circles S_1 and S_2 are determined. The generators (straight lines) are drawn between the points of the spherical evolvents at each stage of the cycle. At the same time, the surface compartment is formed (Fig. 10). The Monge surface is completely inside the sphere with the largest radius (Fig. 11).

Figure 12 shows the types of carved Monge surfaces.

2.2 Helicoids Formation

The helicoids, like the surfaces with identical slopes, have the same angle of inclination α of the generator in all its positions, but not to the normal of the surface [12]. These surfaces have the same angle of inclination to the *x* axis (Fig. 13), except for

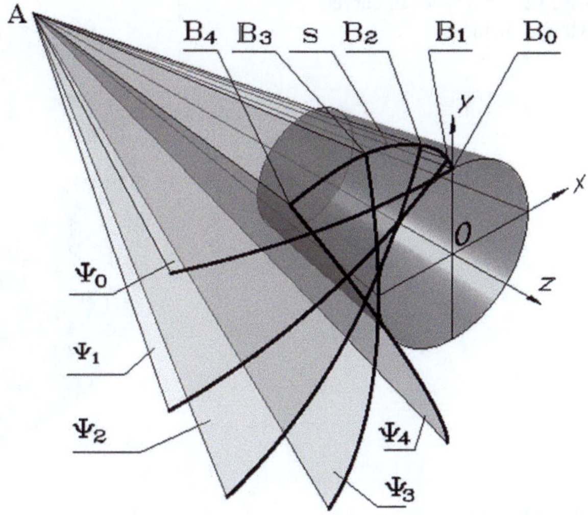

Fig. 10 Monge surface formation

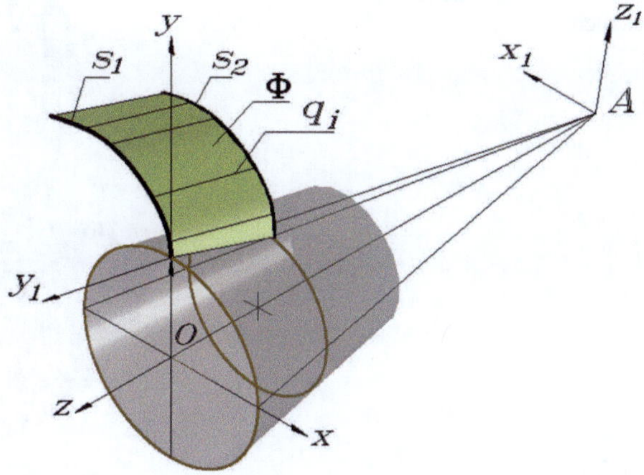

Fig. 11 Monge surface during the formation process

one type of helicoids—developable (evolvent) helicoid (Fig. 14). For the developable helicoid, the projection of the generator is the normal to the evolvent m, which is drawn by the lower end of the generators on the horizontal plane. Moreover, the evolvent m (along with the return edge p) is a guiding line for the generator q. The curve p is the helix. The generator q_i is tangent to the helix and makes a constant angle α with the normals n_i.

The developable (evolvent) helicoid is formed by the motion of the generator q along two helix lines: p and s. The surface is formed in a cycle with the parameter φ. For each value of the angle φ, the coordinates of the point are determined, the coordinate system is transferred to a given point, the x and y axes are set in the vertical plane and the coordinate system rotates by an angle φ. A generator (straight line) is drawn from the point of origin. At each stage of the cycle, a surface compartment is

Fig. 12 Two types of carved Monge surfaces

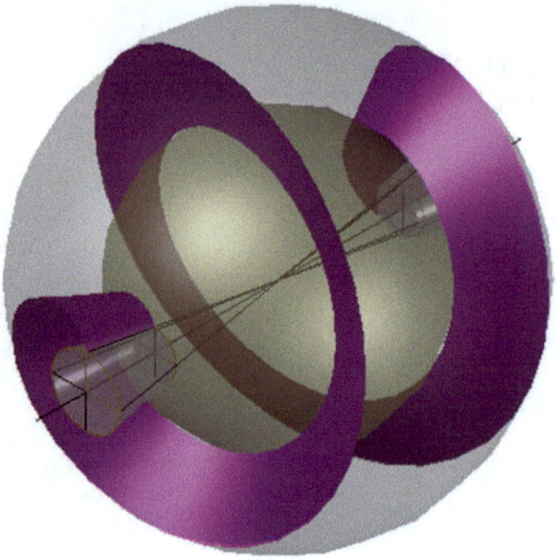

Fig. 13 Initial position of right and oblique helicoids elements

constructed along the generators. At the end of the cycle, a block of compartments is created, that allows the formation of a surface in slow motion by means of the *Form-surface* program.

The cadre of the surface formation from the AutoCAD system is shown in Fig. 15, and the skeleton of the surface is shown in Fig. 16.

Right and oblique helicoids are formed by one user program, in which the angle of inclination of the generator to the x-axis is specified (Fig. 13).

Formation of an array of generators is performed, taking into account the type of motion of this line. At each stage of the cycle, the generator moves translationally along the y axis by an amount ds and rotates about the same axis by an angle $d\varphi$. The formation of the surface compartments of a right helicoid is shown in Fig. 17, the formation of two oblique helicoids are shown in Figs. 18 and 19.

Fig. 14 Two positions of helicoid generator

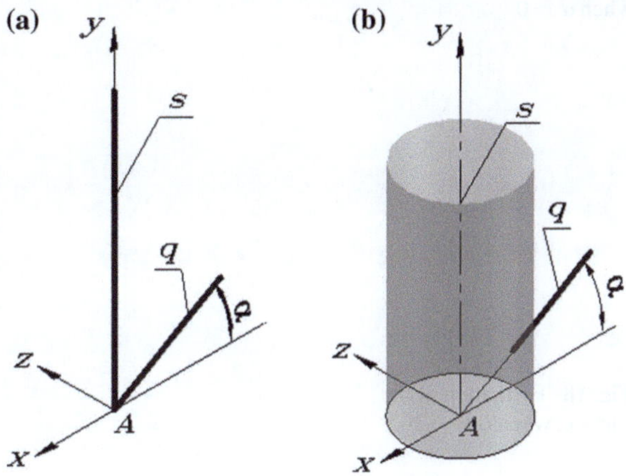

Fig. 15 Formation of developable helicoid

3 Results and Discussion

Various software tools which can be used for surface visualization are known. Some of them are mathematical modeling tools (MathCAD, among others), while others are tools for graphical modeling (AutoLISP, 3Dmax, etc.). The mathematical modeling tools are based on mathematical equations of the surface and allow to obtain the image of the specific surface immediately, but without a clear idea about the way the surface is being built. Some graphical tools allow to create a code for the surface formation and visualization step-by-step, which allow to understand the formation

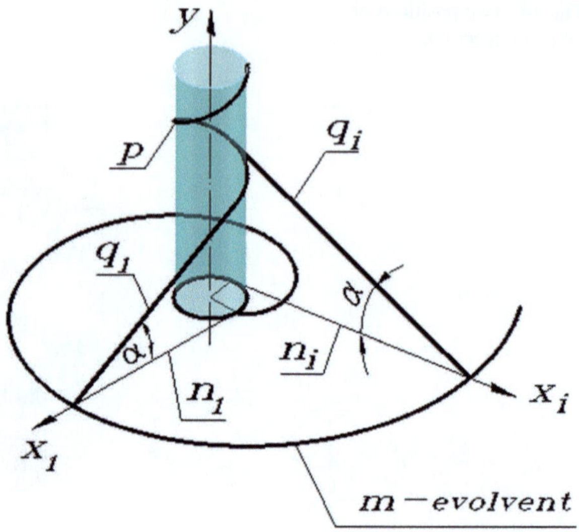

Fig. 16 The skeleton of developable helicoid

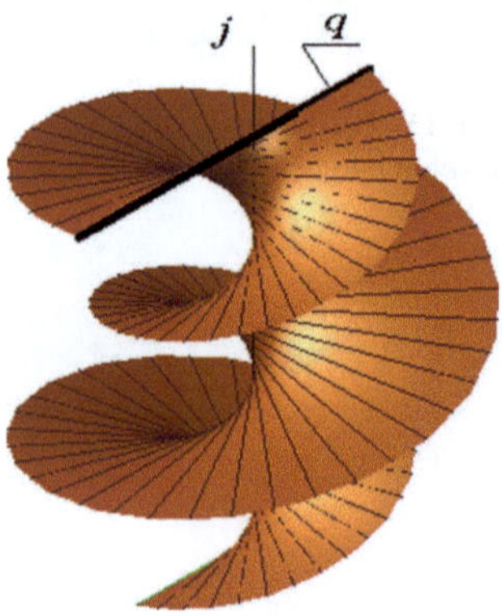

Fig. 17 Surface compartments formation, when $\alpha = 0$

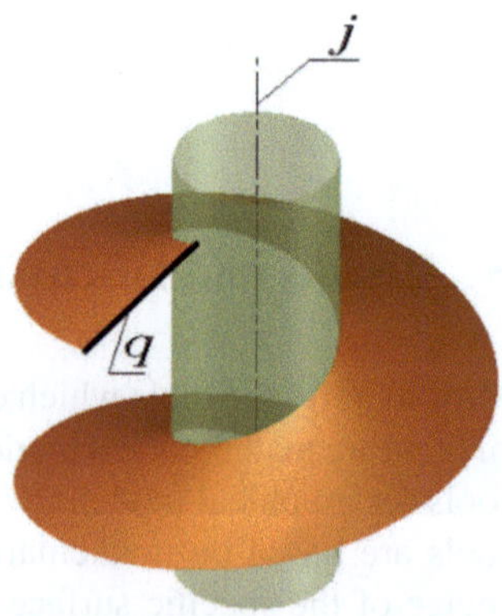

Fig. 18 Formation of the surface, when $\alpha = -20°$

Fig. 19 Formation of the surface, when $\alpha = 20°$

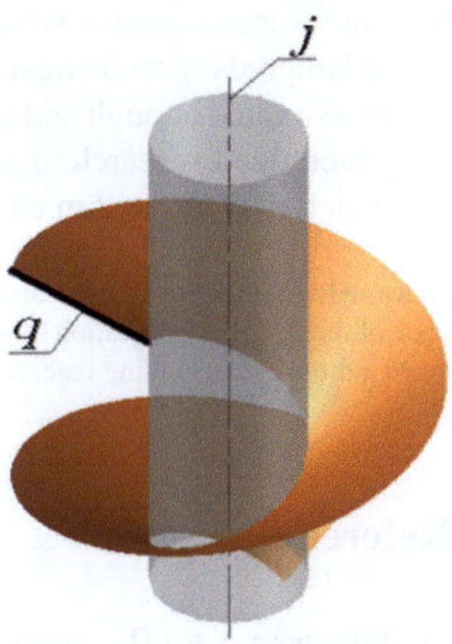

process of the surface. Such an approach may be useful for architects, engineers, teachers and students for deeper analysis of the investigated surface, as well as its possible design and manufacturing.

In this paper, several surface types were presented. There are investigated examples of rather simple surfaces (helicoids and surfaces with the parabola and cosine guide lines) and one example of a complex surface (Monge surface), which is hard to imagine without any step-by-step visualization. Generally, the suggested approach can be used for many other types of surfaces from simple to complex forms which can contribute to popularization and realization of new types of surfaces.

The code which is suggested in this paper is possible to be computed by popular tools (software) and these visualizations do not require too much machine time, capacity, or financial investments.

4 Conclusion

1. Programming in the AutoLisp language (code) provides the formation of analytical surfaces in the AutoCAD system.
2. Programs in the AutoLisp language provide visualization in the AutoCAD system of surface formation process.
3. Using custom programs in the AutoLisp language in AutoCAD system makes it possible to simulate surfaces of different types, to show the process of their formation, which expands the possibilities of work in the AutoCAD system.
4. All above mentioned points may be of interest for industry (engineers and architect as designers of civil structures), research (researchers of similar topics) and educational process (teaching staff and students).
5. Since software surface modeling involves the implementation of their parameterization, this is also a point for future research.

6. For future research it would be interesting to code the visualization of other types of helicoids (pseudo developable and convolute) because these two types are not investigated enough and do not have wide application in industry now. As a more generalized research, authors are going to create a library of various surfaces which may be used in educational process as well as in research and industry.

Acknowledgements This research paper is financially supported by Ministry of Education and Science of the Russian Federation on the program to improve the competitiveness of RUDN University among the world's leading research and education centers in the 2016–2020.

References

1. Romanova, V.A.: The formation of cyclic surfaces with variable radius of generatrix circles in AutoCAD. Str. Mech. Eng. Constr. Build. **3**, 20–24 (2016)
2. Filipova, J., Rynkovskaya, M.: Carved Monge Surfaces as New Forms in the Architecture. In: MATEC Web of Conferences. doi: https://doi.org/10.1051/matecconf/20179517006 (2017)
3. Ivanov, V.N., Rizvan, M.: Geometry of carved Monge surfaces and shell construction. Str. Mech. Eng. Str. Constr. **11**, 27–36 (2002)
4. Ivanov, V., Rynkovskaya, M.: Analysis of Thin Walled Wavy Shell of Monge Type Surface with Parabola and Sinusoid Curves by Variational-Difference Method. https://doi.org/10.1051/matecconf/20179512007 (2017)
5. Krivoshapko, S.N., Rynkovskaya, M.: Five Types of Ruled Helical Surfaces for Helical Conveyers, Support Anchors and Screws. MATEC Web of Conferences https://doi.org/10.1051/matecconf/20179506002 (2017)
6. Hyeng, C.B., Yamb, E.B.: Aplication of cyclic shells in the architecture, machine design, and bionics. Int. J. Mod. Eng. Res. **12**, 799–806 (2012)
7. Winderlich, W.: Kurven Konstanter Ganzer Krummung und Fester Hauptnormalenneigung. Monatsh ath. **77**(2), 158–171 (1973)
8. Cipolla, R.: Visual motion of curves and surfaces. Phil. Trans. R Soc. Lond. A. UK:1–16 (1996)
9. Fahlbusch, K.P., Roser, T.D.: HP PE/SolidDesigner: Dynamic Modeling for Three-Dimensional Computer-Aided Design. H-P J. October (1995)
10. Rachkovskaya, G., Kharabayev, Yu., Rachkovskaya, N.: Development and Computer Graphics of Intricate Geometrical Forms. J. Pol. Soc. Geom. Eng. Graph. **16**, 37–40 (2006)
11. Krivoshapko, S.N., Ivanov, V.N.: Encyclopedia of Analytical Surfaces. Springer, Berlin (2015)
12. Krivoshapko, S.N.: Geometry and strength of general helicoidal shells. Appl. Mech. Rev. **52**(5), 161–175 (1999)

Analytical Method to Analyze Right Helicoid Stress-Strain

Marina Rynkovskaya and Vyacheslav Ivanov

Abstract The authors suggest a simple analytical method to analyze right helicoid stress-strain state based on a corrected and simplified Rekach's method of two Reissner's equations reduced into one eight-order equation with solutions found by the trigonometric Fourier series. In the study of this technique, some inaccuracies were found in the formulas that have a significant impact on the results. A direct approach to the definition of the tangential displacements is also introduced. Improvement of the method allows to simplify the equations and get compact enough formulas which are suitable for programming. It is proven that the numerical results obtained by this modified analytical method are similar to those obtained by finite element method, while the original version of this technique leads to incorrect results.

Keywords Helical surface · Right helicoids · Stress-strain analysis
Analytical method

1 Introduction

A right helicoid is a well known helical surface among the five types of ruled helicoidal surfaces [1], and it is widely used in different fields of mechanical engineering (for anchors, screw conveyors, compressors and presses, bolts [2], augers, heat exchangers, among others), as well as in architecture and civil engineering (helical staircases, ramps, screw piles, among others). It is common to use the shape of the helical surface when designing blades in ship-, airplane- and other engineering industries. The twisted rods, which are also a right helicoid, are mathematical models of a number of crucial elements of engineering structures: operating steam

M. Rynkovskaya (✉) · V. Ivanov
Department of Architecture and Civil Engineering,
Peoples' Friendship University of Russia, RUDN University),
6 Miklukho-Maklaya St, Moscow 117198, Russian Federation
e-mail: marine_step@mail.ru

V. Ivanov
e-mail: i.v.ivn@mail.ru

© Springer International Publishing AG, part of Springer Nature 2019
A. Öchsner and H. Altenbach (eds.), *Engineering Design Applications*,
Advanced Structured Materials 92, https://doi.org/10.1007/978-3-319-79005-3_11

blades, gas and hydraulic turbines, blades of wind power plants, axial expansion joints, helicopter blades, elements of measuring instruments, among others.

The geometry of a right helicoid has been well studied before [3], however in the 21st century there are still several papers on this subject [4]. As for engineering design, there are various calculation methods, the book by Taylor [5] appears to be the first paper devoted to the calculation of the strength of a right helicoid as applied to mechanical engineering. Right helicoids which are used in engineering are usually called rods (strips) with an initial twist. In civil engineering, calculations on the strength of right helicoids were carried out by Solomon [6], Rekach [7], Cohen [8], Koltunov [9], Iura and Masaharu [10] and others.

In practice, right helicoids are mostly calculated by various tools based on numerical methods or experiments. However sometimes it is insufficient for advanced stress-strain analysis [11], because analytical methods make it possible to conduct a deeper analysis of the stress-strain state of a shell [12, 13].

Generally, Iura and Hirashima [10], Krivoshapko [11, 14] and some other scientists suggest developing analytical methods for calculating helical shells to supplement or partially replace expensive calculating software systems based on FEM both in engineering practice and in the educational process. At the same time, several other scientists rely more on numerical methods of calculation [15].

In this paper, an analytical method to analyze right helicoid stress-strain is presented and the obtained results are compared with results from FEM. The authors believe that analytical methods of calculation based on the shell theory are worth developing, especially for engineering structures in the shape of analytical surfaces.

2 Methods and Materials

According to the 2010 Mathematics Subject Classification: a helicoid is a ruled surface described by a straight line that rotates at a constant angular rate around a fixed axis, intersects the axis at a constant angle α, and at the same time becomes gradually displaced at a constant rate k along this axis. The equation of a helicoid in parametric form is $x = \rho \cos t$; $y = \rho \sin t$; $z = \rho \arctan \alpha + k\, t$. If $\alpha = \pi/2$ the helicoid is called straight or right, otherwise it is called oblique. Every straight helicoid is a minimal surface, and the straight helicoid is the only minimal ruled surface.

According to the encyclopedia of analytical surfaces [16], a right helicoid is a helical ruled surface, described by a straight line that intersects the axis of the helicoid at a right angle, rotates at a constant angular velocity around this axis and simultaneously moves translationally at a constant speed along the same axis (Fig. 1). The parametrical equation of right helicoids can be written as:

$$x = u \cos v, \quad y = u \sin v, \quad z = cv,$$

Fig. 1 Right helicoid

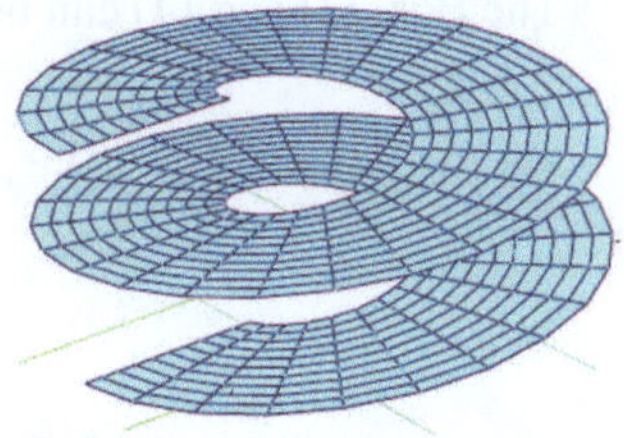

where c is the displacement of the generator AB upon its rotation by 1 rad; u, v are curvilinear coordinates of the point C of helicoids; u is the distance from point C to axis z; v is the rotation angle of generator AB from the plane zOx to the point C.

The equations of combined method which were obtained by Marquere [17] for plates are used:

$$\begin{aligned} D\nabla^2\nabla^2 u_z \pm \nabla_k^2\varphi = Z, \\ \nabla^2\nabla^2\varphi \mp Eh\nabla_k^2 u_z = 0, \end{aligned} \tag{1}$$

where $\nabla^2 \ldots = \frac{1}{AB}\left[\frac{\partial}{\partial\alpha}\left(\frac{B}{A}\frac{\partial\ldots}{\partial\alpha}\right) + \frac{\partial}{\partial\beta}\left(\frac{A}{B}\frac{\partial\ldots}{\partial\beta}\right)\right]$ is a Laplasian;

$\nabla_k^2 \cdots = \frac{1}{AB}\left[\frac{\partial}{\partial\alpha}\left(\frac{B}{A}k_2\frac{\partial\ldots}{\partial\alpha}\right) + \frac{\partial}{\partial\beta}\left(\frac{A}{B}k_1\frac{\partial\ldots}{\partial\beta}\right)\right]$ is a combined operator;

u_z is a function of normal displacement; φ is a special function of stresses; D is cylindrical rigidity; E is Young's modulus; h is helicoid thickness; Z is load.

Equilibrium equations for shallow shells

$$\begin{aligned} &\frac{\partial}{\partial\alpha}(BN_\alpha) - \frac{\partial B}{\partial\alpha}N_\beta + \frac{1}{A}\frac{\partial}{\partial\beta}(A^2S) + ABX = 0, \\ &\frac{\partial}{\partial\beta}(AN_\beta) - \frac{\partial A}{\partial\beta}N_\alpha + \frac{1}{B}\frac{\partial}{\partial\alpha}(B^2S) + ABY = 0, \end{aligned} \tag{2}$$

$$\begin{aligned} k_1N_\alpha + k_2N_\beta + \frac{1}{AB}\Bigg\{\frac{\partial}{\partial\alpha}\frac{1}{A}\left[\frac{\partial}{\partial\alpha}(BM_\alpha) - \frac{\partial B}{\partial\alpha}M_\beta + \frac{1}{A}\frac{\partial}{\partial\beta}(A^2H)\right] \\ + \frac{\partial}{\partial\beta}\frac{1}{B}\left[\frac{\partial}{\partial\beta}(AM_\beta) - \frac{\partial A}{\partial\beta}M_\alpha + \frac{1}{B}\frac{\partial}{\partial\alpha}(B^2H)\right]\Bigg\} - Z = 0, \end{aligned}$$

$$Q_\alpha = \frac{1}{AB}\left[\frac{\partial}{\partial\alpha}(BM_\alpha) - \frac{\partial B}{\partial\alpha}M_\beta + \frac{1}{A}\frac{\partial}{\partial\beta}(A^2H)\right], \tag{3}$$

$$Q_\beta = \frac{1}{AB}\left[\frac{\partial}{\partial\beta}(AM_\beta) - \frac{\partial A}{\partial\beta}M_\alpha + \frac{1}{B}\frac{\partial}{\partial\alpha}(B^2H)\right]. \tag{4}$$

The Eqs. (3) and (4) can be written for helicoid as follows

$$Q_\alpha = -\frac{D}{A}\frac{\partial}{\partial \alpha}\nabla^2 u_z,$$

$$Q_\beta = -\frac{D}{B}\frac{\partial}{\partial \beta}\nabla^2 u_z.$$

Geometrical equations for shallow shells

$$\begin{aligned}
\varepsilon_\alpha &= \frac{1}{A}\frac{\partial u_\alpha}{\partial \alpha} + \frac{1}{AB}\frac{\partial A}{\partial \beta} u_\beta + k_1 u_z,\\
\varepsilon_\beta &= \frac{1}{B}\frac{\partial u_\beta}{\partial \beta} + \frac{1}{AB}\frac{\partial B}{\partial \alpha} u_\alpha + k_2 u_z,\\
\varepsilon_{\alpha\beta} &= \frac{B}{A}\frac{\partial}{\partial \alpha}\left(\frac{u_\beta}{B}\right) + \frac{A}{B}\frac{\partial}{\partial \beta}\left(\frac{u_\alpha}{A}\right),\\
\kappa_\alpha &= -\frac{1}{A}\frac{\partial}{\partial \alpha}\left(\frac{1}{A}\frac{\partial u_z}{\partial \alpha}\right) - \frac{1}{AB^2}\frac{\partial A}{\partial \beta}\frac{\partial u_z}{\partial \beta},\\
\kappa_\beta &= -\frac{1}{B}\frac{\partial}{\partial \beta}\left(\frac{1}{B}\frac{\partial u_z}{\partial \beta}\right) - \frac{1}{A^2 B}\frac{\partial B}{\partial \alpha}\frac{\partial u_z}{\partial \alpha},\\
\kappa_{\alpha\beta} &= -\frac{1}{AB}\left(\frac{\partial^2 u_z}{\partial \alpha\, \partial \beta} - \frac{1}{B}\frac{\partial B}{\partial \alpha}\frac{\partial u_z}{\partial \beta} - \frac{1}{A}\frac{\partial A}{\partial \beta}\frac{\partial u_z}{\partial \alpha}\right).
\end{aligned} \tag{5}$$

Physical equations, when the surface coordinate lines are the lines of curvature α and β, are

$$\begin{aligned}
&N_\alpha = C\left(\varepsilon_\alpha + \nu\varepsilon_\beta\right), \quad N_\beta = C\left(\varepsilon_\beta + \nu\varepsilon_\alpha\right), \quad M_\alpha = -D\left(\kappa_\alpha + \nu\kappa_\beta\right),\\
&M_\beta = -D\left(\kappa_\beta + \nu\kappa_\alpha\right), \quad S_\alpha = S_\beta = S = \frac{1-\nu}{2} C\varepsilon_{\alpha\beta},\\
&M_{\alpha\beta} = M_{\beta\alpha} = H = (1-\nu)\, D\kappa_{\alpha\beta},
\end{aligned} \tag{6}$$

The three static equations from Eq. (2) with geometrical (Eq. 5) and physical (Eq. 6) equations form the system of fifteen equations with fifteen unknown quantities: longitudinal forces N_α, N_β, shear force S, bending moments M_α, M_β, twisting moment H, tangential deformations $\varepsilon\alpha$, $\varepsilon\beta$, $\varepsilon\alpha\beta$, bending deformations κ_α, κ_β, $\kappa_{\alpha\beta}$ and displacements u_α, u_β, u_z.

The quantities may be expressed through the special stress function $\varphi\ (\alpha, \beta)$

$$\begin{aligned}
N_\alpha &= \frac{1}{B}\frac{\partial}{\partial \beta}\left(\frac{1}{B}\frac{\partial \varphi}{\partial \beta}\right) + \frac{1}{A^2 B}\frac{\partial B}{\partial \alpha}\frac{\partial \varphi}{\partial \alpha},\\
N_\beta &= \frac{1}{A}\frac{\partial}{\partial \alpha}\left(\frac{1}{A}\frac{\partial \varphi}{\partial \alpha}\right) + \frac{1}{AB^2}\frac{\partial A}{\partial \beta}\frac{\partial \varphi}{\partial \beta},\\
S &= -\frac{1}{AB}\left(\frac{\partial^2 \varphi}{\partial \alpha \partial \beta} - \frac{1}{B}\frac{\partial B}{\partial \alpha}\frac{\partial \varphi}{\partial \beta} - \frac{1}{A}\frac{\partial A}{\partial \beta}\frac{\partial \varphi}{\partial \alpha}\right).
\end{aligned} \tag{7}$$

For a shallow shell with the middle surface in arbitrary oblique coordinate system u, v, the combined equations are the same as Eq. (1), but differential operators are:

$$\nabla^2 \cdots = \frac{1}{AB\sin\chi}\left[\frac{\partial}{\partial u}\left(\frac{B}{A\sin\chi}\frac{\partial\ldots}{\partial u} - ctg\chi\frac{\partial\ldots}{\partial v}\right) + \frac{\partial}{\partial v}\left(\frac{A}{B\sin\chi}\frac{\partial\ldots}{\partial v} - ctg\chi\frac{\partial\ldots}{\partial u}\right)\right],$$

$$\nabla_k^2 \cdots = \frac{1}{AB\sin\chi}\left[\frac{\partial}{\partial u}\left(\frac{N}{AB\sin\chi}\frac{\partial\ldots}{\partial u}\right) + \frac{\partial}{\partial v}\left(\frac{L}{AB\sin\chi}\frac{\partial\ldots}{\partial v}\right)\right.$$
$$\left. + \frac{\partial}{\partial u}\left(\frac{M}{AB\sin\chi}\frac{\partial\ldots}{\partial v}\right) + \frac{\partial}{\partial v}\left(\frac{M}{AB\sin\chi}\frac{\partial\ldots}{\partial u}\right)\right], \tag{8}$$

where χ is angle between coordinate lines u, v.

We consider further a shallow shell in the shape of right helicoid. The helical shell is shallow if u is much bigger than c and $a \le u \le b$, where a is an inner radius of helicoid, b is an outer radius of helicoid.

Since coefficients of first and second quadratic forms for a shallow right helicoid are

$$A = 1,\ B = u,\ F = 0,\ \chi = \pi/2,\ \cos\chi = 0,\ L = 0,\ M = -c/u,\ N = 0,$$

the operators (Eq. 8) are

$$\nabla^2 \cdots = \frac{1}{u}\left[\frac{\partial}{\partial u}\left(u\frac{\partial\cdots}{\partial u}\right) + \frac{1}{u}\frac{\partial^2\cdots}{\partial v^2}\right],$$

$$\nabla_k^2 \cdots = -\frac{2c}{u^2}\frac{\partial}{\partial u}\left(\frac{1}{u}\frac{\partial\cdots}{\partial v}\right) = -\frac{H}{\pi}\frac{1}{u^2}\frac{\partial}{\partial u}\left(\frac{1}{u}\frac{\partial\cdots}{\partial v}\right),$$

where $H = 2\pi c$, and Eq. (1) for a shallow helicoids are

$$D\nabla^2\nabla^2 u_z = Z + \frac{H}{\pi}\frac{1}{u^2}\frac{\partial}{\partial u}\left(\frac{1}{u}\frac{\partial\varphi}{\partial v}\right),$$
$$\nabla^2\nabla^2\varphi = -Eh\frac{H}{\pi}\frac{1}{u^2}\frac{\partial}{\partial u}\left(\frac{1}{u}\frac{\partial u_z}{\partial v}\right). \tag{9}$$

Equation (9) were derived by Reissner [18] on the basis of the Marquere [17] equations for annulus with large deformations. They are solved for a number of cases: $\varphi = \varphi(u)$ and $u_z = u_z(u)$; $\varphi = kv$ and $u_z = u_z(u)$; $\varphi = \varphi(u)$ and $u_z = kv$ and some others. Rekach [7] proposed some more general cases to solve Eq. (9).

Since Eq. (9) are Euler's equations, it is possible to make a replacement $u = e^t$, and $t = \ln u$, and combine two Eq. (9) into one differential equation of eight order

$$\nabla^8(t, v)\,\Phi(t, v) + p^2\nabla_k^4(t, v)\Phi(t, v) = -e^{4t}Z(t, v)/D, \tag{10}$$

where $\nabla^4(t, v)\cdots = u^4\nabla^4\ldots$,

$\nabla_k^2(t, v)\cdots = u^4\nabla_k^2\ldots$,

$$\nabla_k^2 \cdots = \frac{1}{u^4}\frac{\partial}{\partial v}\left(\frac{\partial}{\partial t^2} - 1\right)\cdots,$$

$$\nabla^4 \cdots = \frac{1}{u^4}\left(\frac{\partial^4 \cdots}{\partial t^4} - 4\frac{\partial^3 \cdots}{\partial t^3} + 4\frac{\partial^2 \cdots}{\partial t^2} - 4\frac{\partial^3 \cdots}{\partial t \partial v^2} + 2\frac{\partial^4 \cdots}{\partial t^2 \partial v^2} + 4\frac{\partial^2 \cdots}{\partial v^2} + \frac{\partial^4 \cdots}{\partial v^4}\right),$$

$$\varphi = \left(\frac{EhH}{\pi}\right)\nabla_k^2(t, v)\Phi(t, v),$$

$$u_z = -\nabla^4(t, v)\Phi(t, v),$$

$$p^2 = \frac{EhH^2}{\pi^2 D} = \frac{12\left(1 - \nu^2\right)H^2}{\pi^2 h^2}.$$

The solution is found in the trigonometric series

$$\Phi\,(t, v) = \sum_{\substack{m=1\\m=0}}^{\infty} \Phi m\,(t) \begin{matrix}\sin mv\\ \cos mv\end{matrix}. \tag{11}$$

The general solution is obtained from the following equation

$$\left[\frac{d^4}{dt^4} - 4\frac{d^3}{dt^3} + 2(2 - m^2)\frac{d^2}{dt^2} + m\,(4m \pm p)\,\frac{d}{dt} + m^2(m^2 - 4) \mp mp\right]\Phi_m\,(t) = 0, \tag{12}$$

where $\Phi_m(t) = A_{mi}e^{\alpha_{mi}t}$, A_{mi} are arbitrary constants of m-member of series, $i = 1$, 2, …, 8.

For a particular solution, we expand the right part of Eq. (10) into a trigonometric series

$$-\frac{e^{4t}}{D}Z\,(t, v) = -\frac{e^{4t}}{D}\sum_{m=1}^{\infty} Z_m\,(t)\sin mv,$$

and for a constant distributed load, $Z(t, v) = q = const$, the solution is found as follows

$\overline{\Phi_m}(t) = Ae^{4t}$ when $0 < v < \pi$,
$\overline{\Phi_m}(t) = -A\,(-1)^{\frac{m+1}{2}}\,e^{4t}$ when $-\frac{\pi}{2} < v < \frac{\pi}{2}$,
where $A = \frac{-4q}{D\pi m[4^6 + m^8 - 40m^6 + 528m^4 - 2560m^2 - 9p^2m^2]}$ [19].

Finally the inner forces and moments are found from the following equations:

$$N_u = \frac{1}{u}\frac{\partial \varphi}{\partial u} + \frac{1}{u^2}\frac{\partial^2 \varphi}{\partial v^2} = \pm \frac{EhH}{\pi}\frac{1}{u}\sum_{m=1}^{\infty}\left(\frac{d}{du} - \frac{m^2}{u}\right)\nabla_k^2(u,m)\,\Phi_m(u)\sin mv,$$

$$N_v = \frac{\partial^2 \varphi}{\partial u^2} = \pm \frac{EhH}{\pi}\frac{d^2}{du^2}\sum_{m=1}^{\infty}\nabla_k^2(u,m)\,\Phi_m(u)\sin mv,$$

$$S = -\frac{\partial}{\partial u}\left(\frac{1}{u}\frac{\partial \varphi}{\partial v}\right) = \frac{EhH}{\pi}\frac{1}{u}\left(\frac{d}{du} - \frac{1}{u}\right)\sum_{m=1}^{\infty} m\nabla_k^2(u,m)\,\Phi_m(u)\sin mv,$$

$$Q_u = -D\frac{\partial}{\partial u}\nabla^2 u_z = D\frac{d}{du}\frac{1}{u^2}\sum_{m=1}^{\infty}\nabla^6(u,m)\,\Phi_m(u)\sin mv,$$

$$Q_v = -\frac{D}{u}\frac{\partial}{\partial v}\nabla^2 u_z = \pm\frac{D}{u^3}\sum_{m=1}^{\infty} m\nabla^6(u,m)\,\Phi_m(u)\sin mv,$$

$$M_u = -D\left[\frac{\partial^2 u_z}{\partial u^2} + \frac{\nu}{u}\left(\frac{\partial u_z}{\partial u} + \frac{1}{u}\frac{\partial^2 u_z}{\partial v^2}\right)\right]$$
$$= D\sum_{m=1}^{\infty}\left[\frac{d^2}{du^2} + \frac{\nu}{u}\left(\frac{d}{du} - \frac{m^2}{u}\right)\right]\nabla^4(u,m)\,\Phi_m(u)\sin mv,$$

$$M_v = -D\left[\frac{1}{u}\left(\frac{\partial u_z}{\partial u} + \frac{1}{u}\frac{\partial^2 u_z}{\partial v^2}\right) + \nu\frac{\partial^2 u_z}{\partial u^2}\right]$$
$$= D\sum_{m=1}^{\infty}\left[\frac{1}{u}\left(\frac{d}{du} - \frac{m^2}{u}\right) + \nu\frac{d^2}{du^2}\right]\nabla^4(u,m)\,\Phi_m(u)\sin mv,$$

$$M_{uv} = -(1-\nu)\frac{D}{u}\frac{\partial}{\partial v}\left(\frac{\partial u_z}{\partial u} - \frac{u_z}{u}\right)$$
$$= \mp(1-\nu)\frac{D}{u}\left(\frac{d}{du} - \frac{1}{u}\right)\sum_{m=1}^{\infty} m\nabla^4(u,m)\,\Phi_m(u)\sin mv, \tag{13}$$

where $\nabla_k^2(u,m)\cdots = m\left(u\frac{d}{du} - 1\right)\ldots$, $\nabla^2(u,m)\cdots = \left(u^2\frac{d^2}{du^2} + u\frac{d}{du} - m^2\right)\ldots$, among others.

Tangential displacements u_u, u_v are to be found through normal displacements u_z and the special stresses function φ. To do this, we solve the system of geometrical equations (Eq. 5) taking into account physical equations (Eq. 6) and obtain analytic expressions for the tangential displacements.

The first three geometrical equations of Eq. (5) for a shallow right helicoid with arbitrary curvilinear oblique coordinates of middle surface are

$$\varepsilon_u = \frac{\partial u_u}{\partial u},$$

$$\varepsilon_v = \frac{1}{B}\frac{\partial u_v}{\partial v} + \frac{1}{AB}\frac{\partial B}{\partial u}u_u = \frac{1}{u}\left(\frac{\partial u_v}{\partial v} + u_u\right),$$

$$\varepsilon_{uv} = \omega_u + \omega_v = u\frac{\partial}{\partial u}\left(\frac{u_v}{u}\right) + \frac{1}{u}\frac{\partial u_u}{\partial v} + 2\frac{c}{u^2}u_z.$$

Physical equations (Eq. 6) in this case are as follows

$$\varepsilon_u = \frac{1}{Eh}\left(N_u - \nu N_v\right) = \frac{1}{Eh}\left(\frac{1}{u}\frac{\partial \varphi}{\partial u} + \frac{1}{u^2}\frac{\partial^2 \varphi}{\partial v^2} - \nu\frac{\partial^2 \varphi}{\partial u^2}\right),$$
$$\varepsilon_v = \frac{1}{Eh}\left(N_v - \nu N_u\right) = \frac{1}{Eh}\left(\frac{\partial^2 \varphi}{\partial u^2} - \nu\left(\frac{1}{u}\frac{\partial \varphi}{\partial u} + \frac{1}{u^2}\frac{\partial^2 \varphi}{\partial v^2}\right)\right),$$
$$\varepsilon_{uv} = \frac{2(1+\nu)}{Eh}S = -\frac{2(1+\nu)}{Eh}\frac{\partial}{\partial u}\left(\frac{1}{u}\frac{\partial \varphi}{\partial v}\right). \tag{14}$$

We integrate physical equations (Eq. 14) which are expressed through normal displacements u_z and the special stress function φ (Eq. 10) and find tangential displacements

$$u_u = \int \varepsilon_u du,$$
$$u_v = \int \left(u\varepsilon_v - u_u\right) dv.$$

3 Numerical Experiments

We consider here the general solution (Eq. 12) and find normal displacements u_z and special stress function φ for a general solution.

If $m = 0$, we have two multiple roots $\alpha_1 = \alpha_2 = 0$ and $\alpha_3 = \alpha_4 = 2$. The solution is simply $u_z = A_{01} + A_{02}t + A_{03}e^{2t} + A_{04}te^{2t}$and similar to well known solutions for an annulus

$u_z = A_{01} + A_{02} \ln u + A_{03}u^2 + A_{04}u^2 \ln u.$

If $m = 1$, we have two roots which are multiple with multiplicity $k = 1$ ($\alpha_1 = \alpha_8 = 1$) and the solution is

$$\Phi(t, v) = \left[\sum_{i=1}^{7}\left(A_{mi}e^{\alpha_i t}\right) + A_{m8}te^{\alpha_8 t}\right]\sin v.$$

If $m > 1$, four roots occur to be single conjugate complex roots ($\alpha_{3,4} = k_3 \pm z_3 i$, $\alpha_{5,6} = k_5 \pm z_5 i$) and the solution is

$$\Phi(t, v) = \sum_{m=2}^{\infty} \big(A_{m1} e^{\alpha_1 t} + A_{m2} e^{\alpha_2 t} + e^{k_3 t} (A_{m3} \cos z_3 t + A_{m4} \sin z_3 t) + e^{k_5 t} (A_{m5} \cos z_5 t + A_{m6} \sin z_5 t) + A_{m7} e^{\alpha_7 t} + A_{m8} e^{\alpha_8 t} \big) \sin mv.$$

The particular solution may be easily included into the equations, for example, the special stress function for m > 1 in this case is written as follows

$$\varphi(u, v) = \left(\frac{EhHm}{\pi} \right) \sum_{m=2}^{\infty} \left(\sum_{i=1}^{8} A_{mi} u^{\alpha_i} (\alpha_i - 1) + 3Au^4 \right) \cos mv.$$

There are also equations for tangential displacements, when m = 1

$$u_u = \frac{H}{\pi} \left\{ \sum_{i=1}^{7} A_{1i} u^{\alpha_i - 1} (\alpha_i - 1 - \nu\alpha_i(\alpha_i - 1)) + Au^3 (3 - 12\nu) + A_{18} \frac{u^{\alpha_8 - 1}}{\alpha_8 - 1} \left[2\alpha_8 - 2 - \nu \left(3\alpha_8^2 - 4\alpha + 1 \right) + (\ln u (\alpha - 2) - 1)(\alpha - 1)(\alpha - 1 - \nu\alpha(\alpha - 1)) \right] \right\} \cos v,$$

$$u_v = \frac{H}{\pi} \left\{ \sum_{i=1}^{7} A_{1i} u^{\alpha_i - 1} [(\alpha_i - 1)(\alpha_i(\alpha_i - 1) + \nu) - \alpha_i + 1] + 3Au^3 (11 + \nu) + A_{18} u^{\alpha_8 - 1} [\ln u (\alpha_8 - 1)(\alpha_8(\alpha_8 - 1) - \nu\alpha_8 + \nu - (\alpha_8 - 2)(\alpha_8 - 1 - \nu\alpha_8(\alpha_8 - 1))) + 3\alpha_8^2 - 4\alpha_8 + 1 - \nu(2\alpha_8 - 1) + \nu - (\alpha_8 - 1)(\alpha_8 - 1 - \nu\alpha_8(\alpha_8 - 1))] \right\} \cos v,$$

and when m > 1

$$u_u = \left(\frac{Hm}{\pi} \right) \sum_{m=2}^{\infty} \left(\sum_{i=1}^{8} A_{mi} u^{\alpha_i - 1} \left(\alpha_i - m^2 - \nu\alpha_i(\alpha_i - 1) \right) + Au^3 \left(4 - m^2 - 12\nu \right) \right) \cos mv$$

$$u_v = \left(\frac{H}{\pi} \right) \sum_{m=2}^{\infty} \left(\sum_{i=1}^{8} A_{mi} u^{\alpha_i - 1} \left[(\alpha_i - 1) \left(\alpha_i(\alpha_i - 1) + m^2\nu - \nu\alpha_i \right) - \alpha_i + m^2 + \nu\alpha_i (\alpha_i - 1) \right] + Au^2 \left(36 + 3m^2\nu - 4 + m^2 \right) \right) \cos mv.$$

4 Load Distribution

We consider here a compartment of a helicoid with arbitrary length. μ is an arbitrary number, not necessarily an integer: $\mu = 1$ is a half-loop, $\mu = 2\ k - k$.

The solution (Eq. 8) will be

$$\Phi\left(t, v\right) = \sum_{\substack{m=1\\m=0}}^{\infty} \Phi m\left(t\right) \frac{\sin mv/\mu}{\cos mv/\mu}.$$

In all the further formulas, m is replaced by m/μ.

We can take into account the arbitrary load in Eq. (7) such as Z(t, v) = q(t) F(v):

$$Z_m\left(t\right) = \frac{2}{v_0} q\left(t\right) \int_0^{v_0} F\left(v\right) \sin m \frac{v}{\mu} dv. \tag{15}$$

When the load is applied in the interval (v_1, v_2):

$$Z_m\left(t\right) = \frac{2}{v_0} q\left(t\right) \int_{v_1}^{v_2} F\left(v\right) \sin m \frac{v}{\mu} dv.$$

If the load does not vary along the helicoid in the interval (v_1, v_2) $F(v) = 1$:

$$Z_m\left(t\right) = \frac{2}{v_0} q\left(t\right) \int_{v_1}^{v_2} \sin m \frac{v}{\mu} dv = \frac{2}{v_0} \frac{\mu}{m} q\left(t\right) \left(\cos m \frac{v_2}{\mu} - \cos m \frac{v_1}{\mu}\right)$$

$$= \frac{2}{\pi m} q\left(t\right) \left(\cos m \frac{v_2}{\mu} - \cos m \frac{v_1}{\mu}\right)$$

If the load is constant (Z(t, v) = q = const) in the interval $(0, v_0)$:

$$Z_m = \frac{4}{\pi m} q.$$

If the load is applied along a line in the section $v = vq$, $Z(t, v) = q(t, vq)$:

$$Z_m\left(t\right) = \frac{2}{v_0} q\left(t, v_q\right) \sin m \frac{v_q}{\mu}.$$

The main idea of the calculation is the same in all types of loading.

5 Test Example

We take as a test example a shell in the shape of a right helicoid with the following physical and geometrical characteristics: Young's modulus $E = 2 \times 10^5$ MPa; Poisson's ratio $\nu = 0.3$; thickness $h = 0.02$ m; screw pitch $H = 0.314$ m; inner radius $r = 5$ m; outer radius $R = 6.7$ m; uniformly distributed load $q = 10^{-2}$ MPa. Since both curvilinear edges are rigidly clamped, we use Dirichle boundary conditions:

$$u_z(r) = 0,\ \partial u_z(r) = 0,\ u_u(r) = 0,\ u_v(r) = 0;$$

$$u_z(R) = 0,\ \partial u_z(R) = 0,\ u_u(R) = 0,\ u_v(R) = 0.$$

The structure is analyzed by two approaches: the user software based on the analytical method suggested above, and popular software based on FEM. Some results are shown in Figs. 2 and 3, and Tables 1 and 2.

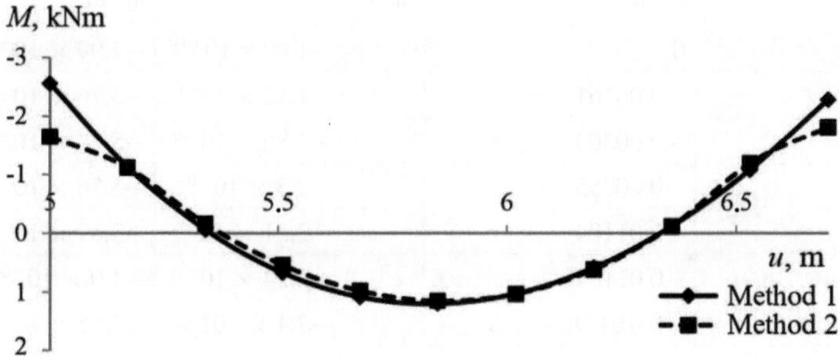

Fig. 2 Results for bending moments obtained by suggested analytical (method 1) and FE (method 2) approaches

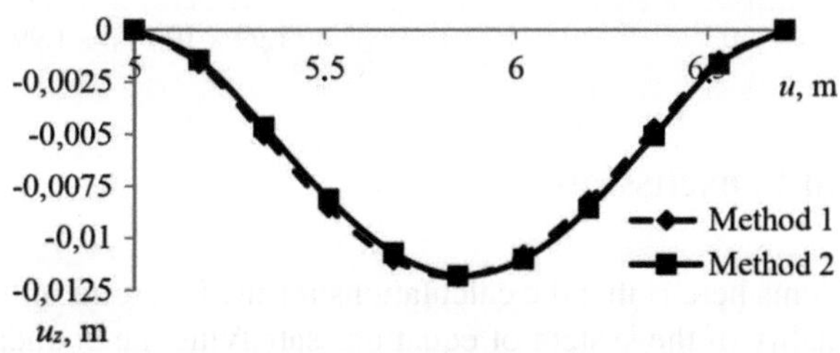

Fig. 3 Results for normal deflections obtained by suggested analytical (method 1) and FE (method 2) approaches

Table 1 Bending moments M_u and normal deflections u_z calculated by analytical (method 1), finite element (method 2), and standard (method 3) approaches of analysis

u, m	Mu, kNm			u_z, m		
	Method 1	Method 2	Method 3	Method 1	Method 2	Method 3
5	−2.5581	−1.650	−2.40833	0	0	0
5.17	−1.1299	−1.120	−1.10783	−0.00161	−0.00143	−0.00169
5.34	−0.0622	−0.164	−0.09633	−0.00503	−0.00463	−0.00534
5.51	0.6692	0.538	0.62617	−0.00855	−0.0081	−0.00920
5.68	1.0837	0.977	1.05967	−0.01104	−0.0107	−0.01203
5.85	1.1979	1.150	1.20417	−0.01184	−0.01180	−0.01305
6.02	1.0267	1.030	1.05967	−0.01079	−0.0110	−0.01203
6.19	0.5830	0.614	0.62617	−0.00817	−0.00856	−0.00920
6.36	−0.1218	−0.118	−0.09633	−0.00469	−0.00505	−0.00534
6.53	−1.0778	−1.190	−1.10783	−0.00147	−0.00163	−0.00169
6.7	−2.2783	−1.810	−2.40833	0	0	0

Table 2 Normal u_z and tangential u_u, u_v deflections ($v = \pi/2$)

u, m	u_z, m	u_u, m	u_v, m
5	0	-1.99×10^{-10}	-1.99×10^{-10}
5.17	−0.00161	-1.12×10^{-6}	-3.76×10^{-6}
5.34	−0.00503	-1.86×10^{-6}	-5.74×10^{-6}
5.51	−0.00855	-2.3×10^{-6}	-5.56×10^{-6}
5.68	−0.01104	-2.51×10^{-6}	-3.69×10^{-6}
5.85	−0.01184	-2.54×10^{-6}	-1.0×10^{-6}
6.02	−0.01079	-2.4×10^{-6}	-1.55×10^{-6}
6.19	−0.00817	-2.09×10^{-6}	-3.18×10^{-6}
6.36	−0.00469	$-1{,}6 \times 10^{-6}$	-3.39×10^{-6}
6.53	−0.00147	-0.9×10^{-6}	-2.17×10^{-6}
6.7	0	-1.99×10^{-10}	-1.99×10^{-10}

6 Results and Discussion

One of the problems here is that the calculations for the first term of the series show a poor conditionality of the system of equations satisfying the boundary conditions at the longitudinal edges of the shell. For the terms of the series $m > 1$, the system of equations for the boundary conditions turns out to be well-conditioned. Similar to the solution for a half-wave of a helicoid, it is easy to obtain a solution for a shell of arbitrary length $0 \leq v \leq v_0$. Calculations show that for v0 $\neq \pi$ no multiple roots occur. It simplifies the equations, and the matrices of the equation system are well-conditioned. It works even with $v0 = 1001\pi$ or $v0 = 0999\pi$, and practically does not affect the calculations. In this regard, it was decided in the author's accounting program to accept these assumptions.

According to the results obtained, the deflections and bending moments in the middle section of a shallow right helicoid per one half-wave ($\nu = \pi/2$) coincide with the deflections and bending moments in the annulus with similar physical and geometrical properties under a uniformly distributed load. Figures 2 and 3 show the results for normal deflections and bending moments obtained by two different approaches: the analytical method (method 1) and FEM (method 2). To check the results, the plate with width of 1 m (with similar physical properties) is also calculated as a beam by standard analytical method of structural mechanics (method 3). The results are shown in Table 1, they are close but there are some differences, especially in bending moments on the edges (boundaries) that can be caused by some problems in boundary conditions and require further investigation. Tangential deflections are insignificant in comparison with normal deflections (Table 2).

For further research, the suggested method and user program can be modified for various boundary conditions and load distributions, as well as for including dynamic loading.

The general results obtained by the suggested analytical method of calculation show that it is still possible to use the analytical approach to the stress-strain analysis of shells in the shape of helical surfaces at least for a preliminary calculation. However, there are also some variances in numerical results of longitudinal forces and tangential displacements which require paying more attention to both FE and analytical methods in order that the correct results will be found. Since the analytical solutions are generally more correct, it would be interesting for future research to investigate these differences and confirm the results by some other method of calculation.

In the case of civil structures like ramps which are large in dimensions and may be designed with a certain safety factor, the calculations represented above seem to be correct enough. But in the case of screws design, where the unification and nomenclature are important, it seems to be necessary to find more precise calculation results for efficient material consumption. On the other hand, the civil structures may require also the effectiveness of the structure during the exploitation. For this, the combined geometric and stress-strain analysis investigation, as well as an experiment may be conducted in the future to obtain the most effective shape of the ramp.

Nowadays, while architects and design engineers need more effective structures for realization [20, 21], there are many instruments for more complicated investigations of shell structures, and future research will obviously involve some ideas of parameterization and optimization of helical surfaces for ramps and screws.

7 Conclusion

Based on Reissner's equations, an analytical method of stress-strain analysis of right helicoid is presented, and the user code is created. It is proven that the numerical results obtained by the analytical approach to the stress-strain are similar to those obtained by finite element method, while the original version of this technique leads to incorrect results.

The user program which is coded based on the suggested analytical method allows to change the dimensions of the structure easily and can be very effective for both an advanced calculation in order to find the proper sizes of the structure (at the stage of preliminary design when architects usually correct the project frequently) and a deeper strain state analysis in order to develop the structural design of the ramp (at the stage of structural design).

The similar analytical approach was checked by Tupikova [22] for the other type of helicoid (oblique helicoid), but the results were finally obtained by a half-numerical method. It would be interesting to develop this method analytically and compare results with other types of analysis.

Surfaces may be analyzed by analytical methods as well as numerical. Both are useful at least because they allow to compare results and deeper analyze the stress-strain state of the shells in the shape of right helicoid.

Acknowledgements This paper was financially supported by the Ministry of Education and Science of the Russian Federation on the program to improve the competitiveness of Peoples' Friendship University of Russia (RUDN University) among the world's leading research and education centers in the 2016–2020.

References

1. Krivoshapko, S.N., Rynkovskaya, M.: Five types of ruled helical surfaces for helical conveyers, support anchors and screws. MATEC Web. Conf. (2017). 10.1051/matecconf/20179506002
2. Abevi, F., Daidie, A., Chaussumier, V., Sartor, M.: Static load distribution and axial stiffness in a planetary roller screw mechanism. J. Mech. Des. **10**(1115/1), 4031859 (2015)
3. Korevaar, N., Kusner, R., Solomon, B.: The structure of complete embedded surfaces with constant mean curvature. J. Diff. Geom. **30**, 465–503 (1989)
4. Edelen, N.: A conservation approach to helicoidal surfaces of constant mean curvature in R3, S3 and H3. arXiv:1110.1068[math.DG] (2011)
5. Taylor, D.W.: The Speed and Power of Ships. Ransdell Inc., Washington D.C (1933)
6. Solomon, L.I.: To the calculation of helicoidal shells. Ph.D. Thesis. Moscow. MISI (1953)
7. Rekach, V.G.: Analysis of shallow helicoidal shells. MISI **27**, 113–132 (1957)
8. Cohen, J.W.: The Inadequacy of the Classical Stress-Strain Relations for the Right Helicoidal Shell. In: Proceedings of the I U T A M Symposium on Shell Theory, pp. 415–433 (1959)
9. Ya, Koltunov S.: To analysis of stress state in limited helicoidal shells. Izvestiya AN SSSR MTT **6**, 149–152 (1980)
10. Iura, M., Hirashima, M.: Geometrically nonlinear theory of naturally curved and twisted rods with finite rotations. Proc. JSCE Struct. Eng./Earthq. Eng. **2**, 107–117 (1985)
11. Krivoshapko, S.N.: Geometry and strength of general helicoidal shells. Appl. Mech. Rev. **52–5**, 161–175 (1999)

12. Dekhtyar, A.S.: Load carrying capacity of helicoidal shell. Str. Mech. Anal. Constr. **6**, 1–6 (2013)
13. Rynkovskaya, M.I.: To the problem of determining the stress-strain state of ruled thin screw shells. Struct. Mech. Eng. Constr. Build. **6**, 13–15 (2015)
14. Krivoshapko, S.N.: Two types of governing equations for shells with the middle surfaces given in arbitrary curvilinear coordinates. Str. Mech. Eng. Constr. Build **1**, 15–22 (2017)
15. Zhao, Y., Su, D., Wei, W., Dong, X.: A meshing principle for generating a cylindrical gear using an Archimedes hob with two degrees of freedom. J. Mech. Eng. Sci. **224–1**, 169–181 (2010)
16. Krivoshapko, S.N., Ivanov, V.N.: Encyclopedia of Analytical Surfaces. Springer, Berlin (2015)
17. Marquere, K.: Zur Theorie der gekrummter Platte prosser Formanderung. In: Proceedings of the Fifth International Congress for Applied Mechanics, pp. 93–101 (1938)
18. Reissner, E.: Small rotationally symmetric deformations of shallow helicoidal shells. J. Appl. Mech. **22–1**, 31–34 (1955)
19. Rynkovskaya, M.I.: Rekach's method of calculation as applied to right helicoids. Struct. Mech. Eng. Constr. Build. **3**, 23–29 (2008)
20. Adriaenssens, S., Block, P., Veennendaal, D., Williams, C.: Shell Structures for Architecture—-Form finding and Optimization. Routledge, UK (2014)
21. Bradshaw, R., Campbell, D., Gargari, M., et al.: Special structures. Past, present, and future. J. Struct. Eng. **6**, 691–701 (2002)
22. Tupikova, E.M.: Investigation of V.G. Rekatch's method of stress-strain analysis of the shell of long shallow oblique helicoid form. Struct. Mech. Anal. Constr. **1**, 14–19 (2016)

Effect of Anisotropy of Masonry on the Behaviour of Unreinforced and Confined Masonry Walls Under Ground Motion

Marija Smilovic, Jure Radnic, Nikola Grgic and Goran Baloevic

Abstract This paper presents the numerical test results of the effect of anisotropy on the behaviour of unreinforced and confined masonry walls under horizontal harmonic ground acceleration. Two-storey masonry walls with different lengths (3, 6, 12 m) are analysed. The wall foundations are supported on a rigid base, with possibility of lifting. A non-linear macro model of masonry is used. Cases of strong masonry (high strength and stiffness) and weak masonry (low strength and stiffness) are analysed. Different coefficients of anisotropy of masonry ca (0.2, 0.4, 0.6, 1.0) are considered. The excitation period is taken to correspond to the fundamental period of the individual elastic wall. Horizontal displacements at the top of the walls, vertical stresses in masonry and concrete at the bottom of the walls and stresses in the longitudinal bars at the bottom of the vertical ring beams are shown for each analysed case. It is concluded that the ultimate bearing capacity and deformability of walls, as well as stresses in masonry, concrete and reinforcement under dynamic excitation significantly depend on the anisotropy of masonry. It is also concluded that, with the same coefficient of anisotropy, the behaviour of masonry walls depends on its type (unreinforced, confined), height to length ratio (0.5, 1.0, 2.0), and the quality of masonry (strong, weak).

Keywords Masonry wall · Ground motion · Numerical model
Effect of anisotropy

M. Smilovic · J. Radnic (✉) · N. Grgic · G. Baloevic
Faculty of Civil Engineering, Architecture and Geodesy, University of Split, Split, Croatia
e-mail: jradnic@gradst.hr

M. Smilovic
e-mail: marija.smilovic@gradst.hr

N. Grgic
e-mail: nikola.grgic@gradst.hr

G. Baloevic
e-mail: goran.baloevic@gradst.hr

© Springer International Publishing AG, part of Springer Nature 2019

A. Öchsner and H. Altenbach (eds.), *Engineering Design Applications*,
Advanced Structured Materials 92, https://doi.org/10.1007/978-3-319-79005-3_12

1 Introduction

Masonry walls generally have anisotropic properties. Even when the masonry units are isotropic, anisotropy of masonry is present due to the different properties of horizontal and vertical joints between the masonry units. Additional significant anisotropy of masonry is often present due to vertical holes in masonry units. Generally, masonry has a significant difference in stiffness and strength in the vertical and horizontal directions, and usually are higher in the vertical direction. In precise numerical analysis of masonry walls, the effect of anisotropy of masonry should be considered, especially for horizontal loads. Likewise, for masonry walls which have vertical and horizontal concrete ring beams, their effect should also be included in the numerical model.

Various numerical models for the simulation of anisotropic masonry are used, and only some of them are shortly presented. The first nonlinear constitutive model for masonry was proposed by Page [1], in which bricks and mortar are considered separately. The failure surface of masonry in the form of three elliptical cones was developed by Dhanasekar et al. [2]. A scalar damage model in a mesoscopic study to assess the need for incorporating non-orthotropic induced anisotropy in macrocopic models was used by Massart et al. [3]. A constitutive model of masonry using three failure criteria was developed by Andreaus [4]. Lourenco [5] defined a limiting surface to model orthotropic characteristics of masonry. De Oliveira Neto et al. [6] presented an application of a Boundary Element Method formulation for anisotropic body analysis using isotropic fundamental solutions. Using third order polynomial to describe the failure surface of masonry was presented by Syrmakezis and Asteris [7]. An ortotropic damage model based on four internal damage parameters was developed by Berto et al. [8]. Plevris et al. [9] used neural networks to approximate the failure surface for masonry as a brittle anisotropic material. The plane stress damage macro-model of ortotropic masonry was used by Pela et al. [10]. The phenomenological failure criteria for orthotropic homogeneous masonry was developed by Lishak et al. [11].

Such numerical models which are simple enough, based on a small number of parameters and at the same time able to adequately describe the actual behavior and the limit bearing capacity of masonry walls are of interest for extensive engineering application. Unfortunately, such numerical models are still rare and further researches in this direction are needed.

In this paper, a previously developed numerical model for static and dynamic analysis of plane unreinforced and confined masonry walls [12] was used. The model is quite simple and can describe the main nonlinear effects of masonry, concrete and reinforcement. The orthotropic model of masonry assumes anisotropy in the horizontal and vertical directions, with different stiffness and strength of masonry. The criteria for the limit bearing capacity of anisotropic masonry are separately defined for normal and shear stresses. By using this model, this paper investigated the effect of anisotropy of masonry on the behavior of two-story unreinforced and confined masonry walls under harmonic ground acceleration. The anisotropy coefficient of

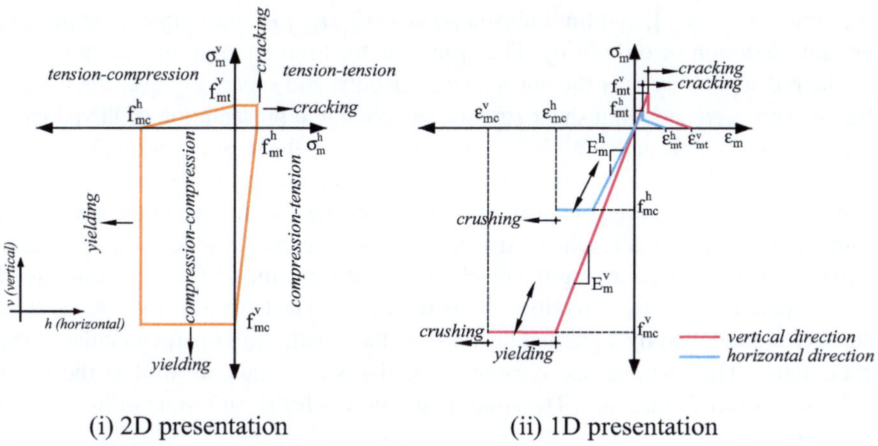

Fig. 1 Adopted orthotropic constitutive model for masonry

masonry was varied for different height to length ratios of walls, as well as for different quality of masonry (strong, weak). It is assumed that foundations are supported on rigid base, with possibility of uplifting. Only some numerical results are presented. The main conclusions of the performed analysis are focused on the effect of anisotropy of masonry on the behaviour and limit bearing capacity of the considered masonry walls.

2 Adopted Numerical Model

A numerical finite element model developed by Radnic et al. [12] was adopted, and is only briefly described here. Spatial discretisation of the structure is approximated by the plane stress state, with 8-node basic elements and 6-node contact elements. The effects of large displacements and formation of the structure in different stages can be modelled. The constitutive models for masonry, concrete and contact elements are only briefly described hereinafter, for validation of results of the performed numerical tests.

2.1 Anisotropic Masonry Model

A simplified macro model for masonry is used, which can simulate its main nonlinear effects. An orthotropic model of masonry is assumed, where the main directions of orthotropy are horizontal (h) and vertical (v). Different compressive strengths $\left(f_{mc}^{h}, f_{mc}^{v}\right)$, tensile strengths $\left(f_{mt}^{h}, f_{mt}^{v}\right)$, modulus of elasticity $\left(E_{m}^{h}, E_{m}^{v}\right)$, limit compressive strains $(\varepsilon^{h}_{mc}, \varepsilon^{v}_{mc})$, and limit tensile strains $(\varepsilon^{h}_{mt}, \varepsilon^{v}_{mt})$ of masonry can be defined for each direction of orthotropy. The criterions for limit strength of masonry were considered separately over the normal stresses only, and separately over the normal stresses and corresponding shear stresses. A schematic presentation of the adopted constitutive model is given in Fig. 1, where σ^{h}_{m}, σ^{v}_{m} are the masonry normal stresses.

An elastic-ideal plastic behaviour with linear unloading is adopted in biaxial compression. The crushing of masonry occurs when the maximum compressive strain from the uniaxial test is reached. An elastic behaviour in biaxial tension is adopted until the tensile masonry strength is reached. Afterwards, smeared cracks are assumed to be perpendicular to the direction of tensile stresses. The tensile stiffness of cracked masonry is modelled by a gradual decrease of the tensile stress perpendicular to the crack plane. The opening and closing of cracks is modelled, as well as the shear stiffness of cracked masonry. The simplified criterion for shear failure of masonry is adopted.

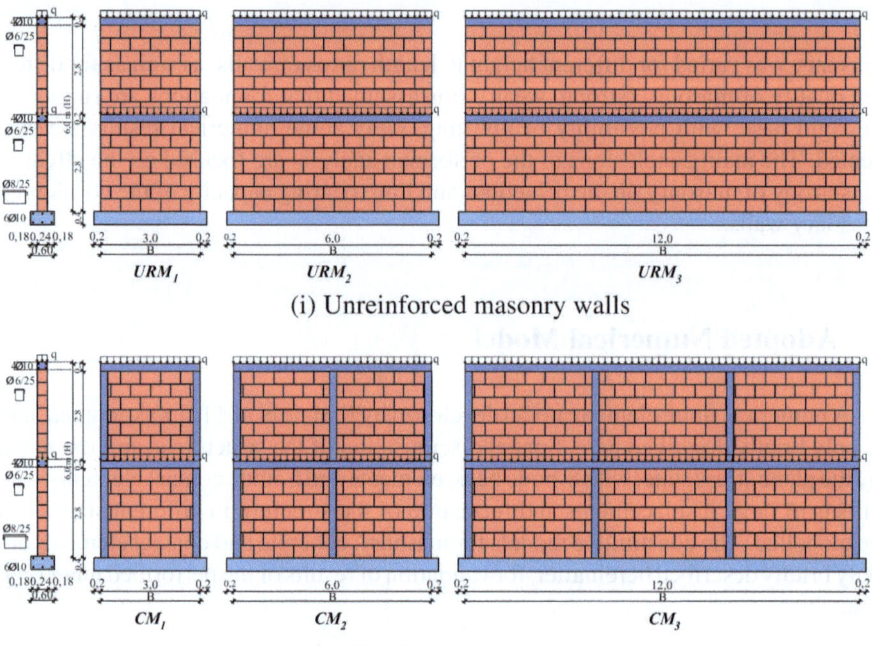

Fig. 2 Analysed masonry walls

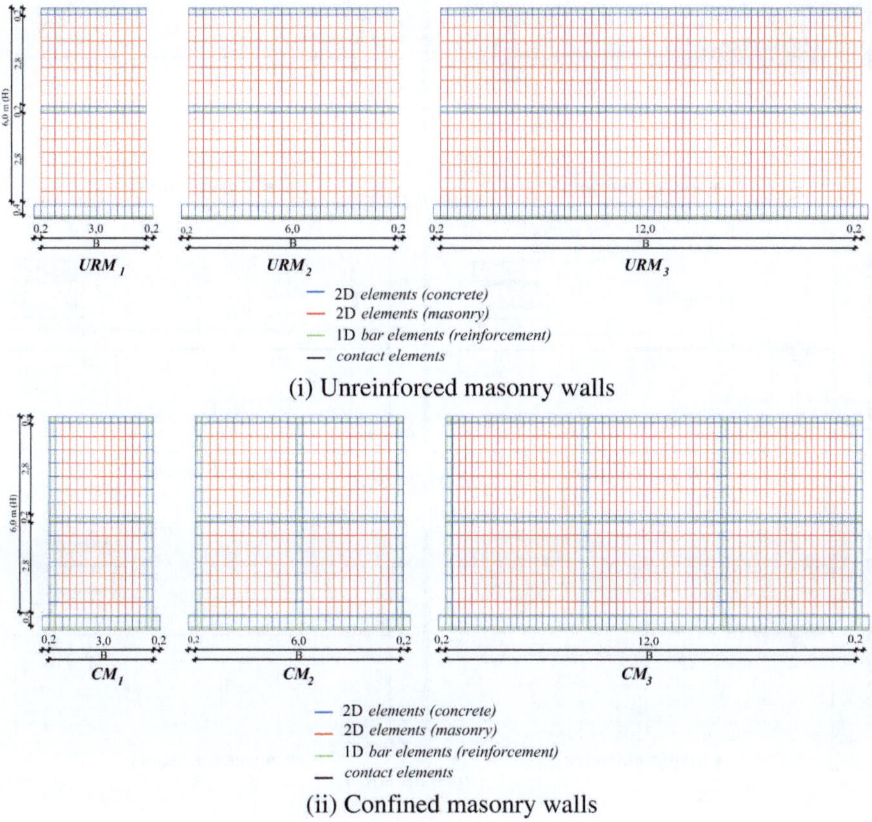

Fig. 3 Finite element discretization of the walls

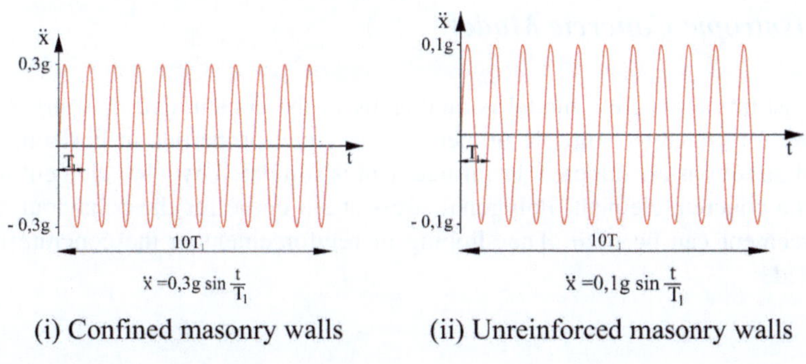

Fig. 4 The adopted horizontal base acceleration

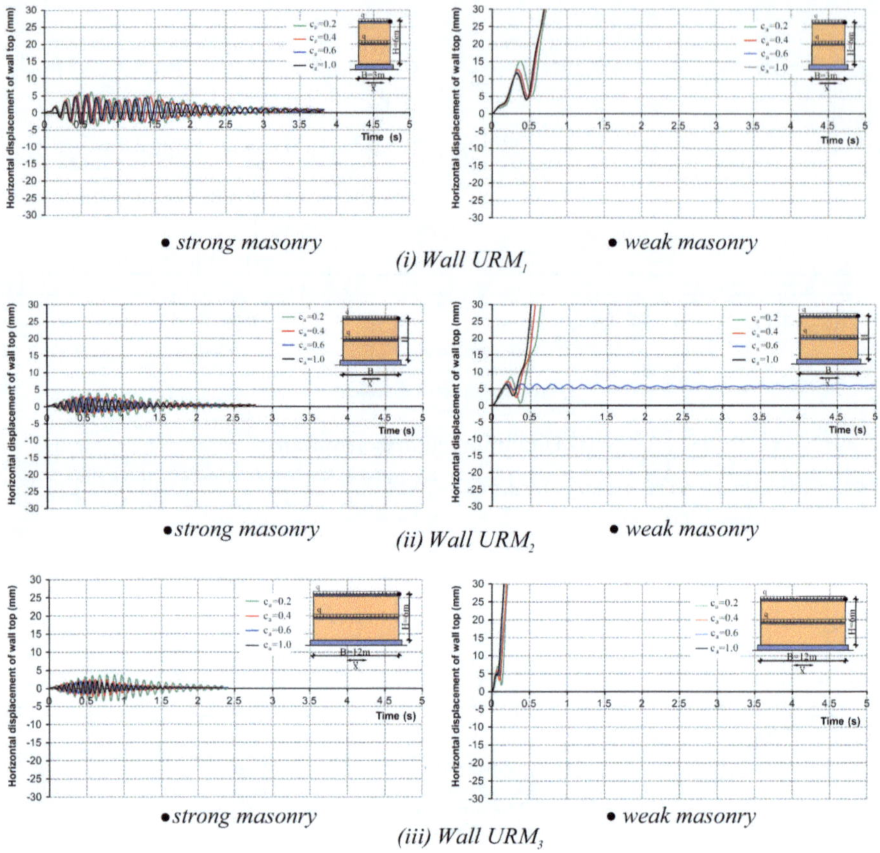

Fig. 5 Horizontal displacements of the top of unreinforced masonry walls

2.2 *Isotropic Concrete Model*

The constitutive concrete model is analogous to the constitutive masonry model, whereby for concrete isotropic properties, i.e. equal properties in horizontal and vertical directions are adopted. Reinforcement is simulated by a bar element within the basic concrete element. Polygonal stress-strain curve for the behaviour of the reinforcement can be used. The slipping of reinforcement in the concrete is not modelled.

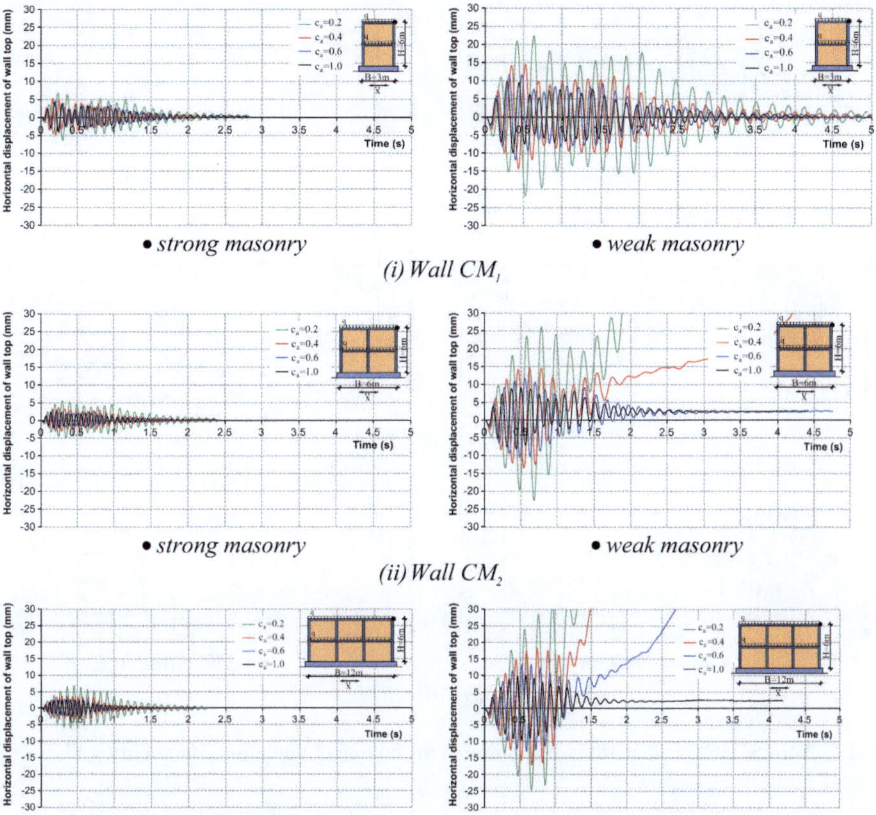

• *strong masonry* • *weak masonry*

(i) Wall CM_1

• *strong masonry* • *weak masonry*

(ii) Wall CM_2

• *strong masonry* • *weak masonry*

(iii) Wall CM_3

Fig. 6 Horizontal displacements of the top of confined masonry walls

2.3 *Model for Contact Elements*

Planar contact elements can transmit normal and shear stresses according to the adopted uniaxial stress-strain diagrams. Penetration and separation on the contact surface can be modelled.

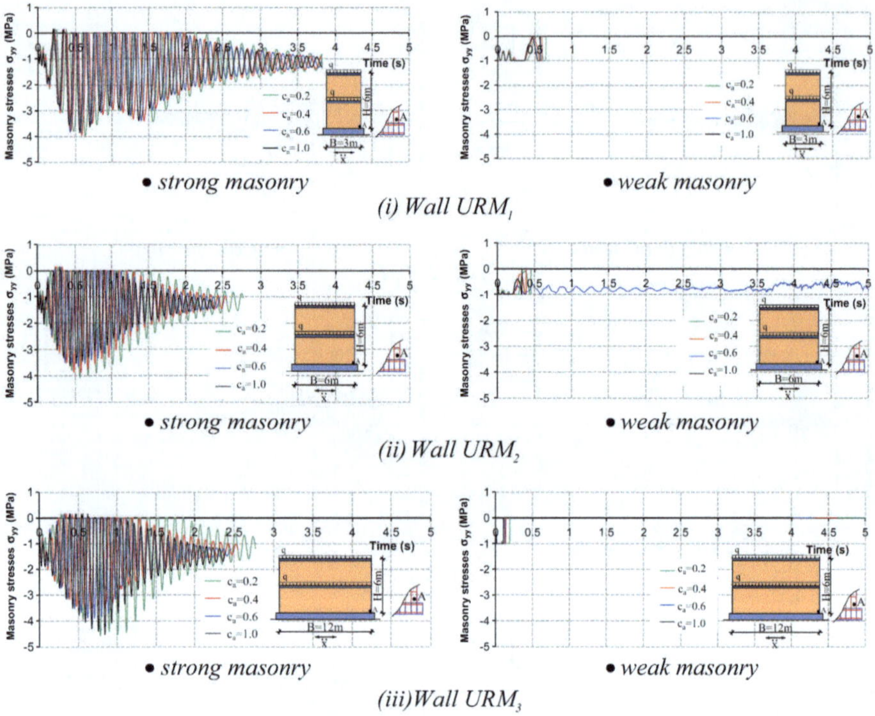

Fig. 7 Vertical stresses in masonry in point A at the bottom of unreinforced masonry walls

3 Numerical Tests of Anisotropy Effects on the Behavior of Unreinforced and Confined Masonry Walls Under Ground Acceleration

3.1 Analysed Masonry Walls

The effect of anisotropy of masonry under dynamic load was considered on the two –story unreinforced and confined masonry walls shown in Fig. 2. The walls are 0.24 m thick. The wall length is varied (3, 6, 12 m) as well as the quality of masonry (strong, weak). The horizontal ring beams are reinforced with 4Ø10 horizontal bars, and vertical ring beams with 4Ø12 bars. The stirrups of all vertical and horizontal ring beams are Ø6 with 250 mm spacing.

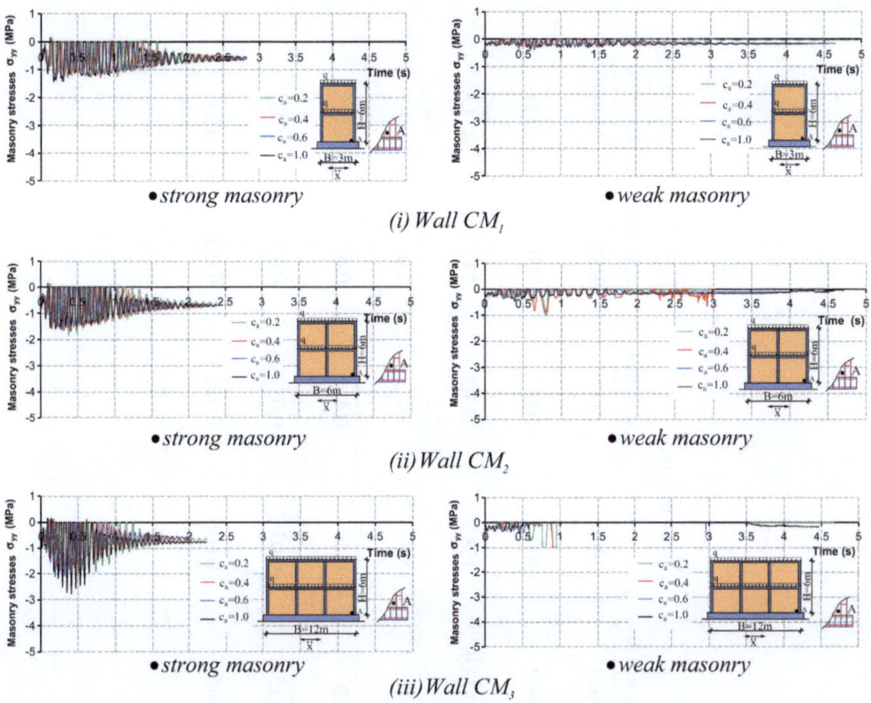

Fig. 8 Vertical stresses in masonry in point A at the bottom of confined masonry walls

3.2 Finite Element Discretization and Adopted Material Parameters

Finite element discretization of the walls is presented in Fig. 3. It is assumed that the wall foundation stands on a rigid base, with the possibility of its uplifting. For this purpose, contact elements are placed below the foundations. Masonry with high strength and stiffness (the so-called strong masonry) and masonry with low strength and stiffness (the so-called weak masonry) are considered. The basic material parameters that were adopted for the masonry, for the considered coefficient of anisotropy ca, are shown in Table 1, where were:

$$c_a = E_m^h / E_m^v = f_{mc}^h / f_{mc}^v = f_{mt}^h / f_{mt}^v ; E_m^v \nu_m^h = E_m^h \nu_m^v \tag{1}$$

where ν_m^h and ν_m^v are the Poisson's ratios in the horizontal and vertical directions. The shear modulus of masonry G_m is given by:

$$G_m = 1 / \left[(1 + \nu_m^h)/E_m^h + (1 + \nu_m^v)/E_m^v \right] \tag{2}$$

Table 1 The adopted basic material parameters for masonry

Parameters	Unit	Strong masonry				Weak masonry			
		$c_a = 0.2$	$c_a = 0.4$	$c_a = 0.6$	$c_a = 1.0$	$c_a = 0.2$	$c_a = 0.4$	$c_a = 0.6$	$c_a = 1.0$
Elasticity modulus E_m^v	MPa	5000	5000	5000	5000	1000	1000	1000	1000
Elasticity modulus E_m^v	MPa	1000	2000	3000	5000	200	400	600	1000
Shear modulus G_m	MPa	793.6	1315.7	1685.4	2173.9	158.7	263.2	337.1	434.8
compressive strength f_{mc}^v	MPa	5.0	5.0	5.0	5.0	1.0	1.0	1.0	1.0
Compressive strength f_{mc}^h	MPa	1.0	2.0	3.0	5.0	0.2	0.4	0.6	1.0
Tensile strength f_{mt}^v	MPa	0.15	0.15	0.15	0.15	0.03	0.03	0.03	0.03
Tensile strength f_{mt}^h	MPa	0.03	0.06	0.09	0.15	0.006	0.012	0.018	0.03

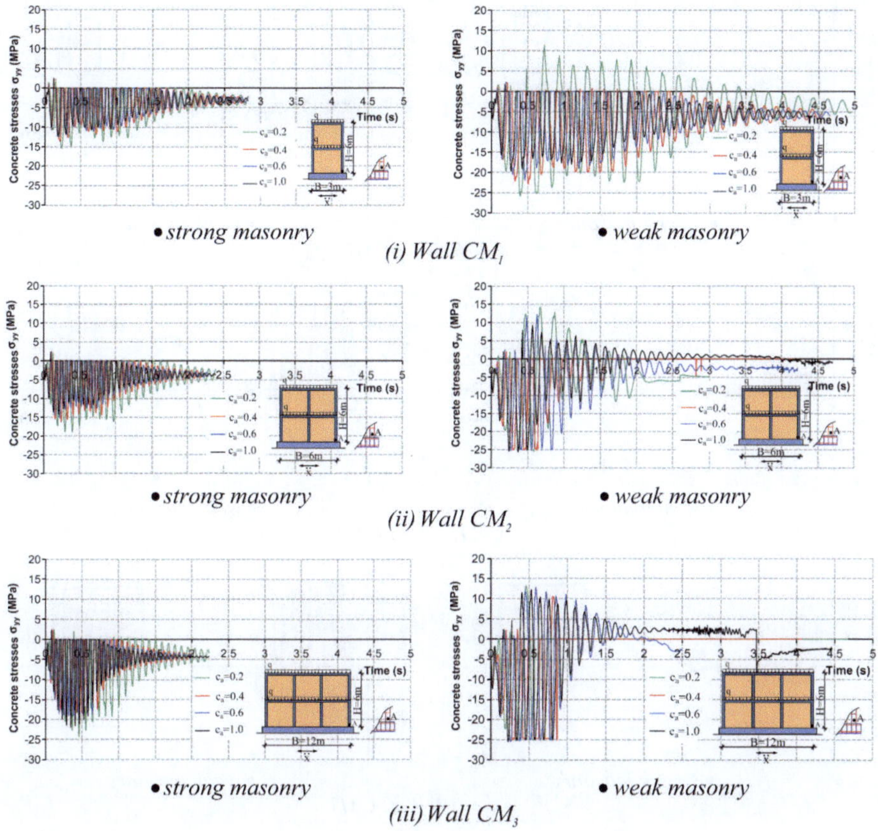

Fig. 9 Vertical stresses in concrete in point A at the bottom of confined masonry walls

Table 2 The adopted basic material parameters for concrete, reinforcement and contact elements

Parameters	Unit	Concrete	Reinforcement	Contact elements
Elasticity modulus	MPa	30,500	210,000	30,500
Shear modulus	MPa	13,260	–	
Compressive strength	MPa	25	560	25
Tensile strength	MPa	2.5	560	0

The effect of shear stresses on the bearing capacity of masonry is not included in the performed analysis of masonry walls, for clarity of presentation of the anisotropy effect on the masonry.

The adopted basic material parameters of concrete, reinforcement and contact elements are shown in Table 2.

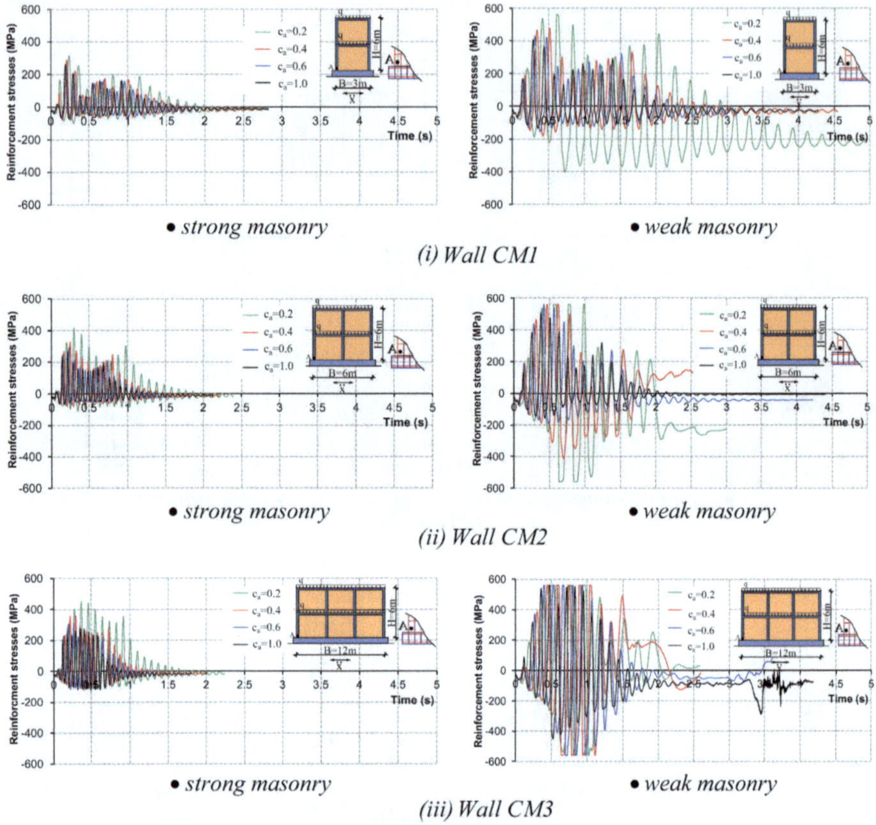

Fig. 10 Reinforcement stresses in point A of confined masonry walls

3.3 Base Excitation

In the numerical analysis, walls were subjected to horizontal harmonic base acceleration according to Fig. 4, where T_1 is the fundamental oscillation period of the elastic wall that includes its anisotropy. For the analysed walls and their corresponding initial stiffness, the eigenvalue problem was solved first [12] and is presented in Table 3.

The maximum base acceleration of 0.3 g is adopted for confined masonry walls and 0.1 g for unreinforced masonry walls. A duration for the base excitation of 10 T_1 and a duration for the numerical analysis of 25 T_1 are adopted.

Table 3 The first (T1) period of free oscillations of elastic walls

Coefficient of anisotropy	Unreinforced masonry wall				Confined masonry wall			
	Strong masonry	T_1 [s]	Weak masonry	T_1 [s]	Strong masonry	T_1 [s]	Weak masonry	T_1 [s]
$c_a = 0.2$	URM_1	0.1528	URM_1	0.3393	CM_1	0.1128	CM_1	0.1987
$c_a = 0.4$		0.1414		0.3139		0.0987		0.1624
$c_a = 0.6$		0.1374		0.3051		0.0936		0.1480
$c_a = 1.0$		0.1343		0.2987		0.0896		0.1357
$c_a = 0.2$	URM_2	0.1107	URM_2	0.2460	CM_2	0.0956	CM_2	0.1884
$c_a = 0.4$		0.0944		0.2097		0.0790		0.1497
$c_a = 0.6$		0.0886		0.1965		0.0728		0.1341
$c_a = 1.0$		0.0839		0.1860		0.0677		0.1207
$c_a = 0.2$	URM_3	0.0961	URM_3	0.2137	CM_3	0.0893	CM_3	0.1868
$c_a = 0.4$		0.0771		0.1710		0.0708		0.1457
$c_a = 0.6$		0.0698		0.1547		0.0636		0.1291
$c_a = 1.0$		0.0638		0.1412		0.0575		0.1146

3.4 Numerical Results

Some results of the performed numerical analysis are presented below. Horizontal displacements of the top of unreinforced masonry walls are shown in Fig. 5, and of confined masonry walls in Fig. 6. Vertical stresses in masonry in point A at the bottom of the unreinforced masonry walls are shown in Fig. 7 and of the confined masonry walls in Fig. 8. Vertical stresses in concrete in point A at the bottom of confined masonry are shown in Fig. 9. Stresses in longitudinal bars at the bottom of vertical ring beam of confined masonry are shown in Fig. 10.

As shown in presented results, the limit strengths and displacements (deformability), as well as the stresses in masonry, concrete and reinforcement, depend on the coefficient of anisotropy ca. As the anisotropy of masonry is higher (smaller ca), the difference in these values for walls with isotropic and anisotropic masonry is also higher. With the same coefficient of anisotropy, the behaviour of the masonry wall depends on the type of the wall (unreinforced or confined), on its height-length ratio and on the quality of masonry (strong, weak).

The effect of anisotropy of masonry on the bearing capacity and deformability of the masonry walls exposed to base acceleration is significant. The effect of anisotropy in the confined masonry walls is slightly higher than for the unreinforced masonry walls. The effect of anisotropy is more significant for walls with weak masonry. As can be seen from the results presented, the largest number of unreinforced masonry walls collapsed at the maximum acceleration $a_{max} = 0.1$ g. Likewise, part of the confined masonry walls with the weak masonry also collapsed under the excitation with $a_{max} = 0.3$ g. In the analysed example, longer walls have shown mostly unfavourable behaviour than the shorter walls.

4 Conclusions

Based on the performed numerical tests in this work, in which the effect of anisotropy of masonry on the behaviour and the ultimate bearing capacity of two storey unreinforced and confined masonry walls with different lengths and stiffness of masonry exposed to ground acceleration is investigated, the most important conclusions are given below.

The ultimate bearing capacity, displacements and stresses in the confined and unreinforced masonry walls exposed to dynamic base excitation depends on the anisotropy of masonry. With higher anisotropy, the difference in the above listed variables is also higher. The behaviour of the walls with low anisotropy of masonry is similar to the behaviour of the walls with isotropic masonry. The effect of anisotropy also depends on the type of wall (unreinforced, confined), the quality of masonry (strong masonry, weak masonry) and height-length ratio of the walls. Confined masonry walls have a great effect of anisotropy than the unreinforced masonry walls. The effect of anisotropy for the strong masonry is significantly greater than for the weak masonry. Longer walls have greater effect of the anisotropy then shorter walls. The use of weak masonry, especially those with higher anisotropy, should be excluded in practice.

References

1. Page AW (1978) Finite element model for masonry. J. Struct. Div—ASCE. 104(8), 1267–1285
2. Dhanasekar, M., Page, A.W., Kleeman, P.W.: Masonry failure criterion under biaxial stress state. Proc. Inst. Civ. Eng. **79**, 295–313 (1985)
3. Massart, T.J., Peerlings, R.H.J., Geers, M.G.D.: Mesoscopic modeling of failure and damage-induced anisotropy in brick masonry. Eur. J. Mech. A Solid J. **23**, 719–735 (2004)
4. Andreaus, U.: Failure criteria for masonry panels under in-plane loading. J. Struct. Div. ASCE **122**(1), 37–46 (1996)
5. Lourenco, P.B., De Borst, R., Rots, J.G.: A plane stress softening plasticity model for orthotropic materials. Int. J. Numer. Methods Eng. **40**, 4033–4057 (1997)
6. de Oliviera, Neto L., Masia, M.J., Taguti, Y., Rigitano, A.C.: An anisotropic linear elastic boundary element formulation for masonry wall analysis. Eng. Anal. Bound. Elem. **37**, 434–440 (2013)
7. Syrmakezis, C.A., Asteris, P.G.: Masonry failure criterion under biaxial stress state. J. Mater. Civ. Eng. **13**(1), 58–64 (2001)
8. Berto, L., Scotta, R., Vitaliani, R.: An orthotropic damage model for masonry structures. Int. J. Numer. Methods Eng. **55**, 127–157 (2002)
9. Plevris, V., Asteris, P.G.: Modelling the masonry failure surface under biaxial compressive stress using neural networks. Constr. Build. Mater. **55**, 447–461 (2014)
10. Pelà, L., Cervera, M., Roca, P.: Continuum damage model for orthotropic materials: application to masonry. Comp. Meth. Appl. Mech. Eng. **200**(9), 917–930 (2011)
11. Lishak, V.I., Yagust, V.I., Yankelevsky, D.Z.: 2D orthotropic failure criteria for masonry. Eng. Struct. **36**, 360–371 (2012)
12. Radnić, J., Matešan, D., Harapin, A., Smilović, M., Grgić, N.: Numerical Model for Static and Dynamic Analysis of Masonry Structures. Springer-Verlag, Berlin, Advanced Structured Materials, Heidelberg (2012)

New Procedure to Construct an Anisotropic Elastic FE-Model Based on Swine Femoral Bones Using Numerical Modeling

Rafael Rodríguez-Martínez, Christopher René Torres-San Miguel, Guillermo Urriolagoitia-Sosa, Beatriz Romero-Ángeles and Guillermo Urriolagoitia-Calderón

Abstract This work presents a new procedure to determine the density of bone tissue from collected data by means of computed axial tomography (CAT) and bone density correlations of cancellous and cortical tissue with their elastic properties which were obtained from structure reconstruction software to determine the apparent density of bone. The main objective is to determine the anisotropic elastic properties in swine femoral bones on two orthogonal directions to produce a most real behavior of bone FE-model. A *TC Brilliance* axial scanner (tomograph) was used to obtain the desired images, which were then processed in the Digital Imaging and Communication in Medicine format using different software packages, including *Scan IP*™, *Scan FE v3.1*™, *and ANSYS vl2* ™. Five *Duroc-Jersey* type swine femur specimens were used for the analysis. These specimens are considered to be equivalent to human specimens of 50 to 55 years of age, since they have been used for eighteen months in experimental processes. The experimental procedure included the processing of 60 tomographic cuts, which allowed the determination of the zone where the density maximum and minimum values of the bone tissue seem to be located. The results obtained displayed the anisotropic elastic behavior for each bone specimen within a

R. Rodríguez-Martínez (✉) · C. R. T.-S. Miguel · G. Urriolagoitia-Sosa · B. Romero-Ángeles
G. Urriolagoitia-Calderón
ESIME Unidad Zacatenco. Sección de Estudios de Posgrado e Investigación,
Instituto Politécnico Nacional, Edif. 5, 2th Piso, Av. Instituto Politécnico Nacional S/N,
Col. Lindavista, C.P. 07738 Cd. De México, Mexico
e-mail: rafarm68@hotmail.com

C. R. T.-S. Miguel
e-mail: napor@hotmail.com

G. Urriolagoitia-Sosa
e-mail: guiurri@hotmail.com

B. Romero-Ángeles
e-mail: romerobeatriz97@hotmail.com

G. Urriolagoitia-Calderón
e-mail: urrio332@hotmail.com

© Springer International Publishing AG, part of Springer Nature 2019
A. Öchsner and H. Altenbach (eds.), *Engineering Design Applications*,
Advanced Structured Materials 92, https://doi.org/10.1007/978-3-319-79005-3_13

voxel unit of precision. The findings of this study could allow a significant advance in the development of customized endo-prostheses by determining the elastic properties of the bone and developing more accurate FE-models. This will contribute to the improvement of the performance of artificial implants as well as to increasing the service life of these prostheses.

Keywords Computer axial tomography · Bone densitometry · Elastic properties Swine specimens · Numerical analysis

1 Introduction

In several medical processes a quantification of human osseous mass is an essential factor. Nowadays one of the most common techniques used for this aim is bone densitometry (BD), which is a radiological method, based on computer axial tomography (CAT), with great sensitivity for diagnosis. Protocols widely used for image digitalization are formatted as DICOM (digital imaging and communication in medicine), DEXA (dual energy x-ray absorptiometry), CT (computed tomography), CAT (computer axial tomography), and MRI (magnetic resonance imaging), among other, are these where DICOM files are generated. Currently, reduced versions of these formats have been developed, which provide a greater resolution and precision in digitalization of clinical radiology [1]. Specialized protocols in the orthopedic field to generate organic tissue models have also been developed. These models are used in the design and development of endo-prostheses for the hip, knee, shoulder, foot, hand, and many others [2].

In order to measure different tissue densities, Hounsfield Units (HU) are used, for which an arbitrary magnitude is assigned, taking for reference the density of water (0 HU) [3]. Based on this method, fat tissue displays negative values (between −70 and −90 HU) whereas soft tissue ranges between +30 and +70 HU, displaying positive values. Tomography images are recorded in the DICOM format and the information is processed in HU. Some of the first researchers who have published work on this subject are mentioned in [4]. They found that bone density is affected by the patients' age and fracture stress of cancellous bone of the vertebrae under compression loading. They also demonstrated that the fracture stress is related to the mineral content of bone, and that it is not dependent on age or sex of the patient. Later, a study [5] was published, relating the apparent density of the bone to the mechanical properties of bone tissue; at the present time, relationship is still being applied. Several studies related to human bone tissue were also performed in the 80's. One of the most remarkable papers [6] demonstrated that the failure caused by tension or compressive loads applied on trabeculae is not dependent on its density or its elastic modulus. Therefore, the failure is fundamentally produced by the strain which the bone is under. Considering the need to define a bio-modeling technique that allows to derive suggestions for customized solutions for bone problems, this work establishes a methodology to determine the elastic properties of swine femoral bone of controlled specimens obtained from computer tomography. Recent results based on alternating methodologies to determine the mechanical properties of bone tissue have been reported [7–12]. However, the reports have only focused on establishing the relation that exists between the apparent bone density or degree of mineralization of the bone matrix (DMB) and its mechanical properties. In this research, the main objective is to determine the mechanical properties of swine bone tissue which will be used as base for the development of 3D numerical models. These models will contribute to propose reliable structural conditions at the bone-implant interface, which in turn, will facilitate the design of customized implants. The proposed model could also increase the effectiveness and service life of prosthetic devices. Furthermore, a simple methodology to determine the correlation between a pixel matrix and mechanical properties of bone has been proposed, using mathematical models that approximate the elastic behavior for these biological materials [13]. Finally, using the commercial software *Scan IP*™, values of cortical and trabecular apparent bone density were obtained and exported to an FE software, in this case *ANSYS v12*™.

2 Materials and Methods

2.1 Image Reconstruction

The software package employed to reconstruct images, based on data sent from the tomography, makes a conversion by means of a Fourier transform and it assigns a value of absorption to each studied volume, called *Voxel*. This value is later represented in the plane by means of Pixels [14–16] (see Fig. 1).

The bone reconstruction processing requires, initially, reproducing an external geometry based on data obtained from the CAT. Figure 2 displays one of the five study cases used in this analysis, where the density of the bone tissue, cortical

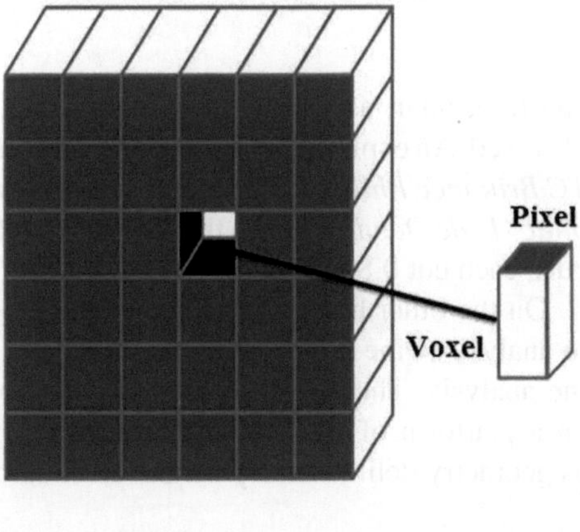

Fig. 1 Geometric structure of a voxel and a pixel

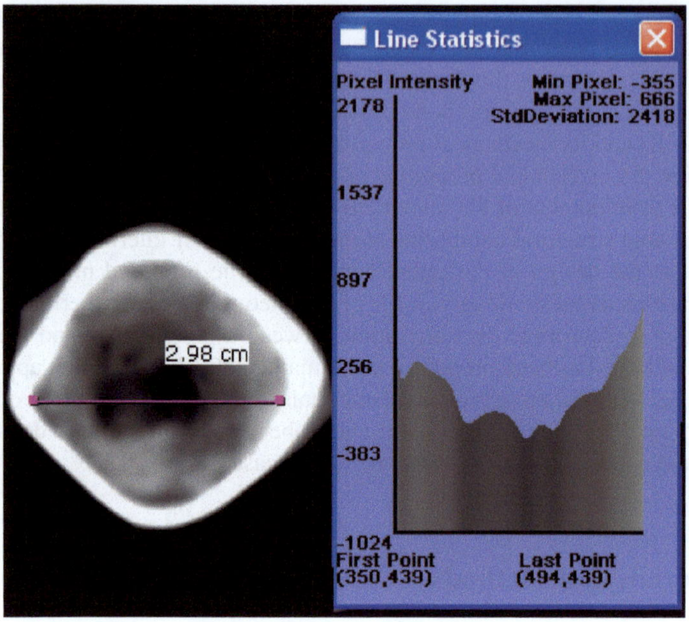

Fig. 2 Image of a transverse cut obtained from the *TC Brillance Phillips* tomograph

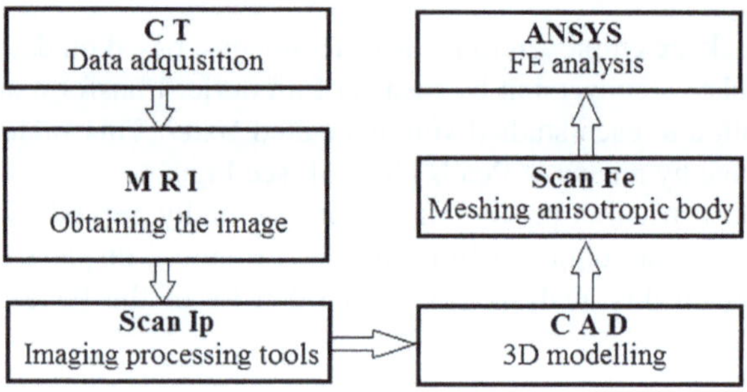

Fig. 3 Modeling stage of swine bony tissue

and trabecular, as well as the relationship of the density versus the pixel can be observed. An exploration protocol for inferior members, which included the use of the *TC Brillance Phillips* tomograph, was followed. The tomograph, located at the *Hospital "1° de Octubre"* from the *ISSSTE* in México City, has the capacity to produce16 cuts, each cut 0.8 mm thick.

On the other hand, Fig. 3 shows the modeled stages for image processing used to analyze swine bone tissue; the mass and volumetric properties are considered in the analysis. The software used in this work for structural regeneration, is based on a platform of basic image processing, which a meshing tool module, as well as geometry definitions by means of CAD capabilities. The proposed methodology in this research requires determining masks that identify tissues in the cortical and trabecular regions [17]; this is achieved by means of *spline* delineation of the required area creating internal and external walls of the swine femur in order to develop a 3D geometry. Once the walls have been defined in each tomography cut, a Stereolithography file is generated.

3 Experimental Analysis

CT Number is defined by the numerical value of pixel attenuation, which is obtained from one voxel, and its magnitude is expressed in HU. This value is selected by the operator in an interactive window of the software by means of a circle (Fig. 4). Once the value is assigned a graphical representation of the relationship, between maximum and minimum values of intensity versus total pixels number, which in turn defines the content of the different biological materials. The procedure used to determine the HU of each tomographic cut was based on the selection of the trabecular tissue means of the greatest circumference, whereas for the cortical tissue a smaller circumference was considered, both displayed in Fig. 4. The process described previously was followed for every slice of every specimen. The mechanical properties of bone tissue are modeled based on the apparent density of bone ρ_{ap}, which is defined as the mineralized tissue mass divided by the total volume, including pores. It is important to notice that the density was calculated using tomography; the mass of other tissues is significant because they are the marrow, fat and blood, which lack the capacity to support loads. A certain value of apparent density is assigned to the maximum density, which corresponds to the medullar phase contained in the cortical tissue, that is 2 g/cm^3. For this study, 30 tomographic cuts of each specimen of swine femur were analyzed in order to establish the apparent density (in HU) corresponding to each zone.

Specimens were taken from pigs of *Duroc-Jersey* type of 18 months of age which are equivalent to human specimens of 50 to 55 years of age [18]. To establish a reference scale, the value of −528 HU was assigned to the apparent density of 0 g/cm^3, while a value of 2275 HU was assigned to 2 g/cm^3. The mathematical model reported by [19, 20] was adapted to be used in the present study, to calculate with:

$$\rho_{ap} = \frac{2}{2654} UH + \frac{1000}{2654} \quad (\text{gr/cm}^3) \tag{1}$$

In agreement with (1), average values for HU were obtained, and the apparent density of the bone tissue is presented in Table 1.

Generally, the apparent density is used as a measurement of the directionality of the microstructure which characterizes anisotropic behavior. Nevertheless, it has been found that numerical models do not always incorporate the anisotropic characteristics of the bone structure due to the difficulty to find the main directions of elastic-constant

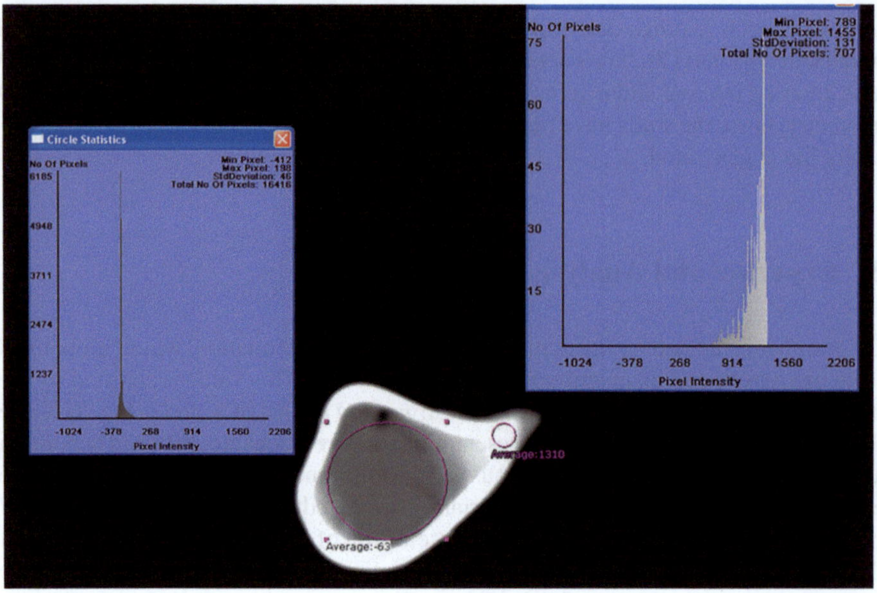

Fig. 4 Interactive window of computational program used to determine HU of cancellous and cortical bone

Table 1 Bone parameters of swine specimens

Specimen	Cortical tissue		Cancellous tissue	
	HU	Apparent density (g/cm^3)	HU	Apparent density (g/cm^3)
1	1952	1.848	163	0.500
2	1820	1.748	186	0.517
3	1983	1.871	176	0.509
4	1790	1.726	193	0.522
5	1963	1.856	145	0.486

tensors, which is variable for each 3D spatial point of bone tissue. On the other hand, cancellous and cortical bones are different from each other, mainly because of their porosity and their density. The law of variation of the elastic modulus for cortical and trabecular bone has been defined in agreement with exponential relations defined by [19, 20]. In the last 30 years, multiple investigations have attempted to determine the properties of in vivo behavior for cortical and trabecular bone and in spite of the mutual relationship between the elastic modulus and compressive strength of the cortical and spongy bones with the apparent density, there are still many difficulties determining the density of each one because of the variations between subjects as well as the close relationship of density with the age of the patient. Tables 2 and

Table 2 Elastic behavior of cortical bone [19]

Properties	Mathematical model
Elastic modulus E_p	$E_p = 2065\rho_{ap}^{3.09}$
Elastic modulus E_t	$E_t = 2314\rho_{ap}^{1.57}$
Poisson's ratio υ_p	$\upsilon_p = 0.3$
Poisson's ratio υ_{pt}	$\upsilon_{pt} = 0.0623\rho_{ap}^2 - 0.2827\rho_{ap} + 0.5249$
Shear modulus G_p	$G_p = 2065\rho_{ap}^{3.09}/[2(1+\upsilon_p)]$
Shear modulus G_t	$G_t = 3892\rho_{ap}^2 - 7385\rho_{ap} + 4606$

Table 3 Elastic behavior of cancellous bone [19]

Properties	Mathematical model
Elastic modulus E_p	$E_p = 1904\rho_{ap}^{1.64}$
Elastic modulus E_t	$E_t = 1157\rho_{ap}^{1.78}$
Poisson's ratio υ_p	$E_t = 1157\rho_{ap}^{1.78}$
Poisson's ratio υ_{pt}	$\upsilon_{pt} = -0.05331\rho_{ap}^4 + 0.1843\rho_{ap}^3 -$ $0.2438\rho_{ap}^2 + 0.1721\rho_{ap} + 0.1675$
Shear modulus G_p	$G_p = 1904\rho_{ap}^{1.64}/[2(1+\upsilon_p)]$
Shear modulus G_t	$G_t = 1157\rho_{ap}^{1.78}/[1.7(1+\upsilon_p)]$

Table 4 Results of elastic properties of swine cortical bone

Specimen	Properties					
	E_p (MPa)	E_t (MPa)	υ_p (adim)	υ_{pt} (adim)	G_p (MPa)	G_t (MPa)
1	13773	6069	0.3	0.2152	5297	4250
2	11598	5561	0.3	0.2211	4461	3589
3	14310	6188	0.3	0.2141	5504	4413
4	11153	5452	0.3	0.2226	4289	3454
5	13958	6110	0.3	0.2148	5368	4306

3 display the mathematical models obtained by [19, 20] for calculation of elastic constants of the cortical and spongy bone tissue, respectively.

4 Results

Tables 4 and 5 present the values obtained for elastic properties of the cortical and trabecular bone, respectively.

Table 5 Results obtained for the elastic properties of swine trabecular bone

Specimen	Properties					
	E_p (MPa)	E_t (MPa)	υ_p (adim)	υ_{pt} (MPa)	G_p (MPa)	G_t (MPa)
1	611	337	0.3	0.2123	235	152
2	645	358	0.3	0.2129	248	162
3	629	348	0.3	0.2127	242	157
4	656	364	0.3	0.2132	252	165
5	583	320	0.3	0.2117	224	145

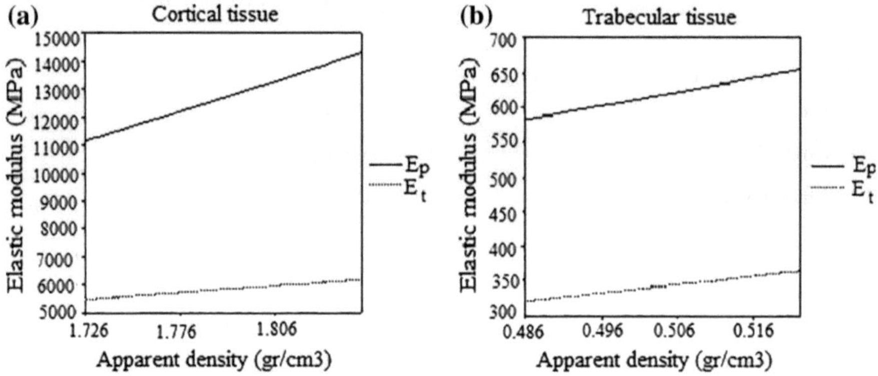

Fig. 5 Behavior of elastic modulus. **a** Difference between values of axial modulus (E_p) and transverse modulus (E_t) of cortical bone; **b** difference between values of axial modulus (E_p) and transverse modulus (E_t) of trabecular bone

The behavior curves of elastic properties for cortical and trabecular bone are displayed in Figs. 5, 6 and 7, where great differences in magnitudes obtained for properties are observed. Furthermore, the construction of the obtained anisotropic model was produced by means of an application (Fig. 8), using the same *Scan IP*™ development of software by inserting the data calculated for the elastic modulus, in axial and transverse load, as well as the different values for the Poisson's ratio.

Figure 9 presents the development of a three-dimensional model for one of the specimens by means of *Scan IP v3.1*™ software. The model was used for the generation of meshes in agreement with values found for cortical and trabecular bone properties.

Finally, the obtained mesh structures were exported to the numerical structural analysis software package *ANSYS v12*™, with which the FE model of the swine femur was produced. The result of the entire process was a matrix that represents the anisotropic FE model for every point considered (Fig. 10). Moreover, its nominal value is given according to the voxel calculated in the same position. This means that for each specific point of the specimen there is an elastic modulus and a Poisson's ratio value different from the one obtained at any other point.

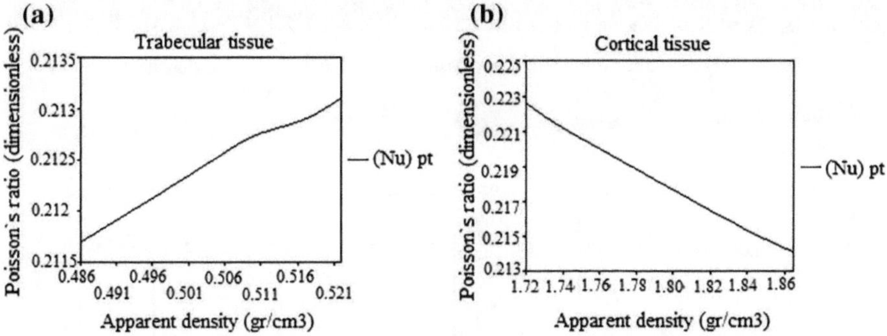

Fig. 6 Behavior of Poisson's ratio. Negative trend of Poisson's ratio of cortical bone (**a**), contrast with positive trend of Poisson's ratio to trabecular bone (**b**)

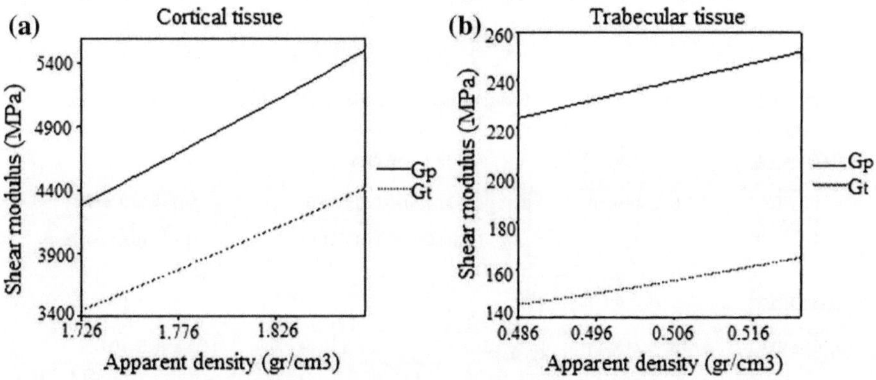

Fig. 7 Behavior of Shear modulus. **a** Difference between values of axial modulus (G_p) and transverse modulus (G_t) of cortical bone; **b** difference between values of axial modulus (G_p) and transverse modulus (G_t) of spongy bone

5 Discussion

The variation of the elastic constants of bone tissue, cortical and cancellous bone, is a function that depends directly on the apparent density of the biological material in a non-linear fashion. Because the properties of the bone tissue are altered by a great number of factors, it would be important to have a database with the average values of these properties for the local population, which would be of great utility to professionals in the medical field and orthopedic implant designers as well as specialist researchers in biomechanics. Results concerning the anisotropic-elastic behavior of bone tissue reported by [9] are similar to those obtained in this work. However, the reported studies have limited results because they merely focused on finding the mechanical properties of the cortical bone, disregarding its possible applications; and although it shows the variation of the properties through three different zones of cross-section of the cortical bone, the results are still limited to

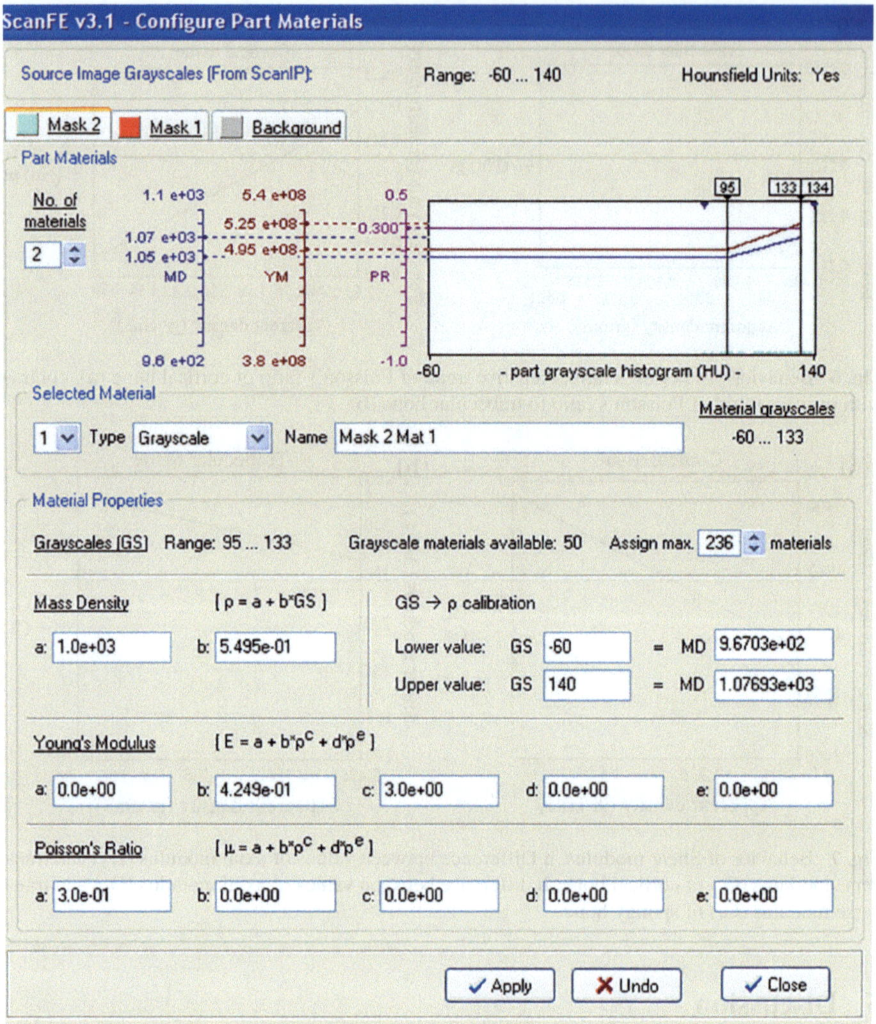

Fig. 8 Graphic interface for the development of anisotropic mesh

the reported values. Additionally, the method was only applied to a human femoral neck bone specimen and the studied zone was very small. The analysis performed by [7], aimed to find a specific parameter, which has the strongest influence on the bulk behavior of human bone tissue, but only one apparent elastic modulus was considered as the main mechanical property. In another recent study [11], the authors predicted analytically the effective elastic constants of cortical bone by modeling its elastic response at different scales, spanning from the nano-structural to meso-structural levels, using micromechanics methods and composite material laminate theories. Their work shows that the results obtained at a lower scale serve as input

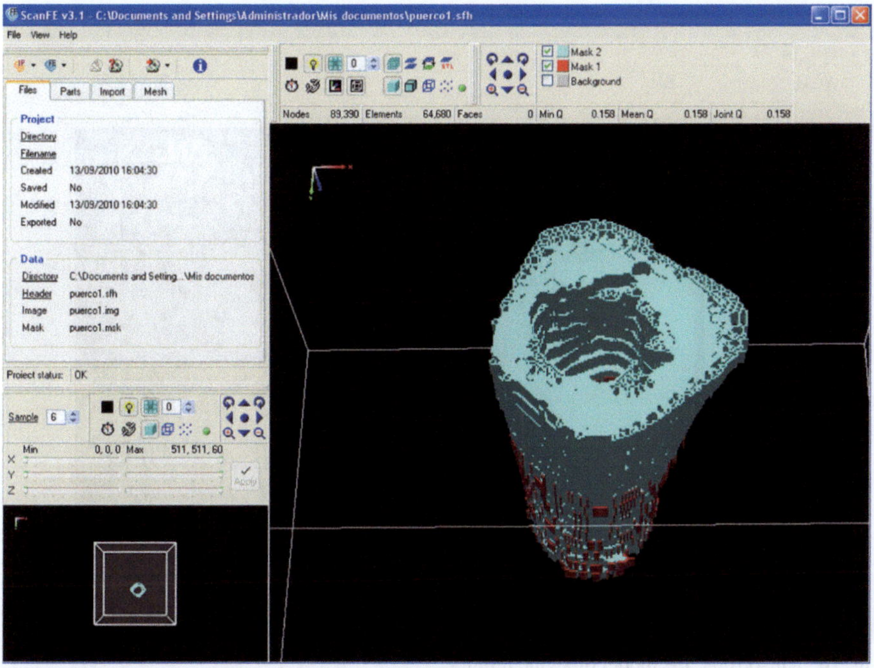

Fig. 9 Mesh of anisotropic elements for swine bone tissue

for the modeling at a higher scale and the predictions are in good agreement with the experimental data reported in the literature. However, the analysis in this study covered mechanical properties extensively to involve six different elastic properties and its magnitudes point by point across the transverse-sectional cut and longitudinal-sectional cut. Furthermore, a complete FE-model to be analyzed structurally was produced. On the other hand, a great amount of voxels (Fig. 10) were observed; this represents the anisotropic elastic properties of the bone tissue and the significant variation in the elastic properties is shown by their values, which go from 5.4 to 13.8 GPa in the cortical bone, while the cancellous bone displayed a gradient of variation going from 320 to 660 MPa. This is due to the diversity of factors that modify the properties of the bone tissue such as the weight, nutrition and age of the specimen, which should be closely, looked wen elaborating anisotropic numerical models. This will improve numerical models rendering them more genuine when compared to the behavior of the bone tissue.

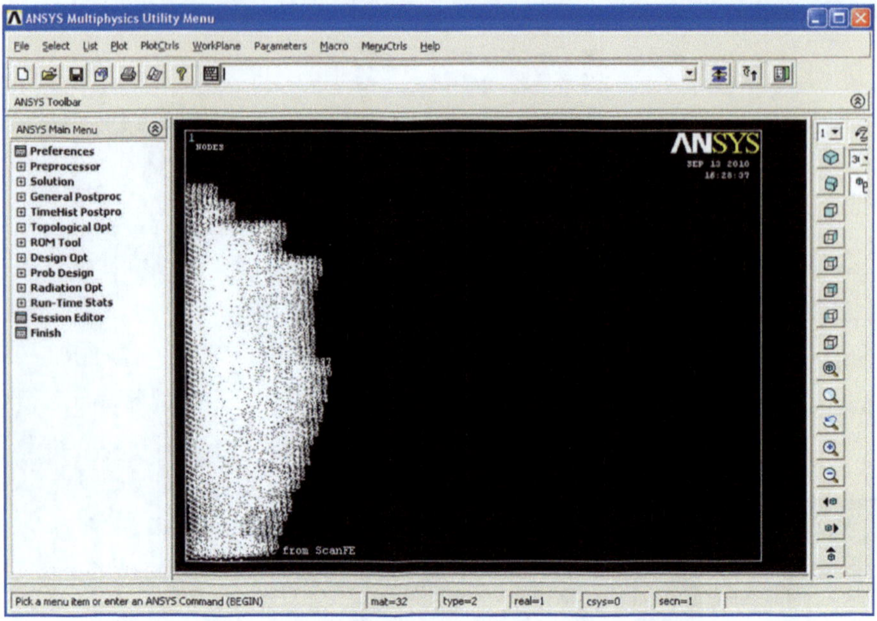

Fig. 10 Anisotropic FE model of swine femur

6 Conclusions

From this work important conclusions have been derived, such as:

1. The most important aspect of this study is that a customized model of bone tissue can be created for each patient, which will help produce an optimized design for required implants.
2. The novel technique proposed in the present work allows producing anisotropic FE-models of the bone, in a rapid and precise manner, to be used in the design of required implants.
3. By utilizing the apparent density, the analysis performed in the present study leads to the determination of the bone quality of the patient. Furthermore, combining this data with the geometry of the bone, elastic behavioral patterns can be established.
4. Knowing the specific properties of the bone for each point could help determine the most suitable structural-contact conditions for implants, which would possibly improve the device service life.
5. It seems possible to reduce the failure of implants if designs are improved by determining the bone tissue characteristics.
6. Subsequently, investigations will be able to produce experimental results on the design of prostheses based on customized anisotropic FE-models for the Mexican phenotype.

7. The development of an anisotropic FE-model of the proximal human femur is possible based on the current knowledge of the bone materials and their properties. However, the validation of these models through experimental analysis has proven difficult because of the present legislation in the matter of human cadaveric organs. This can be resolved with the use of animal bones, like the swine femoral bone tissue, which is very similar to the human bone.

Acknowledgements Authors thank the support provided by the Instituto Politécnico Nacional, Consejo Nacional de Ciencia y Tecnología (CONACyT) and Hospital 1° de Octubre from the ISSSTE for the accomplishment of this research.

References

1. Bessho, M., Ohnishi, I., Matsuyama, J., Matsumoto, T., Imai, K.: Prediction of strength and strain of the proximal femur by a CT-based finite element method. J. Biomech. **40**, 1745–1753 (2007)
2. Herrera, A.: Densitometric and finite-element analysis of bone remodeling further to implantation of an uncemented anatomical femoral stem. Revista Ortopedica Traumatologia **52**, 269–282 (2008)
3. Hofer, M.: Manual Práctico de TC. Introducción a la TC. Ed. Médica Panamericana, Buenos Aires, Arg (2008)
4. Weaver, J., Chalmers, J.: Cancellous bone: its strength and changes with aging and evaluation of some methods for measuring its mineral content. J. Biomech. **8**, 363–367 (1966)
5. Carter, D., Hayes, C.: The compression behavior of bone as a two-phase of porous structure. J. Bone Joint Surg. **59a**, 954–962 (1977)
6. Hvid, I.: Trabecular bone strength patterns at the proximal tibial epiphisy. J. Orthop. **3**, 462–472 (1985)
7. Levasseur, A., Ploeg, H.L., Petit, Y.: Comparison of the influences of structural characteristics on bulk mechanical behavior: experimental study using a bone surrogate. Med. Biol. Eng. Comput. **50**, 61–67 (2012)
8. Rajapakse, Ch.S., Magland, J.F., Wald, M.J., Liu, X.S., Zhang, X.H., Guo,X.E.: Computational biomechanics of the distal tibia from high-resolution MR and micro-CT images. Bone **47**, 556–563 (2010)
9. Sansalone, V., Naili, S., Bousson, V., Bergot, C., Peyrin, F., Zarka, J.: Determination of heterogeneous anisotropic elastic properties of human femoral bone: from nano-scopic to organ scale. J. Biomech. **43**, 1857–1863 (2010)
10. Mc Donnell, P., Harrison, N., Lohfeld, S., Kennedy, O., Zhang, Y., Mc Hugh, P.E.: Investigation of the mechanical interaction of the trabecular core with an external shell using rapid prototype and finite element models. J. Mech. Behav. Biomed. Mater. **3**, 63–76 (2010)
11. Hamed, E., Lee, Y., Jasiuk, I.: Multiscale modeling of elastic properties of cortical bone. Acta Mech. **213**, 131–154 (2010)
12. Hambli, R.: Numerical procedure for multiscale bone adaptation prediction based on neural networks and finite element simulation. Finite Elem. Anal. Des. **47**, 835–842 (2011)
13. Wirtz, D., Schiffers, N., Pandorf, T., Radermacher, K., Weichert, D., Forst, R.: Critical evaluation of known bone material properties to realize anisotropic FE simulation of the proximal femur. J. Biomech. **33**, 1325–1330 (2000)
14. Lee, Y., Seon, J., Shin, V., Kim, G., Jeon, M.: Anatomical evaluation of CT-MRI combined femoral model. Biomed. Eng. Online **7**, 6 (2008)
15. Torres, E.: Desarrollo de un maniquí virtual tipo vóxel a partir de imágenes en formato DICOM. Superficies y Vacío **23**, 90–93 (2010)

16. Mauren, A., Tania, M., Helio, P.: Integrando reconstrucción 3D de imágenes tomográficas y prototipado rápido para la fabricación de modelos médicos. Revista Brasileira de Engenharia Biomédica **19**, 103–115 (2003)
17. Beltrán, J., Hernández, L., Urriolagoitia-Calderón, G., González, A., Urriolagoitia-Sosa, G.: Biomechanics and numerical evaluation of cervical porcine models considering compressive loads using 2-D classic computer tomography CT, 3-D scanner and 3-D computed tomography. Appl. Mech. Mater. **24**, 287–295 (2010)
18. Hernández, L., Beltrán, J., Urriolagoitia-Calderón, G., González, A., Galán, M., Urriolagoitia-Sosa, G.: Biomechanical characterization of a cervical corporectomy using porcine specimens, following an experimental approach. Key Eng. Mater. **478**, 103–111 (2011)
19. Taylor, R., Roland, E., Ploeg, H., Hertixg, D., Klabunde, R., Warner, M.: Determination of orthotropic bone elastic constants using FEA and modal analysis. J. Biomech. **35**(6), 767–773 (2002)
20. Buroni, F.: Determinación de las constantes elásticas anisótropas del tejido óseo utilizando tomografías computadas. Aplicación a la construcción de modelos de elementos finitos. Mecánica Computacional XXIII, 1–24 (2004)

Fatigue Safety Design of Automotive Rear Sub-frame by Using CAE (Computer Aided Engineering) Static Analysis

Kee Joo Kim and Jun-Hyub Park

Abstract Reducing vehicle weight and developing alternative materials, like aluminum and magnesium alloys, are current topics of interest in the automobile industries. Therefore, in the present study, the structure analysis of a rear sub-frame made by aluminum alloys (AA6061), with a tensile strength of 310 MPa, was performed using computer aided engineering (CAE) to investigate the lightweight design process. The reference material for the rear sub-frame was the commonly used SAPH440 steel. The maximum von Mises stresses obtained from the simulated strength analysis were normalized as the fatigue limit, and converted to the weight ratios of the same type as the fatigue safety factor suggested and named like that in present study. From these, it was suggested that the fatigue properties could be predicted from this static CAE simulation. A weight reduction ratio of 25% was achieved between the SAPH440 and AA6061 model. It was a useful method to satisfy the fatigue safety design by using a simple static analysis and weight factors.

Keywords Weight reduction · Sub-frame · CAE · Static analysis · Fatigue

1 Introduction

Reducing vehicle weight and improving fuel efficiency are important issues around the world for the automobile industries [1]. In order to reduce the amount of carbon dioxide produced by automobile exhaust gas, vehicle companies around the world have been investigating technologies to increase fuel efficiency, such as the production of lightweight automobiles and reducing engine friction loss, among others [1, 2]. Designing a vehicle with both a lightweight frame and possessing high stiffness, is a pressing concern in vehicle design, because it is quite difficult to achieve both concurrently. In order to improve fuel efficiency, improvements in engine tech-

K. J. Kim (✉) · J.-H. Park
Department of Mechatronics Engineering, Tongmyong University, Busan 48520, Korea
e-mail: kjkim@tu.ac.kr

J.-H. Park
e-mail: jhpark@tu.ac.kr

© Springer International Publishing AG, part of Springer Nature 2019
A. Öchsner and H. Altenbach (eds.), *Engineering Design Applications*,
Advanced Structured Materials 92, https://doi.org/10.1007/978-3-319-79005-3_14

nology are required, and therefore are a major focus for automobile companies. Fuel efficiency improving technologies can be divided into three parts: the efficiency improvement of the engine and driving system, energy loss reduction and automotive body weight reduction. Because there is a limit to how much fuel efficiency can be improved while preserving the vehicle's efficiency, it more effective to design lightweight vehicles.

The automotive weight reduction research field can be classified into two fields. The first one focuses on improving the unnecessary parts in structural components made by steels while second one focuses on using the light weight materials such as aluminum alloys [1–5]. The former focuses on achieving the weight reduction by using the steel materials which are low cost and recyclable, while the latter is mainly used for luxury passenger vehicle due to the very expensive materials such as aluminum alloys, among others [5–8].

A representative recent technology that is being developed in order to achieve the goals of the vehicle body weight reduction is the usage of ultra lightweight steel bodies or vehicle bodies using lightweight materials such as aluminum alloys or composite materials. In particular, significant increase in the use of aluminum alloys, it has become increasingly commercialized around the casting materials such as cylinder block, cylinder head, piston, links, aluminum wheels, radiator, bumper body housing, etc. In addition, the entire vehicle body made by aluminum alloys has been developed [2, 4–6].

Therefore, in the present study, the structural analysis of the rear sub-frame made of AA6061 aluminum alloys, with a tensile strength of 310 MPa was performed using computer aided engineering (CAE) to investigate the lightweight design process from the reference of the rear sub-frame which was made by the SAPH440 steel materials and was already proven in the commercial market. Especially, the simulated maximum von Mises stresses after strength analysis were normalized as fatigue limit and these were converted to the weight ratios of the same type as the fatigue safety factor suggested and named first in present study. From these, it was suggested that the fatigue properties could be simply predicted only from this static CAE simulation [5, 6].

2 Analysis Method

A rear sub-frame is placed in the rear and under the space of the passenger car as a connection between the rear wheel and the body. Rear wheels are connected to this part by links or arms to ensure a comfortable suspension. The rear sub-frame is composed of front and rear cross members and two side members and 6 kinds of brackets. Figure 1 shows the shape of the rear sub-frame including 2 side members, front, and rear cross members. In the present study, the lightweight design process was conducted from the material change of the rear sub-frame substituted from steel materials to AA6061 aluminum alloys.

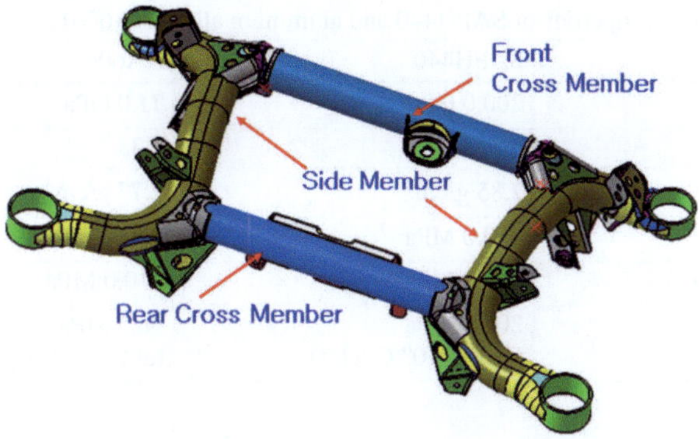

Fig. 1 Geometry of rear sub-frame

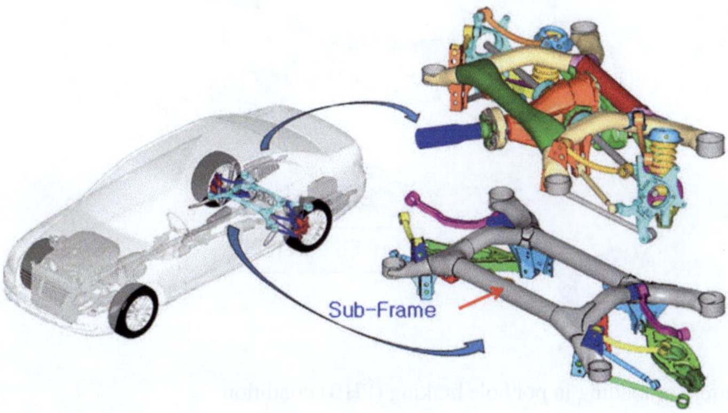

Fig. 2 Rear sub-frame module with assembled automotive components

Figure 2 shows the automotive component parts that are mounted to the rear sub-frame assembled as a module type in the vehicle. As shown in Fig. 2, it plays a role of attaching various kinds of links and axles.

The simulation performed in the present study used the commercially Nastran package (using the 101 static analysis module). Table 1 shows the mechanical properties of SAPH440 steel materials and 6061 aluminum alloys.

Figure 3 shows example boundary conditions simulated in this study. From this example, the PHB (pot hole breaking) loading conditions was represented as shown in Fig. 3 [3]. In the case of the PHB loading condition, when the vehicle is driving on the road of the pot hole portion and give it the breaking action, it represent the loading conditions imparted to the vehicle. At this time, the x-direction, y-direction and z-direction represent the length, width and height direction of the vehicle, respectively. Here, in the case of the rear sub-frame, 1 g denotes half the weight of the vehicle. In addition, RH and LH show where the loads were applied on the vehicle's right hand and left hand side, respectively. The loads are imparted to the wheel center and

Table 1 Mechanical properties of SAPH440 and aluminum alloy (AA6061)

	SAPH440	AA6061
Young's modulus	200.0 GPa	71.0 GPa
Poisson's ratio	0.30	0.33
Density	7.85 g/cm^3	2.77 g/cm^3
Yield strength	150.0 MPa	276.0 MPa
Ultimate strength	290.0 MPa	310.0 MPa
Fatigue limit	208.8 MPa (at 1 × 10^6 Cycles)	96.5 MPa (at 5 × 10^8 Cycles)

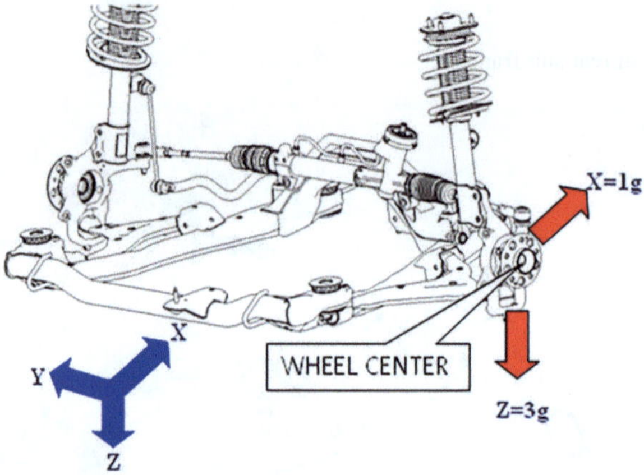

Fig. 3 Example of loading in pot hole braking (PHB) condition

the tire footprint is fixed as shown in Fig. 3. Loading conditions were obtained from the design conditions of the commercial simulation company of ETA Co., Ltd. [3]. Table 2 shows these loading conditions.

3 Analysis Results and Discussion

$$WF = \frac{\sigma}{\sigma_f} \tag{1}$$

From Eq. 1, σ is the von-Mises stress values from the strength analysis results and σ_f is the fatigue limit. The fatigue limit used in the present study is that of SAPH440 steel sheets with 2.3 mm thickness and of AA6061 aluminum sheets with 5.0 mm thickness are 208.8 MPa (at 1,000,000 cycles) and 96.5 MPa (at 5 × 10^8 cycles), respectively. By using Eq. 1, the optimum design criteria regarding fatigue properties can be determined. It is necessary to design the weight factor (WF) of the AA6061

Table 2 Loading conditions of suspension [3]

Load case	Wheel	X	Y	Z (g)
Pothole braking (PHB)	RH	1 g	–	3
	LH	1 g	–	3
Pothole ornering (PHC)	RH	–	1 g	3
	LH	–	1 g	3
Reverse braking (RB)	RH	−1 g	–	1
	LH	−1 g	–	1
Ultimate vertical (UV)	RH	–	–	4
	LH	–	–	4
Oblique curb strike (OCS)	RH	3 g	1.5 g	1
	LH	1 g	–	1
Lateral curb strike (LCS)	RH	–	2 g	1
	LH	–	1 g	1

should be necessary to be as close as possible to that of the SAPH440 steel sheets. For example, from the simulation results of the application of a unit load, the maximum stresses of the SAPH440 and AA6061 materials were calculated to be 100 MPa and 50 MPa, respectively, the WF (σ/σ_f) were normalized as 100/208.8 = 0.479 and 50/96.5 = 0.518 obtained from Eq. 1. Because the WF of AA6061 is larger than that of SAPH440, it should be reinforced and designed so that it does not exceed the WF of the SAPH440 in order to make up a safe sub-frame without failure.

Figure 4a–f show the strength analysis results of the rear sub-frame made by AA6061 and SAPH440 steel materials when given the PHB, PHC, RB, UV, OCS and LCS loading conditions. From these results, it was found that the area causing the maximum stress varied following the variation of the applied loading conditions. Rather, in case of the sub-frame made by 6061 aluminum alloys, it could be found that the maximum stress was reduced due to the material substitution.

Table 3a–e show the comparison of the results for maximum stresses and weight factors of the rear sub-frame made by AA6061 and SAPH440 steel materials when given the PHB, PHC, RB, UV, OCS and LCS loading conditions considering the weight factor as previously represented in Eq. 1. It was found that the weight factors considering fatigue dynamic properties of the AA6061 were not exceed to that of SAPH440 steel made in all loading conditions such as PHB, PHC, RB, UV, OCS and LCS. It could be verified that the sub-frame was designed safely because the weight factors of the sub-frame made by AA6061 were lower than that made by SAPH440. Here, the weight factors are the normalized values of the maximum stress divided by the fatigue limit. From these results, the fatigue dynamic properties could be verified and predicted in the initial development stage by using only simple static analysis without fatigue simulation based on weight factors comparison. This method was verified in the author's many different papers [2, 4–6]. Therefore, it was found that

Fig. 4 Results of maximum stresses and weight factors in various loading conditions **a** PHB, **b** PHC, **c** RB, **d** UV, **e** OCS and **f** LCS

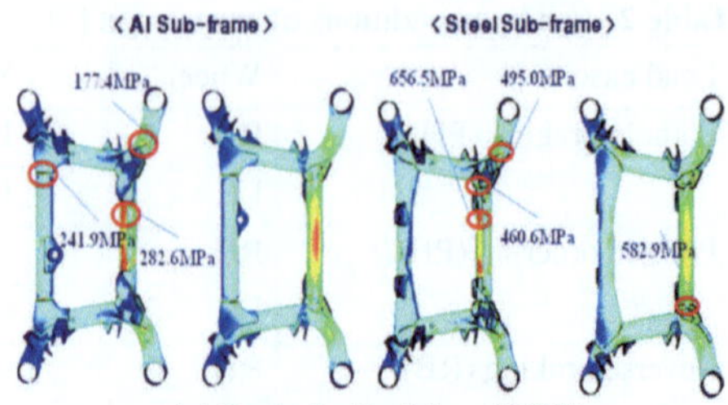

(a) Pothole Braking (PHB)

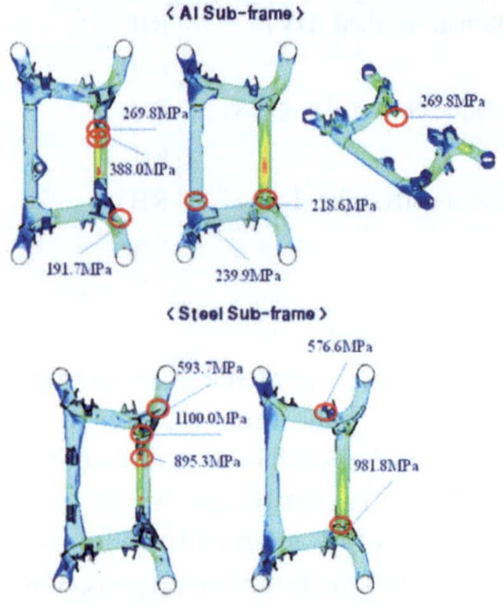

(b) Pothole Cornering (PHC)

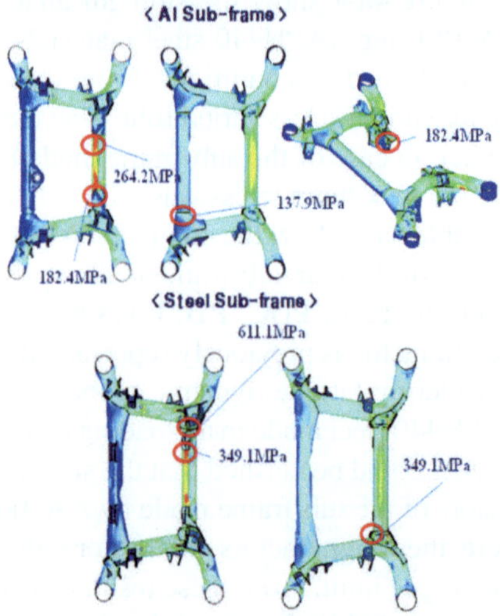

(c) Reverse Braking (RB)

Fig. 4 (continued)

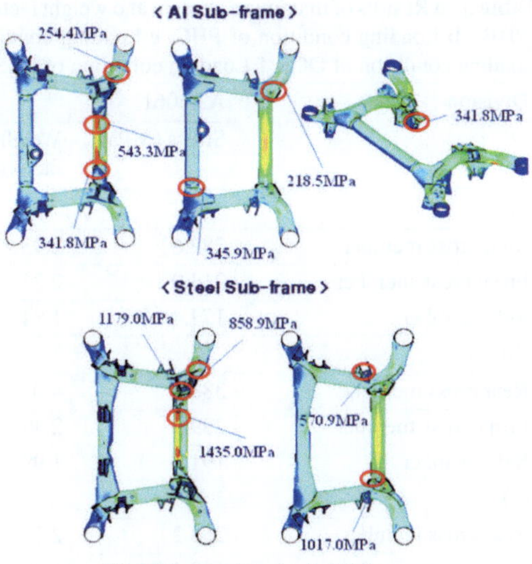

(d) Ultimate Vertical (UV)

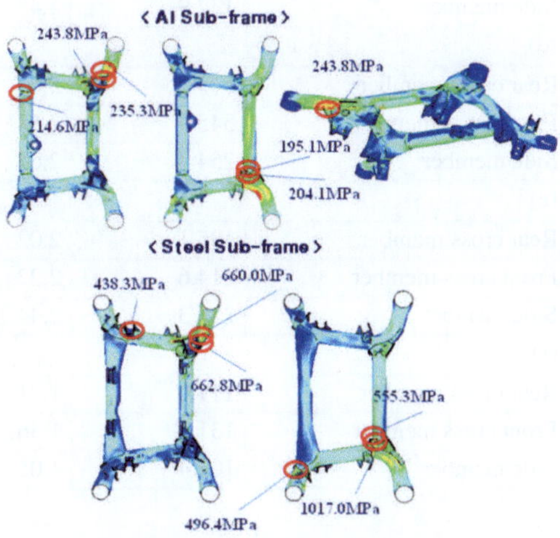

(e) Oblique Curb Strike (OCS)

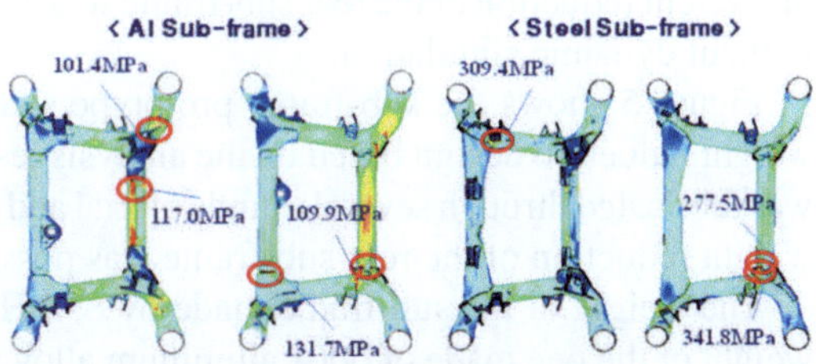

(f) Lateral Curb Strike (LCS)

Table 3 **a** Results of maximum stresses and weight factors in loading conditions of pot hole braking (PHB). **b** Loading condition of PHC. **c** Loading condition of RB. **d** Loading condition of UV. **e** Loading condition of OCS. **f** Loading condition of LCS

Division	AA6061		Steels (SAPH440)	
	Stress (MPa)	Weight factor(σ/σ_f)	Stress (MPa)	Weight factor(σ/σ_f)
(a)				
Rear cross member	282.6	2.93	656.5	3.14
Front cross member	214.9	2.23	582.9	2.79
Side member	177.4	1.84	495.0	2.37
(b)				
Rear cross member	388.0	4.02	1100.0	5.27
Front cross member	239.9	2.49	895.3	4.29
Side member	191.7	1.99	576.6	2.76
(c)				
Rear cross member	264.2	2.74	611.1	2.93
Front cross member	182.4	1.89	551.1	2.64
Side member	137.9	1.43	349.1	1.67
(d)				
Rear cross member	543.3	5.63	1435.0	6.87
Front cross member	345.9	3.58	1179.0	5.65
Side member	254.4	2.64	858.9	4.11
(e)				
Rear cross member	195.1	2.02	496.4	2.38
Front cross member	214.6	2.22	660.	3.16
Side member	235.3	2.44	662.8	3.14
(f)				
Rear cross member	117.0	1.21	309.4	1.48
Front cross member	131.7	1.36	341.8	1.64
Side member	101.4	1.05	236.7	1.13

the weight reduction of the rear sub-frame was possible through simple static analysis without dynamic simulation.

Figure 5 shows the sub-frame prototype shape made by aluminum alloys after weight reduction design based on the analysis results in present study. The prototype was fabricated through several rounds of trial and error and through this the successful weight reduction of the rear sub-frame was possible.

The weight of the sub-frame made by SAPH440 was 19.5 kg, however, and the weight of the one made of 6061 aluminum alloy which was studied and designed for weight reduction through current analysis was 14.7 kg. In addition, the maximum stress was reduced in case of the weight reduction model. The weight reduction ratio of 25% was achieved when comparing the SAPH440 and AA6061 model. It was a

Fig. 5 Prototype of automotive rear sub-frame made by aluminum alloy (AA6061) for light-weight design

useful method to satisfy the fatigue safe design by using simple static analysis and weight factors.

4 Conclusions

The rear sub-frame made by AA6061 aluminum alloys, with a tensile strength of 310 MPa was analyzed using CAE to investigate the lightweight design process from the reference of the rear sub-frame which was made by the SAPH440 steel materials and was already proven in the commercial market [9]. Especially, the simulated maximum von Mises stresses after strength analysis were normalized as fatigue limit and these were converted to the weight functions of the same type as the fatigue safety factor suggested and named first in the present study [9]. From these, it was suggested that the fatigue properties could be simply predicted only from this static CAE simulation. It was found that the weight of the sub-frame made by 6061 aluminum alloy and steel (SAPH440) was 14.7 and 19.5 kg, respectively. Therefore, the weight reduction ratio of 25% was achieved and in case of the maximum stress comparison, the sub-frame made by 6061 aluminum alloy was showed the excellent results better than steel made. This paper presented the dynamic characteristics optimization through the static CAE simulation by suggesting weight factor. From this, it was found to be quite simple and useful way in order to design a lightweight components considering dynamic fatigue characteristics through static analysis.

Acknowledgements This work (Grants No. C0395108) was partially supported by Business for Cooperative R&D between Industry, Academy, and Research Institute funded Korea Small and Medium Business Administration in 2016.

References

1. Kim, K.J., Won, S.T.: Int. J. Automot. Technol. **9**(6), 713–717 (2008)
2. Kim, K.J., Lim, J.H., Park, J.H., Choi, B.-I., et al.: Transaction of KSAE **20**(3), 77–82 (2012)
3. Suspension analysis load input formulas, ETA analysis input report, France, pp. 1–14 (1990)
4. Park, J.H., Kim, K.J., Yoon, J.G.: Trans. KSAE **21**(5), 157–161 (2013)
5. Kim, K.J.: Trans. KSMT **18**(5), 694–699 (2016)
6. Kim, K.J.: Trans. KSAE **41**(6), 564–566 (2017)
7. Kim, J.W., Yu, S.N.: Int. J. Precis. Eng. Manuf. 14(2), 267–273 (2013)
8. Rais-Rohani, M., Solanki, K., Eamon, C.: 11th AIAA/ISSMO Multidisciplinary Analysis and Optimization Conference. Portsmouth, Virginia (2006)
9. Kim, K.J.: MATERIALWISSENSCHAFT UND WERKSTOFFTECHNIK (Materials science and Engineering Technology, to be published) (2018)

Magnetic Field Distribution Around Magnetized Steel Ropes and Its Modulation by Rope Defects

Jaromír Pištora, Michal Lesňák, Jan Valíček, Marta Harničárová and Vladimír Vrabko

Abstract The paper is devoted to the modelling of the magnetic field around magnetized steel rope kinds using the ANSYS software package. The influence of wire cracks on the amplitude distribution of the generated field is specified for two steel rope kinds assuming surface and inner defects. The first rope is characterized by six rope strands where each is completed by 28 wires (two different diameters, 11 thick wires and 15 thin ones), while the second one has a Z-geometry with special cross section of wires located on rope surface. The results show a significant influence of surface defects (more 10 mT) and measurable magnetic flux responses (in the frame some mT) by cracks in the inner wires with 1 mm gaps. The limits for nondestructive testing of demonstrated ropes are described in detail.

Keywords Magnetic field · Steel ropes · Magnetic diagnostics

J. Pištora (✉)
Nanotechnology Centre, VŠB—Technical University of Ostrava, 17. listopadu 15, 708 33 Ostrava—Poruba, Czech Republic
e-mail: jaromir.pistora@vsb.cz

M. Lesňák · J. Valíček · M. Harničárová
Department of Physics, VŠB—Technical University of Ostrava, 17. listopadu 15, 708 33 Ostrava—Poruba, Czech Republic
e-mail: michal.lesnak@vsb.cz

J. Valíček
e-mail: jan.valicek@vsb.cz

M. Harničárová
e-mail: marta.harnicarova@vsb.cz

V. Vrabko
Institute of Mining Engineering and Security, VŠB—Technical University of Ostrava, 17. listopadu 15, 708 33 Ostrava—Poruba, Czech Republic
e-mail: vladimir.vrabko.st@vsb.cz

© Springer International Publishing AG, part of Springer Nature 2019
A. Öchsner and H. Altenbach (eds.), *Engineering Design Applications*,
Advanced Structured Materials 92, https://doi.org/10.1007/978-3-319-79005-3_15

1 Introduction

Steel wire ropes are complex structures. Most wire ropes operate in demanding conditions and must resist crushing, bending fatigue, and abrasion. Consequences of their failures can often be catastrophic [1]. A steel wire rope consists of wires laid into a number of strands which are, in turn, laid around a centre core. They belong to the most critical items in the transport industry. Traffic safety is very important, so ropes are routinely controlled by a suitable type of defectoscope. It is well known that magnetic analysis appeared to offer the greatest promise as a nondestructive method for testing wire ropes. Obviously, this method can only be applied to ferromagnetic materials.

At the Institute of Physics, VŠB-TU Ostrava, we have developed a defectoscope for the inspection of steel ropes and tubes up to a diameter of 60 mm (Fig. 1).

Permanent magnets are used as a source of magnetic field. Hall detectors have been used to measure the magnetic field. Permanent magnets as a magnetic source are very stable and do not need electrical power. Hall probes for measuring magnetic fields have the advantage that it is possible to measure the rope in a static state (without the motion between the rope and defectoscope). The advantage of this

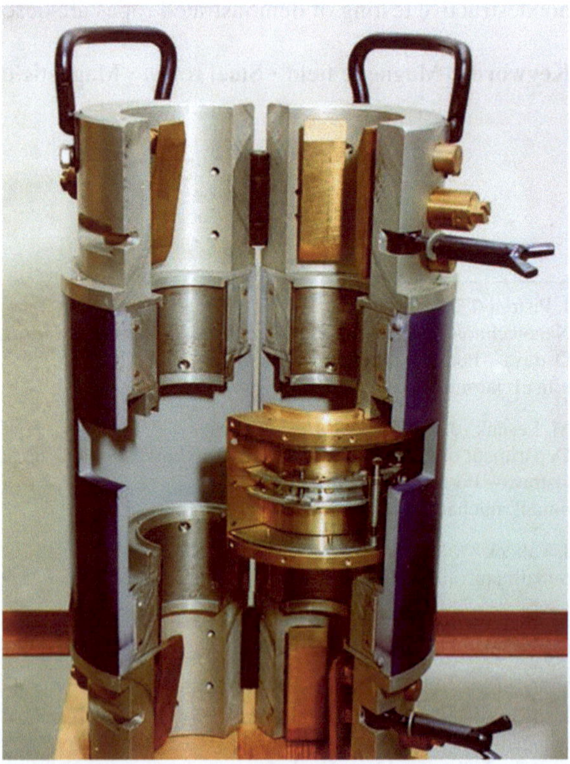

Fig. 1 Magnetic steel rope defectoscope

method is the ability to identify individual defects on the total length of the rope, which normally reaches 1000 m (up to 10,000 m). The method is very sensitive to external defects, while inner defects are more difficult to identify. For this reason, we try to model the responses of individual types of defects for a specific steel rope design [2]. This method allows us to detect not only the surface breakages (broken wires, imperfections, cracks), but also the internal defects. The physical principle of magnetic testing is based on the generation of a secondary magnetic field, which is created by a magnetized body [3]. The flux lines of magnetized object do not all cross the defect area directly [4]. It means generally that the longitudinal and radial components of magnetic field are specified. The solution of the field problem of the magnetized bodies is of great importance in several areas of science especially because of the increasing use of permanent magnets and magnetic circuits [5, 6].

2 Units

The model of the setup can be formulated using a scalar magnetic potential defined as

$$\vec{H} = -\nabla \Phi \tag{1}$$

Two types of the sub-regions can be distinguished: the source-free medium (air) and the magnetic material (steel rope wires). Equation 1 constitutes the formulation governing the scalar magnetic potential Φ solution. Both types of the mentioned sub-regions are considered without free currents, and the result is that the tangential components of the magnetic field intensity along their mutual boundaries are continuous.

The continuity of the normal components of the magnetic flux density leads to the condition expressed as:

$$\mu_0 (-\nabla \Phi_1) \cdot \vec{n} + \vec{n} \cdot \mu_0 \vec{M} - \mu_0 (-\nabla \Phi_2) \cdot \vec{n} = 0, \tag{2}$$

where $\vec{n}$ is the normal unit interface boundary vector. Equation (2) can be treated as governing the differential equation on the domain:

$$L\Phi = f. \tag{3}$$

and together with the appropriate boundary condition it defines a boundary-value problem for the scalar magnetic potential Φ. The solution to this problem can be obtained using the finite element method (FEM). Here, L is the differential operator and f is the so-called excitation or forcing function. The solution to the boundary-value problem can be formulated using a variation approach [6]. Instead of solving Eq. (3) directly, a functional minimization on the domain Ω is used. The functional can be written as:

$$F\left(\tilde{\Phi}\right) = \int_{\Omega} \nabla \cdot \mu_0 \nabla \tilde{\Phi} - \mu_0 \nabla \cdot \vec{M}\, d\Omega = 0, \tag{4}$$

where $\tilde{\Phi}$ is an approximate solution to (3). An approximation of the unknown solution in the e-th element can be expressed as:

$$\tilde{\Phi}^e = \sum_{j=1}^{n} N_j^e \Phi_j^e = \left\{N_j^e\right\}^T \left\{\Phi^e\right\} = \left\{\Phi^e\right\}^T \left\{N^e\right\}. \tag{5}$$

n denotes the number of nodes in the element, Φ_j^e is the value of the magnetic potential at the j-th node and N_j^e is the interpolation function for that node. When the approximate solution is inserted in (4) and the Galerkin's method is followed with the choice of the weighting functions being the same as the functions for the expansion of the approximate solution, the minimization procedure leads to the linear equation system:

$$\sum_{e=1}^{M} \left(\left[K^e\right]\left\{\Phi^e\right\} - \left\{b^e\right\}\right) = 0, \tag{6}$$

where the matrices K^e and b^e for each element are obtained using (4), and where $\{\Phi^e\}$ denotes the column vector of the scalar magnetic potential values for the nodes of the e-th element. The summation over e is related to the total numbers of elements in the domain. Once the system of the elemental equation is assembled, the values of the scalar magnetic potential in the domain can be computed.

The numerical modelling of the experimental array by FEM was carried out using the ANSYS software package. To secure stable results of the computation, the size and the number of the elements in the model was tested. It was found that in order to achieve stable results, models had to contain approximately 2,500,000 elements. Further increasing the element number does not substantially improve the stability increase, and moreover it leads to superfluous growth of computational time.

3 Theoretical Results

The numerical computations of magnetic fields generated by cracks of magnetized steel ropes have been realized on two geometrical constructions of steel ropes. The first one is depicted in Fig. 2, and is characterized by six rope strands where each is completed by 28 wires (two different diameters, 11 thick wires and 15 thin ones). The second rope is in what we have named a Z-geometry (see Fig. 3), with special cross section of wires located on the rope surface.

The fundamental question related to magnetic diagnostics (in general) is the limit responsibility for steel rope testing. This question is connected with the secondary

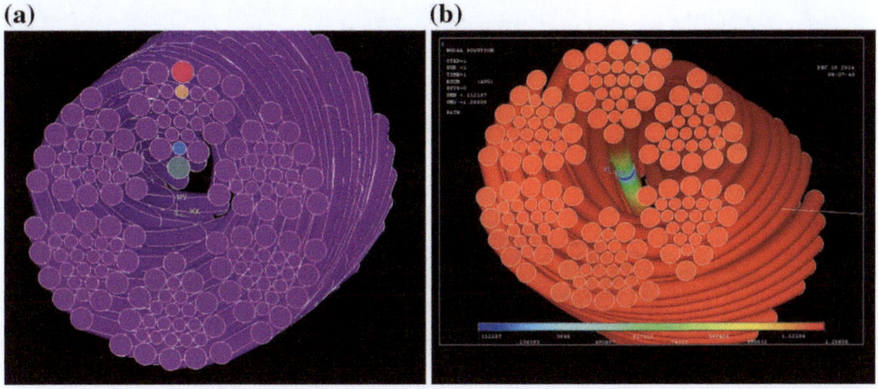

Fig. 2 Six strands rope cross section combining thick and thin wires. The wire defects positions are demonstrated by red, yellow, blue, and green colors (**a**). Figure 1 **b** specifies the geometry of discussed cracks

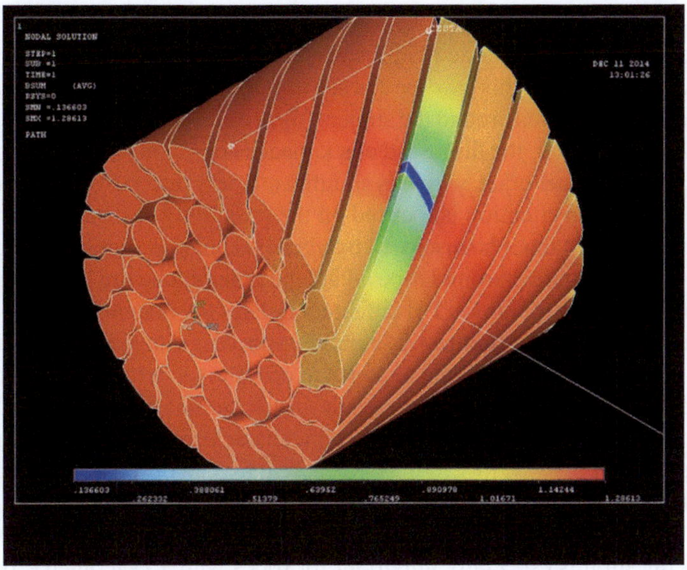

Fig. 3 Z-rope with special cross section of wires located on its surface. The positions of defects are specified by colored fronts of wires

magnetic field magnitude generated by defects of the magnetized rope. In the following we try to describe the magnetic field distribution around magnetized steel ropes with different positions of defects.

The first situation is a thick wire crack (1 mm gap) on the rope's periphery (see Fig. 2a). The corresponding 3D distribution of the secondary magnetic field is demonstrated in Fig. 4. The magnetic flux density peak for this case reaches up to 70 mT.

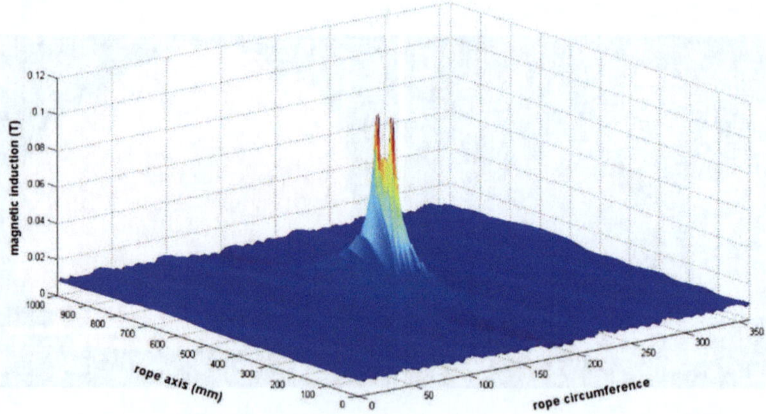

Fig. 4 3D distribution of secondary magnetic field for thick wire crack (1 mm) on the rope periphery (six strands rope)

Figure 5 describes the magnetic field distribution for thin crack wires on the rope's periphery. In the frame of this case we can see importantly can see an important lower peak of the field generated around the subject (the maximum is closely to 15 mT). For the case shown in Fig. 4, the discussed crack means approximately 1% cross section rope change, for thinner crack wire it is significantly less than 0.5%. For nondestructive testing, the main attention is focused on the possibility to determine inner steel ropes defects which cannot be disclosed by visual techniques. Figure 5 describes the magnetic field distribution around rope where the inner thick wire is interrupted by a 1 mm gap (see Fig. 2b). This kind of defect modulates the magnetic flux distribution at the rope's surface in the frame of 2–3 mT. For the case when the inner thin wire is broken (1 mm gap), magnetic field changes in comparison with the perfect rope are about 1 mT. For the magnetized Z-geometry rope, surface breakage (see Fig. 3) is evident, and the surroundings of the defect are characterized by important magnetic flux changes (see Fig. 6). Close to the defect area, we can observe more than 50 mT shifts in the magnetic field amplitude. More attractive for diagnostics is to solve the contribution of inner cracks (defects) to the magnetic field surrounding the tested rope. This question is especially important for ropes with a Z-geometry, because in this case the top wire practically encloses the rope's surface. The theoretical results declare the possibility to use theoretical results show the possibility of using theoretical results show the possibility of using nondestructive magnetic diagnostics to inspect Z-ropes with inner defects. For this case we observed magnetic flux changes more 5 mT (Figs. 7 and 8).

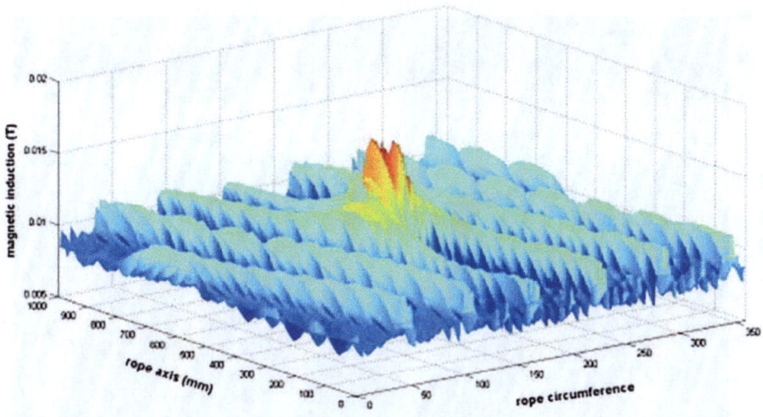

Fig. 5 3D distribution of the secondary magnetic field for thin wire crack (1 mm) on the rope periphery (six strands rope)

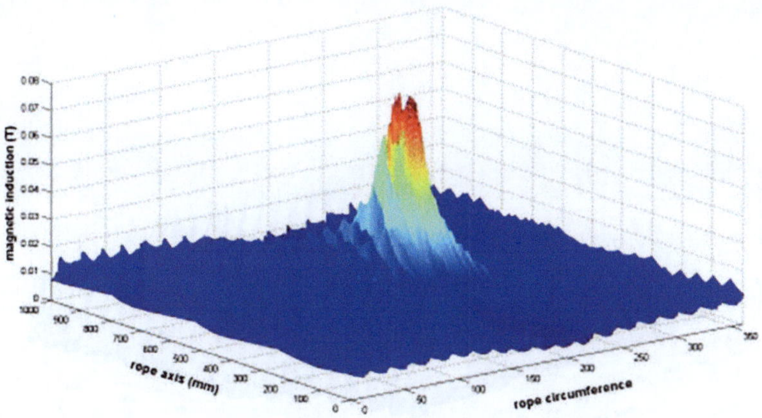

Fig. 6 3D distribution of the secondary magnetic field for surface wire crack (1 mm) on the rope periphery (Z-rope)

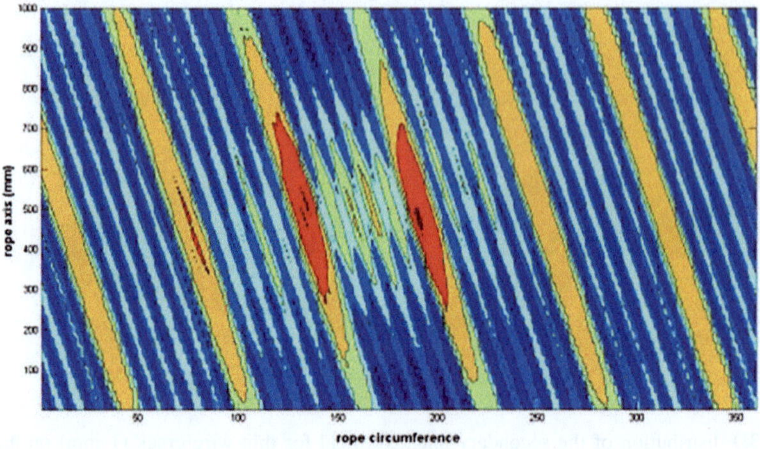

Fig. 7 The distribution of magnetic field around steel rope when inner thick wire is interrupted (six strands rope)

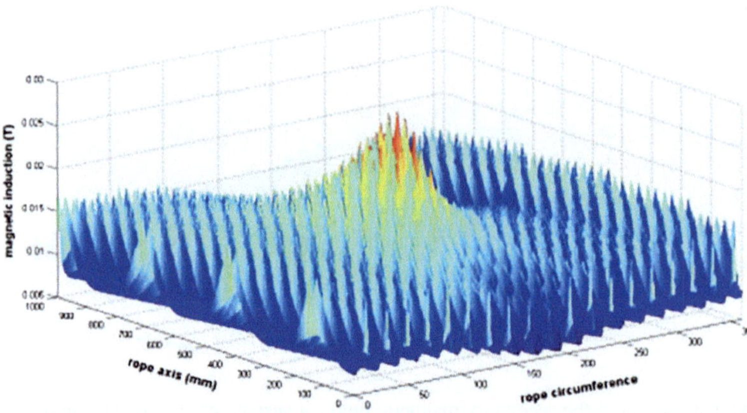

Fig. 8 3D distribution of the secondary magnetic field for inner wire crack (1 mm) on the rope periphery (Z-rope)

4 Conclusions

The achieved results have confirmed the possibility to apply yoke magnetic defectoscope [1] for surface and inner wire cracks diagnostics of ropes characterized by six strands geometry and z-one. The magnetic flux density generated by the magnetized rope with different cracks (surface and inner thick and thin wires) exceeds the defectoscope response limit.

Acknowledgements This work was partially supported by the Grant Agency of the Czech Republic (grant # 15-21547S) and by the Ministry of Education, Youth and Sports: by the National Program of Sustainability (NPU II) project "IT4Innovations excellence in science—LQ1602", and "Regional Materials Science and Technology Centre—Feasibility Program" (# LO1203) and the project SP2017/116.

References

1. Pištora, J., Lesňák, M., Vlček, J., Foukal, J.: Magnetic defectoscope with permanent magnets. J. Mag. Mag. Mat. **196–197**, 283–285 (1999)
2. Collini, L., Degasperi, F.: MRT detection of fretting fatigue cracks in a cableway locked coil rope. Case Stud. Nondestr. Test Eval. **2**, 64–70 (2014)
3. Xiao-yong, Z., Xiao-hong, Z.: Feature extraction and analysis of magnetic non-destructive testing for wire rope. In: Proceedings of Third International Conference on Digital Manufacturing and Automation, pp. 418–421 (2012)
4. Varga, E., Beyer, A.: Magnetic field of a uniformly magnetized hollow cylinder. IEEE Trans. Mag. **34**, 613–618 (1998)
5. Žežulka, V., Pištora, J., Lesňák, M., Straka, P., Ciprian, D., Foukal, J.: Intensity distribution of strong magnetic fields created by opposing linear Halbach assemblies of permanent magnets. J. Mag. Mag. Mat. **345**, 7–12 (2013)
6. Liu, J., Jin, J.M., Yung, E.K., Chen, R.S.: A fast, higher order three-dimensional finite-element analysis of microwave waveguide devices. Micr. Opti. Tech. Lett. **32**(5), 344–352 (2002)

Analysis of the Wear Damage on Offshore Gas Turbine Blades

Luz Yazmin Villagrán-Villegas, Luis Héctor Hernández-Gómez, Miguel Patiño-Ortiz, Juan Alfonso Beltrán-Fernández, Julian Patiño-Ortiz, Juan Rodrigo Laguna-Camacho, Maricela Cuellar-Orozco and Miguel Toledo-Velazquez

Abstract A study of the wear damage of a blade of the seventh-stage of the compressor of a gas turbine was carried out. The manufacturer recommends routine maintenance (30,000 h) to reduce the chances for the shutdown of the gas turbine. In this study, the C40 turbine was analyzed in the field at 24,000 h (shutdown) and out of service after 30,000 h. The turbine was operating in the city of Campeche, Mexico, in a very aggressive environment, where the entry of solid particles is unavoidable. The gas turbine combines high performance operation with rugged industrial construction. This design provides high efficiency, low maintenance and a long service life. For power generation, the components that are in direct contact with the air at different stages of operation are the compressor blades which are exposed to severe

L. Y. Villagrán-Villegas (✉) · J. R. Laguna-Camacho
Facultad de Ingeniería Mecánica y Eléctrica, Universidad Veracruzana, Prolongación Venustiano Carranza S/N, Col. Revolución, Poza Rica 93390, Veracruz, Mexico
e-mail: yvillagran@uv.mx

J. R. Laguna-Camacho
e-mail: jlaguna@uv.mx

L. H. Hernández-Gómez · M. Patiño-Ortiz · J. A. Beltrán-Fernández · J. Patiño-Ortiz · M. Cuellar-Orozco · M. Toledo-Velazquez
INSTITUTO POLITÉCNICO NACIONAL—Escuela Superior de Ingeniería Mecánica y Eléctrica, Sección de Estudios de Posgrado e Investigación Edificio 5. 2do Piso, Unidad Profesional Adolfo López Mateos "Zacatenco" Col. Lindavista, C.P, 07738 Ciudad de México, Mexico
e-mail: luishector56@hotmail.com

M. Patiño-Ortiz
e-mail: mpatino@ipn.mx

J. A. Beltrán-Fernández
e-mail: jbeltranf@hotmail.com

J. Patiño-Ortiz
e-mail: jpatinoo@ipn.mx

M. Cuellar-Orozco
e-mail: mcuellar@ipn.mx

M. Toledo-Velazquez
e-mail: mtoledo@ipn.mx

© Springer International Publishing AG, part of Springer Nature 2019
A. Öchsner and H. Altenbach (eds.), *Engineering Design Applications*,
Advanced Structured Materials 92, https://doi.org/10.1007/978-3-319-79005-3_16

wear damage for the impact of particles, environmental contaminants such as salts, sands and sulphur. Due to this fact, the wear damage on the turbine blades was analysed. The work was developed in two steps. In the first one, an analysis of boroscope images of the surfaces of the blades was carried out in the site, and out of service the study with destructive and non-destructive tests, using tribological characterization were conducted to obtain the chemical composition of the turbine blades, which were in operation, by using energy dispersive X-ray analysis (EDS). Hardness tests were also conducted, using the material that was employed to manufacture the blades. Additionally, optical microscopy and scanning electron microscopy (SEM) were used to identify the wear mechanisms on the surfaces. This allowed obtaining a more complete failure analysis. The wear modes were severe pitting action, large craters similar to those observed in solid particle erosion when the samples are impacted at normal incidence, corrosion and a few irregular scratches similar to ploughing action in abrasive wear.

Keywords Wear damage · Gas turbine blades · Destructive and non destructive testing · Boroscope

1 Introduction

In platforms, turbines operate in highly aggressive environments [1]. The selection of materials in their design plays a key role in the achievement of a greater efficiency, a safer operation and more cost effective operation. The most common faults in gas turbine blades are considered from a metallurgical and mechanical point of view. The lifespan of a blade is reduced due to operating environment and high mechanical and thermal stresses, where two or more factors typically act simultaneously [2]. Degradation is caused by corrosion, hot corrosion, oxidation, erosion, abrasion, particle melting and mechanical degradation. The most common degradation mechanisms are: blade contamination, pitting, clearing between rotor and stator and erosion of the leading and trailing edge of the blade of the hot and cold part of the gas turbine.

Research on turbine blades starts with visual observations, optical microscopy, scanning electron microscopy, fractography analysis, metallography, structural analysis and hardness tests.

Several research works dealing with the failure analysis of turbine blades have developed over the years, for instance, Tabakoff and Hammed [3]. In an experimental analysis with solid particles in an airflow passing through a nozzle, they established that the trajectories and velocities of the particles depend on the inclination along the tangential direction of the profiles of the blades, particle input conditions, particle material density, the angle of attack and the initial location of the collision. Azevedo et al. [4] conducted a failure analysis on steam turbine blades. They were manufactured in forged 12% Cr–NiMoV martensitic stainless steel. The blades belonged to the last stage of a thermoelectric plant steam turbine generator (140 MV A) in Brazil. The results indicated that the failure of the blades was caused by foreign-particle erosion, which attacked the low-pressure side of the lower trailing edge of the blades. On the other hand, Carter [5] carried out an analysis of the typical failures in the blades of the different stages of modern gas turbine engines for aviation applications. It is because the ingestion of solid material such as soft debris (twigs or small birds) and fine abrasive material, as sand, can cause damage by abrasive and erosive wear to blades throughout the engine. Another wear process was corrosion in both compressor and turbine blades, due to ingested air that contained sodium and chlorine, in the form of salt from sea air, from runway treatments or marine environments. Gallardo et al. [6] conducted a failure analysis of the three first-stage blades of a gas turbine that suffered severe degradation after 10,500 h of service. They concluded that the failure was due to a misfit between the rotor and the sectors of the lining, originating from defective mounting. This caused high friction of the blades with some zones of the turbine-case lining, located on the side of minimal clearance. In the C40 Centaur gas turbine, the most common degradation mechanisms are: airfoil (blade) fouling and pitting (loss of flow profile caused by corrosion and surface deposits); opening of blade clearances and leading and trailing edge erosion (caused by mechanical solid and liquid impacts on the rotor blades). Several mechanisms cause the degradation of compressor blades, for example: fouling, corrosion, erosion, abrasion, particle fusing and mechanical degradation [7].

This study proposes a methodology to carry out an analysis of the damage caused by the wear of axial compressor blades of a gas turbine. The damaged blade was located at the 7th stage weighing 22.37 g.

The work was developed in two stages. In the first one, a thorough visual inspection of the turbine blade was carried out. An analysis of boroscope images of the surfaces of the vanes was carried out on site, and microstructural tests were carried out on sample fragments of the blade. The samples of 1 cm^2 were hardness tested to determine the type of material used to manufacture them. Energy dispersive X-ray analysis (EDS) was applied in order to obtain the chemical composition of the material, and the roughness was measured to know the surface degradation of the same. Finally, Scanning Electron Microscopy (SEM) was used to identify the wear mechanisms in several samples. In the second stage of the investigation, EDS and Scanning Electron Microscopy (SEM) were applied to a second blade of the same stage. The analysis was performed on the trailing edge, leading edge and central part of the blade (Fig. 1).

In this study, a description of the wear mechanisms in the blade of an 7th stage axial compressor of the initial stage of a gas turbine Centaur 40 CS was presented using optical and SEM images. These mechanical components (blades) suffered significant damage in this particular section by the direct contact of the intake air mixed with foreign particles. In addition, EDS analysis was performed to compare the chemical changes that originated from the wear damage. The impact and the sliding actions of the mixed inlet particles of different shapes and hardness degraded the surface of the compressor blades. This led to a similar wear mechanism to those mentioned in previous research works such as fatigue erosion on the leading edge in the low and high-pressure sides and corrosion damage on the top part of the low-pressure side

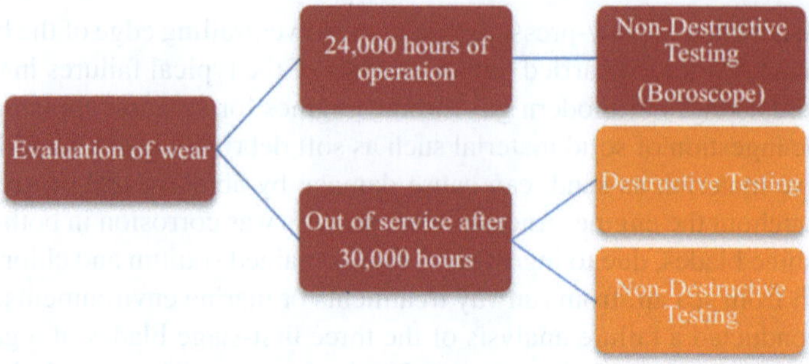

Fig. 1 Evaluation of wear

of the blades. The fatigue erosion was characterized by severe pitting action and the plastic deformation by displaced material, craters and scratches.

2 Visual Examination

Gas boroscope inspection is an internal inspection conducted by a trained specialist who assesses the condition of the gas turbine from the air inlet to the exhaust using an instrument specifically designed to examine the gas turbine. The turbine has four ports, where the boroscope can be introduced (Fig. 2). With a boroscope it is possible to see the first signs of oxidation, erosion, corrosion, cracking and damage results of foreign objects without disassembling the gas turbine.

The greatest benefits of inspection include a notification of equipment condition and effective scheduling of any necessary maintenance interventions. In addition,

Fig. 2 Evaluation of wear

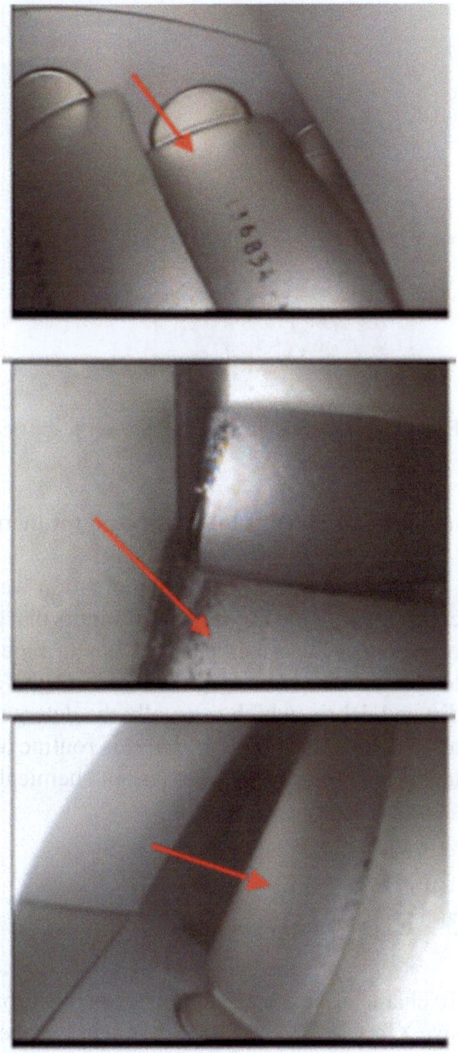

Fig. 3 Images of boroscope

greater reliability and a longer life of the equipment can be achieved, as well as the reduction in the potential of severe equipment failures. The inspector must be able to distinguish between conditions of components that are potentially catastrophic and those that are only insignificant abnormalities.

By design, the gas turbine works exposed to high temperatures and rotate at high speeds. Internal inspections are necessary to determine if thermal erosion or erosion is present. In addition, these inspections will take into account any damage caused by the impact of foreign objects (FOD), or if corrosion has occurred, and allow the

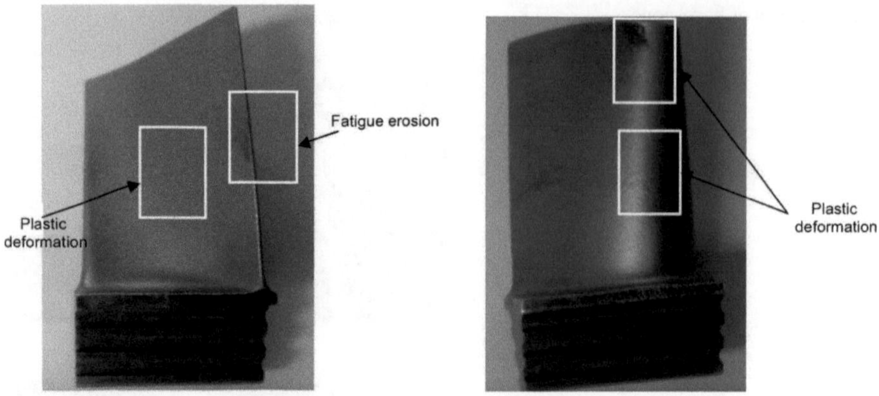

Fig. 4 Wear damage in axial compressor blade (leading and trailing edge)

internal components to be assessed for thermal deterioration, fractures or distortion (Fig. 3).

In the boroscopy test, contamination was observed in the first stage of the axial compressor. It is a mixture of oil vapors that can prevent variable blades from opening or closing, which can lead to a low turbine performance and, even serious damage to the turbines. The stage eleventh of the axial compressor has impregnated with dirt and debris, which is usually the detergent and water used for the washing of the axial compressor (Fig. 4), but this routine activity of maintenance is annual and the technical engineer only carries out chemical analyses of water to detect metals in the samples.

3 EDS Analysis

To characterize the wear damage on the turbine blades firstly was used energy dispersive X-ray (EDS) was used to obtain the chemical composition in an unworn section of the SEM images of the turbine blades using a backscattered electron detector (BED). The characterization was performed on a sample from the blade with 1 cm^2 and at the blade (destructive and non-destructives test). In the first EDS analysis presented in Fig. 4 in a sample of 1 cm^2, the unworn zone of the turbine blade Fig. 5 in the low-pressure side exhibited elements such as chromium (Cr), nickel (Ni) and molybdenum (Mo), with exception of Ni, these elements and their weight percentages are close of superalloy chromium base [8].

The worn regions (Fig. 5a, b) showed different elements from those observed in the unworn surface (chromium, iron, nickel and molybdenum). In zones a and b (Figs. 6 and 7), new chemical elements appeared in the EDS analysis such as carbon (C), oxygen (O), sodium (Na), calcium (Ca), chlorine (Cl) and aluminium (Al), which could be due to the ingested particles (contaminants) travelling in the air intake

(a)

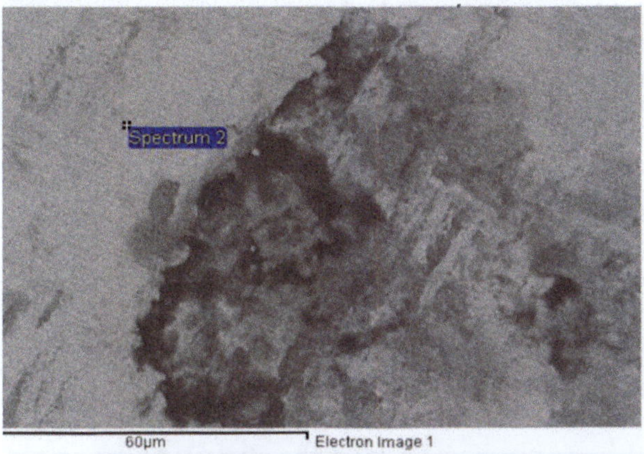

(b)

Element	Wt. %
Chromium (Cr)	11.39
Iron (Fe)	87.35
Nickel (Ni)	1.60
Molybdenum (Mo)	0.34
Totals	100.00

(c)

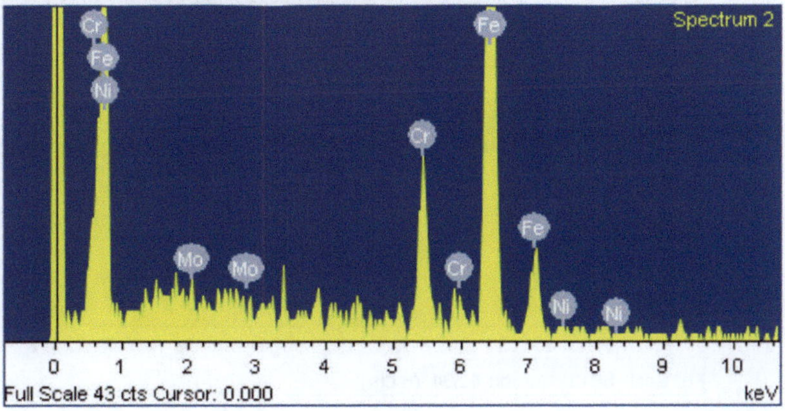

Fig. 5 Chemical analysis of a sample with 1 cm^2 **a** unworn zone **b** chemical elements **c** EDS graph wear damage in axial compressor blade (leading and trailing edge)

(accelerated oxidation damage) due to corrosive deposits. These could be NaCl or surface scales as chromium oxides on the surface (specifically in Fig. 6). Wear debris could also produce new elements. There is evidence of aluminium, chromium and iron oxides in this region [9, 10]. In respect to the contents of chromium (Cr), iron (Fe) and nickel (Ni), these decreased in all wear zones compared to the values obtained in the unworn region. In fact, the nickel element vanished in the EDS analysis of Fig. 4. using the Emission Scanning Electron Microscope (JEOL JSM-7600F).

In the second step the tests were carried out to the whole piece (blade) in the leading and trailing edge with the field emission show in Fig. 7. In the non-worn zone, 10 kV were applied and fourteen elements were found, such as chromium (Cr), nickel (Ni) and molybdenum (Mo) and niobium (Nb). The percentages varied when evaluating a zone without wear, as shown Fig. 9, but in this fourth analysis

(a)

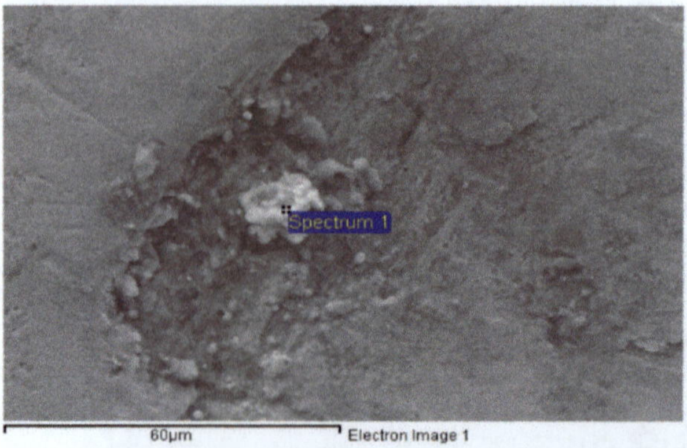

(b)

Element	Wt. %
Oxygen (O)	52.57
Sodium (Na)	2.22
Aluminium (Al)	2.72
Silicon (Si)	2.74
Chlorine (Cl)	2.83
Potassium (K)	0.87
Chromium (Cr)	5.60
Iron (Fe)	28.21
Molybdenum (Mo)	2.23
Totals	100.00

(c)

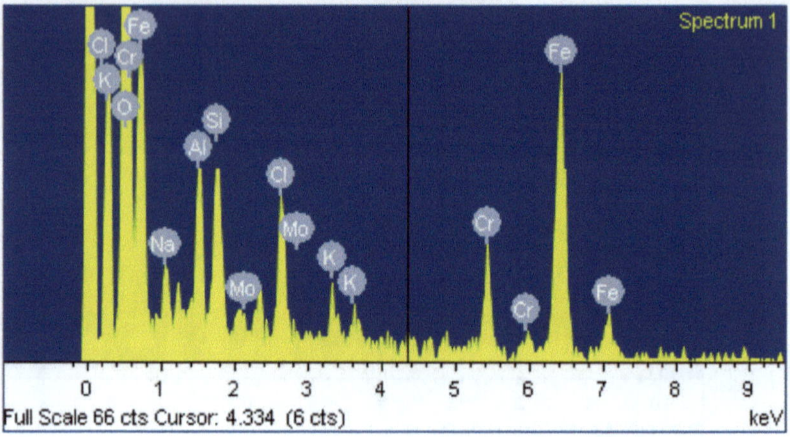

Fig. 6 Chemical analysis of a sample with 1 cm^2 **a** worn zone **b** chemical elements **c** EDS graph

it was possible to identify the type of superalloy with 14 chemical elements in its composition (Figs. 8 and 10).

4 Scanning Electron Microscopy Analysis

The SEM analysis in Fig. 11 was applied to three zones. In SEM images of the unworn (Fig. 12), it is possible to observe wear mechanisms quite similar to those observed in erosion and abrasion wear processes.

With respect to SEM images of zones 1 and 2, shown in Fig. 13 a–c, using two detectors, backscattered electron detector (BED) and namely the low electron detector (LED), respectively, it was possible to observe a ploughing action zone,

(a)

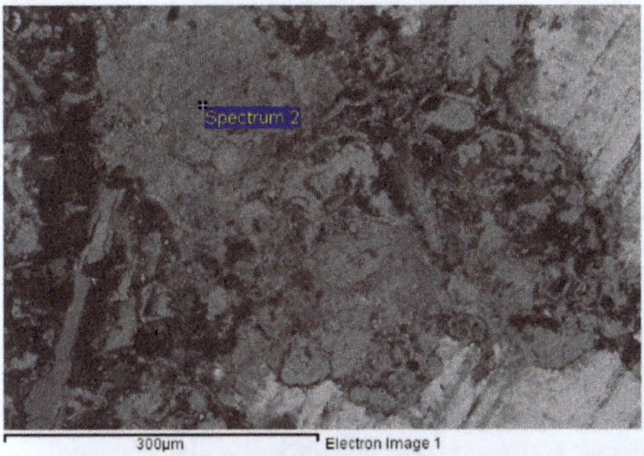

(b)

Element	Wt. %
Oxygen (O)	41.59
Sodium (Na)	0.48
Aluminium (Al)	0.18
Silicon (Si)	0.33
Chlorine (Cl)	4.00
Potassium (K)	0.17
Calcium (Ca)	0.45
Chromium (Cr)	2.61
Iron (Fe)	50.62
Molybdenum (Mo)	0.42
Totals	100.00

(c)

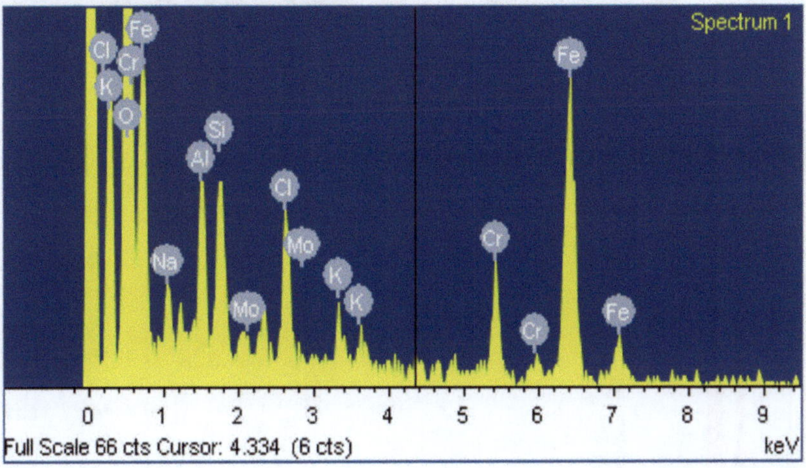

Fig. 7 Chemical analysis of a sample with 1 cm^2 **a** Oxidation in worn zone **b** chemical elements **c** EDS graph

where material was displaced and detached by the impact and sliding actions of foreign particles that were propelled by the air pressure. Here, a fragment (wear debris) was clearly observed on the middle section of the groove. The main wear mechanism was displaced material on the low-pressure side of the compressor blade, however, smeared material on the surface was also observed. The oxidation damage was confirmed with the presence of aluminium and chromium oxides.

Figure 14 shows optical images, which are characterized by large craters and grooves (ploughing action) of around 125–150 μm length with material piled-up (lips) at the sides and in the front of the cavities, which is common in solid particle erosion, as the specimens are impacted at oblique incident angles ($\alpha \leq 45°$) [11, 12]. Some grooves and craters have not piled up material because of the subsequent

Fig. 8 Chemical analysis non-destructive test

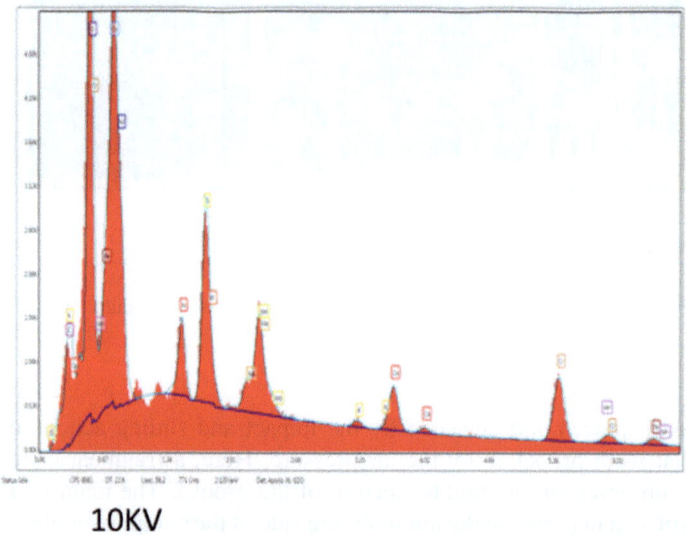

Element	Weight %
C K	7.76
O K	16.02
Fe L	5.23
Co L	21.96
Ni L	8.12
Al K	1.96
Si K	4.77
W M	1.36
Nb L	2.06
Mo L	6.1
K K	0.47
Ca K	4.01
Cr K	19.04
Mn K	1.14

Fig. 9 Chemical analysis in a leading edge (chemical elements and EDS graph)

particle impacts, which caused the detachment of the lips. In addition, a few irregular indentations were identified on the surfaces. In Fig. 15, where possible fatigue erosion damage was indicated, the SEM image confirmed that large craters were formed on the blade surfaces. In addition, irregular cavities, wear debris, foreign particles

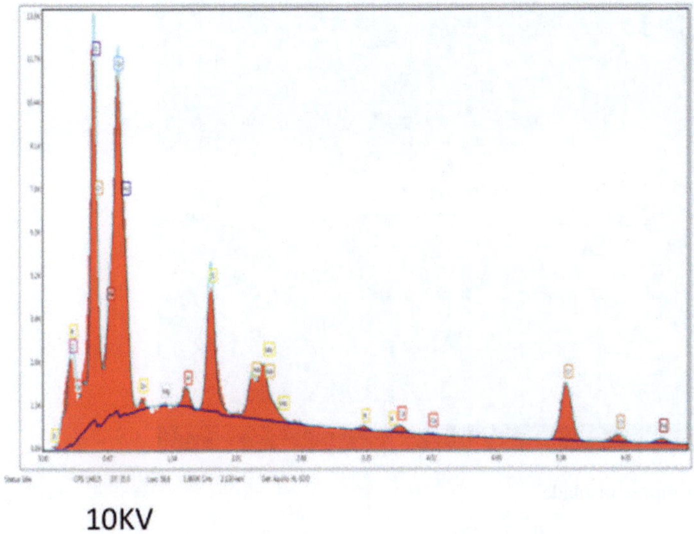

Element	Weight %
C K	10.04
O K	13.12
Fe L	5.15
Co L	24.12
Ni L	8.83
Zn L	0.59
Mg K	0.05
Al K	0.68
Si K	5.11
Nb L	3.95
Mo L	3.72
K K	0.41
Ca K	0.94
Cr K	23.3

Fig. 10 Chemical analysis in a trailing edge (chemical elements and EDS graph)

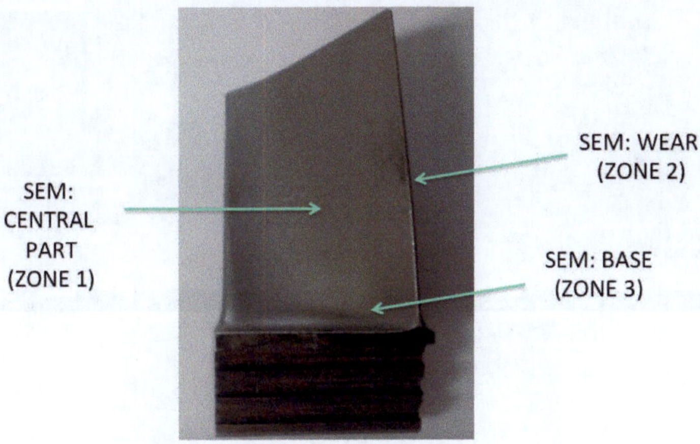

Fig. 11 Zones of SEM analysis

flattened on the surface were other wear mechanisms similar to those presented in erosion and abrasion processes. Due to the appearance of damage zone 3 there is evidence of pitting action [13, 14], it was more related to erosion damage due to the impact of any contaminant or inlet particles.

Fig. 12 Zones 1 of compressor blade

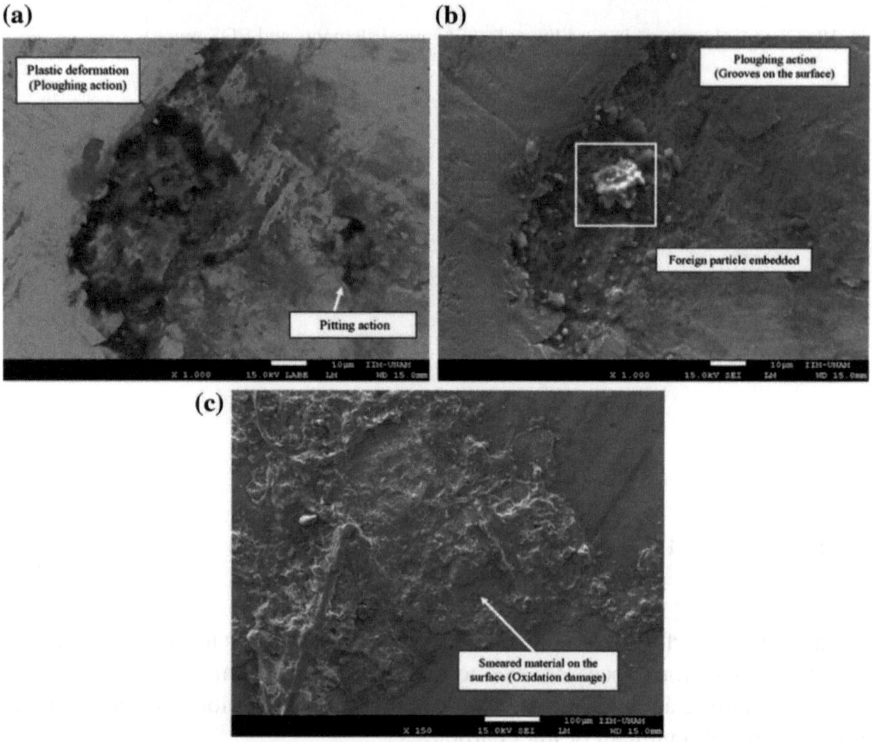

Fig. 13 Wear mechanisms (leading edge) of compressor blade

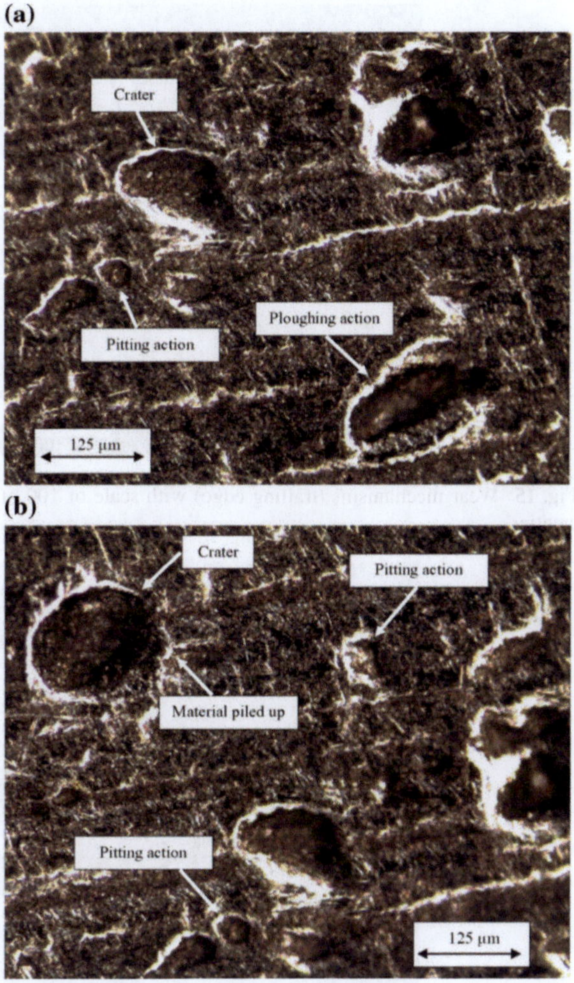

Fig. 14 Wear mechanisms (trailing edge) C with 125 michrometers and show types of damages wear

5 EDS Analysis

The hardness tests were carried out using a Microhardness Tester LECO LM 700 shown in Fig. 16, where the average value was 304 HV with a standard deviation of 34.29. Ten different points were selected to calculate the hardness value in the unworn region and the variation of the values is shown in Fig. 17. The applied load was 100 gf, with an indentation time of 15 s. This hardness value was approximated to the values presented in martensitic stainless steels [15, 16, 17], which correlates well with the EDS analysis where Cr, Mo, Nb and Fe elements appeared in the structure.

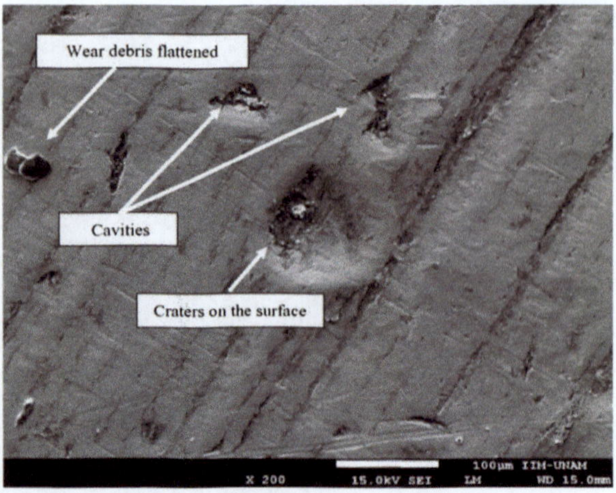

Fig. 15 Wear mechanisms (trailing edge) with scale of 100 michrometers and show craters and cavities

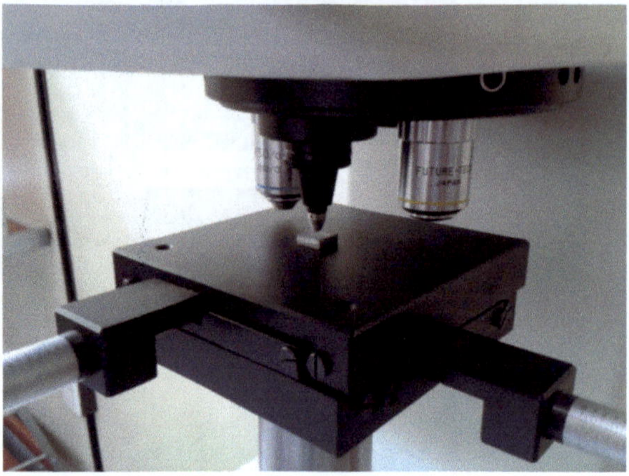

Fig. 16 Microhardness and the sample of 1 cm^2 (unworn zone)

6 Conclusions

Different types of analysis were carried out on the leading and trailing edges of a turbine blade.

The SEM analysis showed that the material was a chromium-based superalloy with chemical elements such as Co, Cr, Ni, Fe, C, Nb and Mo. The degradation of the surfaces took place at the zones of greater pressure. It was generated by air mixed with particles.

In the microstructural tests, wear mechanisms were characterized by large craters and grooves. The wear residues over the surfaces of the blade were identified.

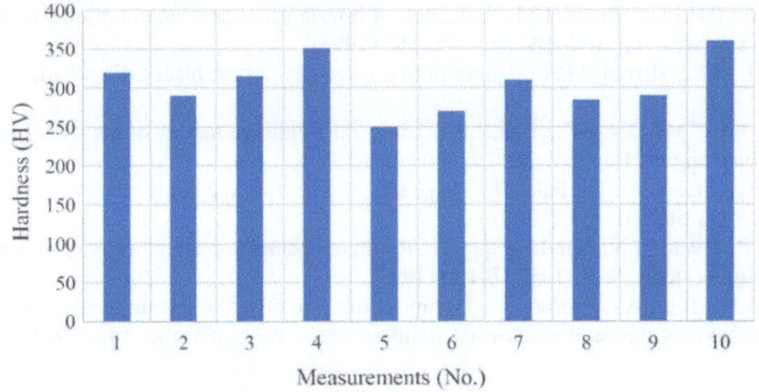

Fig. 17 Hardness values (unworn zone)

Mechanisms such as corrosion damage and irregular cavities were determined.

Foreign particles impacted on the surface of the blade. The parallel grooves showed the trajectory of the solid particle during its operation.

It was demonstrated that there are blades with 30,000 h of operation working in aggressive environments (oil rig) without cracks in the leading and trailing edge of the blade.

Acknowledgements The support given by the Universidad Veracruzana (PRODEP) and the authors kindly acknowledge the grant for the development of the Project 211704. It was awarded by the National Council of Science and Technology (CONACyT).

References

1. Tabakoff, H.S.: Blade deterioration in a gas turbine engine. G. Science Ed. Int. J. Rotating Mach. **4**, 233–241 (1998)
2. Hutchings, I.: Tribology (vol. 1, p. 76). In: Heinemann, B (ed.), Elsevier Science, London, UK (2003)
3. Tabakoff, W., Hamed, A.: Erosion and deposition in turbomachinery. J. Propul. Power **22**, 350–360 (2005)
4. Azevedo, C.R.F., Sinátora, A.: Erosion-fatigue of steam turbine blades. Eng. Fail. Anal. **16**, 2290–2303 (2009)
5. Carter, T.J.: Common failures in gas turbine blades. Eng. Fail. **12**, 237–247 (2005)
6. Gallardo, J.M., Rodríguez, J.A., Herrera, E.J.: Failure of gas turbine blades. Wear **252**, 247–264 (2002)
7. Kurz.: Gas Turbine Degradation. Turbomachinery Laboratory, US (2014)
8. Losertová, M.: Advancved materials (vol. 1). Ostrava: V.S.B. (ed.) (2014)
9. Kumari, S., Satyanarayana, D.V.V., Srinivas, M.: Failure analysis of gas turbine rotor blades. Eng. Fail. Anal. **45**, 234–244 (2014)
10. Eliaz, N., Shemesh, G., Latanision, R.M.: Hot corrosion in gas turbine components. Eng. Fail. **9**, 31–43 (2002)
11. Christman, T., Shewmon, P.G.: Erosion of a strong aluminium alloy. Wear **52**, 57–70 (1979)

12. Shipway, P.H., Hutchings, I.M.: Influence of nozzle roughness on conditions in a gas-blast erosion rig. Wear, pp. 162–164 and 148–158 (1993)
13. Khajavi, M.R., Shariat, M.H.: Failure of first stage gas turbine blades. Eng. Fail. **11**, 589–597 (2004)
14. Qu, S., Fu, C.M., Dong, C., Tian, J.F., Zhang, Z.F.: Failure analysis of the 1st stage blades in gas turbine engine. Eng Fail. **32**, 292–303 (2013)
15. Lapides, L., Levy, A.: The halo effect in jet impingement solid particle erosion testing of ductile metals. Wear **58**, 301–311 (1980)
16. Singh, T., Tiwari, S.N., Sundararajan, G.: Room temperature erosion behavior of 304, 316 and 410 stainless steels. Wear **145**, 77–100 (1991)
17. Mesa D.H., Toro, A., Sinatora A., Tschiptschin.: The effect of testing temperature on corrosion–erosion resistance of martensitic stainless steels. Wear **255**, 139–145 (2003)

Termomechanical Analysis of 3D Printing Specimens (Acrylonitrile Butadiene Styrene)

Juan Atonal-Sánchez, Juan Alfonso Beltrán-Fernández, Luis Héctor Hernández-Gómez, Luz Yazmin-Villagran, Juan Alejandro Flores-Campos, Adolfo López-Lievano and Pablo Moreno-Garibaldi

Abstract In the present work the thermomechanical properties of acrylonitrile butadiene styrene (ABS) were evaluated at ambient and higher temperatures. The test specimens were previously modeled in a computer aided design (CAD) program (SolidWorks) with standardized dimensions according to the D638 Standard of the American Society for Testing Materials (ASTM) (ASTM in D638-10 Standard Test Method for Tensile Properties of Plastics, 2012). These models were exported to computer aided manufacturing (CAM) software for later 3D printing. The 3D printer used is based on fused deposition modeling (FDM) technology. The impressions were configured with a distance of 0.3 mm between layers and angles of 45° and 135° from the horizontal axis. Experimental tests were performed at a speed of 10 mm/min on a universal testing machine (Shimadzu AG-I) with the aid of a support and control system for the band-type electrical resistance, that was used as a heat source, as well

---

J. Atonal-Sánchez (✉) · J. A. Beltrán-Fernández · L. H. Hernández-Gómez
J. A. Flores-Campos · A. López-Lievano · P. Moreno-Garibaldi
Instituto Politécnico Nacional - Escuela Superior de Ingeniería Mecánica y Eléctrica, Sección de Estudios de Posgrado e Investigación Edificio 5. 2do Piso, Unidad Profesional Adolfo López Mateos "Zacatenco" Col. Lindavista, C.P, 07738 Ciudad de Mexico, Mexico
e-mail: juan_atonal@hotmail.com

J. A. Beltrán-Fernández
e-mail: jbeltranf@hotmail.com

L. H. Hernández-Gómez
e-mail: luishector56@hotmail.com

J. A. Flores-Campos
e-mail: jaflores@ipn.mx

A. López-Lievano
e-mail: k_lievano@hotmail.com

P. Moreno-Garibaldi
e-mail: pmg170588@gmail.com

L. Yazmin-Villagran
Facultad de Ingeniería Mecánica y Eléctrica, Universidad Veracruzana, Prolongación Venustiano Carranza S/N, Col. Revolución, 93390 Poza Rica, Veracruz, Mexico
e-mail: yazvillagranv@gmail.com

© Springer International Publishing AG, part of Springer Nature 2019
A. Öchsner and H. Altenbach (eds.), *Engineering Design Applications*, Advanced Structured Materials 92, https://doi.org/10.1007/978-3-319-79005-3_17

as an infrared pyrometer to make temperature measurements during the tensile test. Finite element analyses were performed with the help of the ANSYS software using the SolidWorks generated model, which was imported with the file format*. IGS for further study. The results obtained from the experimental tests and numerical simulations differ, because the Poisson´s ratio and the coefficient of thermal expansion remained constant for the numerical analyses.

Keywords 3D printing · Fused deposition modeling (FDM) · Finite element method (FEM) · Ultimate stress · Modulus of elasticity · Thermal expansion coefficient

1 Introduction

In recent years, many sectors have become great consumers of polymers. One of the most used polymers in engineering applications is ABS. Characterized by easy reprocessing, low cost, high impact and mechanical and chemical resistance [2].

The mechanical properties of solid materials used in engineering designs are determined by destructive mechanical testing [3]. For some years destructive tests have been reduced less and less, at present, numerical simulation based on the finite element method (FEM) has taken more relevance every day, decreasing conventional tests.

There are standardized tests governed by associations, where the dimensions of the specimen are established for each type of test, either of tension, compression, torsion, flexion, among others. Different 3D printing technologies have been developed and continue to evolve for the generation of parts.

The materials that can currently be used for 3D printing are varied and the fact is that they greatly influence the cost of printing [4]. The most widely used thermoplastics in FDM printers is ABS and polylactic acid (PLA) [5]. ABS is the plastic studied in this work.

As is well known, many of these parts are subjected to different states of loading in different applications, so that the proper selection of the material of each of them is based mainly on their mechanical properties.

On the other hand, the ABS characterization depends on the criteria of printing (orientation of fibers, distance between layers, percentage of compaction), these configurations of impression alter the resistance and it is reflected in the performance of the work to which it will be applied.

The mechanical behavior of the materials under different states of loading is still studied and generally thermal fluence experiments are used to analyze and predict the behavior of the materials at different temperatures [6]. Therefore, it is essential to know and understand in detail the thermo-mechanical properties of ABS.

2 Finite Element Method

The FEM has been generally applied structurally, however, advances in this field have been able to include heat transfer for which two types of thermal analyses were developed, one "stable", which does not consider time as a factor in the analysis and the other type is the "transient" analysis, in which the conditions are variable in time.

In the case of thermal analysis, it is necessary to create a discrete model of the parts to be analyzed, so we proceed to a division of geometry into a finite number of elements, which in turn are defined by nodes, and in this way, it is possible to establish the responses generated by each element attached to another through its nodes.

There are many elements in the libraries of each program based on the FEM that adapt to the type of analysis to study, which in the case of ANSYS are the element types beam, solid, shell, among others.

SHELL181 is suitable for analyzing thin or moderately thick structures. It is an element of four nodes with six degrees of freedom in each node: translations in the X, Y, Z, and rotations on the X, Y, and Z axes. The option of triangular elements should only be used as filler elements in the mesh generation.

Figure 1 shows the geometry, location of the nodes, and the coordinate system for this. The element is defined by four nodes: I, J, K, and L.

Temperatures can be introduced as loads to the element at the 1–8 corners. The first corner temperature T1 is set by default to the uniform temperature (TUNIF). The other temperatures, if not specified acquire the temperature T1.

This element has the characteristic of generating layers with orientations that allows simulating the printed specimens in ABS.

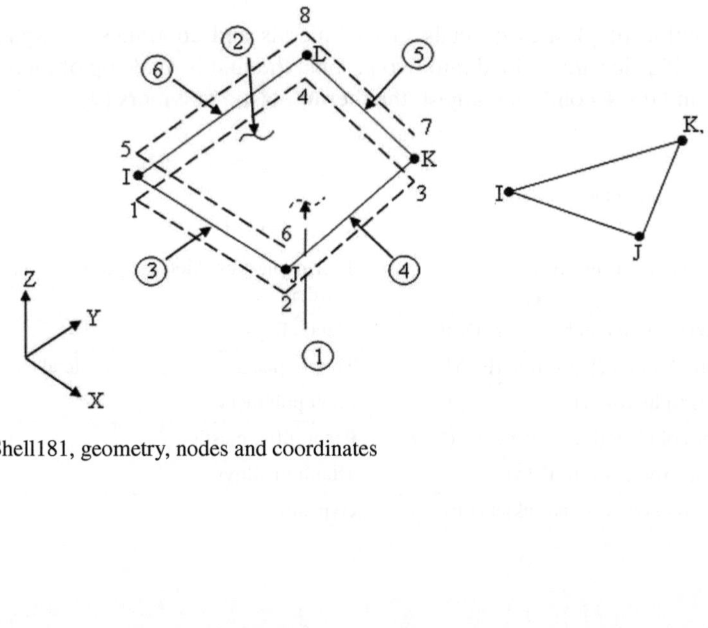

Fig. 1 Shell181, geometry, nodes and coordinates

3 3D Printing

3D printing has gained a lot of popularity in recent years, owing to the interest consumers have had for low-cost 3D printers. Many sectors such as medicine, engineering, architecture, aerospace, the automotive industry and many others are the main areas of application for this innovative printing device [7].

When talking about 3D printing several terms appear to be used interchangeably: rapid prototyping, fast manufacturing or additive manufacturing. These terms used are erroneously because not all printers have the same technology. Rapid prototyping, based on the definition of Ian Gibson, is the process of creating a part or system quickly before its commercialization [8].

There are several technologies used in 3D printers, the main difference is in the way each layer is put to create the piece. Table 1 shows some of these technologies.

For the generation of the used specimens, the tensile tests were produced with a da Vinci 3D printer based on the FDM technology. This 3D printer creates parts layer by layer from bottom to top by heating and extruding the thermoplastic filament. The procedure is as follows:

- The 3D CAD file is placed in the CAM software. The trajectory is then plotted to extrude the thermoplastic material.
- The 3D printer heats the thermoplastic material until it reaches a semiliquid state and deposits it into ultra-fine droplets along the extrusion path.
- The material is removed.

3.1 *Printing Material (ABS)*

The production of plastics exceeds that of metals and continues to expand on a large scale. Plastics are divided into two groups: the first is made up of thermostatic polymers and the second, the largest, the thermoplastic polymers [9].

Table 1 3D printing technologies

Technology	Materials used
Selective laser sintering (SLS)	Thermoplastics, Metallic powders, ceramic powders
Direct sintering in metal by Laser (DMLS)	Metal alloys
Modeling by Molten Deposition (FDM)	Thermoplastics and Eutectic Metals
Stereolithography (SLA)	Photopolymers
Manufacture of objective by laminate (MOL)	Paper, film, plastics
Fused by electronic beam (EBM)	Titanium alloys
3D printing by powders and inkjet (PP)	Gypsum

Fig. 2 ABS filament for 3D printing

Acrylonitrile butadiene styrene is a thermoplastic polymer composed of three types of monomers. Acrylonitrile, butadiene and styrene (see Fig. 2) that when combined they result in one of the most widely used synergies in the world of commercial polymers.

The result of this combination is another polymer with final properties of the base polymers. With this fusion a great chemical resistance is achieved, such as resistance to fatigue, hardness and stiffness thanks to acrylonitrile. On the other hand butadiene, which is an elastomeric component, provides good tenacity and resistance to impact. Styrene for its part provides resistance to heat, increases its processability, hardness and stiffness, and allows the addition of color. Its molecular formula is (C_8H_8–C_4H_6–C_3H_3N) which represents the monomer that constitutes ABS.

Most ABS resins usually contain 21–27% acrylonitrile, 12–25% butadiene and 54–63% styrene. Today more as a product for 3D ABS has been increasingly extended in diverse sectors and introduced for 3D printing for mechanical and biomedical applications in order to evaluate the final models [10–13].

3.2 Heat Source

Although there are a wide variety of devices for heat generation, the one used in this study was an electrical resistance band type. This device that provides energy in the form of heat in various types of processes is generated by cylindrical surface elements.

The band-type resistance has a hollow cylindrical shape; different models were considered according to the specifications of each project. The manufacture of a band resistance is based on the characteristics of the process to be heated, dimensions, assembly and electrical capacities, for this there are three forms of manufacture used:

- Resistance Type Mica Band
- Band-type resistance with mineral insulation
- Band-type resistance with ceramic element.

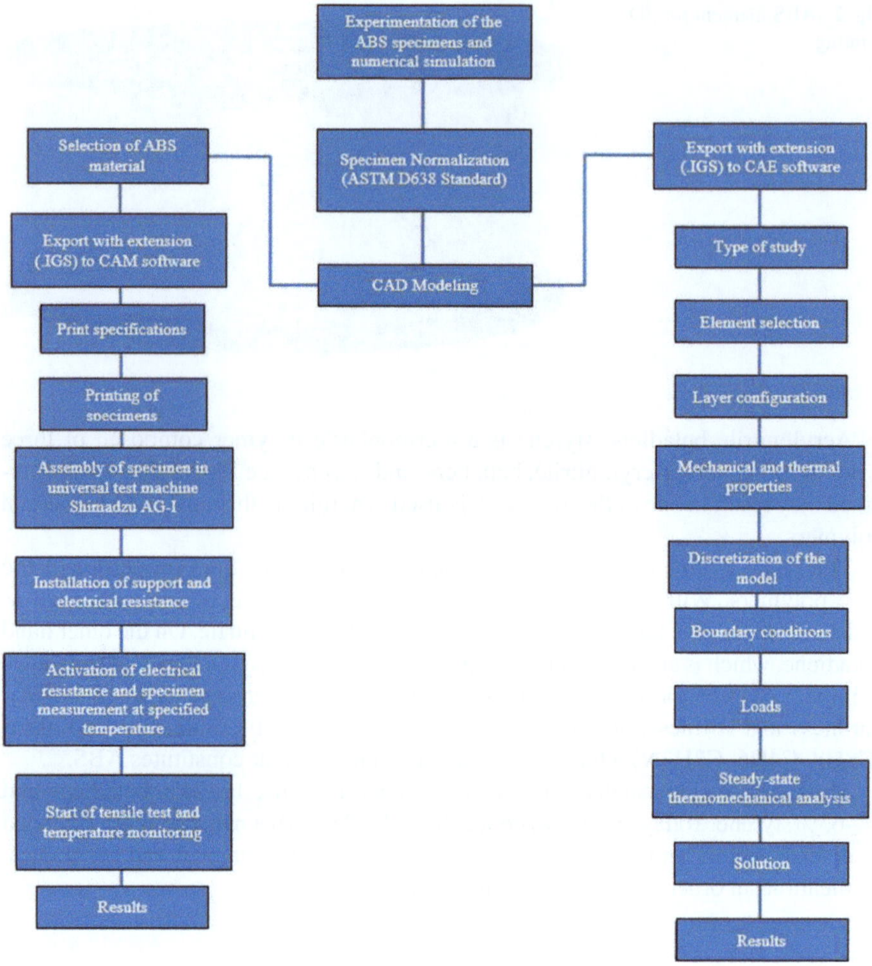

Fig. 3 Stages of experimental and numerical development

4 Experimentation

Before beginning to explain the experimentation developed and the materials used in 3D printing, as well as the auxiliary devices, a diagram (Fig. 3) of the process that was followed during the experimentation is presented.

The specimen model was generated with standardized measurements established by ASTM D638 for tensile tests on flat test specimens (Fig. 4).

The standard establishes the following dimensions for the specimen:

W	Width of the narrow zone	6 mm
L	Length of the narrow zone	33 mm
WO	Width of ends	19 mm
LO	Total length	115 mm
G	Calibrated length	25 mm
D	Distance between jaws	65 mm
R	Radio	14 mm
T	Thickness	4 mm

Defined the dimensions of the specimen the model was generated. In this case, the Solidworks 2015 design program was used for this purpose (Fig. 5).

The part was modeled in SolidWorks, and exported with extension (IGS) to CAM software, which is connected to the da Vinci 1.0 printer with FDM technology. Before

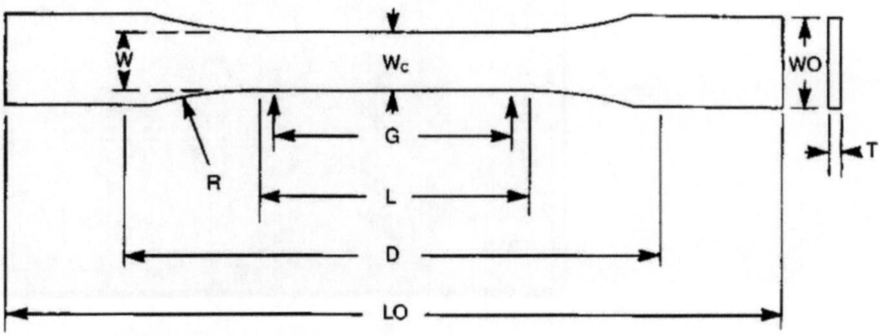

Fig. 4 Standardized specimen according to ASTM D638

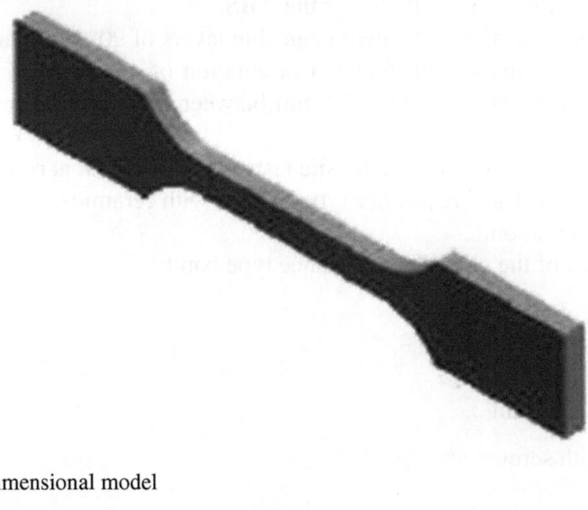

Fig. 5 Three-dimensional model

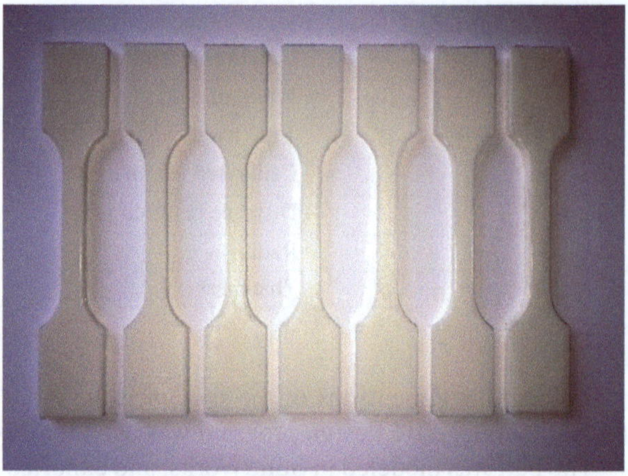

Fig. 6 ABS Printed specimens

Fig. 7 Filament directions 45° and 135°

starting the printing of the specimens, it was necessary to heat the injector at 230 °C and heat the printing base to 110 °C for the ABS.

Since the printer works with interchangeable layers of 90° of separation, the 3D prints (Fig. 6) were made with filament orientation of 45° and 135° (Fig. 7 with respect to the horizontal axis, with 0.3 mm between layer and layer and filling of 90%).

The performance of mechanical tensile tests was carried out at room temperature and higher, so an electrical resistance type of band with ceramic coating (Fig. 8) was used for heat generation.

Specifications of the electrical resistance type band:

Power	950 W
Voltage	120 V
Outer diameter	80 mm
Height	60 mm

Fastening with screws

Fig. 8 Band-type electrical resistance with ceramic coating

As it can be seen, there are machines with accessories to carry out thermo fluency tests, due to the lack of this type of attachments a resistance was designed as a heat source, as well as an adjustable support (Fig. 9) and independent of the universal testing machine to not interfere during the test.

The support was made of low carbon steel (mild steel), round tubular 1 ½ inch outer diameter and thin wall with vertical and horizontal sliding systems, in order to adapt to the height of the test machine.

On the horizontal axis support for the heat source was adapted, which in turn was sliding in a horizontal direction to keep the zone calibrated to the desired temperature during the tension test. Since the jaws of the test machine are very robust (Fig. 10a), an auxiliary jaw system (Fig. 10b) was adapted to not interfere with the heat source.

The jaws of the universal machine were attached to the auxiliary jaws and these in turn to the specimen by means of screws and commercial millimeter nuts.

The temperature monitoring of the ABS test tube was carried out continuously with an infrared pyrometer (Fig. 11) which allows obtaining measurements of the specimens without having to be in direct contact with it.

The tensile tests were carried out based on the standard ASTM D638, the Universal machine Shimadzu AG-I (Fig. 12) of 100 KN was used, at a constant speed of 10 mm/min.

After a series of stress tests (Fig. 13) results were obtained, and were subsequently plotted to observe the thermomechanical behavior of ABS at different temperatures. Figure 14 and Table 2, the ABS has a higher mechanical strength at room temperature, while increasing the temperature to 35 °C this becomes more ductile but a little resistant when compared to the test at room temperature.

In the test at 45 °C the material further increases the ductility, but like the test of 35 °C its resistance decreases, as can be observed, the results are not proportional in decrease or in increment in the case of the elongations.

By increasing the temperature to 70 °C ABS becomes too ductile and its resistance decreases considerably compared to previous tests.

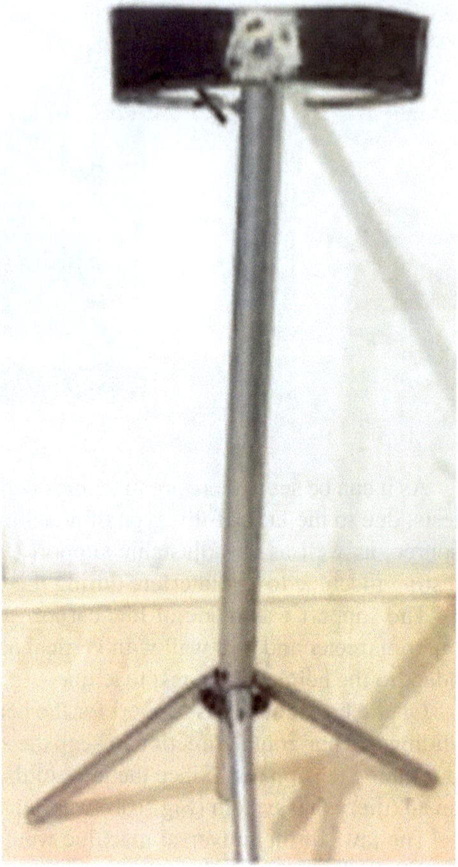

Fig. 9 Independent adjustable support

Table 2 Table of results of S_y, S_u and E

Temperature (°C)	Yield stress (S_y) (MPa)	Ultimate stress (S_u) (MPa)	Modulus of elasticity (E) (MPa)	Elongation at failure (δ) (mm)
25	30.078	32.76	1570	1.1
35	27.56	29.5	1285	1.44
45	21.72	23.41	713	1.71
70	1.17	1.35	0.517	140

(a) (b)

Fig. 10 **a** Original jaw, **b** Auxiliary jaw

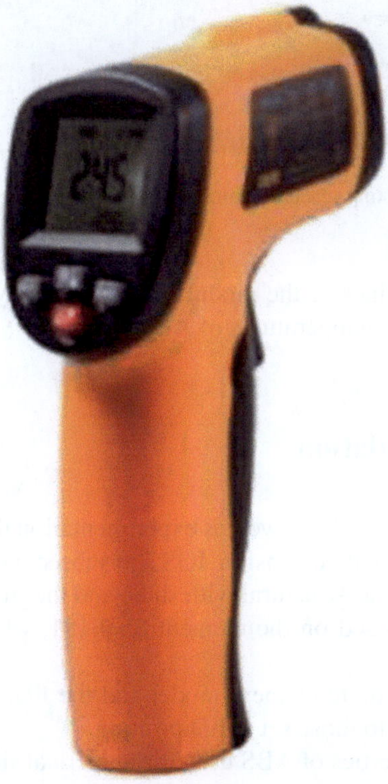

Fig. 11 Infrared pyrometer

Fig. 12 Universal tensile testing machine Shimadzu AG-I

Table 2 shows the values of the modulus of elasticity, yield stress and ultimate stress of tests at different temperatures of ABS.

5 Numerical Simulation

For numerical analysis of ABS as well as experimentation the model was developed in SolidWorks and saved with extension. IGS and subsequently exported to ANSYS.

The type of analysis was structural with static and thermal options in stable state, the discretization was based on the element Shell181, which, has been described previously.

Figure 15 shows the finite element model and the filaments of each layer with orientation angles equal to those set in 3D printing.

The mechanical properties of ABS used for numerical simulation were: modulus of elasticity (1525 MPa), Poisson ratio (0.33) and the thermal expansion coefficient ($8.2 \times 10e{-}5$ m/M °C) [14].

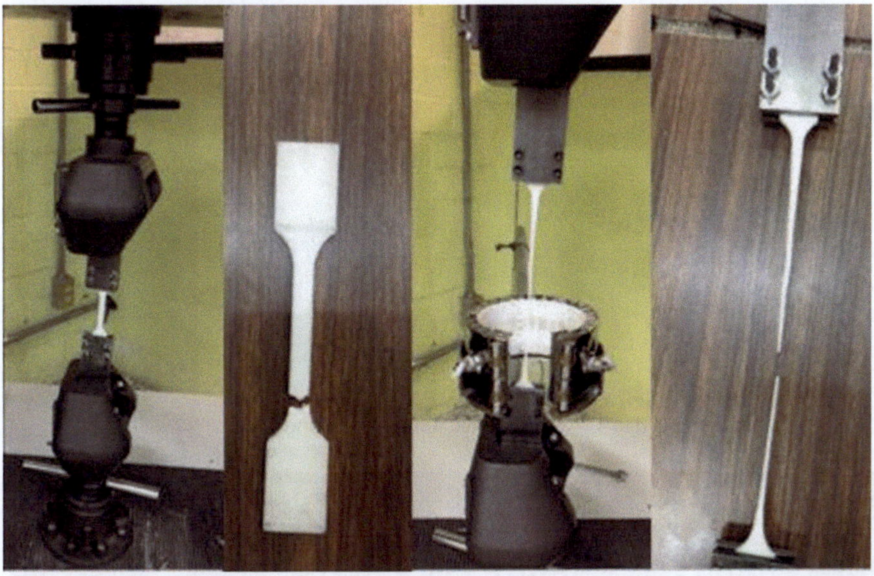

Fig. 13 ABS tensile test

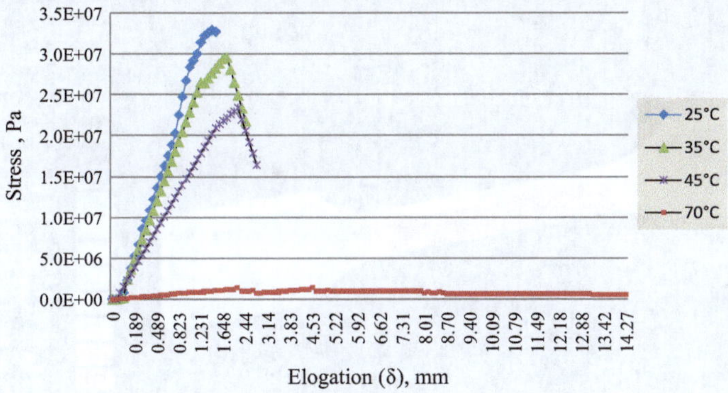

Fig. 14 Thermomechanical behavior at different temperatures

The boundary conditions for the analyses performed were the constraints of displacement and rotations in the X, Y, Z axes.

The displacement generated from each of the experimental tests was introduced in each simulation.

In Figs. 16, 17, 18 and 19, the distribution of Ultimate stress (Pa) after the application of temperature and displacement is observed.

The results of each numerical analysis for each case study are set out in Table 3.

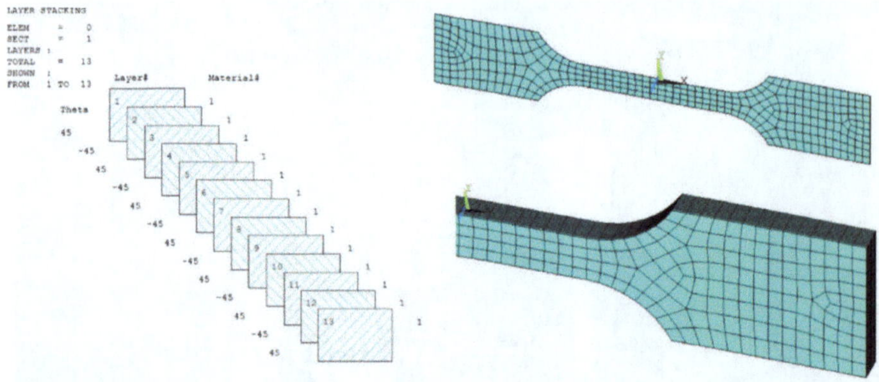

Fig. 15 Filaments oriented to 45° and 135° and finite element model

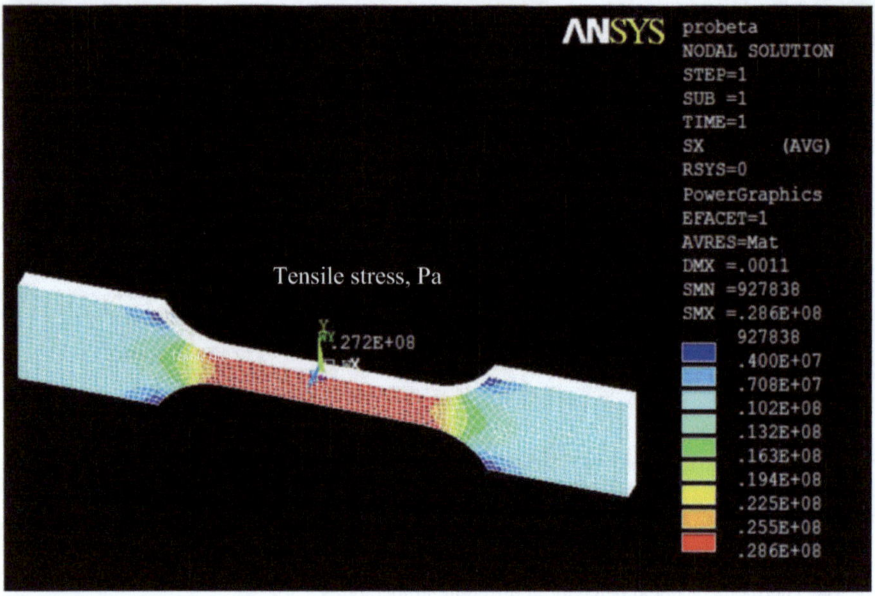

Fig. 16 Ultimate stress (Pa) distribution acting in the x-direction at 25 °C

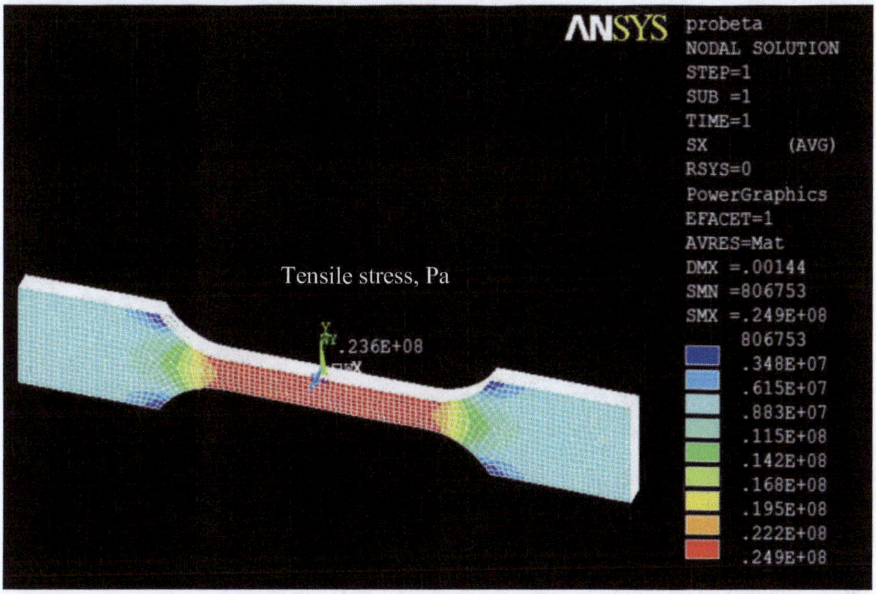

Fig. 17 Ultimate stress (Pa) distribution acting in the x-direction at 35 °C

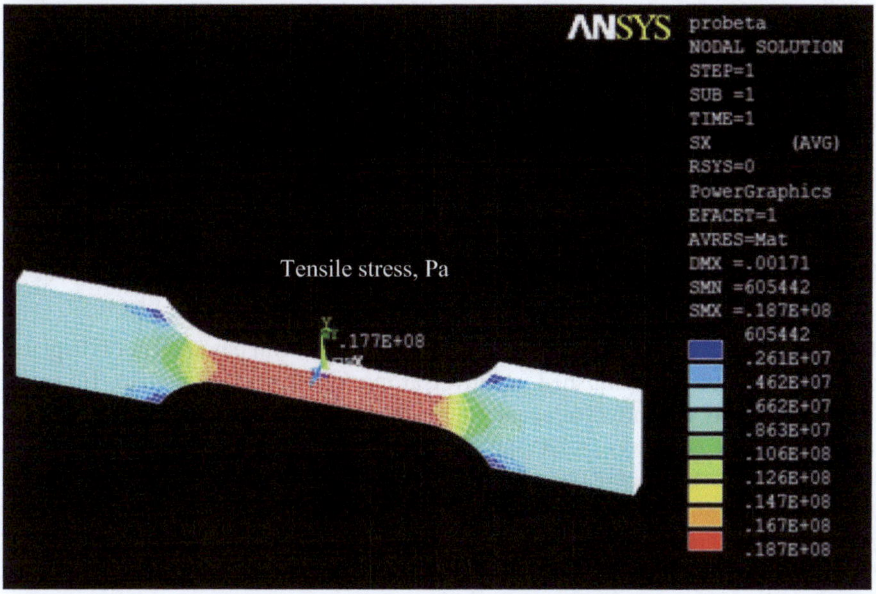

Fig. 18 Ultimate stress (Pa) distribution acting in the x-direction at 45 °C

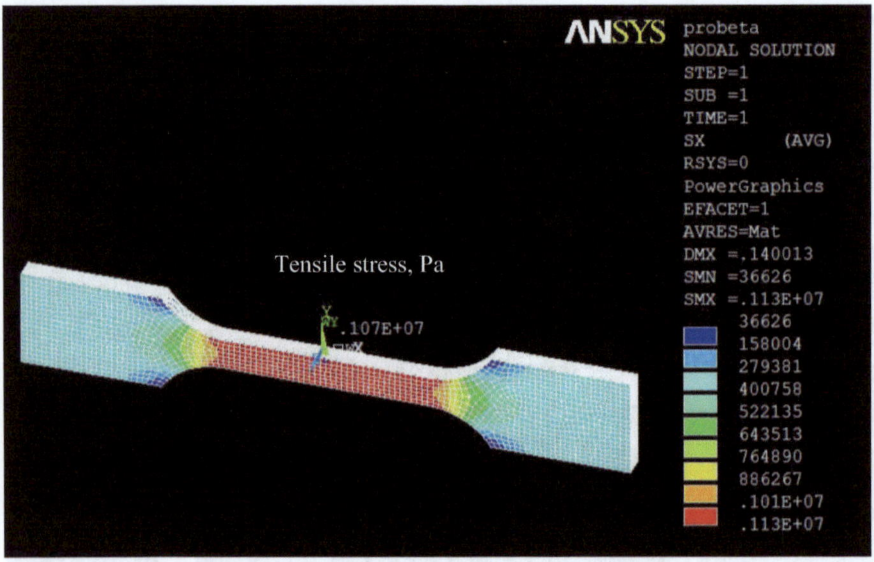

Fig. 19 Ultimate stress (Pa) distribution acting in the x-direction at 70 °C

Table 3 Stresses obtained from numerical simulations

Temperature (°C)	Ultimate stress (S_u) (MPa)	Modulus of elasticity (E) (MPa)	Elongation at failure (δ) (mm)
25	27.2	1570	1.1
35	23.6	1285	1.44
45	17.7	713	1.71
70	1.07	0.517	14

6 Conclusions

The tests performed showed results that were obtained with parameters of displacement of 10 mm/min established in universal machine tests (Shimadzu). However, these results may fall into discrepancy since there is a loss of heat because the resistance does not completely isolate the specimen from the environment.

The ABS printed specimens were based on the configuration of filaments of 45° and 135° according to results obtained in subsequent work, which show that this orientation has greater mechanical resistance.

Submitting the ABS to temperatures above the ambient implies a decrease of its resistance since the material becomes more ductile.

The experimentation of 3D printing specimens of ABS was very helpful because it provides thermo dependent data of the material such as the elastic module, in which it was of vital importance to carry out the numerical simulation.

The results of the numerical analyses compared with the experimental ones present variation, because the coefficient of thermal expansion and the Poisson relationship remained constant in each one of the analyses.

The numerical simulation is a tool of great help to carry out studies of printed pieces, since it decreased the need for prototype generation.

Acknowledgements The support given by the Instituto Politécnico Nacional, COFAA, and CONACyT for the development of this work is kindly acknowledged.

References

1. ASTM.: D638-10 Standard Test Method for Tensile Properties of Plastics, vol. 08.01, June (2012)
2. Guinault, A., Sollogoub, C.: Thermomechanical properties of ABS/PA and ABS/PC blends. Int. J. Mater. Form. **2**(S1), 701–704 (2009). https://doi.org/10.1007/s12289-009-0531-8
3. Artimez, J. M., Belzunce, F.J., Betegón, C., Rodriguez, C.: Estudio del comportamiento mecánico y de los micromecanismos de fractura de aceros DP mediante el uso del ensayo miniatura de punzonado. In XI Congreso Nacional de Propiedades Mecánicas de los Solidos (p. 85). Cadiz (2008)
4. Zukas, J., Zukas, V.: An introduction to 3D printing, 1st edn. First Edition Design Publishing, Sarasota, Florida (2015)
5. Cortés-Cedillo, A.R.: Caracterización mecánica de piezas de PLA fabricadas mediante impresión 3D (Master). Instituto Politécnico Nacional, ESIME Zacatenco (2016)
6. Mosca, H.O., Mastrícola, H.: Termofluencia primaria y secundaria del policloruro de vinilo. In Jornadas SAM 2000—IV Coloquio Latinoamericano de Fractura y Fatiga (pp. 001–1002). Neuquén, Argentina (2000)
7. Hernández-Palafox, E.: Estudio comparativo de las propiedades mecánicas de probetas fabricadas por procesos de manufactura de arranque de viruta y las de impresión en 3D metálicas (Master). Instituto Politécnico Nacional, SEPI ESIME Zacatenco (2016)
8. Gibson, I., Rosen, D., Stucker, B.: Additive manufacturing technologies. Springer, New York (2010)
9. Beltrán-Fernández, J.A.: Beltrán Fernández (Master). Instituto Politécnico Nacional, SEPI ESIME Zacatenco (2002)
10. Relaño-Pastor, A.A.: Estudio comparativo de piezas de ABS y PLA procesadas mediante modelado por deposición fundida (Bachelor). Universidad Carlos III de Madrid, (2013)
11. Beltrán-Fernández, J.A., Martínez-Paredes, J., González-Rebattú, M., Hernández-Gómez, L.H., Ruíz-Muñoz, O.: Customization and Numerical Simulation of a Cranial Distractor Using Computed Axial Tomography (CAT). Properties and Characterization of Modern Materials, 371–398. (2016). http://dx.doi.org/10.1007/978-981-10-1602-8_30
12. Beltrán-Fernández, J.A., Calderón, G., Camacho, N., Escalante, E., Garibaldi, P., Hernández-Gómez, L.H., Saucedo, F.: Design and manufacturing of prosthesis of a jaw for a young patient with articular ankylosis. Adv. Struct. Mater. 73–87. (2015). http://dx.doi.org/10.1007/978-3-319-19470-7_6
13. Beltran-Fernández, J.A., Gonzales-Rebatú, A.I., Hernandez-Gómez, L.H., Martínez-Paredes, J., Rangel-Elizalde, M.: Reconstruccion de fosa ocular con prótesis de hidroxiapatita y malla de titanio. Journal De Ciencia E Ingeniería **6**(1), 53–60 (2014)
14. Harper, C.: Modern Plastics Handbook, 1st edn. McGraw-Hill, New York (2000)

2D Crack Problems in Functionally Graded Magneto-Electro-Elastic Materials

Yonko Stoynov

Abstract Magneto-electro-elastic composite materials have extensive applications in modern smart structures, because they possess good coupling between mechanical, electrical and magnetic fields. This new effect was reported for the first time by Van Suchtelen [1] in 1972. Due to their ceramic structure, cracks inevitably exist in these materials. If these cracks extend, the material may lose its structural integrity and/or functional properties. In this study we consider functionally graded magneto-electro-elastic materials subjected to anti-plane time-harmonic load. The purpose is to evaluate the dependence of the stress concentration near the crack tips on the frequency of the applied external load. The mathematical model is described by a boundary value problem for a system of partial differential equations. A Radon transform is used to derive fundamental solutions in a closed form. Following Wang and Zhang, for the piezoelectric case, the boundary value problem is reduced to a system of integro-differential equations along the crack. For the numerical solution, software code in FORTRAN 77 is developed and validated using available examples in literature. Simulations show the dependence of the stress intensity factors (SIF) on frequency of the incident wave for different types of load, crack dispositions and the magnitude and direction of the material gradient.

Keywords Crack interaction · Anti-plane load · SIF

1 Introduction

Magneto-electro-elastic materials (MEEM) have drawn the interest of researchers in recent years. Their widespread application is due to their ability to convert mechanical energy into electric and magnetic energy, and vice versa. The magneto-electric property in piezoelectric/piezomagnetic composite materials was reported for the first time by Suchtelen [1] in 1972. Unlike single-phase magneto-electric materials,

Y. Stoynov (✉)
Faculty of Applied Mathematics and Informatics, Technical University of Sofia, Studentski grad, bl.58A, 1734 Sofia, Bulgaria
e-mail: ids@tu-sofia.bg

© Springer International Publishing AG, part of Springer Nature 2019
A. Öchsner and H. Altenbach (eds.), *Engineering Design Applications*,
Advanced Structured Materials 92, https://doi.org/10.1007/978-3-319-79005-3_18

composite MEEM possess large magneto-electric effects at room and above, which makes them suitable for application. Owing to this, composite MEEM are potential candidates for fabricating new smart and intelligent structures. Functionally graded MEEM are the next generation of high performance multifunctional materials. Their structural integrity is becoming increasingly important as their use is extended to new frontiers. The conception of functionally graded MEEM (FGMEEM) arose for the first time in 1984 in Japan during the project SPACEPLANE. Structures consisting of several layers have been used in many products. These layered structures create stress, which can cause failures in products made of such materials. Contrary to this, FGMEEM do not have this disadvantage, because their material properties vary in a continuous way. A main disadvantage of the magneto-electro-elastic composites is their brittleness. Cracks or flaws inevitably exist in such materials. These defects provoke regions of high stress and electric/magnetic field concentrations, and may initiate fracture and damage. When MEEM with cracks are subjected to external magneto-electro-mechanical loads, the stress concentration near crack tips may increase high enough to cause crack extension and eventually lead to a serious degradation of the material. To understand the failure mechanism of these materials, an analysis of the behavior of MEEM under applied external loading is necessary. The solution of the general boundary value problem for continuously inhomogeneous MEEM requires advanced numerical tools due to the high mathematical complexity arising from the electro-magneto-elastic coupling combined with the smooth variation of material characteristics leading to the solution of partial differential equation with variable coefficients. The aim of the paper is to propose a new effective boundary integral equation method (BIEM) for solution of the problem. The proposed BIEM technique is based on a frequency dependent fundamental solution of the equation of motion derived analytically by the usage of an appropriate algebraic transformation and Radon transform. To the best of the authors' knowledge the non-hypersingular traction BIEM for solution of the dynamic problem of an exponentially graded MEE composite with multiple cracks has not been developed.

In this paper we study the dynamic behavior of MEEM with two cracks, subjected to an incident time-harmonic anti-plane wave. The numerical results, obtained by the BIEM, show the dependence of the generalized stress intensity factor on the normalized frequency of the applied load, wave propagation direction, direction and magnitude of the material gradient and crack interaction phenomenon.

2 Statement of the Problem

Consider an infinite transversely-isotropic functionally graded linear magneto-electro-elastic solid in a rectangular coordinate system $Ox_1x_2x_3$, with the symmetry axis and poling direction along the Ox_3 axis; in this case Ox_1x_2 is the isotropic plane. The material is subjected to antiplane shear mechanical, and inplane electric and magnetic time-harmonic load. The geometry of the considered problem in the plane

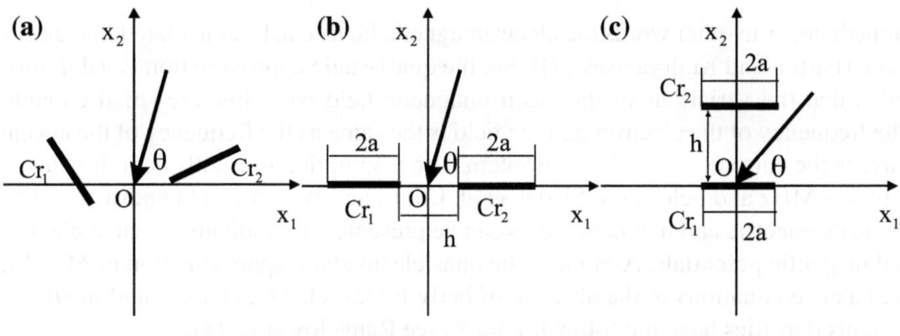

Fig. 1 A magnetoelectroelastic plane with two cracks: **a** general disposition of the cracks; **b** collinear cracks; **c** parallel symmetry cracks. The angle θ shows the wave propagation direction

$x_3 = 0$ is shown on Fig. 1. The problem is two-dimensional, and the anti-plane time-harmonic wave motion with respect to the plane $x_3 = 0$ is considered.

We introduce a generalized tensor of elasticity C_{iJKl}, $i, l = 1, 2; J, K = 3, 4, 5$ in the following way:

$$C_{i33l}(x) = \begin{cases} c_{44}(x), \; i = l \\ 0, \; i \neq l \end{cases}, \quad C_{i34l}(x) = C_{i43l}(x) = \begin{cases} e_{15}(x), \; i = l \\ 0, \; i \neq l \end{cases},$$

$$C_{i35l}(x) = C_{i53l}(x) = \begin{cases} q_{15}(x), \; i = l \\ 0, \; i \neq l \end{cases}, \quad C_{i44l}(x) = \begin{cases} -\varepsilon_{11}(x), \; i = l \\ 0, \; i \neq l \end{cases},$$

$$C_{i45l}(x) = C_{i54l}(x) = \begin{cases} -d_{11}(x), \; i = l \\ 0, \; i \neq l \end{cases}, \quad C_{i55l}(x) = \begin{cases} -\mu_{11}(x), \; i = l \\ 0, \; i \neq l \end{cases},$$

where c_{44} is the elastic module, e_{15} is the piezoelectric coefficient, q_{15} is the piezomagnetic coefficient, ε_{11} is the dielectric permittivity, μ_{11} is the magnetic permeability, d_{11} is the magnetoelectric coefficient. We also introduce a generalized stress tensor $\sigma_{iJ} = (\sigma_{i3}, D_i, B_i)$, $i = 1, 2$, where σ_{i3} is the mechanical stress, D_i and B_i are the components of the electric displacement and magnetic induction respectively in the plane $x_3 = 0$. The constitutive equations for this type of medium are see Sladek et al. [2], Soh and Liu [3]:

$$\sigma_{iJ} = C_{iJKl} u_{K,l} \tag{1}$$

$$div\,\vec{D} = \rho_f, \quad div\,\vec{B} = 0 \tag{2}$$

where the density of free charges is denoted by ρ_f and summation under repeated indexes is assumed, while comma means differentiation. The characteristic frequencies for elastic and electromagnetic processes are $\mathrm{f_{el}} = 10^4$ Hz and $\mathrm{f_{elm}} = 10^7$ Hz, respectively. Thus, if we consider dynamic loadings, with temporal changes corresponding to $\mathrm{f_{el}}$ the changes of the electromagnetic fields can be assumed to be immediate, or in other words the electromagnetic fields can be considered like quasi-static (Parton and Kudryavtsev [4]). So, the quasi-static approximation is valid, provided that the variations of the electromagnetic field with time are small enough. The frequency of the electromagnetic field is the same as the frequency of the elastic wave in the currently considered problem. The assumption is usually valid for elastic waves at MHz and below, see Sladek et al. [5], Li and Wei [6], and Dineva et al. [7]. Then, the electric and magnetic fields can be presented as gradients of scalar electric and magnetic potentials. Assuming the quasielectrostatic approximation of MEEM, the balance equations in the absence of body forces, electric charges and magnetic currents densities have the following form, see Rangelov et al. [8]:

$$\left|\begin{array}{l}(c_{44}(x)u_{3,i})_{,i}+(e_{15}(x)\varphi_{,i})_{,i}+(q_{15}(x)\psi_{,i})_{,i}+\rho\omega^2u_3=0\\(e_{15}(x)u_{3,i})_{,i}-(\varepsilon_{11}(x)\varphi_{,i})_{,i}-(d_{11}(x)\psi_{,i})_{,i}=0\\(q_{15}(x)u_{3,i})_{,i}-(d_{11}(x)\varphi_{,i})_{,i}-(\mu_{11}(x)\psi_{,i})_{,i}=0,\end{array}\right. \tag{3}$$

Here ω is the frequency of the applied load and ρ is the density. In generalized notation this system can be written in the following way:

$$(C_{iJKl}(x)u_{K,l})_{,l}+\rho_{JK}(x)\omega^2u_K=0, \tag{4}$$

where $\rho_{JK}(x)=\begin{cases}\rho(x), J=K=3\\0, J,K=4\,or\,5\end{cases}$. FGM is a kind of material in which the individual material composition varies continuously along certain directions in a controllable way. The modern fabrication technology of FGM allows the possibility to manufacture graded components to meet prescribed gradients in properties. More or less, most mechanical models describing the inhomogeneous material profiles are based on the assumption that material properties vary in a similar manner, which is an idealization. This fact is connected with the available computational tools. For most of them, it is impossible to consider independent variation of the material properties. The solution of the general boundary-value problem for continuously inhomogeneous MEEM requires advanced numerical tools due to its high mathematical complexity arising from the electro-magneto-elastic coupling plus variation of material properties. In this aspect any numerical realization for any type of material gradient is useful because these results can be used as benchmark examples when validating new computational methods. We assume here that all material properties depend on x in one and the same way and describe this by an inhomogeneity function $h(x)$:

$$c_{44}(x)=h(x)c_{44}\quad e_{15}(x)=h(x)e_{15},\quad q_{15}(x)=h(x)q_{15},$$

$$\varepsilon_{15}(x)=h(x)\varepsilon_{15},\quad d_{11}(x)=h(x)d_{11},\quad \mu_{11}(x)=h(x)\mu_{11},\quad \rho(x)=h(x)\rho.$$

We will consider exponentially inhomogeneous materials with the inhomogeneity function: $h(x) = e^{2(k_1x_1+k_2x_2)} = e^{2<k,x>}$, where the inhomogeneity coefficient is denoted by $k = (k_1, k_2)$.

When the incident SH-wave interacts with the cracks a scattered wave is produced. The total displacement and traction at any point of the plane can be calculated by the superposition principle:

$$u_J = u_J^{in} + u_J^{sc}, \quad t_J = t_J^{in} + t_J^{sc} \tag{5}$$

In (5) u_J^{in} and t_J^{in} denote incident wave fields and u_J^{sc} and t_J^{sc} are the displacement and traction of scattering by the crack's wave fields respectively. We impose the following boundary conditions:

$$t_J \left|_{Cr_1 \cup Cr_2} = 0, \quad \text{or} \quad t_J^{sc} \right|_{Cr_1 \cup Cr_2} = -t_J^{in} \left|_{Cr_1 \cup Cr_2} \right., \tag{6}$$

$$u_J(x_1, x_2) \to 0, \quad \text{when} \quad (x_1^2 + x_2^2)^{1/2} \to \infty. \tag{7}$$

The boundary condition (6) means that the cracks are free of mechanical traction and also that they are magnetoelectrically impermeable, the same as one of the extreme models available in the literature, see Sladek et al. [2]. The difference between impermeable and permeable crack models is discussed in details in Dineva et al. [7]. The boundary value problem (4), (6), together with the Sommerfeld-type radiation condition at infinity for the scattered wave field (7) is reduced to an equivalent system of integrodifferential equations along the cracks and then solved numerically.

3 Boundary Integral Equation Method

The fundamental solution of (4) is the solution of the equation:

$$\sigma_{iJM,i}^* + \rho_{JK}(x)\omega^2 u_{KM}^* = -\delta_{JM}\delta(x, \xi), \tag{8}$$

where $i = 1, 2$, i.e. the direction of the applied unit concentrate force, $\delta(x, \xi)$ is Dirac's delta function: $< \delta(x, \xi), \varphi(x) >= \varphi(\xi)$, where δ_{JM} is the Kronecker's symbol. The solution of (8) depends on the value of $\gamma = \frac{\rho\omega^2}{\tilde{a}} - |k|^2$, where $|k|^2 = k_1^2 + k_2^2$, $\tilde{a} = \tilde{c}_{44} + \frac{\tilde{e}_{15}^2}{\tilde{\varepsilon}_{11}}$, $\tilde{c}_{44} = c_{44} + \frac{(q_{15})^2}{\mu_{11}}$, $\tilde{e}_{15} = e_{15} - \frac{d_{11}q_{15}}{\mu_{11}}$, $\tilde{\varepsilon}_{11} = \varepsilon_{11} - \frac{(d_{11})^2}{\mu_{11}}$. We will consider the case when $\gamma > 0$ i.e. the case of wave propagation.

Following Wang and Zhang [9] and Gross et al. [10] the following representation formulae are valid:

$$t_J^{sc}(x) = C_{iJKl}n_i(x) \int\limits_{Cr_1 \cup Cr_2} [(\sigma_{\eta JK}^*(x, y, \omega)\Delta u_{J,\eta}(y, \omega) - \rho_{QJ}\omega^2 u_{QK}^*(x, y, \omega)$$

$$\Delta u_J(y, \omega))\delta_{\lambda l} - \sigma_{\lambda JK}^*(x, y, \omega)\Delta u_{J,l}(y, \omega)]n_\lambda(y)d\Gamma(y) \tag{9}$$

and

$$u_J^{sc}(x,\omega) = \int\limits_{Cr_1 \cup Cr_2} \sigma^*_{\eta MJ}(x,y,\omega)\Delta u_M(y,\omega) d\Gamma(y), \tag{10}$$

where $\Delta u_J = u_J\,|_{Cr^+} - u_J\,|_{Cr^-}$, $Cr = Cr_1 \cup Cr_2$, Cr^+ and Cr^- are the upper and lower bound of the cracks correspondingly. Δu_J are the jumps of the generalized displacement along the crack, also known as crack opening displacements (COD), u^*_{QK} are the fundamental solutions, $\sigma^*_{iJM} = C_{iJK\,l} u^*_{KM,l}$.

The boundary condition (7) is satisfied because of (10) and the limit $\sigma^*_{\eta MJ}(x,y,\omega) \underset{x\to\infty}{\to} 0$. Using boundary condition (6) and representation formula (9) we have:

$$t_J^{in}(x) = -C_{iJKl} n_i(x) \int\limits_{Cr_1 \cup Cr_2} [(\sigma^*_{\eta JK}(x,y,\omega)\Delta u_{J,\eta}(y,\omega) - \rho_{QJ}\omega^2 u^*_{QK}(x,y,\omega)\Delta u_J(y,\omega))\delta_{\lambda l}$$
$$- \sigma^*_{\lambda JK}(x,y,\omega)\Delta u_{J,l}(y,\omega)] n_\lambda(y) d\Gamma(y),\ x \in Cr_1 \cup Cr_2 \tag{11}$$

Since the incident wave field $t_J^{in}(x)$ in (11) is known, a system of integrodifferential equations for the unknown Δu_J is obtained. Once the unknown COD are found we can calculate the traction filed at any point using (9). The stress concentration field near the crack tips is computed using the following formulae:

$$K_{III} = \lim_{\varepsilon\to 0} t_3\sqrt{2\pi\varepsilon}$$

$$K_E = \lim_{\varepsilon\to 0} E_2\sqrt{2\pi\varepsilon},$$

$$K_H = \lim_{\varepsilon\to 0} H_2\sqrt{2\pi\varepsilon},$$

where ε is the distance to the crack tip.

4 Numerical Realization and Results

The Eq. (11) is solved numerically. The cracks are discretisized using 7 boundary elements for each crack. The unknown COD are approximated by parabolic shape functions. The two-dimensional integrals are solved numerically using the Monte-Carlo method. FORTRAN 77 code is created for the numerical solution. The MEEM is the piezoelectric/piezomagnetic composite $BaTiO_3/CoFe_2O_4$. The half-length of the cracks is $a = 5$ mm and the material constants for this composite can be found in Song and Sih [11], Li [12]. The components of the inhomogeneity function are

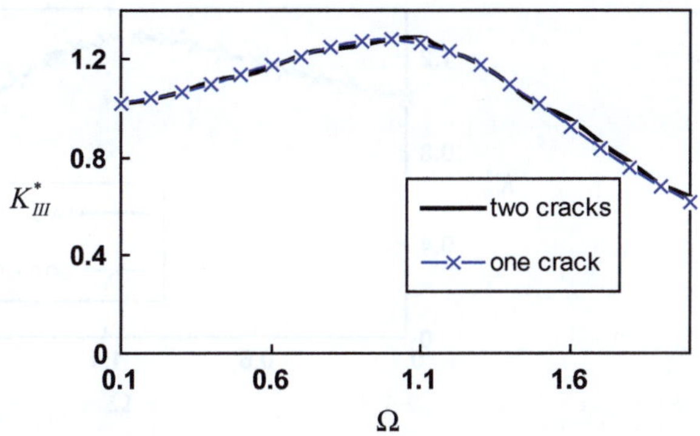

Fig. 2 Comparison between the results for two collinear cracks and one crack

presented in the following way: $k_1 = \beta \frac{\cos\alpha}{2a}$, $k_2 = \beta \frac{\sin\alpha}{2a}$, where α shows the inhomogeneity direction and β is the inhomogeneity magnitude.

4.1 *Validation*

The proposed numerical scheme is validated using the solution for one crack and solutions obtained by the dual integral equation method. In the first example we consider two collinear cracks (see Fig. 1b) when the distance between them is very large. Under this condition, we expect that the result converges to the case with one crack, because crack interaction is minimal. The following normalization of the SIF is used: $K^*_{III} = \frac{K_{III}}{t_3^{in}\sqrt{\pi a}}$ and the normalized frequency is $\Omega = a\sqrt{\rho c_{44}^{-1}}\omega$. The results for normalized SIF versus the normalized frequency at a fixed distance $h = 10a$ are given in Fig. 2. We see good coincidence of the results in the case of this big distance between cracks.

As another example we consider two symmetric parallel cracks (see Fig. 1c) as the distance $\boldsymbol{h}$ between them increases. We expect to obtain the result for one crack, when the distance between them is large enough. The comparison shown in Fig. 3 shows the distance $\boldsymbol{h}$ between the cracks to be $h = 20a$.

Our results have been validated with the results of Zhou and Wang [13], who used a dual integral equation method. The cracks are parallel as shown in Fig. 1c, and the distance $\boldsymbol{h}$ between them is increasing: $h = 0.2a, \ldots, 6.5a$. The external load is static. We present the comparison in Fig. 4. The difference is no more than 1.5%.

Results for the same scenario but in the case of dynamic load also show good agreement, see Fig. 5.

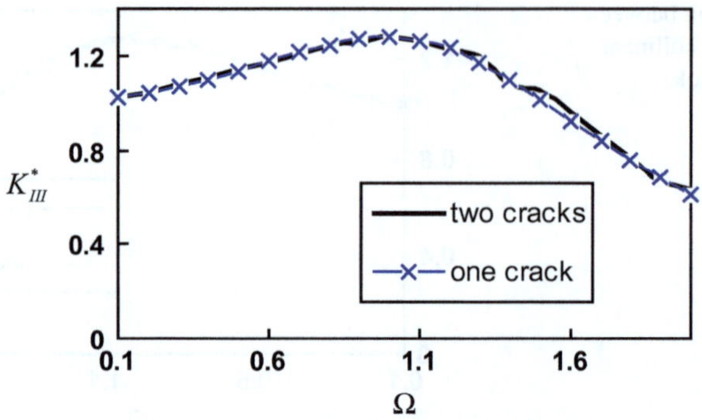

Fig. 3 Comparison between the results for two symmetry parallel cracks and one crack

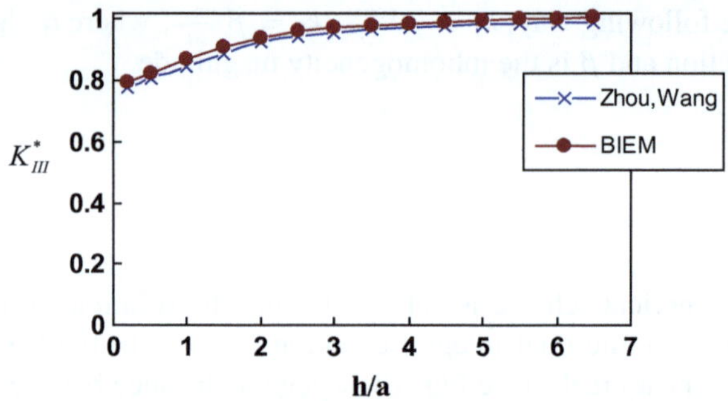

Fig. 4 The normalized SIF versus the ratio $\frac{h}{a}$

Fig. 5 The normalized SIF versus the ratio $\frac{h}{a}$, dynamic load, $\Omega = 0.4$

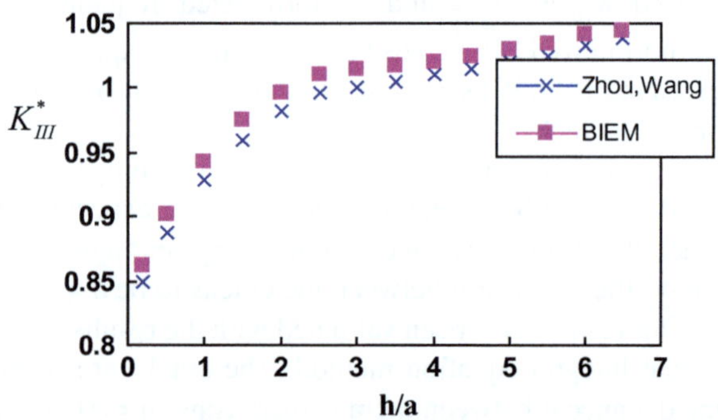

4.2 Parametric Studies

Our aim is to show the sensitivity of the generalized SIF to the wave propagation direction, the distance between the cracks and their geometrical configuration. The

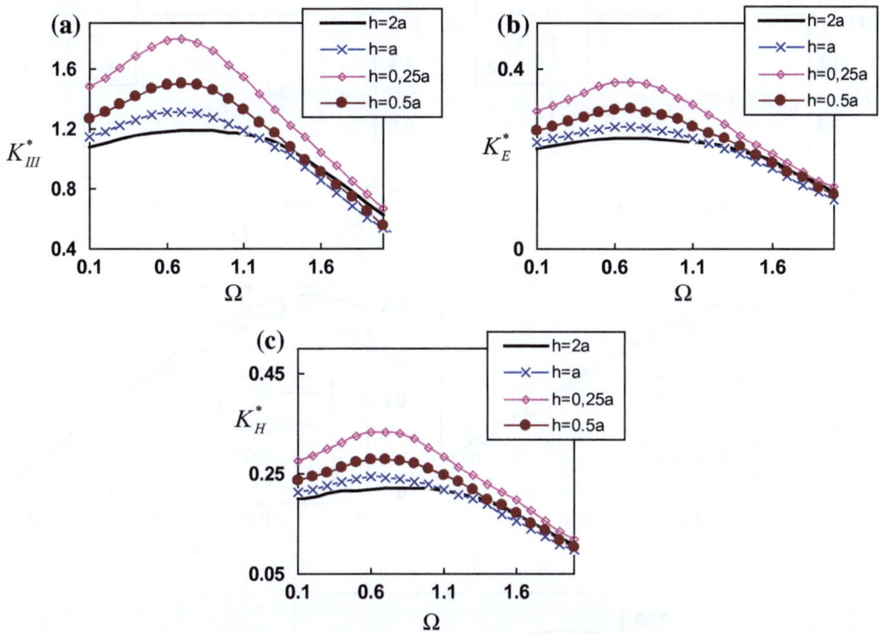

Fig. 6 Normalized SIF, EFIF and MFIF versus the normalized frequency for a normal incident wave

electric field intensity factor (EFIF) and magnetic field intensity factor (MFIF) are normalized as follows: $K_E^* = 10\frac{K_E}{t_3^{in}\sqrt{\pi a}}$ and $K_H^* = 10^4\frac{K_H}{t_3^{in}\sqrt{\pi a}}$.

In Fig. 6 we present the dependence of the generalized SIF on the normalized frequency for a normal incident wave. The cracks are collinear (see Fig. 1b) and the distance between them is: $h = 0.25a,\ 0.5a,\ a,\ 2a$. The SIF is increasing until it reaches its peak at a normalized frequency of $\Omega \approx 0.7$, then it starts to decrease. We see increasing of the SIF when the distance between the cracks is decreasing. The EFIF and MFIF have similar behavior.

Results for two symmetry parallel cracks at incident angles $\theta = \frac{\pi}{2}$ and $\theta = \frac{\pi}{3}$ are given in Fig. 7a, b, where the distance between them is: $h = 6; 9; 12$ mm. SIF increases in the frequency interval [0.7–1.6], when the distance between them increases. This is in opposite to the results shown in Fig. 6 for the case of collinear cracks. This phenomenon, called the crack shielding effect, is discussed in Ratwani [14], Zhou and Wang [13]. Results for SIFs at the right crack-tip of the left crack in the case of two collinear cracks (see Fig. 1b) in an inhomogeneous plane at a fixed distance between them $h = 0.5a$, normal incident wave $\theta = \frac{\pi}{2}$ and at different inhomogeneity magnitudes $\beta = 0.2; 0.4; 0.6$ are given in Fig. 7c. It can be seen that by increasing the inhomogeneity magnitude, the SIF decreases, but this effect is frequency dependent. So the conclusion is that defect driving force can be reduced

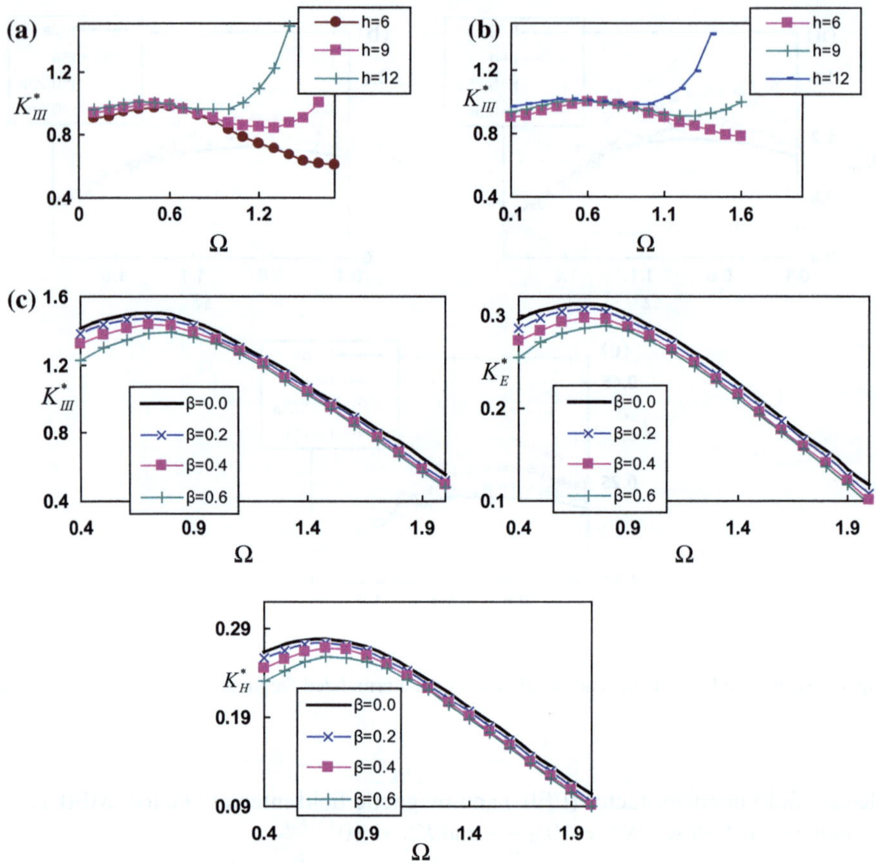

Fig. 7 Normalized SIF: **a** homogeneous material, parallel cracks, normal incident wave; **b** homogeneous material, parallel cracks, incident wave angle $\theta = \frac{\pi}{3}$; **c** inhomogeneous material, collinear cracks, the distance is fixed $h = 0.5a$, $\theta = \frac{\pi}{2}$, the inhomogeneity angle $\alpha = 0$, the inhomogeneity magnitude $\beta = 0.0; 0.2; 0.4; 0.6$

by using the concept for the FGM and the idea to replace the homogeneous material with smoothly inhomogeneous one in the new smart structure technologies works.

5 Conclusion

We present a numerical BIEM solution for a plane of MEEM with two cracks, subjected to incident SH waves. The numerical results show the sensitivity of SIF, EFIF and MFIF to wave propagation direction, distance between the cracks, crack disposition, crack interaction and the material inhomogeneity. The near field results can be applied in computational fracture mechanics, while the information for the scattered wave field can be used for development of new efficient non-destructive test methods for monitoring the integrity and reliability of the multifunctional materials and smart structures based on them.

Acknowledgements The author acknowledges the support of the Bulgarian National Science Fund under the Grant: DFNI-I 02/12.

References

1. Van Suchtelen, J.: Product properties: a new application of composite materials. Phillips Res. Rep. **27**, 28–37 (1972)
2. Sladek, J., Sladek, V., Solek, P., Pan, E.: Fracture analysis of cracks in magneto-electro-elastic solids by the MLPG. Comput. Mech. **42**, 697–714 (2008)
3. Soh, A.K., Liu, J.X.: On the constitutive equations of magnitoelectroelastic solids. J. Intell. Mater. Syst. Struct. **16**, 597–602 (2005)
4. Parton, V.Z., Kudryavtsev, B.A.: Electromagnetoelasticity, piezoelectrics and electrically conductive solids. Gordon & Breach Science Publishers, New York (1988)
5. Sladek, J., Sladek, V., Solek, P., Zhang, Ch.: Fracture analysis in continuously nonhomogeneous magneto-electro-elastic solids under a thermal load by the MLPG. Int. J. Solids Struct. **47**, 1381–1391 (2010)
6. Li, L., Wei, P.J.: Surface wave speed of functionally graded magneto-electro-elastic materials with initial stresses. J. Theor. Appl. Mech. **44**, 49–64 (2014)
7. Dineva, P., Gross, D., Müller, R., Rangelov, Ts.: Dynamic fracture of piezoelectric materials. In: Solid Mechanics and Its Applications, vol. 212, Springer, (2014)
8. Rangelov, T., Stoynov, Y., Dineva, P.: Dynamic fracture behaviour of functionally graded magnetoelectroelastic solids by BIEM. Int. J. Solids Struct. **48**, 2987–2999 (2011)
9. Wang, C.Y., Zhang, C.: 2D and 3D dynamic Green's functions and time domain BIE formulations for piezoelectric solids. In: Yao, Z.H., Yuan M.W., Zhong W.X., (eds.) Proceedings of the WCCM VI in conjunction with APCOM'04 September 5–10, Tsingua University Press & Springer, Beijing, China (2004)
10. Gross, D., Rangelov, T., Dineva, P.: 2D wave scattering by a crack in a piezoelectric plane using traction BIEM. J. Struct. Integ. Durab. **1**, 35–47 (2005)
11. Song, Z.F., Sih, G.C.: Crack initiation behaviour in magneto-electro-elastic composite under in-plane deformation. Theor. Appl. Fract. Mech. **39**, 189–207 (2003)
12. Li, X.F.: Dynamic analysis of a cracked magnetoelectroelastic medium under antiplane mechanical and inplane electric and magnetic impacts. Int. J. Solids Str. **42**, 3185–3205 (2005)
13. Zhou, Z., Wang, B.: Dynamic behavior of two parallel symmetry cracks in magneto-electro-elastic composites under harmonic anti-plane waves. Appl. Math. Mech. (English Edition) **27**, 583–591 (2006)
14. Ratwani, M., Gupta, G.D.: Interaction between parallel cracks in layered composites. Int. J. Solids Struct. **10**, 701–708 (1974)

Nonlinear Dynamic Behaviour of a Rectangular Thin Plate with a Bifurcation Diagram

Sohayb Abdul Karim

Abstract Nonlinear dynamics of rectangular isotropic thin plates with different boundary conditions are investigated. The plate governing equations of motion are nonlinear partial differential equation (PDE). The nonlinearity is caused by large deflection. The finite element method was used to discretize the continuous structures and convert the PDE into a second order ordinary differential equation (ODE).The implicit Newmark's scheme and Newton-Raphson's iteration were used to perform the time integration. Comparisons with the literature are presented in terms of backbone curve and frequency responses for three cases: (1) a simply supported plate with in-plane immovable edges, (2) out-of-plane restrained and in-plane immovable edges, (3) a fully clamped in-plane immovable edge. The model's results showed excellent agreement, particularly for the first case, validating the procedure. Using the same set of results, bifurcation diagrams were extracted as well. The bifurcation diagrams are an important analysis tools extensively used in nonlinear dynamics showing in a glance the boundaries between different oscillatory modes. To the author's knowledge, bifurcation diagrams have not been found for thin plates.

Keywords Dynamics · Geometrical · Nonlinear · Bifurcation · Plates

Abbreviations

APDL	Ansys parametric design language
BCM	Boundary collocation method
BEM	Boundary element method
FFT	Fast Fourier transform
FEM	Finite element method
ODE	Ordinary differential equation
PDE	Partial differential equation

S. Abdul Karim (✉)
Gaziantep University, Gaziantep, Turkey
e-mail: souheibak@yahoo.com; karim@gantep.edu.tr

© Springer International Publishing AG, part of Springer Nature 2019

A. Öchsner and H. Altenbach (eds.), *Engineering Design Applications*,
Advanced Structured Materials 92, https://doi.org/10.1007/978-3-319-79005-3_19

1 Introduction

This paper focuses on solving problems of the nonlinear dynamic response of thin elastic rectangular plates using an approximate numerical method. The von Karman strain-displacement relationships are considered as a source of geometrical nonlinearity. The resulting governing equations of motion are a set of nonlinear partial differential equations, fourth order in space and second order in time. They describe the large lateral deflection of thin plates. An exact analytical solution is impossible. Alternatively, careful selections of approximate numerical schemes are needed to solve the dynamical response.

In this type of family of problems, a large number of publications are found in the literature dedicated to solving this problem using various approximate methods. Most of these methods are capable of solving the problem within very strict conditions in size, boundary conditions, and shape of the plates. In addition, there is another lack noticed in the way of visualizing the nonlinear dynamical behaviour of plates. More specifically, the results, in all reviewed works [1–20], are shown in terms of frequency-response, backbone curves and/or time domain response plots. However, in this work, it was found that the bifurcation diagram could describe the nonlinear dynamical response of thin plates. Bifurcation diagrams by definition are tools widely used in nonlinear dynamics allowing one to obtain a global representation of the different zones that are encountered in nonlinear dynamical systems as a function to a control variable. For more information the reader is advised to read textbooks related nonlinear dynamics such as Gilmore and Lefranc [1].

The adopted procedure, in this work, depends on the finite element method (FEM) to discretise the continuous structure. While the integration process in the time domain was performed by the Newmark scheme beside the Newton-Raphson technique to solve iteratively the nonlinearity by performing the balance between the external applied force and the internal resulted resistance at the end of every time-step. The resulting time-series of the central point of the plate are processed by fast Fourier transform (FFT) to convert the output signal into the frequency domain in order to capture the resonance frequency under external load.

The size of the space discretisation is selected carefully based on the convergence of modal analysis results of the FEM model, while the integration time-step size was selected based on the adequate time-series length in order to obtain a reasonable frequency spacing, because the time-series had to be processed by FFT. Additionally, minimizing the numerical error within acceptable computational costs was another consideration in choosing the appropriate time-step. The numerical errors were estimated in terms of the ratio of the spurious energy loss, due to the numerical integration error, to the total energy of the complete system. The main advantages of using FEM rely on the fact of its capability to model larger size models, with more complex shapes and non-classical boundary conditions. This unlike other methods adopted in the literatures. Just to list a few, the Galerkin technique is used extensively in the literature to solve the space derivative [2–4]. J. Crawford et al. used Galerkin technique and multiple-time-scaling technique to solve the nonlinear vibrations of a flat plate with initial stresses.

The results and comparison with the literature were shown in terms of the frequency response plots [2]. The Galerkin technique, beside the multiple-time-scaling technique, was used by A. Shooshtari to solve the linear, nonlinear vibration of composite and fiber metal laminated rectangular plates. The results were shown in terms of frequency responses [3]. Tanish Dey and L.S. Ramachandra, implemented Galerkin's method to reduce the nonlinear PDE to a nonlinear ODE with an incremental harmonic balance method to obtain the frequency response and time response for the free and forced vibration of laminated composite simply supported circular cylindrical shell [4]. The extended Kantorovich method of variational calculus was used by J. Jones et al. to compute the response of several combinations of simply supported and clamped boundaries for a rectangular plate under linear assumptions [5]. The superposition method was used by R.K. Singal, D.J. Gorman and S.A. Forgues to compare plate results with experimental works for a wide range of configurations [6].

The energy method was used by A.W. Leissa et al. to perform an approximate analysis of the forced vibration of plates. The analysis results were presented by frequency response plots [7]. A FEM quad-8 shear-flexible element was used by M. Ganapathi to solve the nonlinear vibration of laminated composite curved panels, while the Wilson-θ scheme was used to perform the numerical time integration [8]. In order to achieve the equilibrium at the end of each integration time-step, the Newton-Raphson iterations were implemented [8]. Comparatively, Ganapathi used a FEM quad-4 shear-flexible element to solve the same problem for thin circular cylindrical shells problems. In both works, the frequency response plots were used to show the dynamical response [9].

The Shooting and Newton methods were used by P. Ribeiro to solve a dynamic FEM model of isotropic plates in a certain frequency range [10]. The forced response of a simply supported square metallic plate was computed by G. Anlas and O. Elbeyli using the multiple scales method [11]. Identically, the results are introduced and discussed in terms of the frequency response plots. K.N. Saha, D. Misra, S. Ghosal, and G. Pohit solved the forced nonlinear vibration of a rectangular plate by dividing the problem into two parts, a static problem and dynamic one, and used the successive relaxation method as an iterative scheme. The results were documented by backbone curves [12]. P. Ribeiro used p-version hierarchical FEM and harmonic balance and continuation methods to solve the geometric nonlinear vibration of isotropic simply supported plates. Similarly, the results were shown in terms of resonance and backbone curves [13].

The pseudo-arclength continuation method was used to numerically model a rectangular plate [14–17]. M. Amabili used the pseudo-arclength continuation and collocation method which allows a bifurcation analysis, three cases of rectangular plates were analysed. The cases were simply supported movable edges, simply supported immovable edges, and fully clamped edges. The frequency response (backbone curve) and time domain, spectrum of the response, phase portrait diagrams were shown [14]. M. Amabili used the same technique to numerically model a stainless-steel rectangular plate with the simply supported, in-plane moveable and immovable edges. The frequency spectrum, time response, and frequency response plots were compared with the literature and experiments [15]. The same procedure was used again by M. Amabili and S. Carra to solve the nonlinear vibration of a rectangular plate subjected to thermal loads in the presence of fluid-structure interactions, the resultes were demonstrated using frequency response curves [16]. M. Amabili employed the pseudo-arclength continuation method and bifurcation analysis for a rectangular laminated composite plates with three different boundary conditions: a simply-supported movable, a simply-supported immovable and a clamped plate. The results were described in terms of backbone curves [17]. E. Jomehzadeh and A.R. Saidi found the solution for a rectangular simply supported in-plane movable and immovable plate, the results were presented by backbone curves [18].

Gunwoo Noh, and Klaus-Jürgen Bathe, presented a new explicit numerical time integration scheme, and the solution for the wave propagation was presented for rectangular membranes [19]. Awrejcewicz illustrated three feigenbaum type scenarios of transition from a regular to chaotic dynamics for in-plane excited rectangular plate, Poincare map, Fourier power spectrum, and equal deflection curves are plotted in different states of excitation frequencies. In order to show the chaotic transitions, several phase portrait plots for different levels of excitation were presented separately. It was possible to present bifurcation diagrams to show the transition more efficiently [20]. M. Ducceschi employed a modal algorithm for nonlinear plate dynamics. The time-integration was performed using a Störmer-Verlet scheme, and the response showed in the time domain plots [21]. M.M. Banerjeea and J. Mazumdarb reviewed methods for the linear and nonlinear vibration analysis of plates and shells found in literature. The most frequently used approximation method were the Rayleigh-Ritz method, Galerkin Method, and Kantorovich's method (generalization of the Ritz' method). Some other technique was used by modification of equations of motion such as Berger's approach, and constant deflection contour method [22].

2 Formulations

Consider a thin plate of total thickness h with dimensions $a \times b$. The origin of the Cartesian coordinate system is located in the middle of the plate. Since large amplitude motion is assumed, the von Karman type strain-displacement relations are used as [23].

$$\begin{Bmatrix} \varepsilon_x \\ \varepsilon_y \\ \varepsilon_{xy} \end{Bmatrix} = \begin{Bmatrix} \varepsilon_{x0} \\ \varepsilon_{y0} \\ \varepsilon_{xy0} \end{Bmatrix} + z \cdot \begin{Bmatrix} \kappa_x \\ \kappa_y \\ \kappa_{xy} \end{Bmatrix} \tag{1}$$

where

$$\begin{Bmatrix} \varepsilon_{x0} \\ \varepsilon_{y0} \\ \varepsilon_{xy0} \end{Bmatrix} = \begin{Bmatrix} \frac{\partial u}{\partial x} + \frac{1}{2}\left(\frac{\partial w}{\partial x}\right)^2 \\ \frac{\partial v}{\partial y} + \frac{1}{2}\left(\frac{\partial w}{\partial y}\right)^2 \\ \frac{\partial u}{\partial y} + \frac{\partial v}{\partial x} + \frac{\partial w}{\partial x}\frac{\partial w}{\partial y} \end{Bmatrix} \tag{2}$$

$$\begin{Bmatrix} \kappa_x \\ \kappa_y \\ \kappa_{xy} \end{Bmatrix} = \begin{Bmatrix} -\frac{\partial^2 w}{\partial x^2} \\ -\frac{\partial^2 w}{\partial y^2} \\ -2\frac{\partial^2 w}{\partial x \partial y} \end{Bmatrix} \tag{3}$$

Differential equations for large deflections and homogeneous isotropic material thin plates are given by two equations: the equilibrium and the compatibility. They were first introduced in 1910 by von Kármán, as follows.

$$\frac{\partial^4 \phi}{\partial x^4} + 2\frac{\partial^4 \phi}{\partial x^2 y^2} + \frac{\partial^4 \phi}{\partial y^4} = Eh\left[\left(\frac{\partial^2 w}{\partial x \partial y}\right)^2 - \frac{\partial^2 w}{\partial x^2}\frac{\partial^2 w}{\partial y^2}\right] \tag{4}$$

$$\frac{\partial^4 w}{\partial x^4} + 2\frac{\partial^4 w}{\partial x^2 y^2} + \frac{\partial^4 w}{\partial y^4} = \frac{1}{D}\left(q + \frac{\partial^2 \phi}{\partial y^2}\frac{\partial^2 w}{\partial x^2} + \frac{\partial^2 \phi}{\partial x^2}\frac{\partial^2 w}{\partial y^2} - 2\frac{\partial^2 \phi}{\partial x \partial y}\frac{\partial^2 w}{\partial x \partial y}\right) \tag{5}$$

where

q is the applied external load.

ϕ is the stress function related to the in-plane forces as follows

$$N_x = h\frac{\partial^2 \phi}{\partial y^2} \tag{6}$$

$$N_y = h\frac{\partial^2 \phi}{\partial x^2} \tag{7}$$

$$N_{xy} = h\frac{\partial^2 \phi}{\partial x \partial y} \tag{8}$$

An analytical solution is only possible in very special conditions using trial shape function. For example, Ventsel's textbook [23] introduces an approximate solution for simply supported geometrically nonlinear thin plate using Galerkin's method.

In the case of dynamic problems, the external applied loads, q, explicitly include inertia forces in the surface lateral loads according to D'Alambert's principle neglecting the rotational inertia.

$$q(x, y, t) = q - \rho h \frac{\partial^2 w}{\partial t^2} \tag{9}$$

where, ρ is the mass density of the plate material, and h is the plate thickness.

There are several combinations of boundary conditions [23] as follows:

Simply supported edge:

$$\left.\begin{array}{l} w = 0,\ M_x = -D\left(\frac{\partial^2 w}{\partial x^2} + \nu \frac{\partial^2 w}{\partial y^2}\right) = 0 \quad (for\, x = 0, a) \\ w = 0,\ M_y = -D\left(\frac{\partial^2 w}{\partial y^2} + \nu \frac{\partial^2 w}{\partial x^2}\right) = 0 \quad (for\, y = 0, b) \end{array}\right\} \tag{10}$$

It can be written in more simplified form as follows:

$$\left.\begin{array}{l} w = 0,\ \frac{\partial^2 w}{\partial x^2} = 0 \quad (for\, x = 0, a) \\ w = 0,\ \frac{\partial^2 w}{\partial y^2} = 0 \quad (for\, y = 0, b) \end{array}\right\} \tag{11}$$

Plates with all sides clamped (C-C-C-C) are given as follows:

$$\left.\begin{array}{l} w = 0,\ \frac{\partial w}{\partial x} = 0 \quad (for\, x = 0, a) \\ w = 0,\ \frac{\partial w}{\partial y} = 0 \quad (for\, y = 0, b) \end{array}\right\} \tag{12}$$

Plates with free edges are given as follows:

$$\left.\begin{array}{l} \frac{\partial^2 w}{\partial x^2} + \nu \frac{\partial^2 w}{\partial y^2} = 0,\ -\frac{\partial}{\partial x}\left[\frac{\partial^2 w}{\partial x^2} + (2 - \nu)\frac{\partial^2 w}{\partial y^2}\right] = 0 \quad (for\, x = 0, a) \\ \frac{\partial^2 w}{\partial y^2} + \nu \frac{\partial^2 w}{\partial x^2} = 0,\ -\frac{\partial}{\partial y}\left[\frac{\partial^2 w}{\partial y^2} + (2 - \nu)\frac{\partial^2 w}{\partial x^2}\right] = 0 \quad (for\, y = 0, b) \end{array}\right\} \tag{13}$$

Plates with in-plane immovable edges are given as follows:

$$\left.\begin{array}{l} u = 0,\ v = 0 \quad (for\, x = 0, a) \\ u = 0,\ v = 0 \quad (for\, y = 0, b) \end{array}\right\} \tag{14}$$

3 Solution Procedure

There are several approximate numerical methods to solve the nonlinear PDE system of equations. Most numerical methods depend on discretizing the continuous space domain. The discretization process can be performed in direct/indirect methods such as the boundary element method (BEM), boundary collocation method (BCM), FEM, among other. For more details, the reader is referred to the textbook by Ventsel

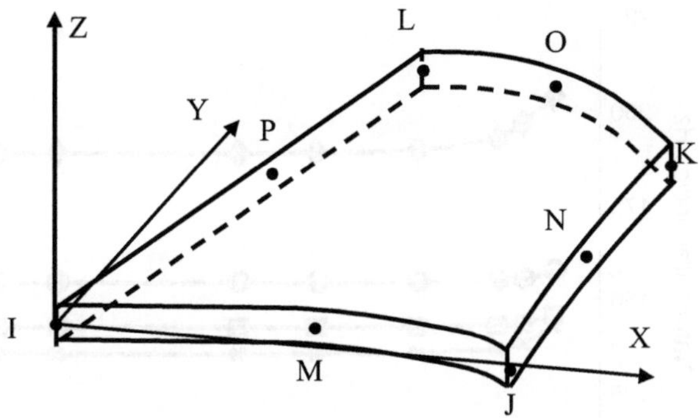

Fig. 1 8-Nodes element with six degree of freedom at each node [25]

[23]. In this work, the finite element method was used for the spatial discretization process while Newmark's method with Newton-Raphson iterative scheme were used numerically to integrate step-by-step in the time domain.

3.1 Spatial Discretization Method

Studies show that the quadratic interpolation of four-edge shell elements with six degrees of freedom show an excellent performance for nonlinear thin plate analysis [24]. This type of element allows the usage of low mesh density, hence, lowering computational costs. The type of element that was used in this study has 8 nodes, as shown in Fig. 1, and six degrees of freedom at each node: nodal translations in x, y, and z directions and nodal rotations about x, y, and z-axes. The deformation interpolations are quadratic in both in-plane directions. The element has a large deflection, and large strain capabilities and implemented by Ansys parametric design language (APDL) [25].

It is well known that the size of the mesh both in the x and y directions greatly affects the performance of the numerical model. The finer mesh allows for increased accuracy, but on the other hand, it has a high computational cost. The computational cost, in particular, becomes more questionable when nonlinear full transient analysis is needed. Since the stiffness matrix has to be updated at the end of every time-step integration in nonlinear structures. This becomes more critical when dynamic phenomena have to be studied on a microsecond time scale. For these reasons, a sensitivity analysis of the space mesh size is necessary.

Accordingly, the impact of the mesh size on the modal analysis results was studied. The results of the first ten modes were included in the observations. The main reason for this was that the model was intended to be subject to different levels of excitation; thereby high modes could be excited. Therefore, the quality of the model was assessed by its capability to accommodate accurately higher possibly excited modes.

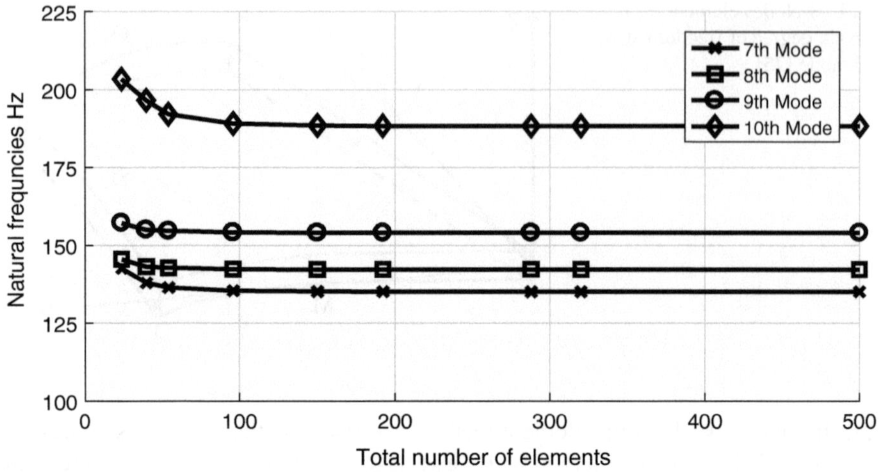

Fig. 2 The FEM resulted fundamental frequency modes versus number of elements

In order to simplify the mesh process, the plate was divided equally in the two directions. The ratio of the elements was made equal and adjusted to obtain nearly square elements. Hence, the total number of elements uniquely defines the mesh steps in x and y directions. In this report, the computations were performed by APDL. This enables writing APDL code to compute fundamental frequencies of the first ten modes at different mesh densities. Figure 2 shows the convergence of the seventh to the tenth modes obtained.

To obtain these results, the following plate data was used: The plate dimensions were $0.5 \times 0.2 \times 0.0008$ [m], material density $\rho = 2840$ [kg/m^3], Poisson's ratio $\nu = 0.35$ and Young's modulus $E = 73E9$ [N/m^2]. According to the plot, the low frequency modes quickly converge with low mesh density. With very few elements, satisfactory results were obtained. The higher frequency modes, specifically from mode number six and higher, require a much finer mesh.

In conclusion, it was found that all modes converge to the final solution at 320 elements. For this reason, in this study, the step size that is compatible with this number of elements was used.

3.2 *Time Integration Newmark's Method*

Newmark's method is used primarily for the analysis of structural dynamics problems. The discretization process reduces the system of nonlinear partial differential equations to a set of nonlinear ordinary differential equations in the following matrix form

$$[M]\left\{\ddot{U}(t)\right\} + [C]\left\{\dot{U}(t)\right\} + \{F_{in}(t)\} = \{F_{ex}(t)\} \tag{15}$$

where

$[M]$ structural mass matrix,
$[C]$ structural damping matrix,
$\left\{\ddot{U}(t)\right\}$ nodal acceleration vector,
$\left\{\dot{U}(t)\right\}$ nodal velocity vector,
$\{F_{in}(t)\}$ internal load vector,
$\{F_{ex}(t)\}$ external applied load vector.

Newmark's method was used to integrate the system of equations in the time domain. Using Taylor's expansion by a quadrature scheme of the displacement and its time derivatives and some algebraic manipulation leads to the standard form for Newmark's equation, as given in Zienkiewicz et al. [26]:

$$U_{t+\Delta t} = U_t + \Delta t \dot{U}_t + \left(\frac{1}{2} - \beta\right) \Delta t^2 \ddot{U}_t + \beta \Delta t^2 \ddot{U}_{t+\Delta t}$$

$$\dot{U}_{t+\Delta t} = \dot{U}_t + (1-\gamma)\,\Delta t\,\ddot{U}_t + \gamma\,\Delta t^2\,\ddot{U}_{t+\Delta t} \tag{16}$$

where, γ and β are parameters associated with the quadrature scheme. This method is conditionally stable for the zero damping method if

$$\gamma \geq \frac{1}{2}, \quad \beta \leq \frac{1}{2} \tag{17}$$

$$\Delta t = \frac{T_{\min}}{2\pi\sqrt{\frac{\gamma}{2} - \beta}} \tag{18}$$

In order to make this method unconditionally stable, the amplitude decay factor, α, was employed as shown in Zienkiewicz et al. [26], and the parameters of Newmark's method were chosen as follows [27]:

$$\alpha = 0.005 \tag{19}$$

$$\gamma \geq \frac{1}{2} + \alpha \tag{20}$$

$$\gamma = 0.505 \tag{21}$$

$$\beta \geq \frac{1}{4}\left(\frac{1}{2} + \gamma\right) \tag{22}$$

with these factors, Newmark's method is unconditionally stable [26], as revealed by the results of the iteration, which always converge. However, the stability of the solution does not mean accurate results. The accuracy is greatly affected by the size of the integration time-step, but it also affects the computational costs.

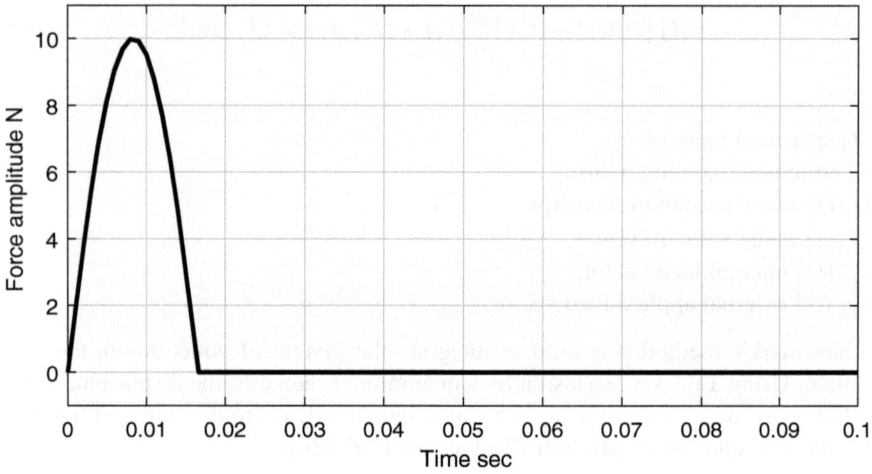

Fig. 3 The external applied half-sine wave force

3.3 *Integration Time-Step Size*

In order to select the size of the time-step suitably, the impact of the time-step size on the overall computation costs and accuracy was studied by performing the full transient analysis at different trial value of time-steps. For every value, the numerical errors were estimated.

The error estimation relied on the fact that numerically integrated dynamical systems tend to lose energy if the integration time-step is not adequate. In other words, the loss of energy increases when the time-step size increase. Therefore, in order to capture the lost energy due to numerical errors only, the damping factors are not considered. Thus, the system becomes theoretically conservative. Hence, any lost energy during the integration process is inevitably caused only by numerical errors.

Thus, the error due to the size of the time-step can be evaluated by estimating spurious energy loss which equals the external provided work, W, minus the total stored energy for both the kinetic energy, K, and the strain energy, S, as follows:

$$Error = \frac{W - (S + K)}{(S + K)} \tag{23}$$

For convenience, in this study the sampling rate in [Hz], which is the inverse of the time step, was used instead of the time step itself.

To plot the errors versus the sampling rates, a full nonlinear transient analysis was done using APDL code for a clamped-free-clamped-free plate several times. The number of elements was taken to be 16 $\times$ 20 elements. The force amplitude was 10 [N], the analysis time was done for only 0.1 s. The applied force was taken to be a half-sine-wave as illustrated in Fig. 3.

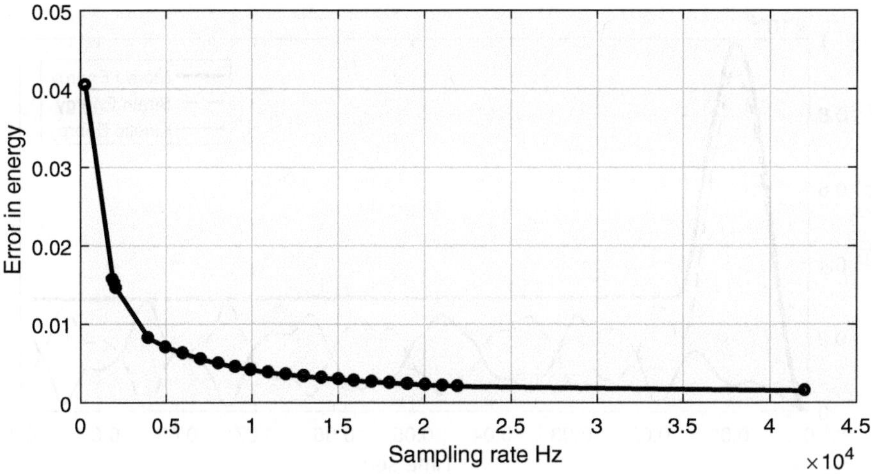

Fig. 4 The numerical integral error in terms of energy loss versus sampling rate

The results of the full transient analysis were performed several times at different sampling rates. For every run, the error was estimated according to Eq. (23). Thus, the errors as a function of the sampling rate are plotted as shown in Fig. 4. The curve shows that the error decreases dramatically when the sampling rate is increased in the first range of sampling, at less than 5 kHz. It also shows that the slope of the curve becomes less as the sampling rate is increased. In other words, at high sampling rates the improvement in the error is not considered.

For clarification purposes, the time histories of three types of energies are shown. The work, strain, and kinetic energies are plotted at a sampling rate of 5000 [Hz] in Fig. 5. The graph compares the variation of the two types of energies: strain and kinetic energy. The stored energy, which is the sum of kinetic and strain ones, is nearly fixed after the transient solution. However, a very small decline exists but it is not noticeable.

In order to visualize the energy decline due to the numerical error during the integration, the additional plot of the energies time-history is presented in Fig. 6, which shows three curves for three different sampling rates, at 800, 5000, and 15,000 [Hz]. The curves show a severe decline in the stored energy at 800 [Hz], while it shows nearly preserved energy at 15,000 [Hz] in contrast for the 5000 [Hz] case where the energy curve shows slight decline. Thus, this example shows that a sampling rate as high as 6000 [Hz] is enough for the analysed range of plates.

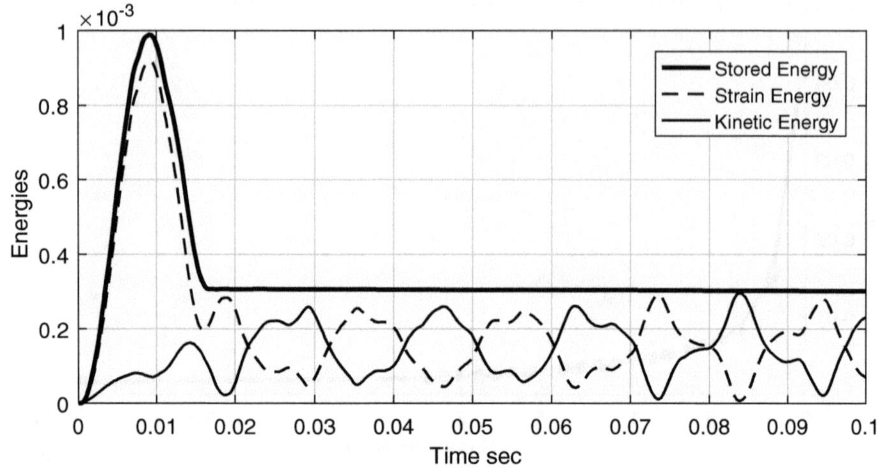

Fig. 5 The time histories of stored, kinetic and strain energies at 5000 Hz

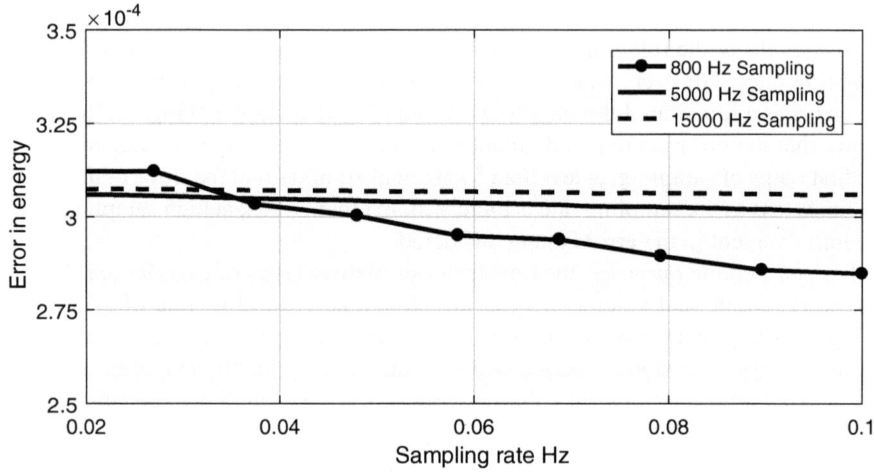

Fig. 6 The time histories of the stored energy under different sampling rates

4 Model Numerical Results

4.1 Model Verification (Backbone Curve)

The case of a simply supported square plate with immovable edges was studied by Amabili [14], with the dimensions: 0.3 × 0.3 × 0.001 [m], and the following material properties: material density $\rho = 2778$ [kg/m^3], Poisson's ratio $\nu = 0.3$, Young's modulus $E = 70E9$ [N/m^2] [14].

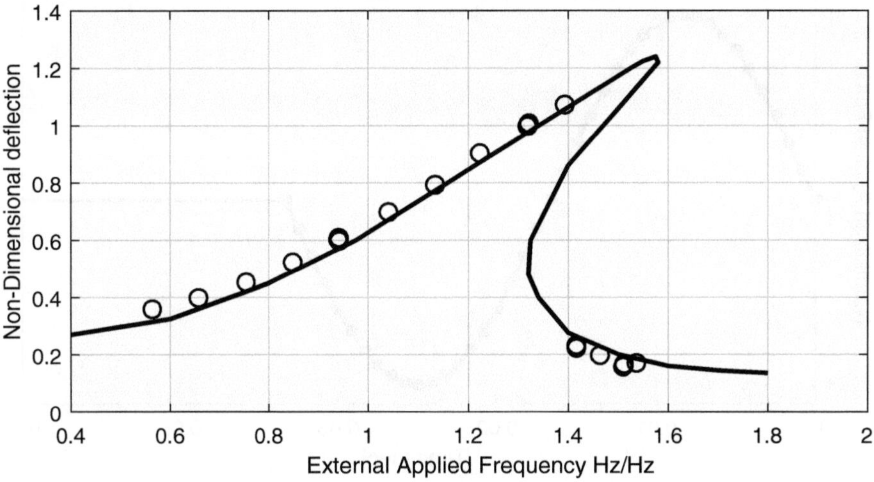

Fig. 7 Response of the simply supported plate with immovable edges; o present results,—Amabili [14]

For verification purposes, the plate according to the last data was studied using the same adopted procedure. The first fundamental frequency was found to be 52.902 [Hz], which was in agreement with that of Amabili (53.02 [Hz]) [14]. In order to excite the plate to obtain a different level of response, the frequency of the sinusoidal excitation was selected in the range from 40 to 90 [Hz]. The overall simulation time was 0.5 s and the sampling rate was 5000 [Hz] for all cases. The simulation time was sufficient to obtain a stable solution and the sampling rate was adequate for convergence and smooth response as well.

Taking the amplitude of the steady state response of the central node for all cases yields the variation of the plate response against the excitation frequencies and compared with results of Amabili [14] as plotted in Fig. 7.

Figure 7 shows excellent agreement of the current study with published results. However, there are small discrepancies due to differences in methodology. In Amabili's model, the plate was simulated with 16 dofs using a trigonometric series. In the present model, the plate was modelled using a second order finite element method with 6 degrees of freedom. (The difference in the fundamental frequency is also due to the same reason.)

In conclusion, the comparison gave full validation of the model used. The model is applicable for other cases, which only require the changing of the input parameters and the boundary conditions.

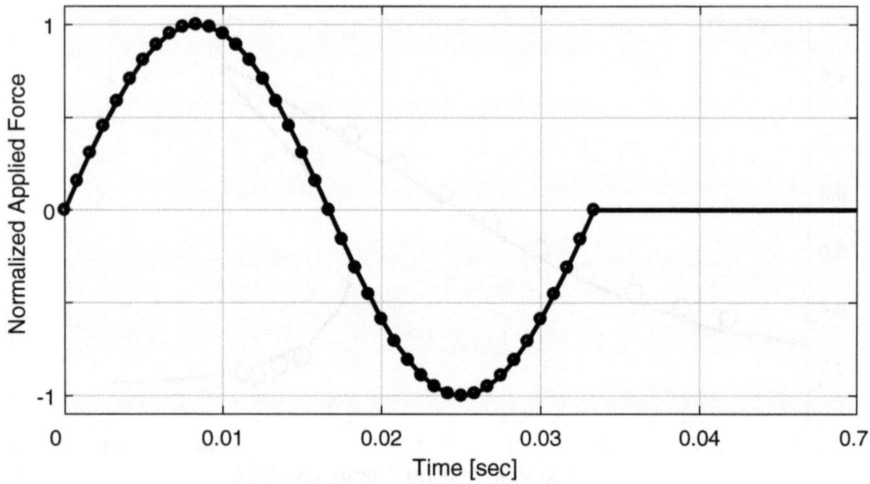

Fig. 8 The excitation signal: one sinusoidal wave 30 [Hz] unity amplitude

4.2 Bifurcation Diagrams (Three Cases)

In this chapter, the numerical results for three different boundary conditions are compared with those available in the literature: (1) simply supported and in-plane immovable edges, (2) clamped and in-plane immovable edges, (3) restrain all degrees of freedom at the edges. Additionally, the same set of results were used to extract bifurcation diagrams.

The sampling rate is $Fs = 6000$ [Hz]. Hence, the time step size equals, $dt = 1/Fs = 1.67\text{E}{-4}$ [s]. The time length of the analysis is 0.7 [s]. The plate dimensions are 0.3×0.3 [m]. The thickness of the plate was taken depending on the boundary conditions under analysis, the thickness values was selected in a way to keep the value of the resonance within the frequency range [20–25 Hz]. Hence, it is possible to use the same sampling rate and spectrum analysis routines without affecting the accuracy.

The plate is harmonically excited by one-sinusoidal-wave uniform pressure; the excitation value was 30 [Hz] which is selected in the vicinity of the first fundamental frequency. Thus, the time length of the wave equals to 0.033 [s] as shown in Fig. 8. The force sampling rate was 1200 [Hz]. The values between the sampled pointed was linearly interpolated.

4.2.1 Bifurcation Diagram for Simply Supported Immovable Edge

The boundary conditions for this case are simply supported and in-plane immovable edge expressed mathematically by Eq. (14). This case is compared to the results shown by Leissa [28]. The plate thickness was selected to be 0.4 [mm]. The modal

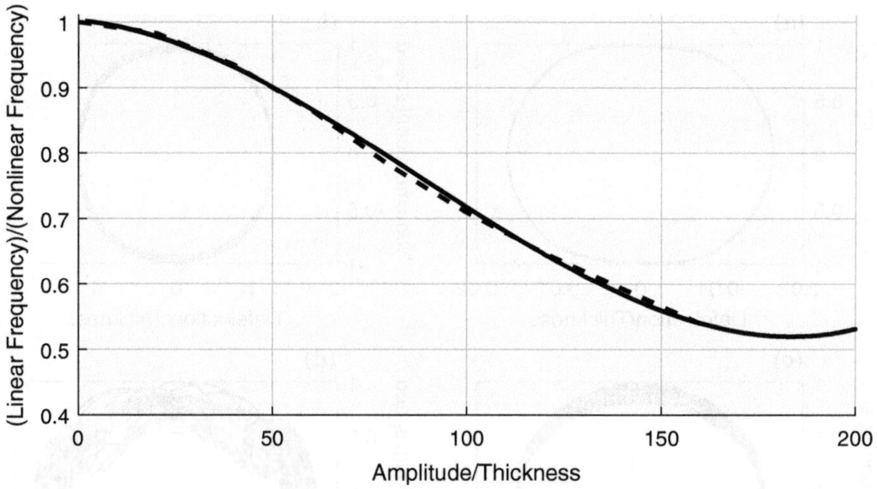

Fig. 9 Frequency response simply supported in-plane immovable edge,--: Leissa [28],—FEM results

analysis results for this case were 21.092, 52.881 [Hz] for the first and the second bending mode respectively. The full transient analysis for the plate was performed 300 times. For every run, the amplitude of the distributed force was varied according to the formula $0.0032 \times i$ [N] where $i = [1,2, \ldots, 300]$. The time histories of the deflection of the central node were recorded. Then the FFT was computed in order to extract the corresponding nonlinear resonance frequency, which is the peak in the spectrum plot. Thus it was possible to figure out the resonance at every level of excitation, and the results were compared with Leissa [28]. The comparison, as plotted in Fig. 9, shows an excellent agreement between the FEM results of the current study and the Leissa results. This gives another verification of the methodology followed in this report.

Using the resulting time-series of the central node of the plate, the phase portrait was plotted. Figure 10 shows phase portraits for six arbitrary chosen solutions corresponding to different excitation amplitudes. The phase portraits were plotted in terms of normalized velocity versus the normalized lateral deflection. The velocity was normalized by the maximum value in the time series, while the deflection was normalized by the plate thickness. Figure 10 shows a clean single loop at a low level of excitation as a sign of a periodic linear behaviour. At a higher excitation level, the loop starts to deviate to show a strand of orbits, the behaviour was transiting from periodic to aperiodic due to rising nonlinearity effects.

In order to show the transition clearly, the bifurcation diagram is extracted from the phase portraits and plotted as shown in Fig. 11. The plotting process was made by selecting the number of records periodically. The diagram shows the limit, where the linearity assumption is no more valid and the onset of the chaotic behaviour as a result of dominating the geometrical nonlinearity.

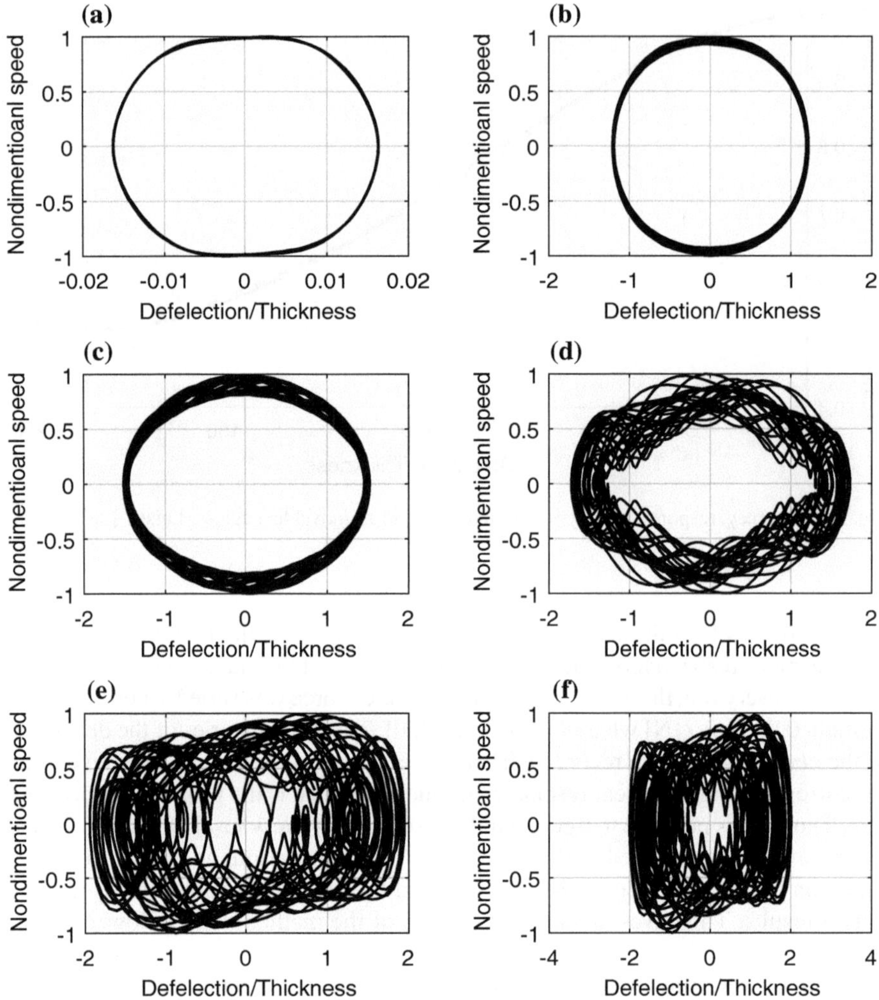

Fig. 10 Case i, the phase portraits of the central node under different level of excitation. Excitation amplitude **a** 0.0032 [N], **b** 0.0032 × 100 [N], **c** 0.0032 × 150 [N], **d** 0.0032 × 200 [N], **e** 0.0032 × 250 [N], **f** 0.0032 × 300 [N]

4.2.2 Bifurcation Diagram for Laterally Clamped in-Plane Immovable Edges

In this exercise, the boundary conditions are laterally clamped as expressed mathematically by Eq. (12) and immovable edge expressed by Eq. (14). To make the numerical analysis available for this case, parameters were taken to be same as before, otherwise it will be listed hereafter, where the plate thickness was 0.25 [mm], and the resulting fundamental frequencies are 24.189, 49.465 [Hz] first and second mode

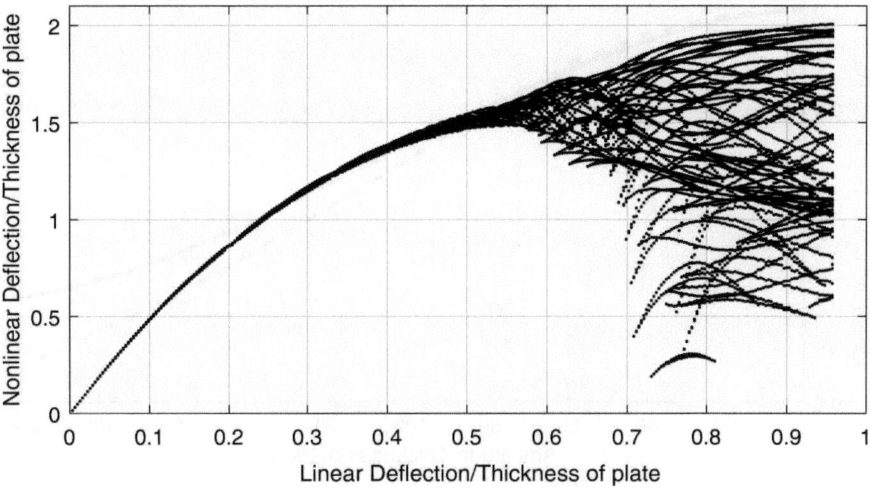

Fig. 11 Bifurcation diagram of immovable simply supported plate

respectively. The full-transient analysis was run 100 times. For every run, the level of excitation was taken from the range of 3.2e−03 to 3.2e−01 [N]. The frequency response for this case was extracted similarly to the previous case, and it was compared to the case-b in Leissa [28]. The plot in Fig. 12, shows acceptable agreement when the ratio of the amplitude to the thickness is less than 100%. In the higher range, the discrepancies start to increase, when the ratio becomes 2.00 and the discrepancy becomes 8%. The trends of the curves show that the difference could be bigger at higher ratio. The reasons could be due to the differences in methods, in spite of the fact that the finite element method is an approximation method. Nevertheless, it gives results that are more accurate when compared with analytical ones, particularly if the analytical trial shape function is chosen based on satisfying the clamped boundary condition, which is difficult to satisfy.

Similarly, the phase portraits were plotted for six levels of excitation as shown in Fig. 13. The graph shows a single orbit at low excitation where it becomes several loops at a higher excitation as an indicator of stochastic behaviour and the transition from periodic to aperiodic.

Similar to the method followed above, the bifurcation map was extracted for clamped immovable edges as shown in Fig. 14. The graph shows that the transition clearly happens when the normalized deflection becomes greater than 0.75. Below this value, the behaviour is linear and the oscillation is periodic, and if higher than this value, the behaviour is chaotic and the oscillation is aperiodic.

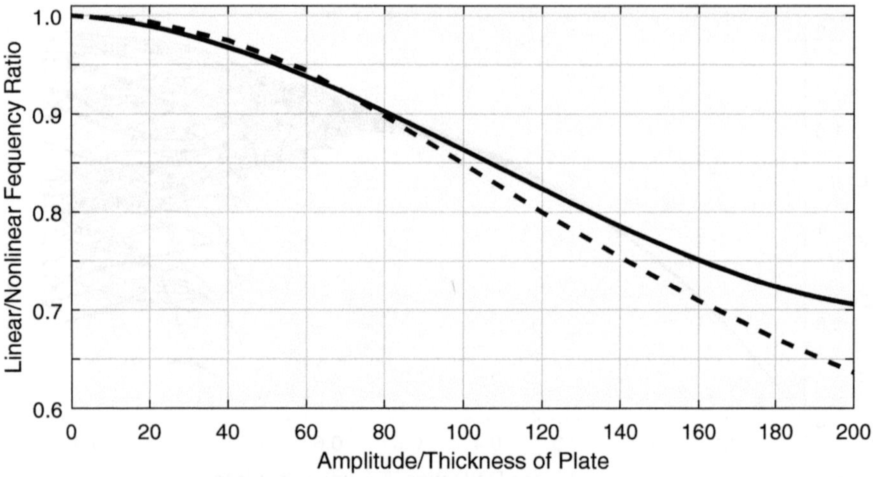

Fig. 12 Laterally clamped in-plane immovable edge,--: Leissa [28] Case b,—FEM results

4.2.3 Bifurcation Diagram for Fully Restrained Edges

In this case, all the degrees of freedom of the edge nodes are restrained ($u = v = w = Rxy = Rxz = Ryz = 0$). The thickness was selected to be 0.2 [mm]. It is less than in the previous cases in order to keep the fundamental frequency within the selected range of frequency. The modal analysis gave the fundamental frequency as 24.245, 49.680 [Hz] for the first and second mode respectively. The full transient analysis was made 300 times. For every run, different levels of excitation were applied in the range from 3.2e−04 to 300 × 3.2e−04 [N]. Similar to the previous cases, the frequency response was plotted and compared to the same case in the previous section. This is because the same boundary conditions are not covered by Leissa [28]. The comparison is plotted in Fig. 15. It shows that for fully restrained edges analysed by the FEM is stiffer than when analysed by analytical methods at a level where the amplitude is less than the thickness (Amplitude/Thickness ratio < 100) where the solution is softer at a higher ratio. The maximum discrepancies are not more than 8%. An extra investigation is suggested here to make more comparisons.

For this case, the phase portraits were plotted at different values of excitations as depicted in Fig. 16. The phase portraits again show the transition from the linear to nonlinear response and the orbit change from single loop to several loops.

Similarly, the phase portraits were condensed in one plot to show the bifurcation diagram for this case as shown in Fig. 17. Again, this diagram shows the onset of chaotic dynamics. It can be considered a design plot to avoid some sort of instability that arises due to the geometrical nonlinearity. The plots shows that the behaviour is linear when the ration of the amplitude to thickness is nearly lower than 0.5; it is nonlinear when the ratio becomes bigger.

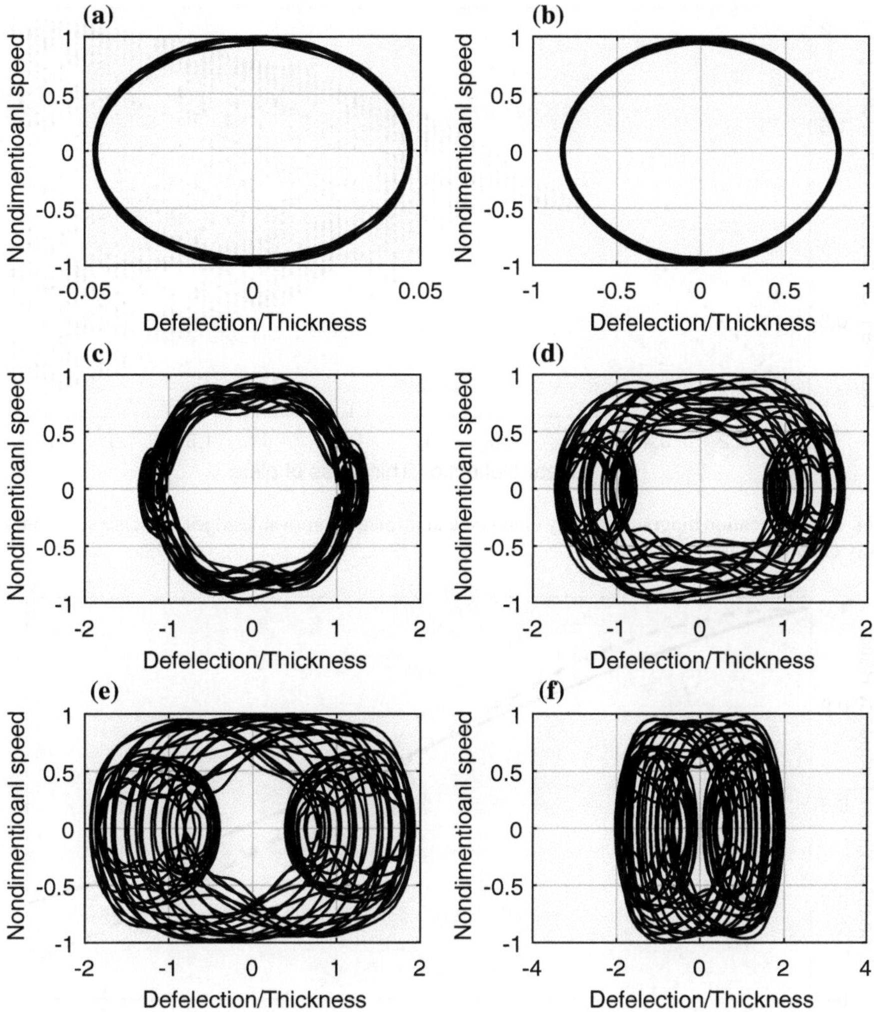

Fig. 13 Phase portrait for 6 levels of excitation for laterally clamped and in-plane immovable edges. Excitation amplitude **a** 0.0032 [N], **b** 0.0032 × 100 [N], **c** 0.0032 × 150 [N], **d** 0.0032 × 200 [N], **e** 0.0032 × 250 [N], **f** 0.0032 × 300 [N]

5 Conclusions

The nonlinear dynamics analysis for isotropic rectangular thin plates was investigated. The plate nonlinearity is due to large deflections. The finite element method was adopted for the spatial discretization process while the numerical time integration was made by Newmark's method in conjunction with the Newton-Raphson iteration technique to solve the nonlinearity. The modelling process was success-

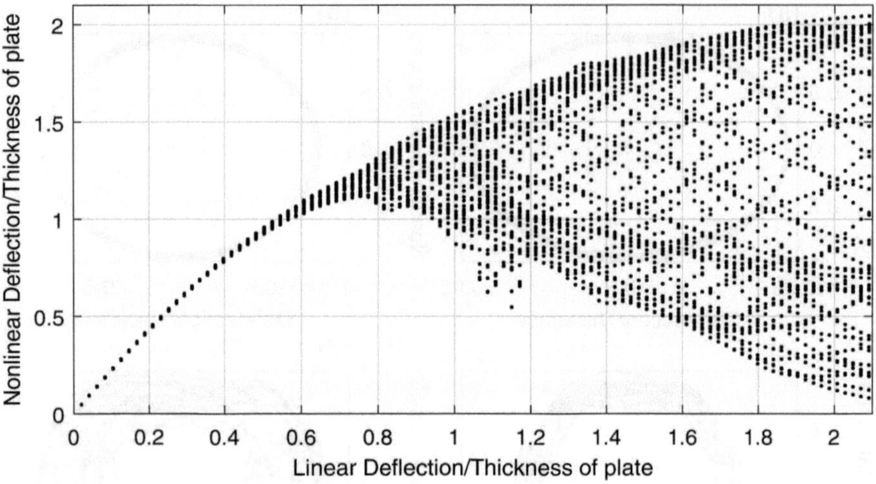

Fig. 14 Bifurcation diagram laterally clamped and in-plane immovable edges Poincare section for fully clamped plate

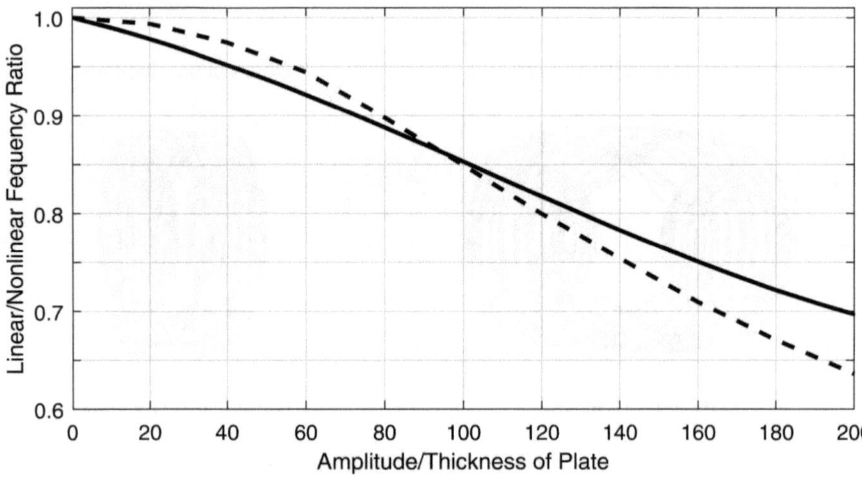

Fig. 15 Frequency response for a fully restrained edges plate—FEM results compared to clamped immovable edges case b--: Leissa [28]

fully verified by comparing with published results. In an advanced step, bifurcation diagrams were extracted as a diagnostic tool for a nonlinear dynamical system. The diagram clearly showed the periodical response, where the linearity assumption is valid and the chaotic response, where the nonlinearity is dominating the response and the transition between these two regions. Correspondingly, the results showed that the geometrical nonlinearity arises when the lateral defection of the plate is greater than one third of the plate thickness. Additionally, the frequency response curves

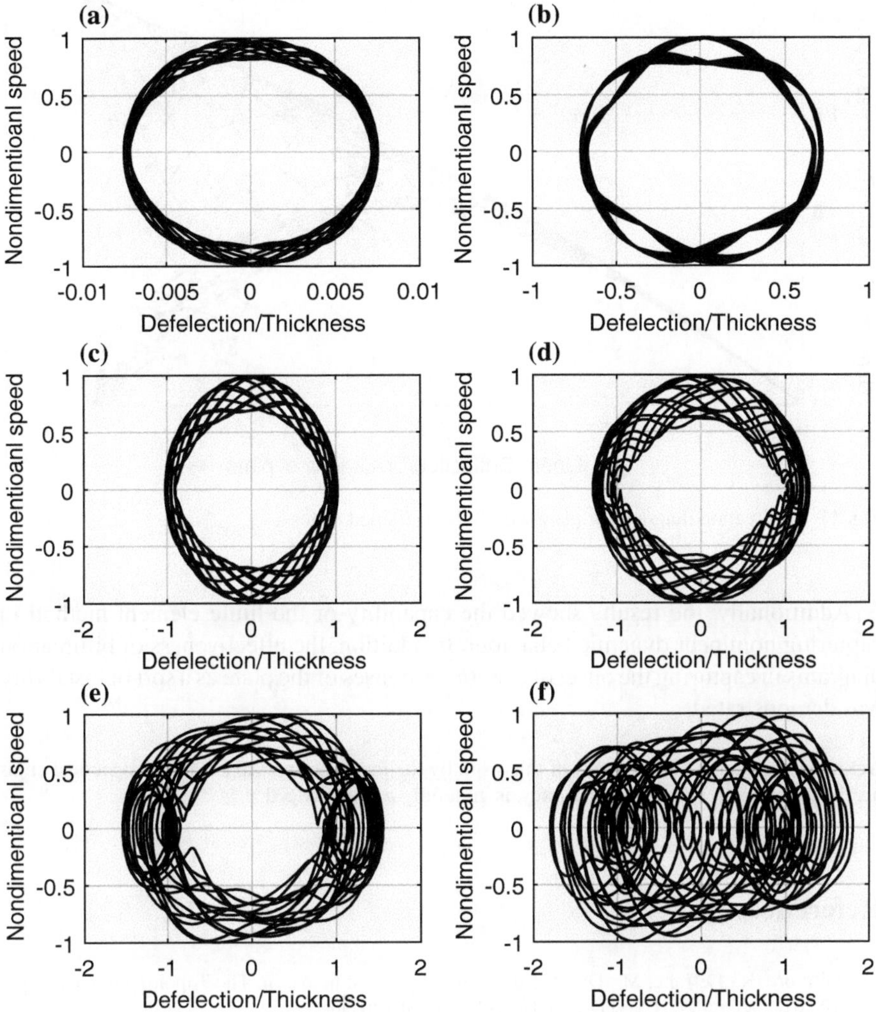

Fig. 16 Phase portrait of the fully restrained edges, excitation by one sinusoidal wave in vicinity of the first fundamental frequency with different level. Excitation amplitude **a** 3.2e−4 [N], **b** 3.2e−4 × 100 [N], **c** 3.2e−4 × 150 [N], **d** 3.2e−4 × 200 [N], **e** 3.2e−4 × 250 [N], **f** 3.2e−4 × 300 [N]

showed that the modal analysis could be used to predict the resonance frequency of loaded plates. More precisely, the resonance frequency of a simply supported immovable edge plate under lateral load, deflected to 100% of the plate thickness, is 17% greater than that computed by modal analysis. This ratio jumps to 43% for clamped immovable edges.

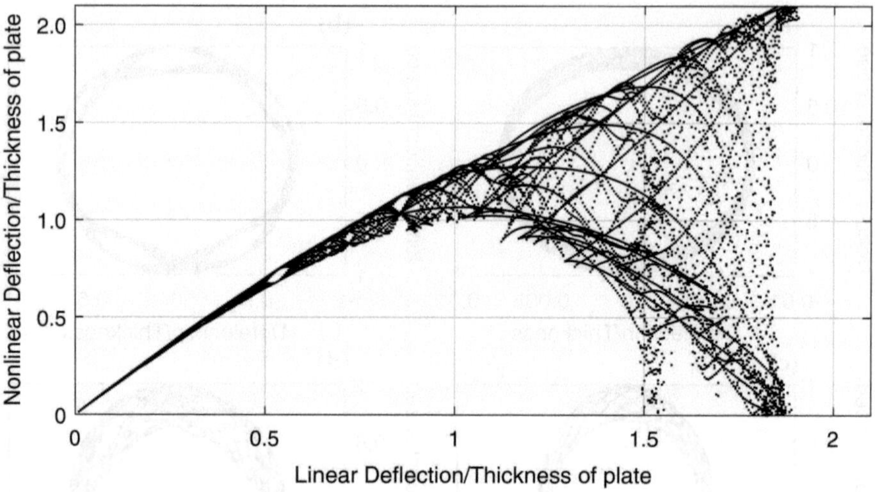

Fig. 17 Bifurcation diagram for plate with fully restrained edges

Additionally, the results showed the capability of the finite element method in capturing nonlinear dynamic behaviour. In addition, the effectiveness of bifurcation diagrams in capturing the onset of chaotic responses of the plate as a sort of instability was demonstrated.

Acknowledgements The support of this work by the faculty of aerospace and aeronautical engineering, Gaziantep University in Turkey, is gratefully acknowledged.

References

1. Gilmore, R., Lefranc, M.: Discrete dynamical systems: maps. In: The Topology of Chaos, pp. 19–104. Wiley-VCH Verlag GmbH & Co. KGaA (2011)
2. Crawford, J., Atluri, S.: Non-linear vibration of a flat plate with initial stresses. J. Sound Vib. **43**, 117–129 (1975)
3. Shooshtari, A., Razavi, S.: A closed form solution for linear and nonlinear free vibrations of composite and fiber metal laminated rectangular plates. Compos. Struct. **92**, 2663–2675 (2010)
4. Dey, T., Ramachandra, L.S.: Non-linear vibration analysis of laminated composite circular cylindrical shells. Compos. Struct. **163**, 89–100 (2017)
5. Jones, R., Milne, B.J.: Application of the extended Kantorovich method to the vibration of clamped rectangular plates. J. Sound Vib. **45**, 309–316 (1976)
6. Singal, R.K., Gorman, D.J., Forgues, S.A.: A comprehensive analytical solution for free vibration of rectangular plates with classical edge conditions: experimental verification. Exp. Mech. **31**, 21–23 (1992)
7. Leissa, A.W., Chern, Y.-T.: An approximate analysis of the forced vibration response of plates. J. Vib. Acoust. **114**, 106–111 (1992)
8. Ganapathi, M.: Large-amplitude free flexural vibrations of laminated composite curved panels using shear-flexible shell element. Def. Sci. J. **45**, 55–60 (1995)

9. Ganapathi, M., Varadan, T.K.: Large amplitude vibrations of circular cylindrical shells. J. Sound Vib. **192**, 1–14 (1996)
10. Ribeiro, P.: Periodic vibration of plates with large displacements. AIAA J. **40**, 185–188 (2002)
11. Anlas, G., Elbeyli, O.: Nonlinear vibrations of a simply supported rectangular metallic plate subjected to transverse harmonic excitation in the presence of a one-to-one internal resonance. Nonlinear Dyn. **30**, 1–28 (2002)
12. Saha, K.N., Misra, D., Ghosal, S., Pohit, G.: Nonlinear free vibration analysis of square plates with various boundary conditions. J. Sound Vib. **287**, 1031–1044 (2005)
13. Ribeiro, P.: Nonlinear vibrations of simply-supported plates by the p-version finite element method. Finite Elem. Anal. Des. **41**, 911–924 (2005)
14. Amabili, M.: Nonlinear vibrations of rectangular plates with different boundary conditions: theory and experiments. Comput. Struct. **82**, 2587–2605 (2004)
15. Amabili, M.: Theory and experiments for large-amplitude vibrations of rectangular plates with geometric imperfections. J. Sound Vib. **291**, 539–565 (2006)
16. Amabili, M., Carra, S.: Large-amplitude vibrations of rectangular plates in air or coupled to free surface liquids. In: 6th EUROMECH Conference ENOC 2008 (2008)
17. Amabili, M., Karagiozis, K., Khorshidi, K.: Nonlinear vibration of plates and higher order theory for different boundary conditions. Int. J. Struct. Stab. Dyn. **11**, 673–695 (2011)
18. Jomehzadeh, E., Saidi, A.R.: The small scale effect on nonlinear vibration of single layer graphene sheets. World Acad. Sci. Eng. Technol. **5**, 235–239 (2011)
19. Noh, G., Bathe, K.J.: An explicit time integration scheme for the analysis of wave propagations. Comput. Struct. **129**, 178–193 (2013)
20. Awrejcewicz, J., Krylova, E.Y., Papkova, I.V., Krysko, V.A.: Regular and chaotic dynamics of flexible plates. Shock Vib. **2014** (2014)
21. Ducceschi, M., Touzé, C.: Simulations of nonlinear plate dynamics: an accurate and efficient modal algorithm. In: Proceedings of the 18th International Conference on Digital Audio Effects (DAFx-15), pp. 1–8, Trondheim, Norway (2015)
22. Banerjee, M.M., Mazumdar, J.: A review of methods for linear and nonlinear vibration analysis of plates and shells. In: Procedia Engineering, 12th International Conference on Vibration Problems, ICOVP 2015, pp. 493–503. The Author(s) (2016)
23. Ventsel, E., Krauthammer, T.: Thin Plates and Shells Theory, Analysis, and Applications. CRC Press (2001)
24. Abdul Karim, S.: Nonlinear Dynamic Behaviour of Joined Lightweight Structures. http://d-nb.info/101505823X/34 (2011)
25. Padt, I., Young, S., Strain, J., Miller, E.: Introduction to the ANSYS Parametric Design Language (APDL). CreateSpace Independent Publishing Platform (2013)
26. Zienkiewicz, O.C., Taylor, R.L., Taylor, R.L.: The Finite Element Method For Solid And Structural Mechanics (2005)
27. Pica, A., Wood, R.D., Hinton, E.: Finite element analysis of geometrically nonlinear plate behaviour using a mindlin formulation. Comput. Struct. **11**, 203–215 (1980)
28. Leissa, A.W.: Vibration of plates. Scientific and Technical Information Division, National Aeronautics and Space Administration. [For sale by the Supt. of Docs., U.S. Govt. Print. Off.] Washington (1969)

Topological and Contact Force Analysis of a Knee Tumor Prosthesis

T. De la Mora Ramirez, M. A. Doñu Ruiz, I. Hilerio Cruz, N. López Perrusquia and E. D. García Bustos

Abstract In the present work, a topological analysis of a knee tumor prosthesis using grade 5 biomedical titanium alloy material (Ti6Al4V) and an ultra-high molecular weight polyethylene insert (UHMWPE) was performed. A three-dimensional finite element analysis (FEM) model was used, taking into account the von Mises stress and the contact forces (CPRESS), using the finite element software ABAQUS that analyzes the mechanical contact between two surfaces. The contact pressure due to the contact force and shear stress from friction, especially in the contact areas in the UHMWPE and Ti6Al4V parts, to be considered occur during the following activities: (1) At the critical point of the walking cycle at 15° of femur flexion with respect to the tibia with a load of 3.3 times the body weight (BW), (2) Climbing stairs with a bend angle of 83° and 3.5 BW and (3) Squatting with an angle of 130° and 5.6 pcs (Alsamhan in J King Saud Univ Eng Sci 25:49–54, 2012, [1]; Brassard in Complicaciones de la artoplasia total de rodilla. Insall, México, 2007, [2]; Buehler in J Arthroplasty 15:698–701, 2000, [3]). The Analysis of the knee tumor was performed by scanning the parts, digitizing them in STL format and debugging the images using CAD software. We performed the simulation of these three conditions using finite element analysis software, obtaining a maximum stress of 229,497 MPa in the squat condition and a displacement of 20.14 μm in this same condition. We used the von Mises failure criterion equation (Jhonson in Mecánica de contacto. Cambridge University, México, 1985, [4]). For the analysis of contact forces in the joints we used the results of CPRESS and were validated with an analytical contact Hertz study (Ludema in Friction Wear Lubrication. CRC Press, Michigan, 1996, [5]). The topology indicates that there are material savings of 80%, mainly on the titanium alloy of the knee. As reference, a male Mexican patient with a weight of 70 kg and 1.70 m of height was taken.

T. De la Mora Ramirez (✉) · I. Hilerio Cruz
Universidad Autónoma Metropolitana Unidad Azcapotzalco, Azcapotzalco, Ciudad de México, Mexico
e-mail: tommora077@yahoo.com

M. A. Doñu Ruiz · N. López Perrusquia
Universidad Politécnica del Valle de México, Tultitlán de Mariano Escobedo, Mexico

E. D. García Bustos
CUCEI, Universidad de Guadalajara, Guadalajara, Jalisco, Mexico

© Springer International Publishing AG, part of Springer Nature 2019

A. Öchsner and H. Altenbach (eds.), *Engineering Design Applications*,
Advanced Structured Materials 92, https://doi.org/10.1007/978-3-319-79005-3_20

Keywords Prosthesis · Finite element analysis · Hertz contact · UHMWPE

1 Introduction

The knee joint is the most complex joint of the human body, due to the critical loads experienced in the various movement conditions [6]. At present, nutritional choices along with a sedentary lifestyle are provoking high rates of obesity, which have increased the incidence of joint diseases [7, 8], and in the specific case knee tumor diseases, requiring total knee replacement surgery [1]. Currently, the main problems of knee prostheses are: (1) The costs of recovery and intervention are very high. (2) The average lifetime of its components, mainly of the pieces made with ultra-high molecular weight polyethylene (UHMWPE) which lasts 15 years under conditions of adequate surgery and passive use [3, 9].

Knee tumors are a frequent cause of disability, due to the magnitude of the total replacement of the knee joint. It is cataloged as an extremely painful, incapacitating disease, and, in addition to generating socio-economic problems and disability due to the high costs of proper prosthesis, surgery, recovery and loss of patient employment. The price of a titanium prosthesis is around 25 thousand pesos, and is of foreign manufacture, and adding the cost of surgery, the cost increases to about 45 thousand pesos [2], which does not make them affordable for public health institutions.

There is a great variety of prosthesis options for transfemoral amputations, because of their low cost. The main use in public institutions is the passive type knee prosthesis. After surgery, the patient is instructed to lower the pace of their sporting activity and social activities to the extent of changing jobs or making it impossible to perform any labour.

For this reason, this research seeks to develop a low-cost knee prosthesis for widespread use in public health institutions in Mexico by conducting a topological analysis of the contact forces of a knee tumor prosthesis made of titanium alloy grade 5 (Ti6Al4V) material and ultra-high molecular weight polyethylene insert (UHMWPE), using a three-dimensional model of finite element analysis, taking into account von Mises efforts and contact stresses, especially in the zones of greater wear that occur in the pieces of polyethylene material of ultra-high molecular weight in everyday activities such as walking, climbing stairs and squatting.

2 Modeling of the Prosthesis

The tumor knee prosthesis is parameterized by means of a digital camera of the David 3D brand to guarantee the reproducibility of the precise dimensions of the commercial model, digitizing it in STL format and debugging the images using CAD software.

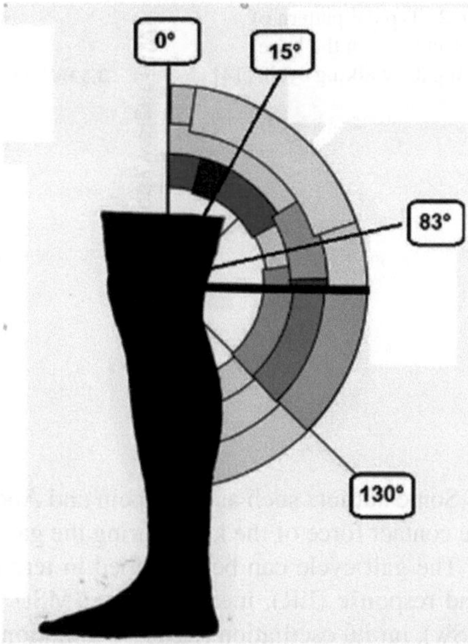

Fig. 1 Degrees of flexion of the knee, in the 0°, 83° and 130° positions with reference to the sagittal plane. *Source* Aaron Swanson

Table 1 Main activities of an individual with their respective degree of flexion and applied load [10, 11]

Activity	Flexion in the knee (°)	Body weight (x BW)
Hike cycle	0–67	3.0
Climbing stairs	0–83	3.5
Squatting	0–130	5.6

2.1 Study Cases

A static analysis was performed at the point where the forces are acting during the critical activity of the cycles of walking, climbing stairs, and squatting, due to the maximum strengths that occur in each of the activities. For the present study, it was necessary to know the ranges of movements generated by the different knee joints, in order to obtain the concentration of efforts in each of the stages.

The sagittal plane is used to obtain the angles generated by the knee flexion between the femur and the tibia. A healthy person has a 0° flexion when the leg is fully extended and a 135° angle when in a squatting position (Fig. 1).

Table 1 shows the degrees of knee flexion during the activities to be analyzed and their load applied with respect to the sagittal plane from [10, 11].

During the walk cycle the maximum force was taken, which occurs at 15° with a value of 2500 N as shown in Fig. 3 [12].

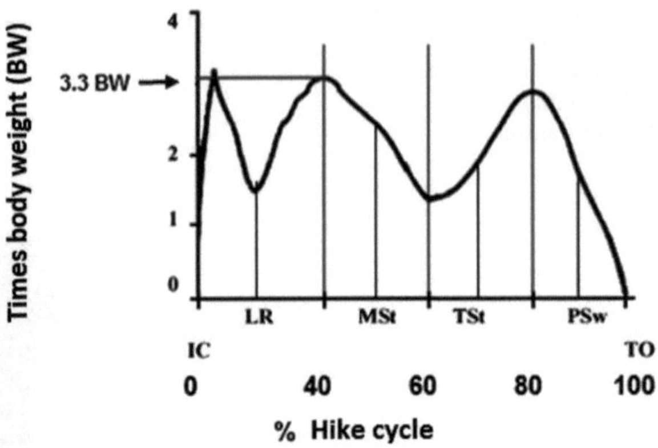

Fig. 2 Typical pattern of contact force in the knee during the walking cycle [14]

Some authors such as Schippein and Andriacchi developed the typical pattern of the contact force of the knee during the gait phase, describing it as follows:

The gait cycle can be described in terms of each phase as: initial contact (IC), load response (LR), mean position (MSt), terminal position (TSt), pre-oscillation (PSw), initial oscillation (ISw)—oscillation (MSw) and terminal oscillation (TSw). The posture period consists of the first five phases: initial contact, load response, middle position, terminal posture and pre-oscillation (see Fig. 2).

To perform the analysis of the tumor prosthesis, a male Mexican patient weighing 70 kg and 1.75 m of height was taken as a reference. This weight was taken because a patient undergoing this type of surgery is subjected to a diet to reach their ideal weight [13].

Schipplein described the gait cycle as a function of body weight (BW), having the maximum load between the body weight response period and the mean position applied to the knee joint. Figure 2 shows the profile of the walking cycle, in the lapse of zero degrees, the profile of the period of response of load presents a peak of maximum force of 3.3 times the body weight of the individual.

Taking the information of the weight of the person as 70 kg and matching the two previous criteria of Mulholland and Shipplein, which indicate that the maximum flexion is found when the flexion of the femur with respect to the tibia is of 15°. In that point we have an axial load value of 2250 N (Fig. 2a).

Figure 3 is the typical pattern of contact force in the knee during the walking cycle [14].

Using Eq. (1) to obtain the approximate value of body weight (BW) taking the criterion of Shippein [14].

$$Load = (x)\,BW \tag{1}$$

where: x is the number of times body weight.

BW: Patient weight [N].

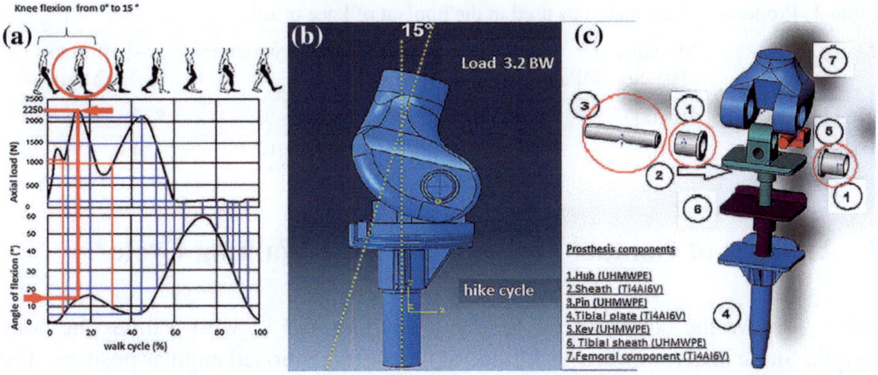

Fig. 3 Conditions for the running cycle, **a** human gait cycle with their respective flexion angles [12], **b** Angle for the gait cycle between femur and tibia and the number of times of loading applied **c** components of the prosthesis

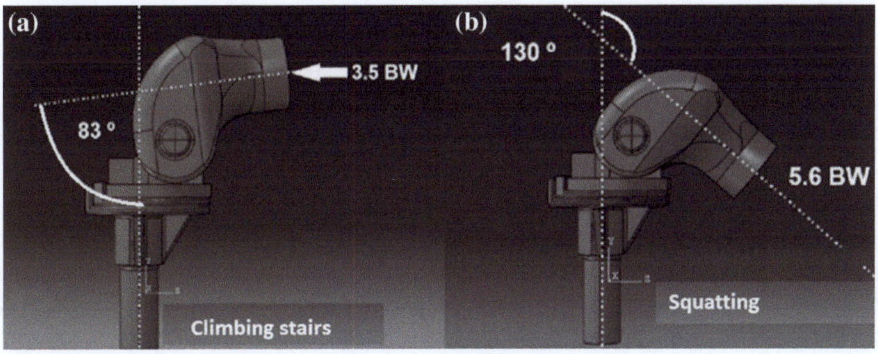

Fig. 4 Angle conditions between femur and tibia, **a** climbing stairs, **b** squatting

Substituting the value of 2250 N for the magnitude of the axial load in the Mulholland gait cycle to Eq. 1, and taking the value of the transverse area of the prosthesis in the area of the femur to be 641.30 mm^2 as shown in Fig. 5, solving for x the value is 3.27 times body weight, a value that agrees with the authors Schippein and Andriacchi although they are different criteria.

Mow, Mulholland, and David, obtained values for climbing up stairs and squatting conditions [12]. Table 1 shows the degrees of knee flexion during activities that were taken as reference.

The scheme shown below (Fig. 4) illustrates the positions of the tumor prosthesis during activity requiring more body weight.

Table 2 Properties of the materials used in the implant of knee prosthesis

Material	Modulus of elasticity [MPa]	Poisson constant	Density [tn/mm^3]	Last effort to creep [MPa]
Ti6Al4V	105,000	0.342	4.43e−9	827
UHMWPE	1080	0.4	9.7e−10	24

3 Analysis of Contact Forces During the Walking Cycle

Table 1 shows the condition for the cycle running at 15°, with 3 times the body weight. Static analyses were performed at the aforementioned angular position. The properties of the materials to be used in the simulation are shown in Table 2.

Modulus of elasticity values were obtained from stress tests, properties of the Poisson constant, density and ultimate stress at creep were obtained from literature [15–17]. The above data was inserted into the software in the material properties module.

3.1 Meshing and Constraint

For the meshing of the piece, C3D8R and linear C3D4 tetrahedral elements were used. Interactions were used between each of the pieces of the prosthesis that was in contact, using the method of surface-to-surface discretization with a coefficient of dynamic friction lubricated between the parts of titanium alloy of 0.275 [18] and the pieces of the polyethylene with those of titanium alloy 0.056 [19]. The load was applied in the form of pressure in the upper part of the knee and a fixed restriction in the lower part of the tibial plate.

4 Results

A simulation was first performed in the critical condition of the walking cycle with a load of 3.5 BW, mainly analyzing the efforts of von Mises, the displacements in the area between the knee and the articulated sheath, as well as the contact pressures (CPRESS) on the bolt and bushing components. Figure 5 shows the reference points where the tensions were taken between the knee and the hinged sheath under the conditions studied. In the knee and hinged sheath, it is important that contact is avoided because friction between the parts made of the biomedical titanium alloy material will be avoided.

The prosthesis without the loading application has a gap between the points shown in Fig. 5 of 0.5 mm.

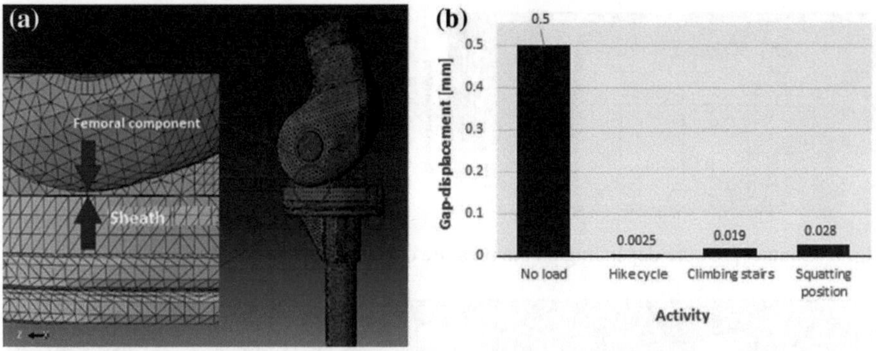

Fig. 5 **a** Separation space between knee and titanium alloy hinged sheath (Ti-6Al-4V), **b** Graph of displacements in the activities analyzed

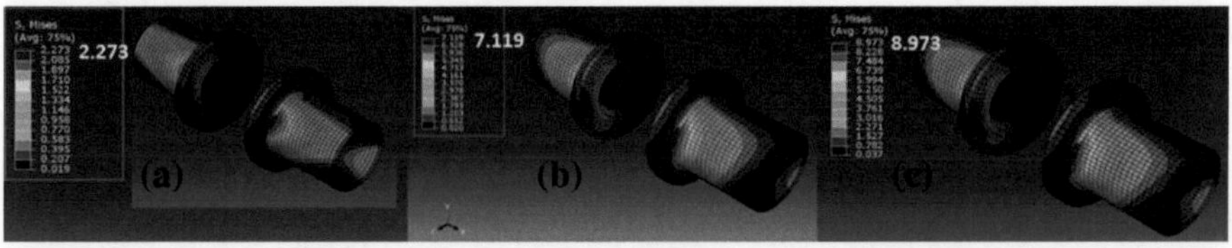

Fig. 6 Von Mises effort on hubs. **a** Cycle walk, **b** Climbing stairs, **c** Squatting

The pieces of ultra-high molecular weight polyethylene material in the prosthesis are the key, and the bushings and the tibial sheath, the ultimate stress to the fluence of the Polyethylene is 24 MPa [16, 17], as shown in Table 2.

The displacements are shown in Fig. 5b, and the graph shows that the largest displacement is in the squatting position with a value in the y-axis direction of 0.028 mm.

The part of interest in the investigation are the bushings due to the contact between the titanium and the polyethylene, and are the ones that support the entire weight of the patient, limiting the useful life of the prosthesis. They present a maximum von Mises effort of 8973 MPa and a minimum of 2273 MPa in the squat and cycle conditions respectively (see Fig. 6). Figure 7 shows the graphs corresponding to the voltage contour diagrams in the three conditions, against the distance of the bushing length.

Figure 7 shows the graphs of contact profile from one end to the end of the bushing. There is a slight upward trend, reaching the maximum values on the right side of the bush zone in contact with the knee. The maximum pressure contact values are not significant during the walk, but are elevated to a large extent in the squatting position, causing possible failures due to fatigue during loading and unloading cycles.

The knee has very little contact pressure because it has the bushings that considerably reduce it. This is another result to be taken into account for the topological analysis of the knee.

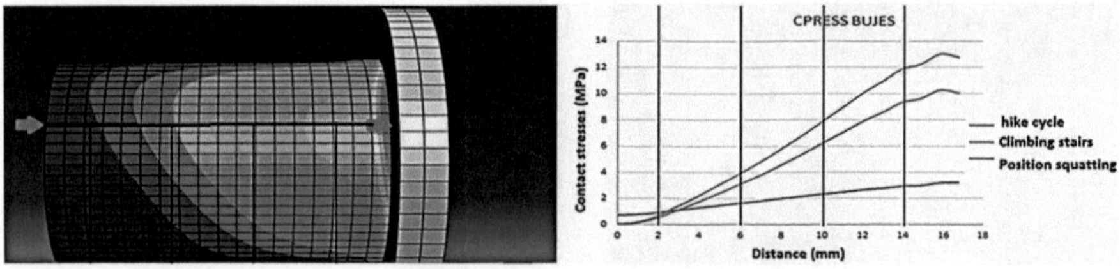

Fig. 7 Contact pressure depending on the pressure contact in the bushing

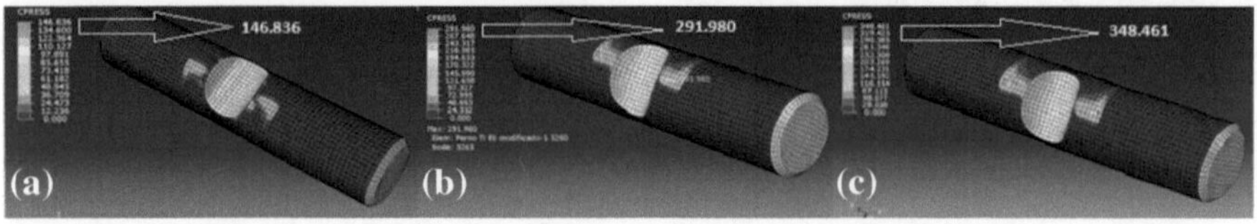

Fig. 8 Stress of pin contact pressures. **a** Cycle walk, **b** Climbing stairs, **c** Squatting

Concluding that the critical part by the previous results is in the bolt with a maximum contact force and therefore the bush with which it is assembled.

4.1 Activity Climbing Stairs

In this condition, the maximum von Mises stress in the polyethylene bushing was 7199 MPa and a minimum of 2693 MPa in the key, just as the non-overrun cycle activity passes the ultimate stress to the UHMWPE creep.

In the titanium alloy bolt the maximum effort is 65,497 MPa in the areas located in the part where it is assembled with the hub.

The most important contact conditions in the running cycle are on the bolt as can be seen in Fig. 8.

4.2 Squat Activity

In this activity, von Mises' maximum stresses, displacements and contact stresses on the polyethylene and titanium alloy parts are presented, without failing to take into account von Mises' failure criterion. The contact force for all three conditions is presented on the bolt with a critical value of 348,461 MPa squatting condition (see Fig. 8). This value will be taken as a reference for the analysis of Hertz forces, covering the loads of previous activities.

Table 3 Results table of von Mises stresses and contact pressures on the bolt, bushings and knee of the prosthesis

Activity	Parts of the prosthesis	Von Mises stress [MPa]	Maximum contact force [MPa]
Hike cycle	Hub	2.27	5.25
	Pin	41.48	146.83
	Knee	18.62	4.48
Climbing stairs	Hub	7.11	15.38
	Pin	65.49	291.98
	Knee	30.07	13.36
Squatting position	Hub	8.97	19.28
	Pin	80.43	348.46
	Knee	41.19	18.97

Table 3 shows a summary of von Mises maximum stress and contact pressure in the pieces of the knee tumor prosthesis, with the squatting position requiring greater concentration of stresses and contact pressures.

4.3 *Topological Analysis*

Based on the results obtained for the knee in the squat position, the topological analysis was performed as an objective function to reduce the volume. The initial mass of the knee is 359.14 g, the Ti-6Al-4V has an elastic limit between 875 and 995 MPa and a last yield stress of 795–843 MPa, values well above the results of the maximum effort of Von Mises of 41.19 MPa and maximum contact pressures of 18.97 MPa.

Figure 9 shows the proposed knee with a material saving of 77%, reducing the material's cost considerably.

4.4 *Hertz Contact Forces*

The mechanics of contact is a fundamental tool for studies in the area of tribology and solid mechanics because most interactions between machine components are made through their surfaces. The forces resulting from these interactions between the surfaces of the elements generate contact stresses. In many cases, the contacts take place in very localized regions of the surfaces of the machine components, high contact stresses being achieved at such sites and their neighborhoods. The knowledge of the location and the magnitude of these efforts is necessary for the realization of a good design.

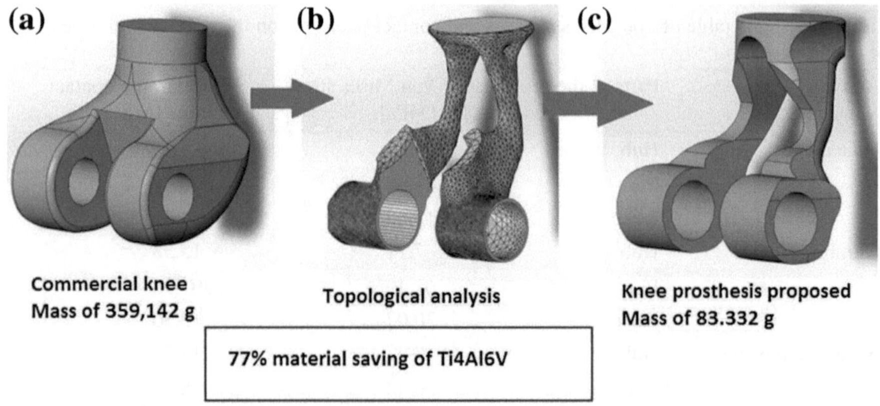

Fig. 9 Knee prosthesis topology. **a** Commercial knee, **b** result of the topological analysis, **c** proposed knee geometry

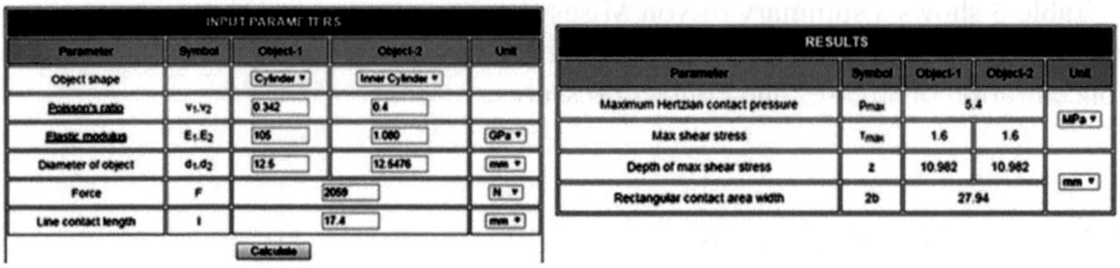

INPUT PARAMETERS				
Parameter	Symbol	Object-1	Object-2	Unit
Object shape		Cylinder	Inner Cylinder	
Poisson's ratio	v_1, v_2	0.342	0.4	
Elastic modulus	E_1, E_2	105	1.080	GPa
Diameter of object	d_1, d_2	12.5	12.5476	mm
Force	F	2059		N
Line contact length	l	17.4		mm

Calculate

RESULTS				
Parameter	Symbol	Object-1	Object-2	Unit
Maximum Hertzian contact pressure	P_{max}	5.4		MPa
Max shear stress	τ_{max}	1.6	1.6	
Depth of max shear stress	z	10.982	10.982	mm
Rectangular contact area width	2b	27.94		

Fig. 10 Hertzian contact voltage calculator data entry and results

Using the Hertz maximum contact pressure equation in the present work it is intended to develop a general method for the solution of elastic contact problems modelable in two dimensions, characterized by a bushing-bolt deformation state. In general, in this method the following variables are considered: intensity and shape of the load distribution, stress states, displacements and deformations of the points located in the contact zone and their neighborhoods. In addition, the method developed to obtain a numerical solution to the problem of the Hertzian contact of two long aligned cylinders is applied, which is contrasted with the analytical solution given by Hertz [9].

The contact configurations were calculated using a "Hertzian contact voltage calculator" [16–21]. The input values of the parameters and results shown in Fig. 10.

According to the Hertz's theory, the maximum pressure in the contact zone occurs in the center of the contact area and its value can be calculated through the following expression, expressed by Eq. (2) [20]:

$$P_{\max} = \frac{2F}{\pi b l^2} \tag{2}$$

Table 4 Variables for calculating Hertz contact forces

Parameter	Variable	Pin	Hub	Units
Poisson constant	ν	0.342	0.4	
Modulus of elasticity	E	105000	1080	[MPa]
Diameter	D	12.5	12.5476	[mm]
Force	F	9817		[N]
Length	L	17.399		[mm]

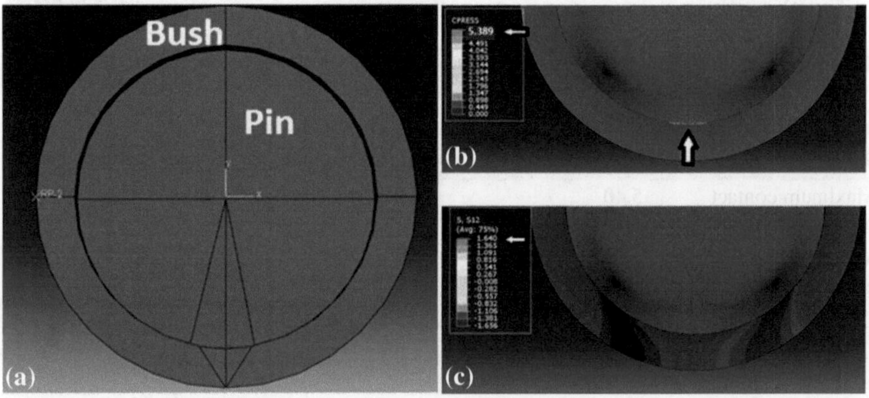

Fig. 11 2D **a** drawing of bushing and pin for obtaining Hertz contact, **b** CPRESS contact force between bushing and bolt and **c** the Effort in the direction S12 between the bolt and hub

To carry out the analysis of wear only the configuration of bush-bolt was taken. The knee prosthesis-bushing dimensions are given in Table 4.

The results obtained are as follows: Maximum Hertz contact force of 5.4 MPa, Maximum shear stress of 1.6 MPa and a rectangular contact area (bl^2) of 27.9 mm^2

4.5 Hertz Contact Analysis by Finite Element

For the simplification of the contact analysis, the only parts involved were the bush and the bolt. The 2D drawings were used to optimize computational resources and to speed up the results (Fig. 11). The maximum contact pressure (CPRESS) equivalent to the Hertz contact force, is 5389 MPa. This is located in the lower part of bushing and bolt contact (Fig. 11b). The maximum shear stress is given by the variable S12 of the finite element results giving a value of 1640 MPa (see Fig. 11c).

The displacement is obtained by measuring the contact distance in the nodes involved.

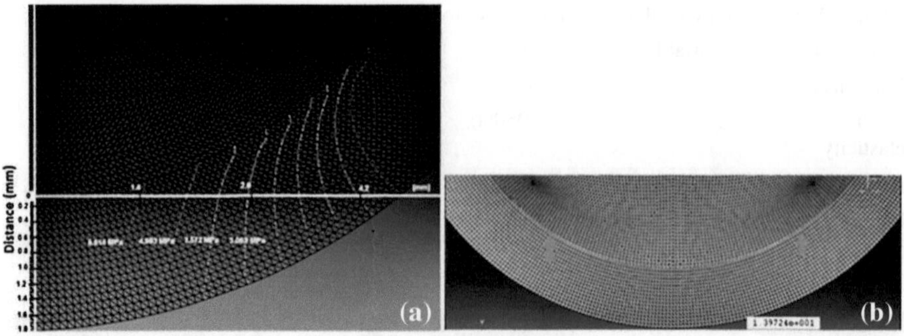

Fig. 12 **a** Isobaric tension stress lines and **b** Distance b, of the contact width between pieces

Table 5 Hzcontact results with different methodologies for obtaining results

Parameter	Calculator data	Finite element data	Error rate %
Maximum contact pressure Hertz [MPa]	5.40	5.39	0.2
Maximum cutting effort [MPa]	1.60	1.64	2.5
Contact width [mm]	27.94	13.97 (2) = 27.94	0

The contact width between the bolt and the hub is 2b, if the value of b is 13.97 mm^2, 2b = 27.94 mm^2. Figure 12b shows the isobaric lines of the tension lines as a function of the transverse distance, indicating that the maximum pressure is at a distance of 1.4 mm, as it is a symmetrical figure the total distance of maximum pressure is 2.8 mm.

Table 5 shows a summary of the results obtained by the online calculator, as well as the finite element analysis, giving a maximum error rate of 2.5% in the maximum shear stress.

The results obtained by the finite element analysis with the Hertzian calculator in line were obtained using the values of maximum contact pressure and contact width given in Table 5. It can be seen from Fig. 12 that the maximum contact pressure occurs at the lower contact portion between the hub and bolt, 5.39 MPa. This contact pressure is directly related as a function of the value obtained in the contact width (b) and length of contact giving the contact area. Therefore, the lower the contact area, the lower the contact pressure, which leads to a longer period of time in the part and therefore reduces its useful life. The results obtained from the maximum contact pressures in the UHMWPE bushings indicate not to modify the joint part between the bolt and the hub in the new topology, in addition to concluding that the result is below the yield material of the hub which is 21–28 MPa.

5 Conclusions

- ➢ The maximum stresses do not exceed the yield stress of either of the two materials, so that the material does not present any of its conditions in the analysis performed with the commercial knee.
- ➢ The topological analysis of the knee gave us a material saving of approximately 80%.
- ➢ The comparison of results with Hertzian in-line contact voltage calculator and the MEF gives us a percentage error of 2.5%
- ➢ By means of the maximum contact pressures in the bushings of UHMWPE we are given an idea not to modify the joint part between the bolt and the hub in the new topology, in addition to concluding that the result is below the limit of Material flow of the hub which is 21–28 MPa.
- ➢ The effort is not very high because there is no punctual contact between the bolt and the bushings, giving a contact width 2b of 27.94 mm, which ensures that the generated tension is within the elastic limit (33–35 MPa) of the ultra-high molecular weight polyethylene.
- ➢ The isobaric curves will serve as reference for the modification of the surface texture of the UHMWPE bushings.

References

1. Alsamhan, A.M.: Rationale analysis of human artificial knee replacements. J. King Saud Univ. Eng. Sci. **25**, 49–54 (2012)
2. Brassard, M.F.: Complicaciones de la artoplasia total de rodilla. Insall, México (2007)
3. Buehler, K.O.: The press-fit condylar total knee system: 8- to 10-year results with a posterior cruciate-retaining design. J. Arthroplasty **15**, 698–701 (2000)
4. Jhonson, K.: Mecánica de contacto. Cambridge University, México (1985)
5. Ludema, K.: Friction Wear Lubrication. CRC Press, Michigan (1996)
6. Kumbhalkar, M.A.: Modeling and finite element analysis of knee prosthesis with and without implant. Univ. J. Comput. Math. **1**, 56–66 (2013)
7. Lübbeke A., Duc S.: BMI and severity of clinical and radiographic signs of hip osteoarthritis. Obesity **17**, 14414–1419 (2010)
8. Teichtahl, A.J., Wang, Y.: Obesity and knee osteoarthritis: new insights provided by body composition studies. Obesity **16**, 232–240 (2008)
9. Baena, J.C., Wu, J.: Wear performance of UHMWPE and reinforced UHMWPE composites in arthroplasty applications: a review. Lubricants **3**, 413–436 (2015)
10. David, D.: In vivo tibial force measurement after total knee arthroplasty. Bioengineering 37–55 (2007)
11. Mow, V.C.: Basic Orthopedic Biomechanics. Reven Press, New York (1991)
12. Mulholland, S.J.: Activities of daily living in non-Western cultures: range of motion requirements for hip and knee joint implants. Int. J. Rehabil. Res. **24**, 191–198 (2001)
13. Sánchez, M.M.: Cirugía Ortopédica. Artrosis de Rodilla. McGraw-Hill, México (2004)
14. Schipplein, O.D.: The influence of initial horizontal weight placement on the lumbo-sacral spine moment. Presented at the Orthopaedic Research Society, 907–912 (1991)
15. Collings, E.W.: Materials Properties Handbook: Titanium Alloys. ASM International, USA

16. Yang, R.-S.: Contact stress on polyethylene components of a new rotating hinge with a spherical contact surface. Clin. Biomech. **16**, 540–546 (2001)
17. Veiga, D.: Properties and applications of titanium alloys: a brief review. Titanium Alloys **32**, 133–148 (2012)
18. Hamrock, B.J.: Diseño de elementos de Máquinas. McGraw-Hill, México (2000)
19. Park, J.: Bioceramics, Properties, Characterizations and Applications. Springer, USA (2008)
20. Jorge, Castrejón: Diseño y contol para una protesis de rodilla tipo policéntrica. Tesis UNAM, México (2013)
21. Solutions, A.M.: Advanced Mechanical Engineering Solutions. Obtenido de http://www.amesweb.info/HertzianContact/HertzianContact.aspx (2016)

Wettability, Interfacial Tension (IFT) and Viscosity Alteration of Nanofluids Under Electromagnetic (EM) Waves for Enhanced Oil Recovery (IFT) Applications

Keanchuan Lee, Muhammad Adil, Hasnah Mohd. Zaid, Beh Hoe Guan, Hassan Soleimani and Martin Weis

Abstract Zinc oxide (ZnO) nanoparticles (NPs) were prepared and synthesized via sol-gel method, by using citric acid as a precursor. The annealing temperatures were 500 and 800 °C which yielded average particle sizes of 56 and 117 nm, respectively. ZnO nanofluids of two different particle sizes (56 and 117 nm) were prepared using 0.1 wt% nanoparticles that were dispersed into brine (3 wt% NaCl) along with SDBS as dispersant. The impact of EM waves on wettability, viscosity and IFT of nanofluid was investigated. Results showed further reduction in wettability under EM influence with increased viscosity and reduced IFT. These results reveal a potential way to employ water-based ZnO nanofluids for enhanced oil recovery purposes at a relatively high reservoir temperature.

Keywords Nanofluid · EM waves · Wettability · Viscosity · IFT

K. Lee (✉) · M. Adil · H. Mohd. Zaid · B. H. Guan · H. Soleimani
Fundamental and Applied Sciences Department, Universiti Teknologi PETRONAS, 32610 Seri Iskandar, Perak, Malaysia
e-mail: lee.kc@utp.edu.my

M. Adil
e-mail: MuhammadAdil86@hotmail.com

H. Mohd. Zaid
e-mail: hasnamz@utp.edu.my

B. H. Guan
e-mail: beh.hoeguan@utp.edu.my

H. Soleimani
e-mail: hassan.soleimani@utp.edu.my

M. Weis
Organic Electronics Lab, Institute of Electronics and Photonics, Slovak University of Technology, Ilkovicova 3, 81219 Bratislava, Slovakia
e-mail: martin.weis@stuba.sk

© Springer International Publishing AG, part of Springer Nature 2019
A. Öchsner and H. Altenbach (eds.), *Engineering Design Applications*, Advanced Structured Materials 92, https://doi.org/10.1007/978-3-319-79005-3_21

1 Introduction

Nanotechnology has developed rapidly in decades. Scientists and researchers focus on this field since a lot of finding showed that nanotechnology can provide alternatives and solutions to various kinds of problems encountered in research. Nanotechnology is the science of materials in a range very close to molecular dimensions (1–100 nm) which has changed our viewpoint in many scientific aspects and has shown novel solutions for unsolved old problems found in previous technologies [1–9]. Therefore, nanotechnology is beneficial to humankind.

Wettability plays a main role in oil recovery [1–5] where in general water wet rocks will show better recovery efficiency compared to those oil wet rocks. However, several studies also showed that intermediate wet formation, i.e. oil and water mixture, showed even better recovery. In order to alter the wettability, metal oxide NPs might be good candidates. It has been reported that various metal oxide NPs have been used to improve the oil recovery [6, 7]. Results showed the size of the particles might have direct impact on the recovery efficiency due to the change of viscosity, surface tension etc.

NPs have the ability to change certain factors in the formation and in oil properties and have been developed to promote wettability alteration, reduction of viscosity ratio, stabilization of foam or emulsion and interfacial tension reduction—in some cases involving application of an external electric or magnetic field. With these, EOR application will be getting better and the production of hydrocarbon can be increased. By injecting a nanofluid into formations and studying its effects on oil recovery, oil and gas Industry can get a lot of benefits if the idea is proven. According to Ogolo et al. [10], some selected types of nanoparticles that are likely to be used include oxides of aluminium, zinc, magnesium, iron, zirconium, nickel, tin and silicon. Here, the authors would like to focus on the study regarding nanoparticles and nanofluid of zinc oxide (ZnO).

Current surfactant and chemicals use in EOR is not really compatible with High Pressure and High Temperature (HPHT) condition. Therefore, more studies need to be conducted in order to check the stability of the nanoparticles under HTHP condition. In this study, we will not go for compatibility studies of ZnO nanoparticle in HTHP condition but we believe that this study will create possibilities for nanoparticles and nanofluids to be used in EOR. In this study, laboratory experiments were designed to establish the relationship between governing parameters and EM-assisted oil recovery due to nanofluids. The crude oil/brine/rock (COBR) interaction was observed using contact angle on quartz plates, which were used as the solid-phase, to reveal the possible oil displacement mechanism under EM waves. In addition, fluid-fluid interactions (oil/brine/nanofluid) were evaluated using interfacial tension results, along with viscosity ratio between nanofluids and crude oil.

2 Methodology

ZnO NPs were synthesized using sol-gel auto-combustion method [11, 12], and were used after calcining at 500 and 800 °C having an average particle size of 56 and 117 nm, respectively. In this paper, these particles were denoted as ZnO@500 and ZnO@800, respectively. The analytical grade of sodium dodecylbenzenesulfonate (SDBS) from Sigma Aldrich was used as stabilizer, without further purification. Deionized water (with $\sigma = 18\ M\Omega$) was used as a solvent. Sodium chloride (NaCl) obtained from Fisher Scientific, was employed as salt to prepare brine of a concentration of 30,000 ppm (equivalent to sea water concentration). The pH value of system was adjusted with hydrochloric acid (HCl) and sodium hydroxide (NaOH) solution by precise pH meter (Mettler Toledo, FE20-Basic).

The NPs were dispersed in brine as the base fluid and magnetically stirred for 1 h to produce nanoparticles suspension. Then, the appropriate amount of sodium dodecylbenzenesulfonate (SDBS) was added to the suspensions. These suspensions were agitated in an ultrasonic bath at ambient temperature for an optimum period, to attain the required concentration of nanofluids. The anionic surfactant, SDBS, was chosen as a stabilizer based on our previous stability tests; where the surfactant concentrations were selected using critical micelle concentration (CMC) determination methods. The pH value of the system was also adjusted by using HCl and NaOH solution in order to improve the quality of dispersion. These pH values were monitored by precise pH meter (FE20-Basic) from Mettler Toledo. After the samples are prepared, viscosity and density tests were carried out to check the viscosity and density for each sample. The IFT was measured by using a Dataphysics Surface Tensiometer. The wettabliity was investigate in terms of contact angle measurement. The above measurements were carried out at ambient condition and under EM waves. The EM field was generated at a designed frequency of 18.8 MHz, and a fixed voltage of 3.5 V.

3 Results and Discussions

Table 1 depicts fluid properties measurement of brine, crude oil and 0.1 wt% ZnO nanofluid concentration at ambient condition and under EM waves. We can observe that the viscosity increases when NPs are introduced into the solution. The viscosity of the nanofluid is further enhanced from 1.05 to 1.10 cP for ZnO@500 NF and 1.07 to 1.14 cP for ZnO@800 NF when under EM waves. Dielectrics (such as ZnO, Al_2O_3) are of great interest as metal oxides for EEOR application due to their high dielectric loss that render the particles as the surface-active agent [13]. The surface-active agents suspended in a liquid undergo polarization when an external electric field is applied, providing the possible oil recovery mechanism including (i) the deformation of oil droplet which increases the surface area for more particles' adsorption and consequently to a reduction in interfacial tension; (ii) the increment in

Table 1 Fluid properties at ambient condition and under EM waves

Fluid	Density (g/cm^3)	Viscosity (cP)
Brine, NaCl 3 wt%	1.0197	1.01
Crude oil	0.8021	7.50
SDBS 0.025 wt%	1.0194	1.02
ZnO@500 NF	1.0204	1.05
ZnO@800 NF	1.0205	1.07
ZnO@500 NF (with EM)	1.0204	1.10
ZnO@800 NF (with EM)	1.0205	1.14

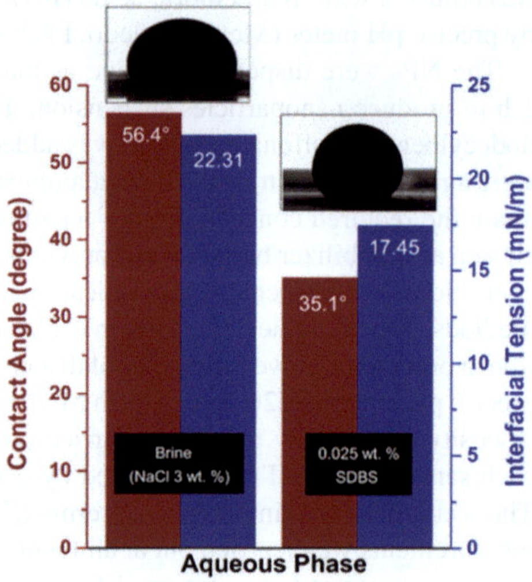

Fig. 1 Interfacial tension and contact angle measurement for crude oil against brine and 0.025 wt% SDBS at ambient condition

rate of wettability alteration which lead to an increase in surface free energy; (iii) the improvement of mobility ratio due to electro rheological effect (ER) which increases the viscosity of nanofluids.

In addition, high temperature leads to the reduction of oil viscosity. This leads to the decrease in viscosity ratio (at 10 s^{-1}) of oil and nanofluids from 3.73 to 2.13 and 3.28 to 2.1 for ZnO@500 and ZnO@800, respectively; since ZnO NFs are proven to be relatively stable at high temperature [14]. This ultimately reduced the mobility ratio which will divert injected NFs into unswept areas, and consequently resulting in a favorable oil recovery.

Figure 1 shows the reduction of interfacial tension between crude oil and surfactant solution, along with wettability alteration at room temperature. However, at high temperature SDBS tends to degrade [15] which reduces the interaction between the surfactant and crude oil. Consequently, this reduced the oil recovery.

In Fig. 2, the introduction of NPs into the fluid leads to higher contact angle compared to the one only with SDBS. The contact angle is further increased when the fluid is under EM waves. The increment in contact angle under EM waves can

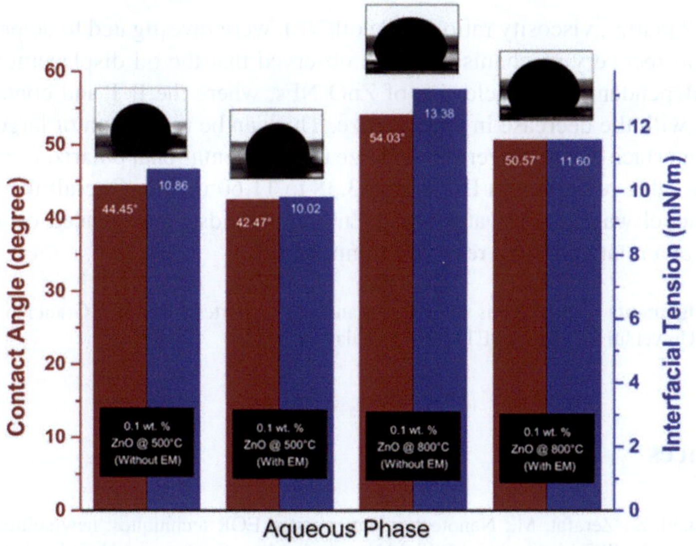

Fig. 2 Interfacial tension and contact angle measurement for crude oil against ZnO NFs/SDBS under EM waves

be explained by different mechanism as follows: (1) the dielectric polarization of nanoparticles causes the shape of oil drops to deform which increases the surface area for more particles' adsorption, and consequently to a reduction in interfacial tension (Fig. 2); (2) the rate of wettability alteration increases, leading to an increase in surface free energy (Fig. 2); (3) the improvement of mobility ratio, due to electrorheological effect (ER) which increases the viscosity of ZnO nanofluids. However, the increasing particle size (i.e. ZnO@800) will decrease will result in lower contact angle, viscosity and IFT [16–19]. This will decrease the incremental oil recovery.

It is suspected that incremental oil at high temperature was recovered not only due to the decreased mobility ratio, but also the decrement in IFT with increasing temperature since the molecular interactions between the liquids are weaker. Another possibility is the intensity of Brownian movement increases with the temperature of the medium and with the reduction of its viscosity and the particle size. Since Brownian motions might be one of energies that drive oil displacement due to nanoparticles, the force will increase. Overall, the combination of ZnO NPs and dispersant (Nano-EOR) seems greatly promising in higher temperature, e.g. reservoir temperature.

4 Conclusion

A comprehensive study on effects of ZnO NPs on fluid properties with and without EM has been demonstrated. Relationship between the IFT (crude oil/NF), contact angle (NF/quartz), viscosity ratio (crude oil/NF), were investigated to determine the possible oil recovery mechanism. It was observed that the oil displacement mechanism is dependent on particle size of ZnO NFs, where the IFT and contact angle decreases with the decrease in particle size. This can be the reason of larger disturbance in interface due to bigger particle size under orientational polarization, providing a noticeable reduction in IFT from 13.38 to 11.60 mN/m. Overall these results reveal a novel way to use water-based ZnO nanofluids for enhanced oil recovery purposed at a relatively high reservoir temperature.

Acknowledgements A part of this work was financially supported by YUTP (Grant No. 0153AA-E17) from Universiti Teknologi PETRONAS Malaysia.

References

1. Ayatollahi, S., Zerafat, M.: Nanotechnology-assisted EOR techniques: new solutions to old challenges. In: SPE International Oilfield Nanotechnology Conference, Noordwijk, Netherland (2012)
2. Duhan, S., Aghamkar, P., Lal, B.: Influence of temperature and time on Nd-doped silica powder prepared by the sol-gel process. J. Alloy. Compd. **474**, 301–305 (2009)
3. Patil, S.R., Hameed, B.H., Skapin, A.S., Stangar, U.L.: Alternate coating and porosity as dependent factors for the photocatalytic activity of sol-gel derived TiO_2 films. Chem. Eng. J. **174**, 190–198 (2011)
4. Vafaee, M., Ghamsari, M.S., Radiman, S.: Highly concentrated zinc oxide nanocrystals sol with strong blue emission. J. Lumin. **131**, 155–158 (2011)
5. Yavarina, N., Alveroglu, E., Celebioglu, N., Siklar, U., Yilmaz, Y.: Effect of temperature, stirring velocity and reactant concentration on the size and the optical properties of ZnO nanoparticles. J. Lumin. **135**, 170–177 (2013)
6. Broido, A.: A simple, sensitive graphical method of treating thermogravimetric analysis data. J. Polym. Sci. **7**(10), 1761–1773 (1969)
7. Albertsson, J., Abrahams, S.C., Kvick, A.: Acta Crystallography. Struct. Sci. **45**, 34 (1989)
8. Pei, L.Z., et al.: Single crystalline ZnO nanorods grown by a simple hydrothermal process. Mater. Charact. **60**, 1063–1067 (2009)
9. Pierre, A.C.: Introduction to Sol-Gel Processing. Kluwer Academic Publisher (2002)
10. Oglo, N.A., Olafuyi, O.A., Onyekonwu, M.O.: Enhanced oil recovery using nanoparticles. In: SPE Saudi Arabia Technical Symposium and Exhibition, Al-Khobar, Saudi Arabia (2012)
11. Adil, M., Zaid, H.M., Chuan, L.K., Alta'ee, A.F., Latiff, N.R.A.: Microscopic evolution of dielectric nanoparticles at difference calcination temperatures synthesized via sol-gel auto-combustion. AIP Conf. Proc. **1669**, 020016 (2015)
12. Lee, K., Saipolbahri, Z.A., Guan, B.H., Soleimani, H.: Organic sol-gel method in the synthesis and characterization of zinc oxide nanoparticles. Am. J. Appl. Sci. **11**, 1250–1254 (2014)
13. Adil, M., Zaid, H.M., Chuan, L.K., Latiff, N.R.A.: Effect of dispersion stability on electrorheology of water-based ZnO nanofluids. Energy Fuel **30**, 6169–6177 (2016)
14. Zaid, H.M., Adil, M., Chuan, L.K., Alta'ee, A.F., Latiff, N.R.A.: Stability and electrorheology of ZnO nanofluids in the presence of anionic surfactants. AIP Conf. Proc. **1787**, 050007 (2016)
15. Manousaki, E., Psillakis, E., Kalogerakis, N., Mantzavinos, D.: Degradation of sodium dodecylbenzene sulfonate in water by ultrasonic irradiation. Water Res. **38**, 3751–3759 (2004)
16. Soleimani, H., Yahya, N., Baig, M.K., Khodapanah, L., Sabet, M.: Synthesis of carbon nanotubes for oil-water interfacial tension reduction. J. Oil Gas Res. (2015). http://dx.doi.org/10.4172/ogr.1000104

17. Soleimani, H., Ahmad Latiff, N.R., Yahya, N., Sabet, M., Khodapanah, L., Kozlowski, G., Chuan, L.K., Guan, B.H.: Synthesis and characterization of Yttrium Iron Garnet (YIG) nanoparticles activated by electromagnetic wave in enhanced oil recovery. J. Nano Res. **38**, 40–46 (2016)
18. Soleimani, H., Yahya, N., Ahmad Latiff, N.R., Demiral, B.M.R.: Novel enhanced oil recovery method using $Co^{2+}xFe^{2+}1-xFe^{3+}2O_4$ as magnetic nanoparticles activated by electromagnetic waves. J. Nano Res. **29**, 105–113 (2014)
19. Mohd Zaid, H., Ahmad Latiff, N.R., Yahya, N., Soleimani, H.: Application of electromagnetic waves and dielectric nanoparticles in enhanced oil recovery. J. Nano Res. **26**, 135–142 (2014)

Design and Characterization of a Mandibular Prosthesis Prototype by Hemimandibulectomy

Juan Alfonso Beltrán-Fernández, Iliana Picco-Díaz, Itzel Bantle-Chávez, Carolina Alvarado-Moreno, Luis Héctor Hernández-Gómez, Juan Luis Cuevas-Andrade, Mauricio González-Rebattú, Adolfo López-Lievano, Juan Alejandro Flores-Campos and Pablo Moreno-Garibaldi

Abstract The design and development of highly accurate prostheses for individual cases according to the patient's condition is one of the current challenges in the fields of medicine and engineering. The use of tomographic archives uploaded to a special-

---

J. A. Beltrán-Fernández (✉) · I. Bantle-Chávez · C. Alvarado-Moreno · L. H. Hernández-Gómez · J. L. Cuevas-Andrade · A. López-Lievano · J. A. Flores-Campos · P. Moreno-Garibaldi
Instituto Politécnico Nacional SEPI - ESIME Zacatenco, Unidad Profesional "Adolfo López Mateos", Edificio 5, 3er Piso, Colonia Lindavista, Gustavo A. Madero, 07738 Mexico, CDMX, Mexico
e-mail: jbeltranf@hotmail.com

I. Bantle-Chávez
e-mail: itzi.bantle@gmail.com

C. Alvarado-Moreno
e-mail: aly-oroshiwa@hotmail.com

L. H. Hernández-Gómez
e-mail: luishector56@hotmail.com

J. L. Cuevas-Andrade
e-mail: jl1z9c88@gmail.com

A. López-Lievano
e-mail: k_lievano@hotmail.com

J. A. Flores-Campos
e-mail: jaflores@ipn.mx

P. Moreno-Garibaldi
e-mail: pmg170588@gmail.com

I. Picco-Díaz · M. González-Rebattú
Hospital Regional 1° de Octubre-ISSSTE-Av. Instituto Politécnico Nacional No. 1669, Lindavista, Gustavo A. Madero, 07300 CDMX, Mexico
e-mail: ilianapicco@yahoo.com.mx

M. González-Rebattú
e-mail: mauriciorebattu@yahoo.com.mx

© Springer International Publishing AG, part of Springer Nature 2019

A. Öchsner and H. Altenbach (eds.), *Engineering Design Applications*,
Advanced Structured Materials 92, https://doi.org/10.1007/978-3-319-79005-3_22

ized computer program for its rendering (creation of a 3-D model) allows specialists decide on the surgical approach to take, by creating a model which possesses precise dimensions, corresponding to the exact anatomy of the patient. The use of computational programs (CAD/CAM programs) not only allows the development of better designs, but also the evaluation of the proposed models through a force moment analysis in conjunction with the assignment of the endorsed materials for its posterior development by setting the prototypes to the Maximum Shear Stress principle and analyzing the results of its displacement's and the Stress von Mises' tests, specialists have the opportunity to know if the final design will be able to support the muscular tensions and occlusion forces caused by the elongation of the muscular fibers.

Keywords Tomography · 3-D printing · Models made in alternative materials Digitalization · Creo Parametric/Simulate · Numeric force evaluation

1 Introduction

A mandibulectomy is a surgical technique consisting of the extraction of the mandibular bone. Generally, it is used for the treatment of cancer in the lower jaw. However, it is also used to treat diseases such as osteomyelitis (inflammation of the bone caused by an infection), osteoradionechrosis (bone death by radiation), or ameloblastoma (a benign tumor). This type of procedure involves a loss of facial contour for the patient, phonation and swallowing disorders, as well as aesthetic defect.

Beginning 1970 in Japan, the replacement of that section of the mandible with a part of bone from another part of the body, the fibula being most commonly used for this purpose, since it is the smallest of the lower bones.

Osteonecrosis is an illness characterized by gradual loss of blood flow, and as a consequence, the interruption of the nutrient supply to osseous tissues [1].

Most of the reported cases show a higher frequency of this suffering for the lower jaw (63–80%) due to its lesser blood supply in comparison to the maxilla (14–38%) [2]. The death of the bone components (osteocytes) and the bone marrow is provoked by ischemia (low blood irrigation) and hypoxia (low oxygen supply). The debilitation of the osseous tissue allows the uncontrolled growth of microbiota species commonly found in the oral cavity, such as *E. coli* or Bacteroides (gram-negative bacteria), which leads to the collapse of the tissue's structure.

As the suffering develops, multiple tiny fractures appear around the bone's surface. Pain often comes with the gradual collapse of the affected area due to the pressure increase caused by the natural forces produced by muscular occlusion movements. Moreover, another consequence is teeth detachment and in the most severe cases, mandibular cancer [1].

Due to the high complexity of these cases, from medical and engineering point of view, the highly accurate replacement of anatomic structures needs a geometric and superficial evaluation process of the affected area's morphology, together with an analysis of the forces (e.g. mastication forces) which naturally influence the lower jaw.

Likewise, anatomical data must be taken into consideration for the design and development of the prosthesis. The bite force produced during mastication is generated by multiple muscles that work simultaneously and in conjunction on the lower jaw's surface. In order to determine the influence of each muscle during the production of a certain force, an anatomical model is used for the analysis of the strain distribution [3]. A coordinate system with its origin set on the projection of the center of the mandible's coronoid process in relation with the symmetric plane is used in order to establish the direction and degree of the force vectors. The x, y, z axis are projected along the intercondilar axis. The muscles that act over the jaw, the masseter, temporal, internal pterygoid, external pterygoid and digastric, create the forces Fm, Ft, Fi, Fe, Fo, respectively. Balancing the forces and force moments on the lower jaw, three equations are obtained for the force's magnitude of the five muscles, altogether with the magnitude and direction of the articulation's force reaction [3].

The following table shows the values of the force's magnitude accordig to the transversal cross area obtained from the calculation of the equations. The muscle's forces will be used for the force's simulation in the computational programm Creo Simulate on the articulated model in order to make a numerical analysis of an accurate anatomical model.

2 Clinical Case

Patient presents with a gradual infection in the lower jaw caused by a post-surgical infection that culminates in osteonecrosis and the weakening of the interior and exterior bone's surface of the lower jaw, together with the surrounding tissues. Due to the seriousness of the area's condition, the patient's lower jaw is completely removed by a surgical procedure known as total mandibulectomy, which consists of its extraction and disarticulation.

In order to support the development and design of the prosthesis digital and modeling techniques are applied, and also financial support for the acquisition of necessary materials for the development and design processes of the prototypes is provided. Different prototype models will be evaluated by maxillofacial specialists and surgeons in charge of the surgical intervention for the patient.

3 Objectives

I. Generation of negative models of the mandibular prosthesis using as a base a 3-D printed model made from alternative materials (Acrylnitril-Butadien-Styrol-Copolymere plastic, knows as ABS), which was obtained from the digital archive of the patient's tomography.

II. Construction of the prosthesis' biomodel using dental, self-healing acrylic, trademark Nic Tone©. Acquisition of the part which will be prepared for digitalization.
III. Application of modeling techniques and analysis processes making use of the following specialized computer programs: Atos, Autodesk NetFabb Standard 2017, Da Vinci XYZ Ware and Creo Parametric, in order to design two prototypes.
IV. Performance of the numeric analysis of the forces acting on the articulated prosthesis using the program Creo Simulate in order to test the mechanical properties of endorsed materials for its posterior construction.

4 Methodology

For the drawing up process of the damaged area's tridimensional image, a specialized computer program such as SCANP IP must be used. One of the main tools of this program allows the opening of the tomographic archive in the *.sip* format, which allows the user to display the tomography in three planes (coronal, lateral and antero-posterior) as multiple layers (tessellation). Through the union of all the painted layers, a three-dimensional image of the anatomical zones that have to be analyzed can be created.

For its realization each layer must be painted (the conjunction of layers are seen as masks by the program). For this particular case two different masks were created in order to distinguish between the maxilla and the skull (presented in a green color) and the inferior jaw (dyed in a blue color), which is also the affected zone, as shown in Fig. 1.

Once the interested zones were painted using the Confidence connected region growing tool, the three-dimensional figure is converted into a .stl archive (Standard Tesselation Language archive) that can be read and modified in any CAD program. The conversion conserves the real state of the patient's anatomical structure and also respects the colors that represent the different parts of the 3-D figure. The final result can be seen in Figs. 2 and 3.

After the rendering process of the pieces which will be evaluated (skull with maxilla and lower jaw) and the export into the .stl format, the archives are uploaded in the DaVinci program to begin the 3-D printing process.

The *3D System Printer* builds the model in ABS plastic based on the digital archive analyzed by the DaVinci program. As a result, the digital archive of the lower jaw is extracted and set over a platform. The program allows the user to modify the configuration of the piece that will be printed in order to save material and time. The prototype's configuration is set as followed: set to have a coat height of 0.4 and a 3-D density of 10%.

The printer proceeds to dispose the plastic filaments over one another, reconstructing the natural morphology of the lower jaw. The interior of the new piece consists of multiple supports that have to be cut one by one, taking 5 h and 43 min for the skull and the maxilla and 2 h and 21 min for the lower jaw. The ABS plastic was chosen as

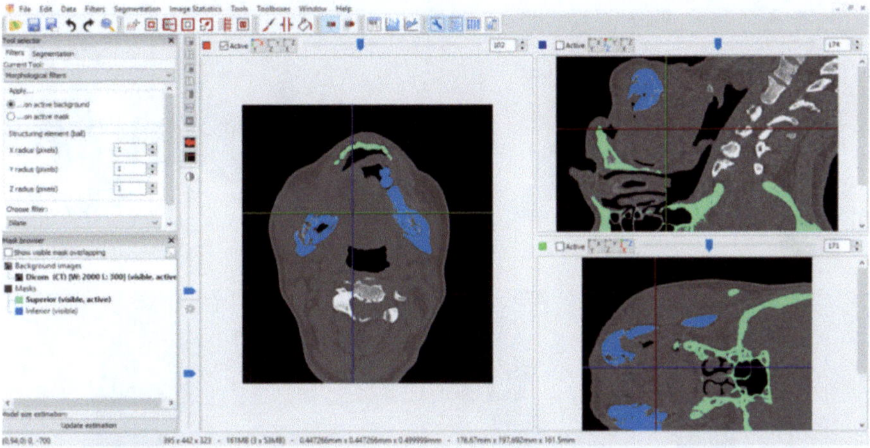

Fig. 1 Tomography's evaluation in the coronal plane (first, superior, left picture), lateral plane (second, superior, right picture) and antero-posterior plane (first, inferior, left picture) using the SCAN-IP computer program. Layers 17–322 coronal, 39–330 lateral, 42–422 antero-posterior layers

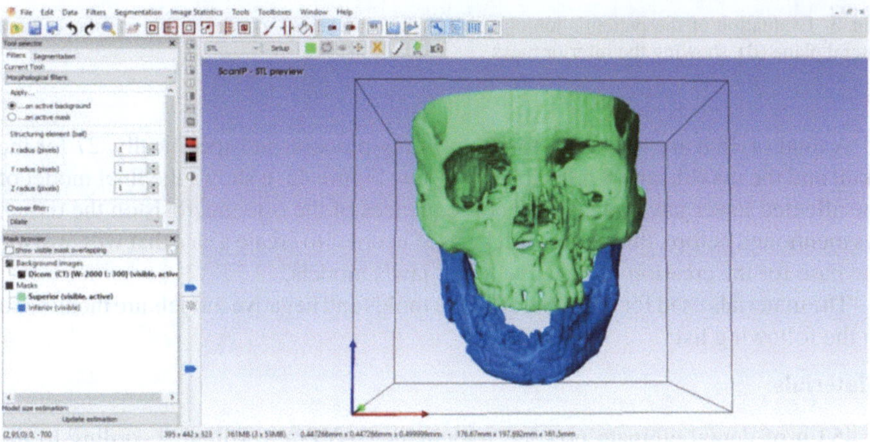

Fig. 2 Evaluation of the patient's tomography as seen in the anterior plane with both masks: superior (dyed in green) and inferior (dyed in blue), exposing the current state of the patient's lower jaw

the building material because of its chemical-mechanical properties. It particularly presents a wide range of characteristics which are optimal for the manual modeling process, such as, a low humidity absorption capacity and its high dimensional stability. Both properties will allow to introduce the final piece in dental alginate to create molds. Taking into account that the alginate is a cold material, the ABS plastic will not change the surface of the printed model by absorbing humidity and, as a consequence, start to swallow.

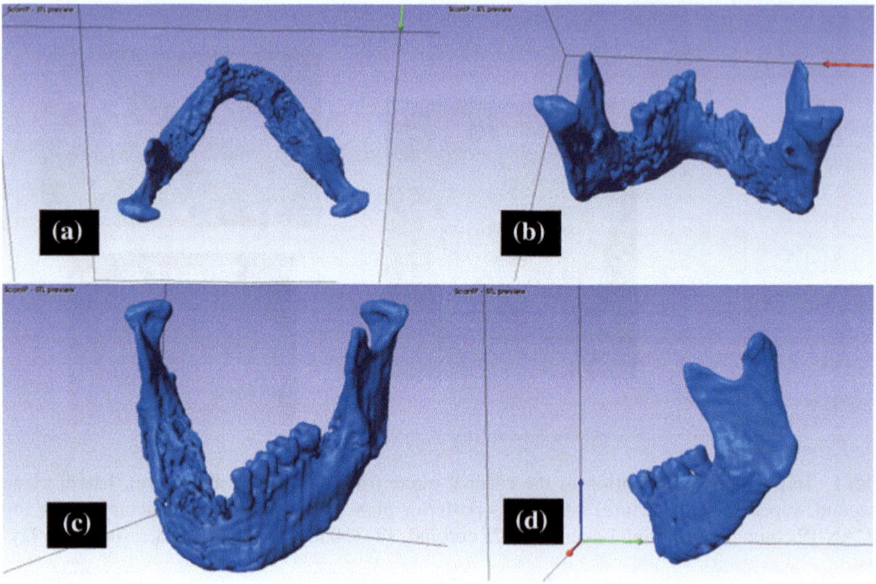

Fig. 3 Evaluation of the patient's tomography in the coronal (**a**), posterior (**b**), angular (**c**) and lateral plane (**d**), showing the inferior mask

A total of 38 h are necessary for the printing process, and specifically, 27 for the skull and the maxilla and 8 h for the lower jaw. Figure 4a, b shows the final model of the affected lower jaw and also the consequences of the osteonecrosis on the tissue. As mentioned before, this model will be used in order to create a cast that will provide the base for the creation of negative lower jaw's models.

The materials used for the creation of the molds and negative models are mentioned in the following list:

Materials:

- 453 g of dental alginate powder, trademark GAYZ©, certified according to the norm ISO 7405:2008.
- 254 g of ultra-quick drying stone plaster, type III, trademark Whip Mix©, certified according to the norm ISO 7405:2008 and endorsed by the *American Dental Association (ADA).*
- 2 l and 200 ml of water.
- 4 plastic bowls.
- 2 plastic shawls for mixing.
- Sandpaper for wood, grain 80, 600, 1000, 1500 and 2000.
- 1 cutter.
- 250 ml of acrylic resin, trademark Nic Tone©.
- 35 g of catalyst for the acrylic resin, trademark Nic Tone©.

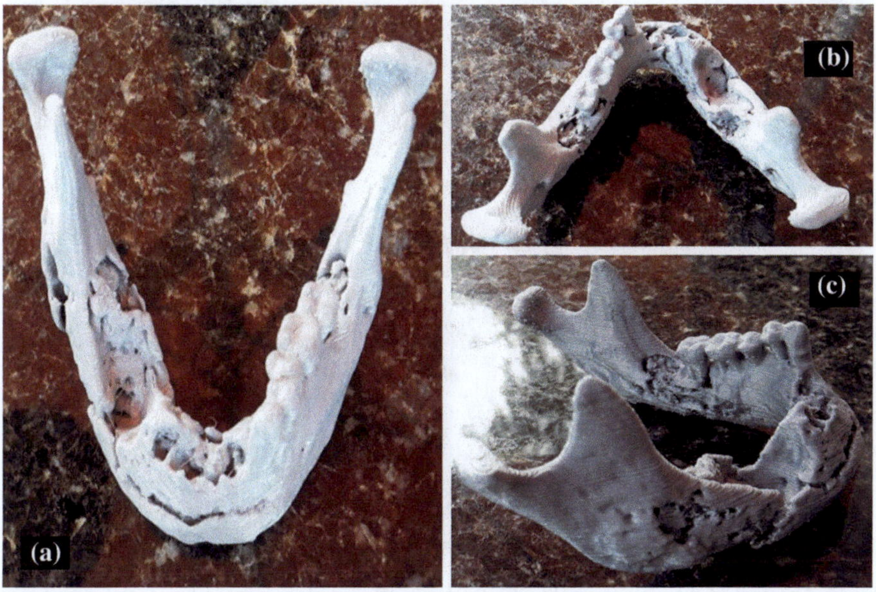

Fig. 4 Lower jaw affected by osteonecrosis, printed in ABS plastic, appreciated from above (**a**), from behind (**b**) and laterally (**c**)

First, the lower jaw printed in ABS plastic is used as a model to create the main mold, with which the negative models in stone plaster and acrylic resin will be created. The surface of the plastic jaw must be completely covered with liquid soap (release agent) in order to allow the easy extraction of the plastic jaw and also to avoid its adhesion to the mold's walls. The 453 g of dental alginate powder are poured in the plastic bowl and 2 L of water are added to prepare a colloid mixture that will be the base over which the surface of the model will be printed. Then the plastic model of the lower jaw is introduced in the alginate mixture and is left to stand approximately 15 min. Subsequently, when the mold is completely dry, it is opened by being cut in half using a cutter in order to extract the plastic model without damaging its printing. The process can be seen in Fig. 5a–c.

The next step consists of closing the mold by adding more of the mixture and covering the cut on the mold's borders. Using the cutter, a tiny hole is made on the upper portion of the mold in order to add a mixture of 254 g of stone plaster and 200 ml of water. Once more the mixture is left to stand inside the mold for two hours to guarantee its drying. The same extraction process is performed and having the first negative model of the mandible, it is necessary to reduce the surfaces of the body, angles and border of the stone plaster's model by using sandpaper of grain 80. Originally the ABS plastic model presented a swallowed body with a rugged surface due to the cellular activity during the development of the illness. As a consequence, the thickness of the stone plaster model must be reduced to a width of 10 mm according to the maxillofacial specialist's indications and all the irregularities have

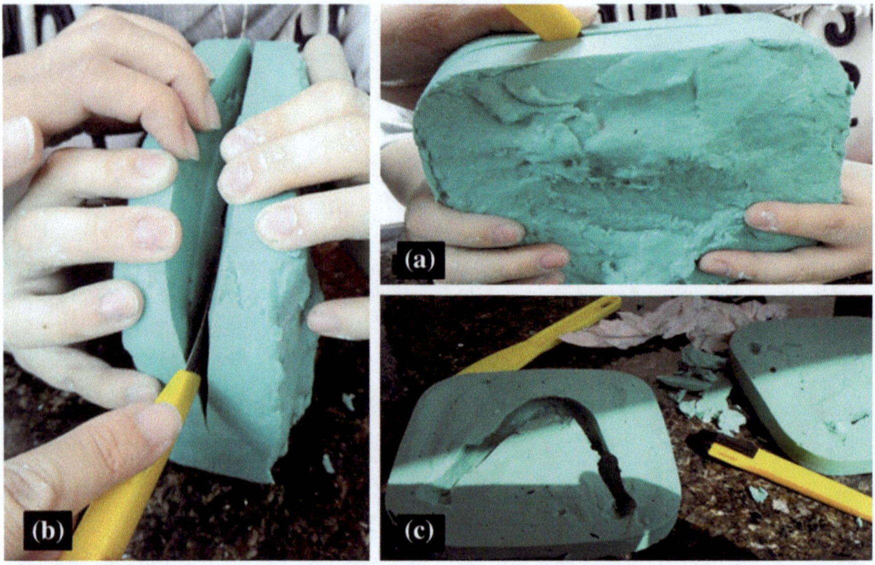

Fig. 5 Creation of the alginate mold with the lower jaw's surface print, **a** horizontal cut as seen from above during the cutting process, **b** horizontal cut made each time more deeply until the cutter touches the different borders and components of the plastic jaw (inferior mandibular border, body, mandible's coronoid process, etc.), **c** print of the mandible on the alginate's surface as seen from above

to be refined using the sandpaper, being careful of preserving the natural curvature of the lower jaw and also the form of the mandible's condylar processes. This process can be appreciated on Fig. 6a–c. The finished piece will be newly used to create a new alginate's mold for the shaping of a second negative model made in acrylic resin (a biomodel that will be digitalized once finished).

Having the alginate model, the mixture for the dental acrylic resin is prepared employing the following admixtures: transparent methyl methacrylate acrylic polymer powder and the self-healing monomer, trademark Nic Tone© (catalyst) in a proportion 2:1 to obtain the final piece shown in Fig. 7a, b. The consolidation takes approximately 15–30 min.

The next step consisted in the digitalization of the acrylic piece using the Scanner Aio 1.0 Da Vinci, obtained as a result from the digital archives shown in Fig. 8. Many points that in conjunction create the lower jaw's surface were not perceived by the scanner because of the transparent color of the acrylic piece which reflected the scanner's photonic beam in different angles out of the scanner's detection range. As a consequence, part of the information got lost and the image could not be completed.

A solution to the problem the piece was painted with watercolors and dyed in a dark blue tone so that the photonic beam of the scanner would not be reflected out of the detector's range. Likewise, another and more potent scanner was used for the piece's digitalization, specifically the ATOS 3D Scanner, equipped with the

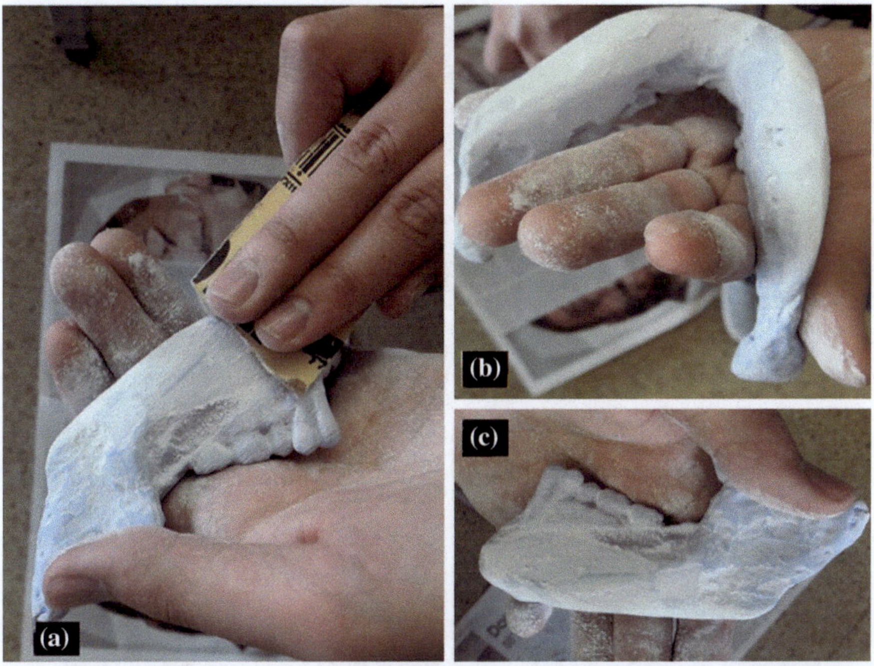

Fig. 6 Reduction of the body's surface and the mandibular branches using sandpaper, grain 80, **a** upper view of the lower jaw's left, lateral body, **b** smoothening of the inner body's surface and the mandible's base, removal of the anatomically deformed areas which altered the natural mandible's geometry, **c** lateral view of the mandible's body after 5 min of smoothening

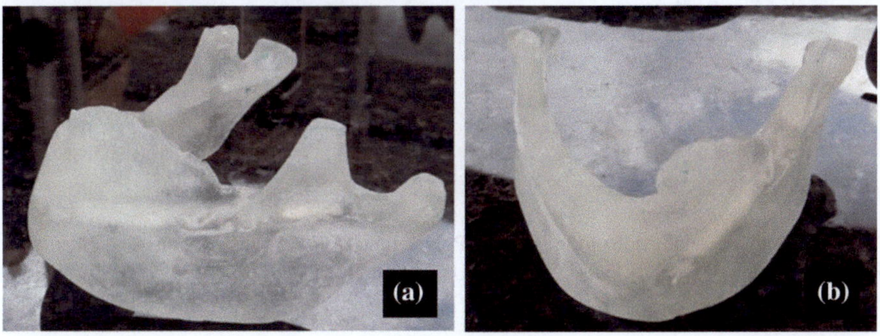

Fig. 7 Final piece in acrylic resin after being extracted form the alginate's mold (**a** lateral view, **b** anterior view)

Atos Core program, trademark GOM that is shown in Fig. 9. This highly specialized equipment needs a variety of steps for its calibration, an important process that prevents the digital archive of contenting noise and other kinds of alteration during the digitalization process.

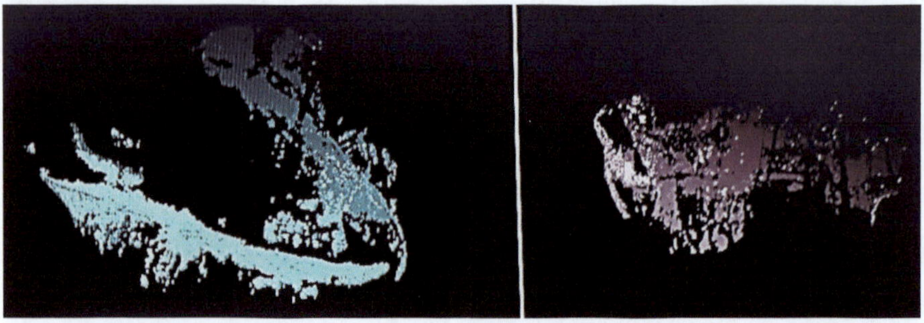

Fig. 8 Digital archives of the acrylic resin's mandible, obtained through the Scanner Aio 1.0 Da Vinci. Several irregularities and a significant lack of surface points can be observed

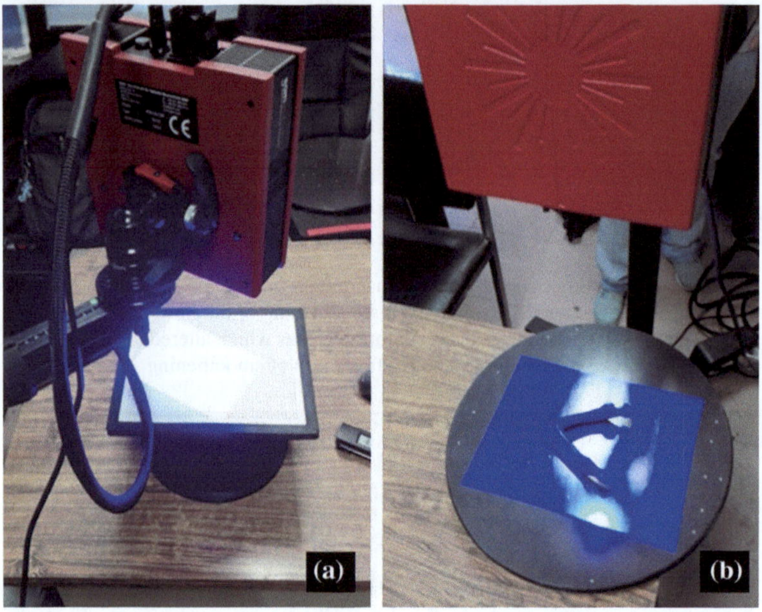

Fig. 9 Equipment used inside of the National Polytechnic Institute's installations, specifically the Scanner ATOS 3D, trademark GOM while using its computational interface Atos Core. **a** Calibration process of the platform, **b** Detection of the lower jaw's points that create the mandible's surface

For the calibration process these steps must be followed:

1. The temperature of the environment which is in contact with the scanner and the calibration panel must be measured using a digital thermometer and specified in the programs platform.
2. The scanner will indicate a range of 18–20 steps (positions) in which the equipment must be placed to ensure the communication between the sensor's camera and its interface in a range that can be detected. The steps consist of

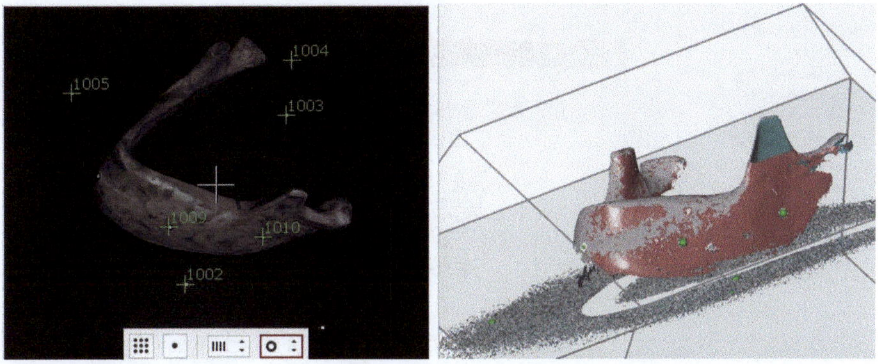

Fig. 10 Interface's change of coloration that indicates with a cherry color the entrance of new information and with a grey color the information previously captured

a placement's change of the sensor along the "X", "Y" and "Z" axis and the rotation of the calibration platform 25° each time.

After finishing the calibration process, several reference points (minimum three) are placed on the piece's surface in order to allow the program to recognize the mandible's space position on the rotatory platform. Likewise, those points are necessary due to the platforms 360° rotation in intervals of 25° to 45° each time the piece is scanned to obtain the superior and inferior views of the piece and also all the surface's details. More reference points must be placed on the rotatory platform so that the program can eliminate the zones more easily the zones that the scanner captured but do not belong to the mandible's surface. By using the tool "*New Measurement Series*" the first scan that corresponds to the upper part of the lower jaw can be initiated. For the second scan (of the lower part of the mandible), the program will start to show the new information of the geometric figure that is detected during each scan and also every time the piece's angle is changed with the help of the tool "*Scan Measuring Object Space*".

In the same scan, multiple captures of the piece's image can be obtained in order to let the sensor detect new information (cluster of points) that will be grouped over the first information and also appear with a different coloration of the piece, specifically showing a cherry coloration for the entrance of new information and a grey color for the previous information (see Fig. 10).

After finishing the first scan, a new scanning process in conjunction with the placement of the reference points using the program's tool "ATOS Measurement Series" is started. The main objective of the second scan is to capture the inferior surface of the mandible by putting the piece upside down and scanning it. Having both scanned archives, the file "Transform by Common Ref. Point" is used to unite them and acquire a complete digital piece. The last step consists of using the tool "*Close Holes Interactively*" in order to select the borders of the spaces without information and fill them according to the size that corresponds to their dimensions. Finally the

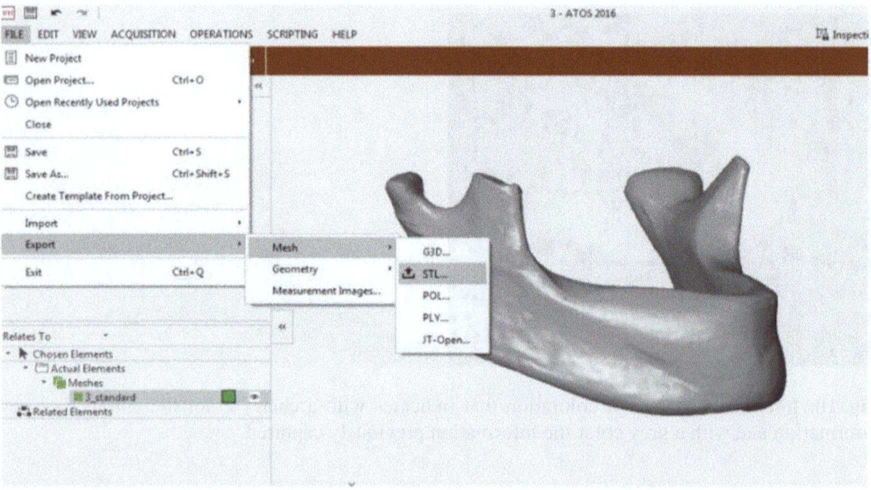

Fig. 11 Digital archive obtained from the digitalization process in its .stl format

digital archive of the whole piece is obtained and exported to a .stl format, as shown in Fig. 11.

To make the archive's manipulation more easily, the lower jaw was divided in two halves which correspond to the right and left side of the mandible. The division was made using the computer program Autodesk Netfabb Standard 2017, which allows the user to create a plane that can be placed freely where the object is intended to be divided or cut in multiple pieces (see Fig. 12).

The digital archives obtained from the Atos Core program are .stl archives that can not be modified in the Creo Parametric, because the pieces considered by the program to be manageable "objects". This fact leads to the conversion of the digitalized object to a piece through the program SolidWorks, which allows the transformation of a .stl archive to an IGES format (in other words from a cluster of points in .stl format to a solid structure for its posterior modification).

5 Results

The archives obtained in the .igs (IGES) format through the computational program SolidWorks were transformed successfully. Unfortunately both archives ended with a digital weight of 1.7 and 1.4 GB, for the right half (equivalent to one billion of bytes) corresponding to the right half and left half of the lower jaw, respectively. Originally with the use of the scanner Atos 3D, the number of facets (triangles that form the piece) was not limited in order to assure the highest resolution of the digital archives and, as a consequence, the program SolidWorks transformed the archives in their

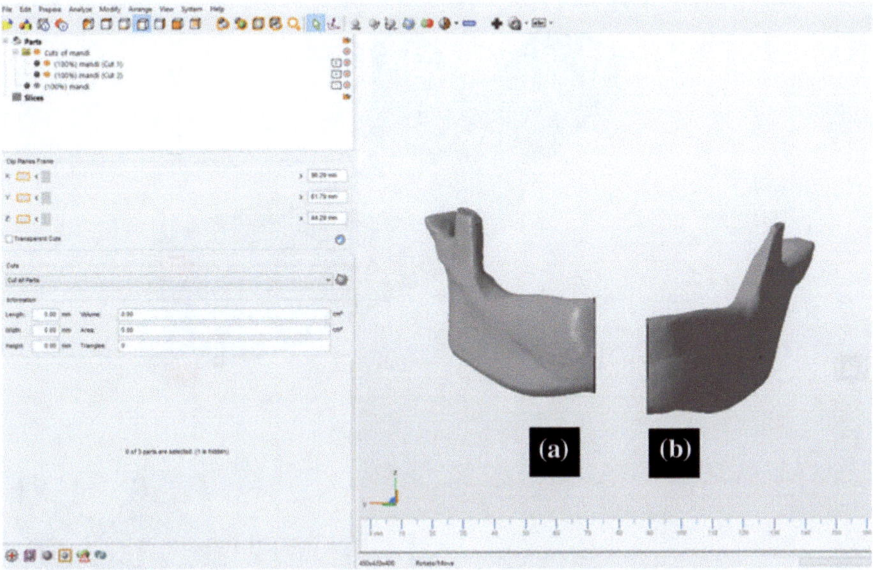

Fig. 12 Placement of the imaginary plane on the digital archive in .stl format for the lower jaw's division, creating two individual pieces that can be saved separately in the same format, (**a** Right side, **b** left side)

highest definition, making their usage and manipulation in conventional computers practically impossible due to their weight.

The team proceeded to perform the conversion of the .stl archives to a .prt format ("shrinewrap" format) through the computational program Creo Parametric 3.0, which also transforms the archives to a manipulable object. After defining the prosthesis' design, the manipulation and modification processes of the digital archives were performed. For the first model, a male-female union for the assembly of both halves was chosen. Taking into account the width of approximately 12.715 mm and the length of 30.634 mm of the internal face of both hemimandibles, the sketch of a rectangle, with 5 mm width and 15 mm in length, was drawn over the face's surface (as seen in Fig. 13b, dyed in purple). Initially the sketch of the rectangle was created with a distance of 7.0628 mm from the superior central point of the right hemimandible's edge and a distance of 7.923 mm from the inferior central point. After defining distances and lengths, the rectangle was extruded 5 mm to the outside in order to create a tridimensional surface. At a distance of 1.5 mm from the superior surface of the central rectangle, a 5 mm width and 4 mm length rectangle is sketched (as shown in a red color in Fig. 13b), which is also extruded 5 mm, but in this case, to the inside of the piece in order to create an assembly zone in which the rectangles of the other hemimandible will enter. The same procedure is followed when the inferior rectangle is created. For the other lower jaw's hemimandible, the same process is

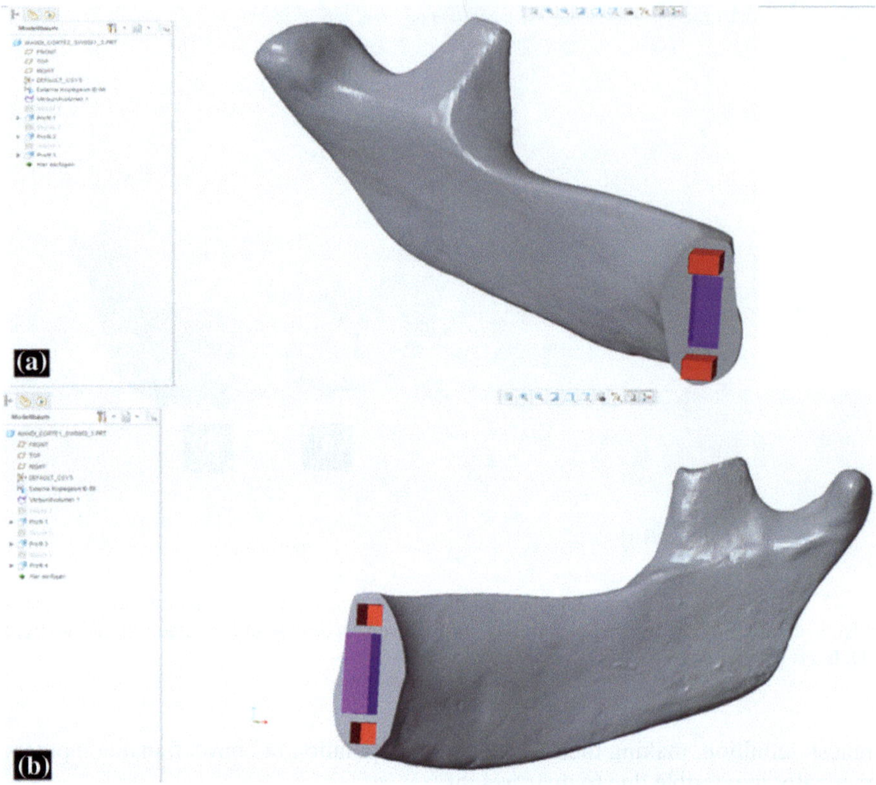

Fig. 13 Design of a model with a male-female union. **a** Right hemimandible with a central 15 mm length and 5 mm width rectangle, internal extrusion of 5 mm, shown in a purple color and two 5 mm width and 4 mm length rectangles with an external extrusion of 5 mm, dyed in a red color. **b** Left hemimandible with a central 15 mm length and 5 mm width rectangle, external extrusion of 5 mm, shown in a purple color and two 5 mm width and 4 mm length rectangles with an internal extrusion of 5 mm, dyed in a red color

followed but the extrusions are made the other way round to allow the assembly of both halves.

After finishing the piece's design, the assembly simulation of both pieces in order to evaluate possible global interferences was performed. One of the pieces must be opened as an assembly piece in the computer program Creo Parametric 3.0 in the modality "by default", while the second must be opened in the modality "automatic" in order to place one surface over another. The next step is to select both faces of the rectangles that have to come together until both halves are properly assembled. Finally, using the program's "analysis" and the "global interference" tool, the user can indicate the program to scan the whole piece's surface and borders in search of overlapping regions or structures that do not fit in each other. The first model did not show signs of any kind of interference, only a warning was given by the program

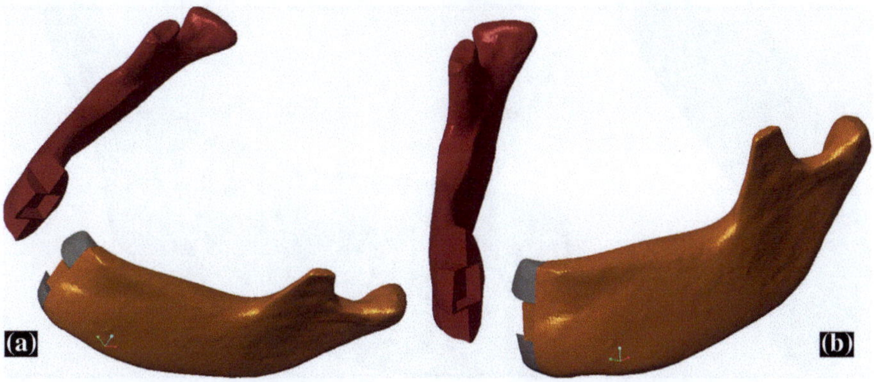

Fig. 14 Ilustration of both hemimandibles, the left hemimandible can be appreciated with the piece of surface extracted from the right hemimandible and also the extrusions made to the inside and outside of the piece that conform the male-female union

due to the difficulty of finding the difference between the distance of the triangles that conform to the pieces. This means there is a minimal volume that can not be calculated by the program, but that does not influence the interference calculation.

For the second design, part of the right hemimandible was separated and bounded to the border of the left hemimandible in order to create a conjunction of pieces that will be assembled as a single piece with the right hemimandible. The process consisted of creating a plane that will allow the removal of the selected material, in this case, the whole hemimandible with the exception of the pieces that will be assembled over the complementary hemimandible. Subsequently, over the face of the right hemimandible's piece, a 10 mm wide and 10 mm long square is sketched. Its distance in relationship with the superior edge of the piece is of 12.9261 mm. Likewise the distance of the inferior square's border and the inferior edge of the hemimandible's border is of approximately 10.073 mm. The internal extrusion of the square was 5 mm. In each model edge blends were avoided in order to optimize the numerical evaluations. The edges of the hemimandible present variations along the curvature. The stability of each design is given by the male-female union, which also will assure the assembly of both halves. For the left hemimandible, an assembled piece is created through the union of the left half with the pieces extracted from the right hemimandible. In its center a new square with similar measures was created, but the extrusion of 5 mm was made to the outside. Then both finished halves are assembled in order to search for global interferences or any irregularity that could provoke the instability of the union. The design of both pieces is shown in Fig. 14a, b.

For the design of the third proposition, the model with the mayor stability and maneuverability according to the maxillofacial specialist's opinion was modified. After corroborating the resistance of the male-female union, the mandible's coronoid process was retired, as according to the specialist's indications so that he would be

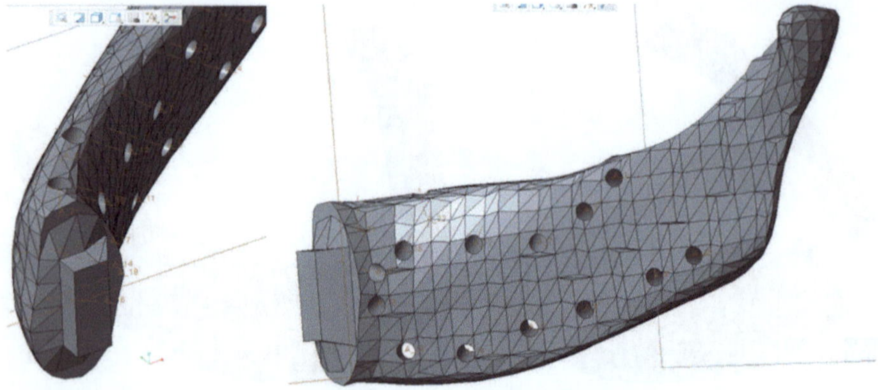

Fig. 15 Proposal for the assembly of both hemimandibles with perforations along the mandible's body and branches with a diameter of 3 mm, including the base for microimplants in case a dental prosthesis needs to be added. Also the removal of the mandible's coronoid process can be appreciated

able to collocate the prosthesis without causing damage to the oral tissues. Likewise, because of the necrosis, part of the internal tissues are severely damaged and the strongest muscles (main muscles that control and move the lower jaw) need to be inserted in the zones of the body and the mandible's branches with the use of sutures planted through holes with a diameter of 3 mm.

In order to make the model's design more accurate to the specialist's indications, several perforations along the mandible's body and branches are sketched and extruded in order to remove material, and further, also the coronoid process is retired, being careful of preserving the progressive curvature of the mandible. Something important to consider is not leaving borders or peaks that could hurt the patient by injuring adjacent tissues or muscles, as shown in Fig. 15.

For the resistance analysis both hemimandibles needed to be transformed with the usage of the "shrink-wrap" tool, so that the archives could be evaluated after assigning them a certain material and also distributing over their surfaces force vectors that simulate the natural occlusion forces on the mandible and also the tensions provoked by the muscles. Although this tool reduces the definition of the archive, the precision of the piece is not lost due to the transformation performed by the GOM program. Only one functional hemimandible was obtained, this means, only on one side the force vectors could be placed and analyzed. According to the force distribution's law, if one side is subjected to a certain force magnitude in a specific direction, the other side will react opposing to that force with a similar magnitude and in the opposite direction. For the resistance analysis this means that if one half of the lower jaw is capable of resisting the forces evaluated on its surface, the other half should be capable of resisting them too. Taking the maximal forces as the reference for the force that the main muscles which control the lower jaw can achieve, the evaluation process could be initiated. For instance, the digital archives were exported in the ._xt (parasolid) format so that the program Creo Simulate 3.0 (a force simulation and

Table 1 Maximal allowed muscle forces, study realized by Q. S. Kang, D. P. Updike, Eric P. Salathe, Institute of Biomedical Engineering and Mathematical Biology, Pennsylvania, USA

Muscle	Muscle's cross sectional area in cm^2	F max. (Maximal force) in Newton (N)
Masseter	3.4	408
Internal pterygoid	1.9	228
External pterygoid	2.1	252
Anterior temporal muscle	2.6	312
Posterior temporal muscle	1.6	192
Total temporal muscle	3.9	468
Digastric	1.0	120

extension program of Creo Parametric 3.0) could recognize the assembled pieces as a single one. The first step was to define the direction and magnitude of the force vectors through the program's "force moment" tool.

Once selected, a new window is opened in which the magnitude can be specified in Newton and also the direction of the force vectors along the y, x or z axis. Following the values of Table 1, the forces are placed over the superior border of the mandible. This is the occlusion zone and is oriented downward (y axis) with a value of 500 N. The value depends of the forces that muscles such as the masseter (408 N) and the temporal (468 N) execute on the zone. In the superior zone of both mandible's branches, near the condylar process, force vectors of a magnitude of 300 N were oriented to the outside (along the x-axis) in order to simulate the internal and external pterygoid's tension (228 and 252 N, respectively). Subsequently, through the "displacement" tool several bases along the inferior border of the lower jaw are placed so that the program may recognize the area against which the forces will be simulated. The process can be appreciated in Fig. 16.

Several shrinewrap archives were tested in order to restructure and consolidate the piece. At the beginning several shrinewrap levels were tested in a scale from one to ten, the last level being the one that preserves the highest resolution. Finalizing the tests of different pieces that conform the model 1, 2 and the complete mandible, the following functional archives were obtained:

Archive "complete mandible", functional in its shrinewrap conversion, level three.
Archive "left hemimandible, model 1", functional in its shrinewrap conversion, level five.
Archive "right hemimandible, model 2", functional in its shrinewrap conversion, level five.

Hereunder, the results obtained in each test with three different materials frequently used for the construction of multiple prostheses and implants are shown in Tables 2, 3 and 4. The three simulations which were performed are "Displacement", "Stress Von Mises" and "Maximum Shear Stress" (see Fig. 17).

Table 2 **First simulation**: Modified pieces, (A) Mandible, (B) Model 1: male-female union and (C) Model 2: guide union with the reinforcement of a male-female union

Materials	Healthy bone			ABS plastic		
Simulation	A	B	C	A	B	C
Displacement (mm)	Min. 0.00000	Min. 0.00000	Min. 0.0000	Min. 0.00000	Min. 0.00000	Min. 0.00000
	Medium 0.27434	Medium 1.00000	Medium 0.30535	Medium 2.50000	Medium 8.10852	Medium 2.7214
	Max. 0.54869	Max. 1.84668	Max. 0.61071	Max. 4.79077	Max. 16.2170	Max. 5.44283
Von Mises Stress (MPa)	Min. 0.00284	Min. 0.00098	Min. 0.01479	Min. 0.00353	Min. 0.00104	Min. 0.00675
	Medium 176.895	Medium 500.000	Medium 372.073	Medium 207.349	Medium 371.862	Medium 285.793
	Max. 353.787	Max. 919.245	Max. 744.130	Max. 345.579	Max. 743.723	Max. 571.580
Maximum Shear Stress (MPa)	Min. 0.00159	Min. 0.00054	Min. 0.00786	Min. 0.00203	Min. 0.00056	Min. 0.00383
	Medium 100.000	Medium 250.000	Medium 200.942	Medium 120.000	Medium 192.400	Medium 157.549
	Max. 188.433	Max. 495.651	Max. 401.876	Max. 193.147	Max. 384.800	Max. 315.095

(continued)

Table 2 (continued)

Materials	Stainless steel 316L			PEEK			Ti6Al4V		
Simulation	A	B	C	A	B	C	A	B	C
Displacement (mm)	Min. 0.0000	Min. 0.00000	Min. 0.00000	Min. 0.00000	Min. 0.00000	Min. 0.00000	Min. 0.00000	Min. 0.00000	Min. 0.0000
	Medium 80.0348	Medium 0.11699	Medium 0.11699	Medium 1.76522	Medium 5.96325	Medium 2.00111	Medium 0.06063	Medium 0.20385	Medium 0.06775
	Max. 0.06974	Max. 0.23399	Max. 0.07738	Max. 3.53045	Max. 11.9265	Max. 4.00223	Max. 0.12126	Max. 0.40770	Max. 0.13551
Von Mises Stress (MPa)	Min. 0.00322	Min. 0.00098	Min. 0.01465	Min. 0.00337	Min. 0.00105	Min. 0.00755	Min. 0.00255	Min. 0.00099	Min. 0.01504
	Medium 180.091	Medium 500.000	Medium 387.621	Medium 170.687	Medium 374.201	Medium 294.801	Medium 173.632	Medium 430.478	Medium 352.423
	Max. 360.178	Max. 937.813	Max. 775.227	Max. 341.371	Max. 748.402	Max. 589.594	Max. 347.262	Max. 860.955	Max. 704.831
Maximum Shear Stress (MPa)	Min. 0.00175	Min. 0.00054	Min. 0.00767	Min. 0.00192	Min. 0.00056	Min. 0.00437	Min. 0.00146	Min. 0.00053	Min. 0.00813
	Medium 100.000	Medium 250.997	Medium 209.024	Medium 100.000	Medium 193.176	Medium 161.917	Medium 120.000	Medium 250.000	Medium 190.823
	Max. 190.757	Max. 501.994	Max. 418.041	Max. 189.382	Max. 386.351	Max. 323.830	Max. 185.507	Max. 460.195	Max. 381.639

Table 3 Second simulation: Modified pieces, (A) Mandible, (B) Model 1: male-female union and (C) Model 2: guide union with the reinforcement of a male-female union

Materials	Healthy bone		ABS plastic		Stainless steel 316L		PEEK		Ti6Al4V	
Simulation	A	B	A	B	A	B	A	B	A	B
Displacement (mm)	Min. 0.00000	Min. 0.00000	Min. 0.00000	Min. 0.00000	Min. 0.0000	Min. 0.00000	Min. 0.00000	Min. 0.00000	Min. 0.00000	Min. 0.00000
	Medium 1.13488	Medium 0.40479	Medium 10.0257	Medium 3.58896	Medium 0.14435	Medium 0.05144	Medium 7.37762	Medium 2.63969	Medium 0.25297	Medium 0.08966
	Max. 2.26977	Max. 0.80957	Max. 20.0513	Max. 7.17791	Max. 0.28669	Max. 0.10288	Max. 14.7552	Max. 5.27939	Max. 0.50593	Max. 0.17932
Von Mises Stress (MPa)	Min. 0.00049	Min. 0.00495	Min. 0.00070	Min. 0.00236	Min. 0.00045	Min. 0.00455	Min. 0.00067	Min. 0.00329	Min. 0.00053	Min. 0.00532
	Medium 800.00	Medium 1048.04	Medium 577.529	Medium 1222.01	Medium 862.888	Medium 1022.1	Medium 482.211	Medium 968.261	Medium 815.721	Medium 1083.12
	Max. 1854.36	Max. 2620.10	Max. 1155.06	Max. 2444.01	Max. 2157.22	Max. 2555.47	Max. 1205.53	Max. 2420.65	Max. 2039.30	Max. 2707.79
Maximum Shear Stress (MPa)	Min. 0.00028	Min. 0.00203	Min. 0.00039	Min. 0.00136	Min. 0.00026	Min. 0.00242	Min. 0.00038	Min. 0.00184	Min. 0.00030	Min. 0.00300
	Medium 400.000	Medium 581.356	Medium 316.049	Medium 698.288	Medium 465.164	Medium 564.389	Medium 262.830	Medium 552.074	Medium. 440.453	Medium 604.343
	Max. 998.828	Max. 1453.39	Max. 632.098	Max. 1396.58	Max. 1162.91	Max. 1410.97	Max. 657.074	Max. 1380.18	Max. 1101.13	Max. 1510.85

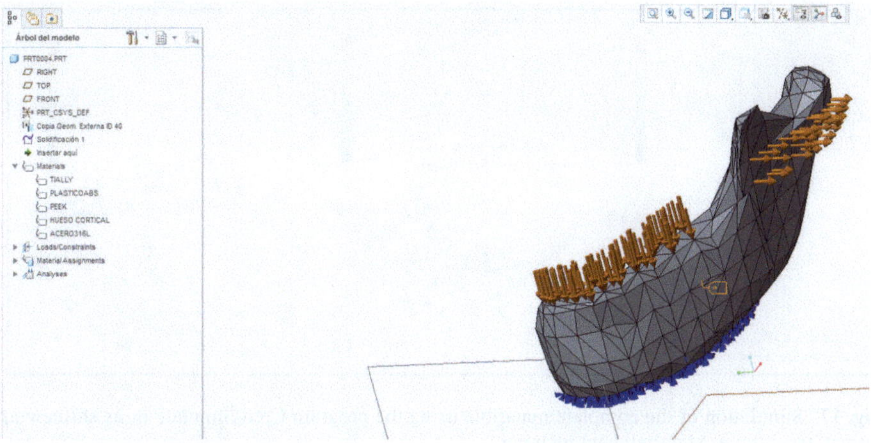

Fig. 16 Placement of the maximal forces specified on Table 1

Table 4 **Third simulation**: Proposed model using the most optimal material

Material	Ti6Al4V	
Simulation	A	
Displacement (mm)	Min.	0.00000
	Medium	0.29002
	Max.	0.58005
Von Mises Stress (MPa)	Min.	0.00234
	Medium	393.401
	Max.	786.799
Maximum Shear Stress (MPa)	Min.	0.00134
	Medium	210.097
	Max.	420.193
Material	Ti6Al4V	
Simulation	A	
Von Mises Stress (MPa)	196.7005	
Maximum Shear Stress (MPa)	210.0965	

6 Discussion

In order to select the right material, the maximum shear stress principle is applied. For instance, the evaluation of the obtained results concerning the test known as von Mises Stress are compared against the mechanical tensile strength of each material. Likewise the test known as Maximum Shear Stress indicates the Maximum Shear Stress localized in a determined area and, as a consequence, it is taken as a base for the material's evaluation. This means if the material is resistant enough to withstand the muscular forces.

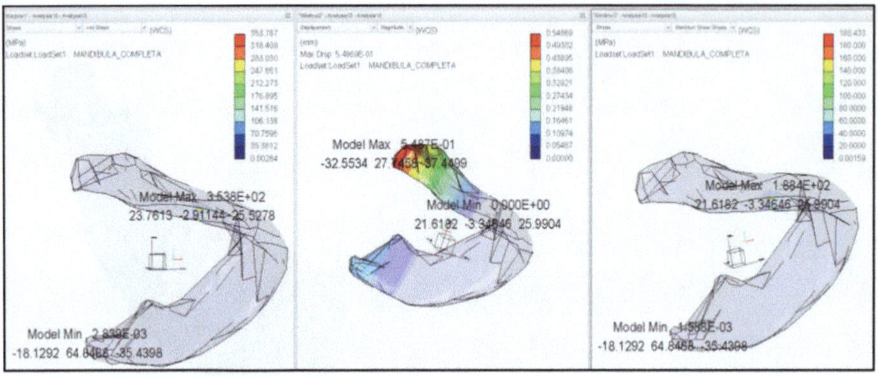

Fig. 17 Simulation of the complete mandible using the program Creo Simulate in its shrinewrap format, level three. Material: healthy bone

Hence the values of the medium values set on Table 2 must be used, which also were selected to form the green area of the simulation tests performed through the program Creo Simulate and reflect the material's response after being subjected to perpendicular forces over the mandible's surface of a magnitude of 500 N in conjunction with forces of 300 N working on the condylar processes' area (region of the mandible's branches). The values are shown in Table 5. To analyze these values, the mechanical tensile strength of each material must be taken into account and also be compared against the medium values obtained of the von Mises Stress test. First the values of the test must be inferior to the limits of the mechanical tensile strength, because all the values found in a range of values inferior to the limit are indicators of an optimal material's resistance. It also states that the material must not lose its original form by deformation and should be considered as an option of which the prosthesis should be built.

Table 5 shows the values of both tests for the stainless steel 316L [4] and also organizes them in three cases for the different models, case A being the simulation of the forces over the complete mandible, case B over the model assembled through a male-female union and C the model with a guide assembly in conjunction with a male-female union. Considering the first test, the values of the von Mises Stress simulation are compared against the inferior limit (490 MPa) and the superior limit (520 MPa) of the mechanical tensile strength. For the first model (A) the medium value of the von Mises Stress test is equal to 180.091 MPa, which means that between the first limit (490 MPa) and that value, a difference of 63.246% is obtained. Then it can be concluded that the complete mandible would easily resist the occlusion forces if the prosthesis is built in stainless steel. Likewise, when compared against the superior limit, the medium value of the von Mises Stress test lays approximately a 65.367% under the limit's value. For the evaluation of the second test the values of the Maximum Shear Stress are compared against the medium value of the stainless steel's mechanical tensile strength. Once again, the value of the test is inferior to the

Table 5 Comparative values of the three models described on Table 2 for the von Mises Stress test against the mechanical tensile strength values of stainless steel 316L and the values of the Maximum Shear Stress against the medium values of the mechanical tensile strength

Comparison			
Stain steel 316L [MPa]			
Model	A	B	C
Von Mises Stress Medium Values	***180.091***	***500.00***	***387.621***
Mechanical Tensile Strength Lim. 1	490.000	490.000	490.000
Mechanical Tensile Strength Lim. 2	520.000	520.000	520.000
Maximum Shear Stress	***100.000***	***250.997***	***209.024***
Mechanical Tensile Strength Lim. 1/2	245.000	245.000	245.000
Mechanical Tensile Strength Lim. 2/2	260.000	260.000	260.000

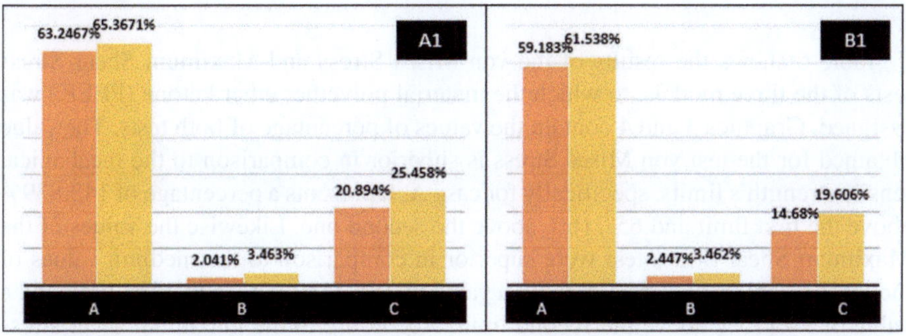

Graphics 1 and 2 Percentages obtained for the tests of the models with stainless steel 316L assigned as the building material. The graphic A1 shows the comparison of both limits of the mechanical tensile strength for the Von Mises Stress test (First limit 490 MPa and second limit 520 MPa) and the graphic B1 corresponds to the values obtained from the Maximum Shear Stress test (First limit 245 MPa and second limit 260 MPa). For both cases the columns with the darkest color indicate the percentage that is above the first limit, in this case, model B is approximately 2.041% above the limit of 490 MPa, but 3.463% below the second limit (520 MPa). Likewise, the model B presents a value of 2.447% above the first limit of the Maximum Shear Stress test (245 MPa), but 3.462% below the second limit (260 MPa)

medium values of both limits, being the percentage that lays below the first limit of 59.184 and 61.538% below the second limit. The same principle is used for the other models and tests performed in different materials and the results (percentage above or below the limits) are organized in the following tables and graphics (Graphics 1 and 2).

Table 6 Comparative values of the three models described on Table 2 for the von Mises Stress test against the mechanical tensile strength values of PEEK and the values of the Maximum Shear Stress against the medium values of the mechanical tensile strength

Comparison			
PEEK [MPa]			
Model	A	B	C
Von Mises Stress	***170.687***	***374.201***	***294.801***
Mechanical Tensile Strength Lim. 1	70.00	70.00	70.00
Mechanical Tensile Strength Lim. 2	103.000	103.000	103.000
Maximum Shear Stress	***100.000***	***193.176***	***161.917***
Mechanical Tensile Strength Lim. 1/2	35.00	35.00	35.00
Mechanical Tensile Strength Lim. 2/2	51.50	51.50	51.50
Von Mises Stress/2	***85.3440***	***187.101***	***147.401***

Table 6 shows the results of the von Mises Stress and Maximum Shear Stress tests of the three models, to which the material polyether ether ketone (PEEK) was assigned. Graphics 3 and 4 contain the values of percentage of both tests. The value obtained for the test von Mises Stress is superior in comparison to the mechanical tensile strength's limits, specifically for case A, it presents a percentage of 143.839% above the first limit and 65.716% above the second one. Likewise the values of the Maximum Shear Stress test were superior in comparison to the medium values of the mechanical tensile strength with a value of 185.714% above the first limit and a value of 94.175% above the second limit. According to the maximum shear stress principle if the values are superior to the medium value of the material's mechanical tensile strength, the piece will behave plastically and suffer structural changes due to deformation until it reaches the maximum tension's point (breaking point).

Finally the titanium alloy was analyzed based on the results of Table 7. The values obtained in the von Mises Stress test are below both mechanical tensile strength's limits (Graphics 5 and 6).

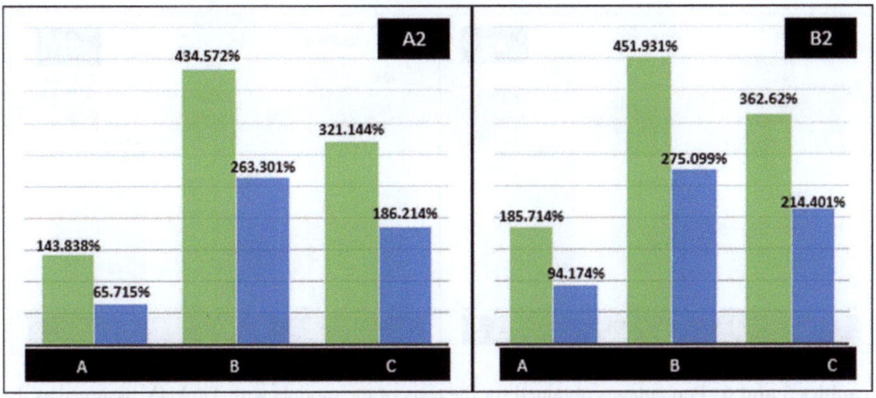

Graphics 3 and 4 Percentages obtained for the tests of the models with PEEK assigned as the building material. The graphic A2 shows the comparison of both limits of the mechanical tensile strength for the Von Mises Stress test (first limit 70 MPa and second limit 103 MPa) and the graphic B2 corresponds to the values obtained from the Maximum Shear Stress test (first limit 35 MPa and second limit 51.5 MPa). For this particular case all the percentage values are above the limits

Table 7 Comparative values of the three models described on Table 2 for the von Mises Stress test against the mechanical tensile strength values of Ti6Al4V and the values of the Maximum Shear Stress against the medium values of the mechanical tensile strength

Comparison			
Ti6Al4V [MPa]			
Model	A	B	C
Von Mises Stress	***173.632***	***430.478***	***352.423***
Mechanical Tensile Strength Lim. 1	800.000	800.000	800.000
Mechanical Tensile Strength Lim. 2	1450.00	1450.00	1450.00
Maximum Shear Stress	***120.000***	***250.000***	***190.823***
Mechanical Tensile Strength Lim. 1/2	400.000	400.000	400.000
Mechanical Tensile Strength Lim. 2/2	725.000	725.000	725.000
Von Mises Stress/2	***86.816***	***215.239***	***176.212***

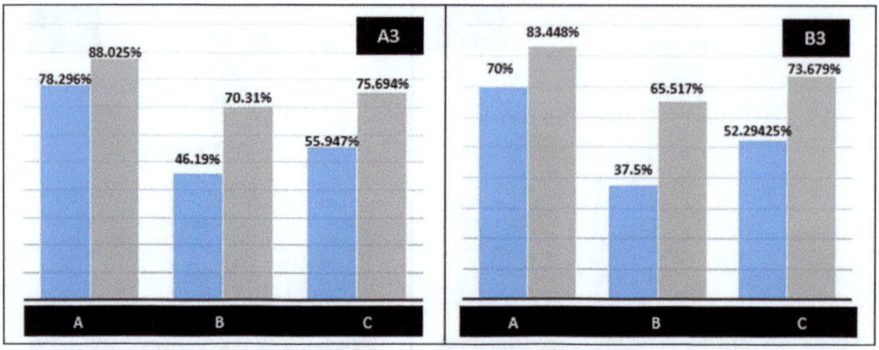

Graphics 5 and 6 Percentages obtained for the tests of the models with Ti6Al4V assigned as the building material. The graphic A3 shows the comparison of both limits of the mechanical tensile strength for the Von Mises Stress test (first limit 800 MPa and second limit 1450 MPa) and the graphic B3 corresponds to the values obtained from the Maximum Shear Stress test (first limit 400 MPa and second limit 725 MPa). For this particular case all the percentage values are below the limits

Table 8 Comparative values of the two models described on Table 3 for the von Mises Stress test against the mechanical tensile strength values of stainless steel 316L and the values of the Maximum Shear Stress against the medium values of the mechanical tensile strength

Comparison		
Stainless steel 316L [MPa]		
Model	A	B
Von Mises Stress Medium Values	***862.888***	***1022.190***
Mechanical Tensile Strength Lim. 1	490.000	490.000
Mechanical Tensile Strength Lim. 2	520.00	520.00
Maximum Shear Stress	***465.164***	***564.389***
Mechanical Tensile Strength Lim. 1/2	245.000	245.000
Mechanical Tensile Strength Lim. 2/2	260.000	260.000
Von Mises Stress/2	***431.444***	***511.095***

6.1 Analysis of the Models with Circular Perforations Along the Body and the Branches (Length of 10 mm and a Diameter of 3 mm)

Table 8 shows the values obtained for both tests for the model assembled through a male-female union (A) and the guide union with the reinforcement of a male-female union (B) (Graphics 7, 8, 9 and 10, Table 9).

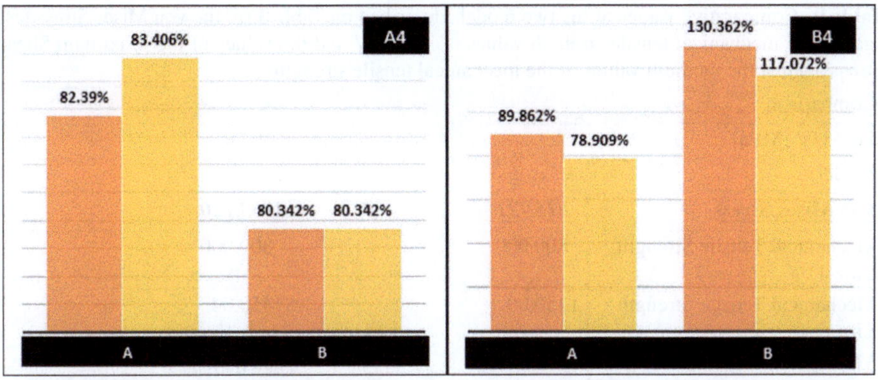

Graphics 7 and 8 Percentages obtained for the tests of the models with stainless steel 316L assigned as the building material. The graphic A4 shows the comparison of both limits of the mechanical tensile strength for the Von Mises Stress test (first limit 490 MPa and second limit 520 MPa) and the graphic B4 corresponds to the values obtained from the Maximum Shear Stress test (first limit 245 MPa and second limit 260 MPa). For this particular case all the percentage values are above the limits

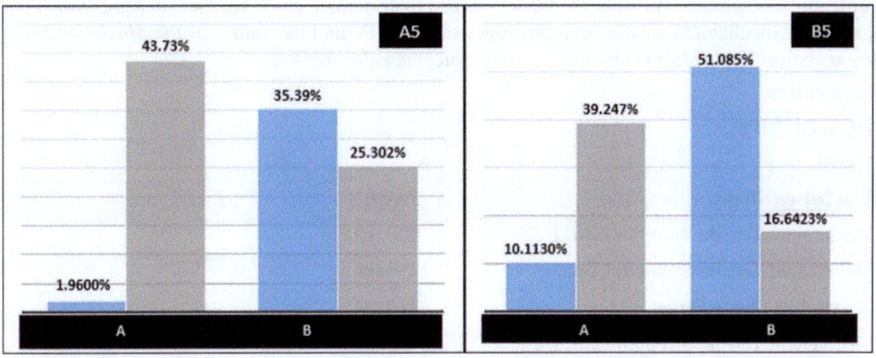

Graphics 9 and 10 Percentages obtained for the tests of the models with Ti6Al4V assigned as the building material. The graphic A5 shows the comparison of both limits of the mechanical tensile strength for the Von Mises Stress test (first limit 800 MPa and second limit 1450 MPa) and the graphic B5 corresponds to the values obtained from the Maximum Shear Stress test (first limit 400 MPa and second limit 725 MPa)

For both cases, the graphics with the darkest color indicate the percentage that is above the evaluated limit, while those dyed in gray refer to the percentage below the material's mechanical tensile strength's limit.

Table 9 Comparative values of the two models described on Table 3 for the von Mises Stress test against the mechanical tensile strength values of Ti6Al4V and the values of the Maximum Shear Stress against the medium values of the mechanical tensile strength

Comparison		
Ti6Al4V [MPa]		
Model	A	B
Von Mises Stress	***815.721***	***1083.120***
Mechanical Tensile Strength Lim. 1	800.000	800.000
Mechanical Tensile Strength Lim. 2	1450.00	1450.00
Maximum Shear Stress	***440.453***	***604.343***
Mechanical Tensile Strength Lim. 1/2	400.000	400.000
Mechanical Tensile Strength Lim. 2/2	725.000	725.000
Von Mises Stress/2	***407.8605***	***541.560***

Table 10 Comparative values of the model with a male-female union for the *von Mises Stress test* against the mechanical tensile strength values of Ti6Al4V and the values of the *Maximum Shear Stress* against the medium values of the mechanical tensile strength

Comparison	
Ti6Al4V [MPa]	
Model	A
Von Mises Stress	***393.401***
Mechanical Tensile Strength Lim. 1	800.000
Mechanical Tensile Strength Lim. 2	1450.00
Maximum Shear Stress	***210.097***
Mechanical Tensile Strength Lim. 1/2	400.000
Mechanical Tensile Strength Lim. 2/2	725.000
Von Mises Stress/2	***196.701***

6.2 Analysis of the Third Model, to Which the Optimal Material was Assigned

For the final part of the analysis, the third model is subjected to the simulation of the forces. The only difference that it presents in comparison to the other models is the perforations placed along the mandible's body and part of the ascendant branches. Likewise, it does not possess the mandible's coronoid process (Graphics 11 and 12, Table 10).

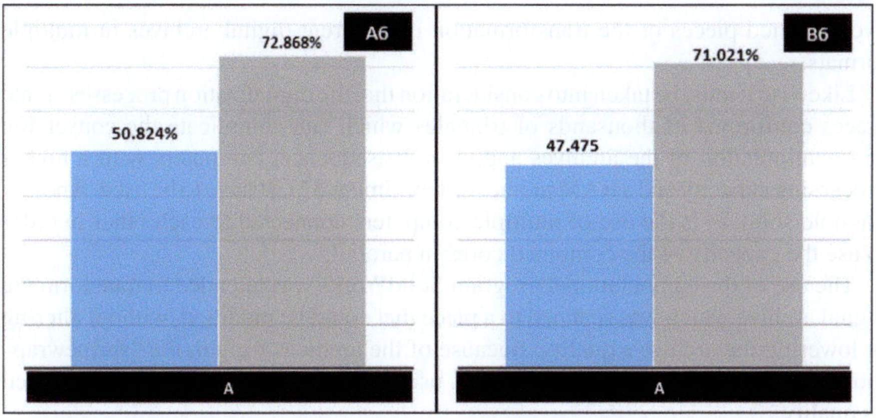

Graphics 11 and 12 Percentages obtained for the tests of the models with Ti6Al4V assigned as the building material. The graphic A6 shows the comparison of both limits of the mechanical tensile strength for the Von Mises Stress test (first limit 800 MPa and second limit 1450 MPa) and the graphic B6 corresponds to the values obtained from the Maximum Shear Stress test (first limit 400 MPa and second limit 725 MPa). For this particular case all the percentage values are below the limits, indicating that the titanium alloy is the most optimal material for the prosthesis design

7 Conclusions

The tomographic analysis of the anatomic regions in conjunction with the reproduction of the tridimensional models through digital processes allows the creation of the base for the physical construction of 3-D models with the use of 3-D printers [5–11]. This is particularly efficient for the reproduction of accurate anatomic structures in individual cases.

The computer and physical analysis of the mandible's structure allows the specialists to detect defects or zones that have to be modified, for example, rugged zones affected by the necrosis.

For tridimensional printing, the most optimal material was the ABS plastic due to its dimensional stability. This property gives the material the surface's resistance against cold materials or fluids, such as the dental alginate.

Easy to use and low cost materials such as stone plaster and dental alginate allow the user to create similar reproductions of the patient's anatomic structures in less time.

The process of rendering, tridimensional printing, model reproduction in stone plaster, sanding and reconstruction are vital for the creation of models which are the base for the development of the model in acrylic material and the following scanning and modeling processes.

Although the process was not performed according to the program which was originally selected (SolidWorks), it is necessary to know how to use another CAD program that fulfills the needed functions. This means the design of different elements over scanned pieces or the transformation of different digital archives in multiple formats.

Likewise it must be taken into consideration that the digitalization processes create pieces conformed of thousands of triangles which can complicate the conversion and manipulation of the archives and, as a consequence, computers with a higher processing capacity and a RAM memory of minimum 32 GB have to be used. Another possible solution is the use of multiple computers connected to each other in order to use the capacity of the combined cores in parallel.

The use of the computational program SolidWorks was intended to transform the digital archive which was scanned to a piece that could be modified, without altering or lowering the archive's quality. Because of the archives weight, the "shrinewrap" tool of the program Creo Parametric was used and also the archive did not lowered its quality.

All the created models are subject to posterior modifications specified by the surgical specialists. Likewise all the models have to be reinforced with screws made of stainless steel and titanium plates to guarantee a stable union.

The force's analysis of the occlusion zone is based on the maximal forces that the muscles can reach, concretely for mastication movements and not for other kind of functions like phonetic movements because the tensions of the muscles tend to be softer.

In all the cases in which the percentage was found to be below the medium value of the mechanical tensile strength, the material is considered optimal for the construction of the prosthesis. This means it is capable of resisting the forces that influence the lower jaw.

Due to the specialist's indications, the mandible needed to be separated in two halves. All the models can be built in stainless steel 316L or titanium Ti6Al4V.

Through the simulation of the forces over the model with perforations, the male-female union presents a mayor resistance when the material titanium is assigned, because the values acquired in the Stress von Mises and Maximum Shear Stress tests were lower than the values of the mechanical tensile strength.

Nowadays, titanium prostheses arise to provide an alternative in fields of application such as dentistry, where they have shown good results due to an extraordinary biocompatibility. Thus, titanium prostheses result in a simplified surgical procedure by reducing surgical time and avoiding a second procedure for obtaining a graft. In addition, the modeling minimizes the risk of contamination of the receptor site, reducing its morbidity.

References

1. Osteonecrosis: Clinic DAM Madrid. Spain. https://www.clinicadam.com/salud/5/007260.html. Accessed 21 Jan 2017
2. Fernández López, R.G., Arellano Flores, A.M., Velázquez Serrano, S.N.: Mandibular ostenecrosis linked to bisphosphates—a clinical study. http://www.scielo.org.mx/scielo.php?script=sci_arttext&pid=S1870199X2013000100008 (2013). Accessed 24 Jan 2017
3. Kang, Q.S., Updike, D.P., Salathe, E.P.: Theoretical Prediction of Muscle Forces on the Mandible During Bite. Institute of Biomedical Engineering and Mathematical Biology, Pennsylvania (1990). Accessed 24 Jan 2017
4. Stainless Steel—Grade 316L—Properties, Fabrication and Applications (UNS S31603). http://www.azom.com/article.aspx?ArticleID=2382. Accessed 21 Jan 2017
5. Beltrán-Fernández, J.A., et al.: Assessment of the structural integrity of C3–C5 cervical porcine vertebrae model based on 2D classic CAD, 3D scanner and 3D computed tomography. In: Ochsner, A., da Silva, L.F.M., Altenbach, H. (eds) Analysis and Design of Biological Materials and Structures. Advanced Structured Materials, vol. 14, pp. 3–17. Springer (2012)
6. Beltrán-Fernández, J.A., et al.: Numerical stress analysis of C3–C5 cervical porcine vertebrae under compressive loading using 3-D scanner and 3-D computed tomography. In: Mechanics of Biological Structures. Springer, Germany (2011)
7. Beltrán-Fernández, J.A., et al.: Biomechanical characterization of a cervical corporectomy using porcine specimens, following an experimental approach. J. Key Eng. Mater. Trans. Tech. Inglaterra **478**, 103–108 (2011)
8. Beltrán-Fernández, J.A., et al.: Biomechanics and numerical evaluation of cervical porcine models considering compressive loads using 2-D classic computer tomography CT, 3-D scanner and 3-D computed tomography. Appl. Mech. Mater. **24–25**, 287–295 (2010)
9. Beltrán-Fernández, J.A., et al.: Biomechanical evaluation of a corporectomy in porcine lumbar specimens using fexible polimer belts. J. Phys.: Conf. Ser. **181**, 12015 (2009)
10. Beltrán-Fernández, J.A., et al.: Modelling of a cervical plate and human cervical section C3–C5 under compression loading conditions using the finite element method. Appl. Mech. Mater. **13–14**, 49–56 (2008)
11. Moreno-Garibaldi, P., Beltrán-Fernández, J.A., Hernández-Gómez, L.H., López-Saucedo, F., Corro-Valdez, N., López-Liévano, A.: Caracterización mecánica de material compuesto enfocado hacia aplicaciones maxilofaciales, DYNA Ing. Ind. **94**(4), 461–465 (2017)

Development of a Sensitive System to Fixed Prosthesis in the Lumbar and Cervical Orthopedic Area

Rodrigo Vázquez-Machorro, Juan Alfonso Beltrán-Fernández, Alexander Reyes-Cruz, Luis Héctor Hernández-Gómez, Juan Atonal-Sánchez, Adolfo López-Lievano, Juan Alejandro Flores-Campos and Alejandro González Rebattu y González

Abstract Substitution for the loss of human members due to different natural consequences has been a reality for many thousands of years. Over time, the evolution of knowledge and technologies, just as the discovery of new materials, gave rise to inventions in the fields of robotics, in particular biomechanics, that have provided human beings with complementary limbs that are improved upon every day. Based on the advances of electronic components, more efficient systems are developed for patient monitoring, which allows for more accurate control and monitoring of the prosthesis, as well as the vital signs of the patient. With the help of technology, efficient

R. Vázquez-Machorro (✉) · J. A. Beltrán-Fernández · A. Reyes-Cruz · L. H. Hernández-Gómez · J. Atonal-Sánchez · A. López-Lievano · J. A. Flores-Campos
Instituto Politécnico Nacional ESIME-SEPI. Unidad Profesional "Adolfo López Mateos", Edificio 5, 3er Piso, Colonia Lindavista, Gustavo A. Madero, 07738 Mexico D.F., Mexico
e-mail: rodrigovazmach_5@hotmail.com

J. A. Beltrán-Fernández
e-mail: jbeltranf@hotmail.com

A. Reyes-Cruz
e-mail: alex_reyes1983@hotmail.com

L. H. Hernández-Gómez
e-mail: luishector56@hotmail.com

J. Atonal-Sánchez
e-mail: juan_atonal@yahoo.com.mx

A. López-Lievano
e-mail: k_lievano@hotmail.com

J. A. Flores-Campos
e-mail: jaflores@ipn.mx

A. G. R. y. González
Hospital Regional 1ro de Octubre-ISSSTE, Av. Instituto Politécnico Nacional, No. 1669, Colonia Lindavista, Gustavo A. Madero, CDMX, 07300 Mexico, Mexico
e-mail: alexrebattu@hotmail.com

© Springer International Publishing AG, part of Springer Nature 2019
A. Öchsner and H. Altenbach (eds.), *Engineering Design Applications*,
Advanced Structured Materials 92, https://doi.org/10.1007/978-3-319-79005-3_23

treatments are applied or to recover bodily parts as far as possible. This paper presents the design for a prototype of a monitoring system for spinal prosthesis, which an system Arduino UNO®, an infrared sensor, an accelerometer and a micro SD memory are used. In addition, computational tools are used for the development of the algorithm, which allow the monitoring of the prosthesis. As a result, measurements are obtained in the Arduino® system for the behaviour and movement of the prosthesis such as the displacement of the vertebra and angular velocities, which are stored in a micro SD. The application of new technologies in spinal prostheses is also obtained in order to make a diagnosis less costly and available in real time.

Keywords Orthopedic area · Prosthesis · Sensitive system · Lumbar vertebra Cervical vertebra

1 Introduction

Today, invasive techniques can be so small at the size of a half-inch incision, that the recovery time is practically one night, and because of this, the doctor can decide whether an invasive spinal surgery is appropriate after diagnosing the situation and determine whether or not there is a high potential for success. There are several procedures available for different conditions [1].

The spine is composed of vertebrae, and between each of them there is a disc, which is called the intervertebral disc, which gives the column mobility and flexibility. The vertebrae are called differently, depending on the area they occupy in the spine: cervical, thoracic and lumbar [2].

2 Spine and Vertebrae

Before adolescence, the spine is made up only of 33 bones, because both the five sacral bones and the four bones of the coccyx do not work until the stage of adolescence arrives.

Each vertebra of the spine is composed of several important parts that are [3, 4] (see Fig. 1):

- *The body*: this is the main region that supports the weight of a vertebra and that constitutes the thickness of the mass of the bone.
- *Transverse process*: extends from the neck as thin bone columns pointing to the right and left sides of the body.
- *Spinous process*: in this case, it extends from the ends of the transverse process in the opposite direction.
- *Vertebral hole*: it is located between the transverse process and the spinous process. Basically, it is a hollow space containing the spinal cord and meninges.

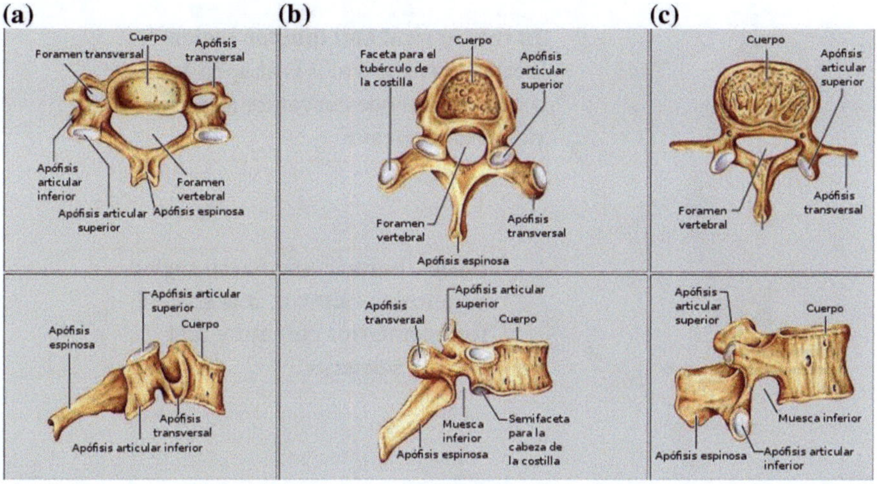

Fig. 1 Types of vertebrae: **a** cervical; **b** dorsal and **c** lumbar

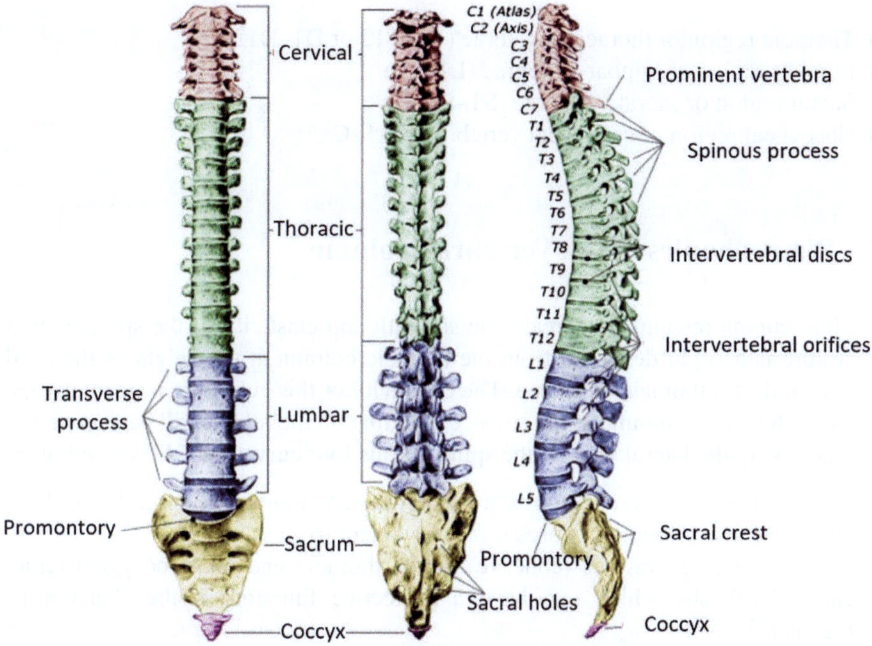

Fig. 2 Regions of the spine

These regions of the body are five and are of great importance in the spine; Fig. 2 shows the locations of each region of the spine [2, 3].

- Cervical region or cervical vertebrae (C1–C7).

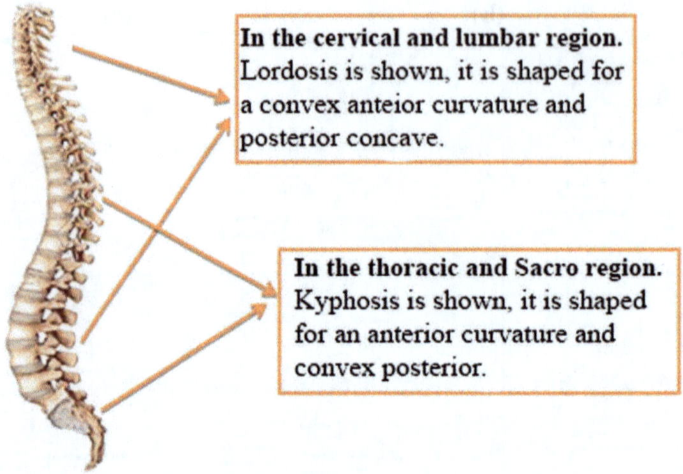

Fig. 3 Curvatures of the vertebral column

- Thoracic region or thoracic vertebrae (T1–T12 or D1–D12).
- Lumbar region or lumbar vertebrae (L1–L5).
- Sacral region or sacral vertebrae (S1–S5).
- Coccygeal region or coccygeal vertebrae (Cx1–Cx5).

3 Biomechanics of the Vertebral Column

Sagittal curvatures aim to increase the strength and elasticity of the spine. Lateral curvature is a slight deviation from the thoracic column at the height of the third, fourth, and fifth thoracic vertebrae. The concavity of this curvature is generally oriented to the left. In summary, the spine, depending on the region will have particular curvatures. In the lateral view of the spine reveals four curvatures [4, 5] (see Fig. 3):

- Two anterior (convex) curvatures in the cervical and lumbar region called lordosis, which the spine commonly uses it for movement [6].
- Two posterior curvatures (concave) in the thoracic and sacrococcygeal region called kyphosis, which can have a protective function in the skeleton for organs [6].

3.1 Column Mobility

The sum of the limited movements between adjacent vertebrae allows an important degree of mobility to the spine as a whole as seen in Fig. 4. The different movements of the spine are [5, 7]:

- Bending
- Extension
- Inclination
- Rotation.

Table 1 shows the degrees of articulation for each region of the spine according to the movement [8]. These movements are more easily given in the cervical and lumbar spine due to the following reasons:

- In these sectors, the intervertebral discs are thicker.
- The spinous process in these regions are shorter and more separated from each other.
- There is no fastening to the rib cage.

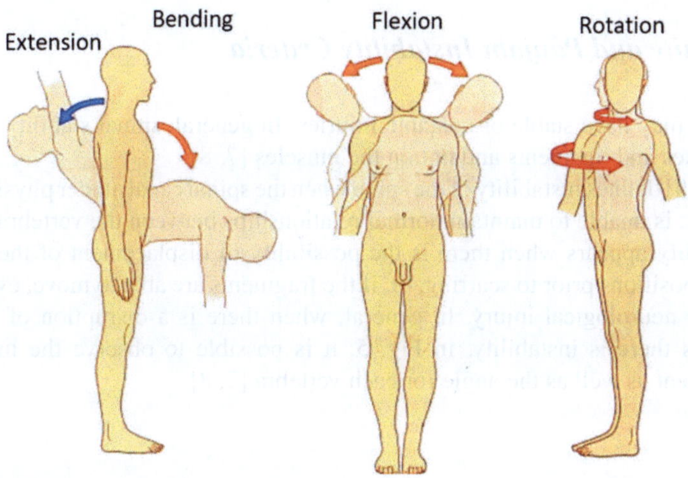

Fig. 4 Movements of the spine

Table 1 Summary of the movement of the spine

Movimiento	Cervical	Dorsal	Lumbar
Flexión	40°	20°	60°
Extensión	75°	25°	35°
Inclinación lateral	(30°–45°)[a]	20°	20°
Rotación	(45°–65°)[b]	35°	5°

[a]30° in the lower cervical region and 10°–15° in high cervical region
[b]25° in the atlo-axoid joint region and 25° in the lower cervical region

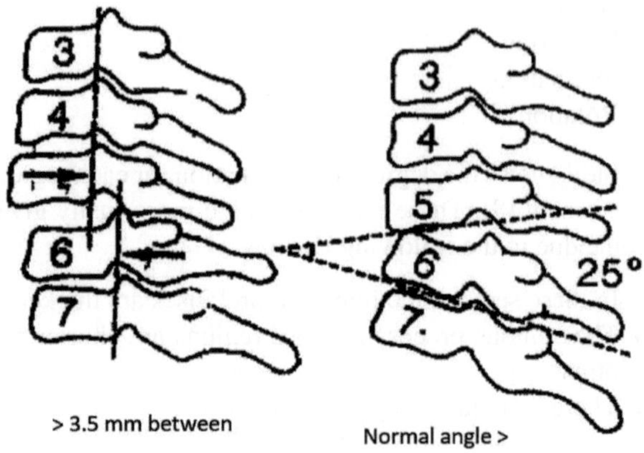

Fig. 5 Maximum displacement and angle of a vertebra

3.2 White and Panjabi Instability Criteria

The spine may have stable or unstable injuries. In general, spinal stability depends on the bones and ligaments and not on the muscles [7, 8].

White [9] defines instability of the spine when the spinal canal, under physiological conditions, is unable to maintain normal relationships between the vertebrae.

Instability appears when there is the possibility of displacement of the relative vertebral positions prior to scarring, or, if the fragments are able to move, eventually leading to neurological injury. In general, when there is a disruption of 2 of the 3 columns there is instability. In Fig. 5, it is possible to observe the maximum displacement as well as the angle for each vertebra [7, 8].

4 Materials and Method (Methodology)

The following are the guidelines of the methodology for the assembly of the sensing system, as well as the acquisition of specimens for experimental analysis [10, 11] (see Figs. 6 and 7).

From the available range of animals, perhaps the most representative is th pig because of the great similarities that it has from the biological point of view with the human. This is why several implants show very good acceptance in the human body. Rephrase maybe; obviously in the literatura regarding the use of porcine specimens the pig is the most widely used animal model especially in research relating to the study of the human spine [9, 12, 13].

Another reason why porcine specimens were chosen, is because in Mexico it is difficult to carry out the experimental analysis in humans due to the legal and health difficulties that accompany the use of obituary material [9, 12, 13].

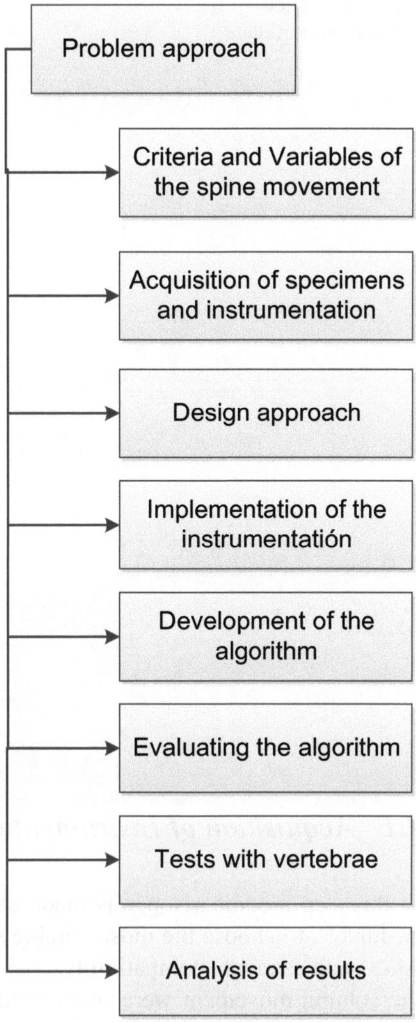

Fig. 6 Stages of the methodology

The experimental analysis was divided into 2 parts for each region (specimens), in the first part the sensing system was installed using an ultrasonic sensor [10] in each one of the specimens in a single plane and in the second part the reflective optical sensor was used with each of the specimens. This work only focused on the analysis with the optical sensor reflective.

Specimens from the cervical region (C3, C4 and C5) and lumbar region (L3, L4, and L5) were used for analysis. Once the specimens were prepared, according to Fig. 7, and the sensing system was assembled according to the methodology of Fig. 6 from the acquisition and instrumentation.

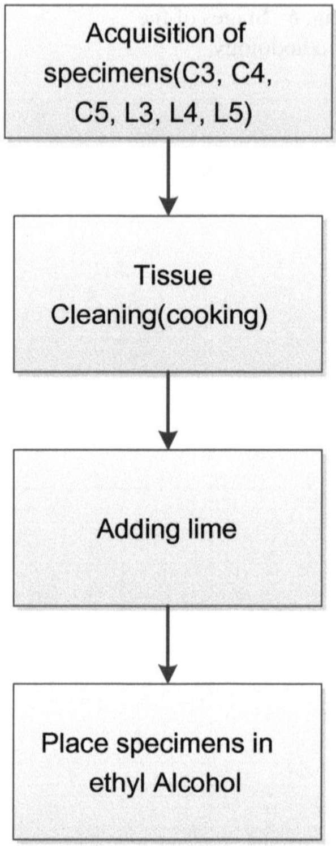

Fig. 7 Specimen preparation procedure

4.1 Acquisition of Instrumentation

In this step a comparison was made of the wide range of components offered by Arduino®, to choose the most suitable sensing system, this according to characteristics such as range, compatibility, cost, among others. The criteria and variables of the column movement were considered. The comparison of some components for the sensing system [14–16] can be seen in Tables 2 and 3.

4.2 Design Approach

Here the compatibility of the components with the Arduino® Board are considered, as well as the pins to be considered for the feeding of the same, and see if these same are inputs or outputs of the board. The selection of instrumentation for the sensing system is also considered. Table 4 shows the components defined to carry out the sensing system [14–16].

Table 2 Sensor comparison

Model	Voltage (V)	Distance (cm)
GP2Y0A41SK0F	5	4–30
GP2Y0A21YK	3.1–0.04	10–80
GP2Y0A02YK	5	20–150
GP2Y0A710K0F	5	100–500
HY-SRF05	5	3–300
HC-SR04	5	1–500
LV-EZ0	5	40–600
IS471F	5	1–15
QRD1114	5	0–3

Table 3 Comparison of accelerometers

Name	Range	Axes	Interface
MMA7361	±1.5 g, ±6 g	3	Analógica
ADXL335	±3 g	3	Analógica
MMA8452	±2 g, ±4 g, ±8 g	3	Digital
ADXL345	±16 g	3	Digital
MMA7455	±2 g, ±4 g, ±8 g	3	Digital
MMA7660	±1.5 g	3	Digital

Table 4 Components of the sensing system

Component	Image	Component	Image
Arduino one R3		Accelerometer ADXL335	
Ultrasonic sensor HC-SR04		Bluetooth HC06	
Reflective optical sensor IR QRD1114		Micro SD reader	

Fig. 8 Connecting the components

4.3 Implementation of the Instrumentation

In this step the necessary physical connections of each component to the plate are made (with support of the data sheets of each one of these components), always taking into account the pins to be used in it, because these will be used for the development of the algorithm. Figure 8 shows the connection of the components of the sensing system.

4.4 Development of the Algorithm

Of the code pins were taken into account to be used in each component, as well as the download of extra libraries for communication with the instrumentation. It should be taken into account that the Arduino® program must first be configured, as this will allow the code to be uploaded to the microcontroller and thus a correct acquisition of the sensor data is achieved (see Fig. 9).

4.5 Evaluating the Algorithm

Once the development of the algorithms for each of the components has been completed, the Arduino® program is responsible for verifying the syntax of each one of the commands used. If there is an error the program automatically points to the line where the error is possible. Until the error is corrected, the program will not compile and load the code to the microcontroller. Otherwise the program will be compiled and loaded. It will be ready to do the sensing in this case. In Fig. 10 you could see a flowchart of the execution of a program or algorithm [14].

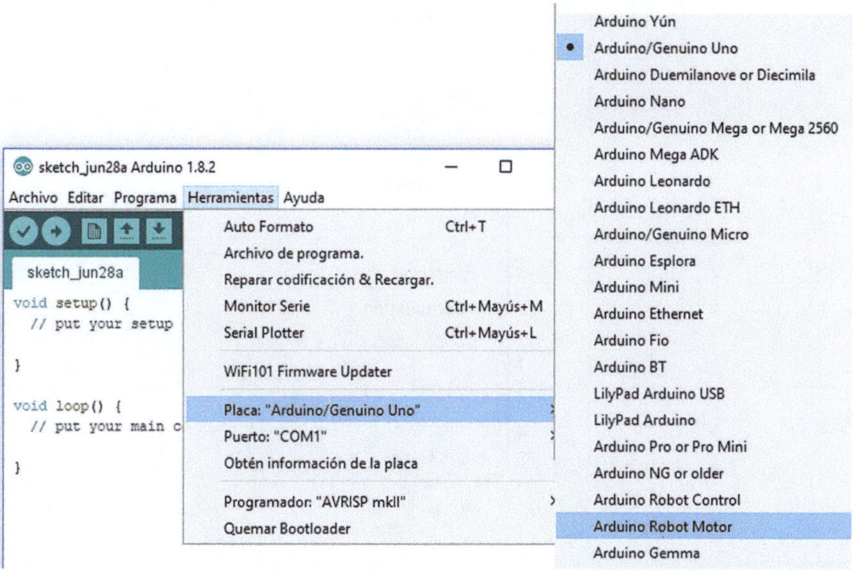

Fig. 9 Arduino® board configuration

4.6 *Tests with Vertebrae*

In this step tests were performed with cervical and lumbar specimens. The experimental analysis was divided into two parts. In the first part, sensing of the cervical specimens is made and in the second part sensing with the lumbar specimens is made.

First, specimen 1 was placed in front of the sensor horizontally. Then, the Arduino board was connected to a PC to power the board, which contains the code developed and evaluated to operate all components of the sensing system.

Once the system was powered, the Arduino® program was opened and the serial console or monitor was activated, which allows to see the results obtained from the sensor when the vertebra moves. Figure 11 shows the window to display the results.

This same process was performed on the remaining specimens (C2 and C3). In Fig. 12, the placement of the specimen in the sensing system can be observed.

The previous procedure was performed for the other specimens (C2 and C3). For the second part the previous steps were followed, but this time for the specimens of the lumbar region (L3, L4 and L5). Figure 13 shows the placement of the specimen in the sensing system. The same was done for the remaining 2 lumbar specimens.

Finally, Fig. 14 shows the operation diagram of the reflective optical sensor in front of each specimen and in Fig. 15 shows the placement of the sensors for the 3 axes. As mentioned earlier in this document only the sensor application is shown on a single axis.

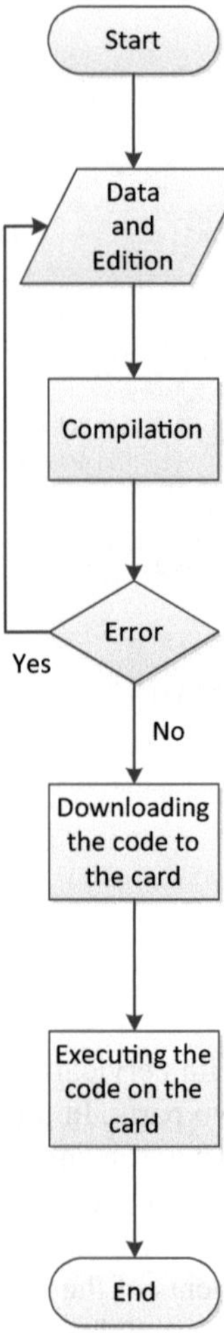

Fig. 10 Diagram of execution for the program

4.7 *Analysis of Results*

Once each of the specimens was sensed, the results of each specimen were obtained. The displacement results of specimens C3, C4 and C5 are observed in Tables 5, 6 and 7.

The displacement results obtained from the lumbar region specimens are shown in Tables 8, 9 and 10.

Table 5 Results of the specimen C3

Frontal displacement (mm)	Lateral displacement (mm)	Superior displacement (mm)
2.7	2.5	3.3
2.6	2.4	3.1
2.3	2.5	3.3
2.3	2.4	3.5
2.1	2.3	2.1
2.4	2.6	2.5
2.5	2.8	3.0
2.6	3.2	3.2
3.0	3.5	3.3
3.5	3.3	2.8

Table 6 Results of specimen C4

Frontal displacement (mm)	Lateral displacement (mm)	Superior displacement (mm)
2.5	2.1	2.2
1.9	2.0	2.0
2.0	2.1	2.1
2.1	2.2	2.2
2.2	2.3	2.4
2.3	2.4	2.6
2.5	2.4	2.7
2.6	2.6	2.9
2.7	2.8	3.1
3.0	3.3	3.4

Table 7 Results of specimen C5

Frontal displacement (mm)	Lateral displacement (mm)	Superior displacement (mm)
2.1	2.1	2.4
1.9	2.0	2.0
1.9	2.2	1.9
2.1	1.9	1.9
2.2	1.9	2.1
2.3	2.5	2.2
2.6	2.6	2.3
2.7	3.0	2.4
2.9	3.4	2.5
3.0	2.7	2.6

Table 8 Results of the L3 specimen

Frontal displacement (mm)	Lateral displacement (mm)	Superior displacement (mm)
2.2	3.0	2.4
2.2	2.2	2.0
2.6	2.0	2.1
2.9	2.1	1.9
3.2	2.2	1.9
2.6	2.3	2.0
3.0	2.5	2.1
2.7	2.8	2.1
2.8	3.1	2.3
2.4	3.2	2.6

Table 9 Results of the L4 specimen

Frontal displacement (mm)	Lateral displacement (mm)	Superior displacement (mm)
3.3	2.4	1.9
2.3	2.2	2.0
2.1	2.1	2.1
2.3	2.2	2.2
2.5	2.5	2.3
2.3	2.6	2.4
2.7	2.7	2.5
3.3	2.9	2.6
3.1	2.8	2.7
2.8	3.0	2.9

Table 10 Results of the L5 specimen

Frontal Displacement (mm)	Lateral Displacement (mm)	Superior Displacement (mm)
2.3	2.4	3.3
2.2	2.0	2.3
2.9	2.1	2.1
2.8	2.2	2.3
2.7	2.3	2.5
2.4	2.4	2.6
2.5	2.5	2.7
2.2	2.6	2.6
2.6	2.7	3.1
3.1	2.8	2.9

Fig. 11 Serial monitor to visualize results of sensing system

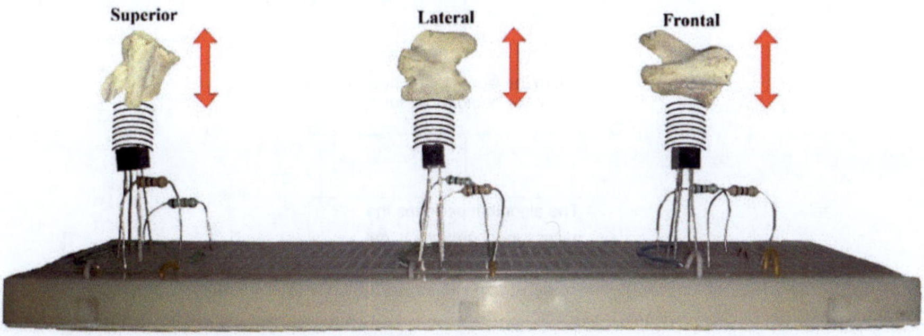

Fig. 12 Sensing of the cervical specimens

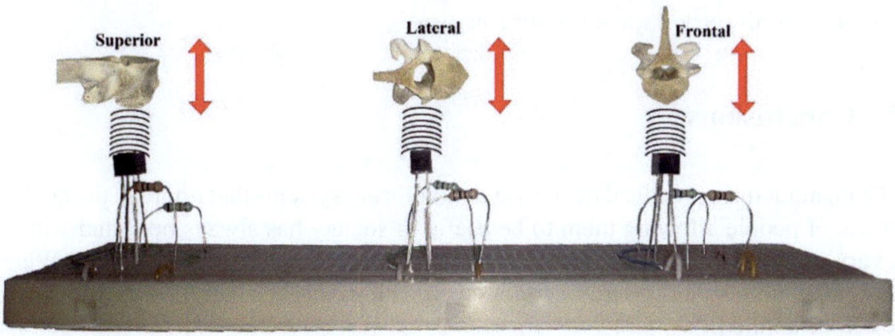

Fig. 13 Sensing of lumbar specimens

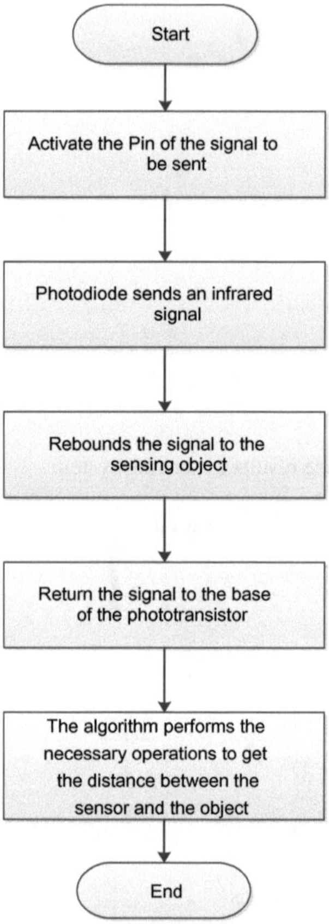

Fig. 14 Reflective optical sensor operating diagram

5 Conclusions

The manufacturing medical devices and monitoring systems that improve the quality of life of people allowing them to be useful in society has always been and will be a very important part of achieving an complete development of the country. When designing monitoring systems, certain specifications must be met according to where it will be applied.

The use of sensors in prostheses allows to increase the complexity and precision of the final models. However, the requirements for controlling the stability of these implants should be clear and the end result may be associated with human health. The criteria of White and Panjabi in orthopedics are necessary to design these systems.

In this paper, a monitoring system was presented to evaluate the displacement of the cervical and lumbar vertebrae according to the stability criteria of White and Panjabi. The integrated equipment allows acquiring and analyzing data of the

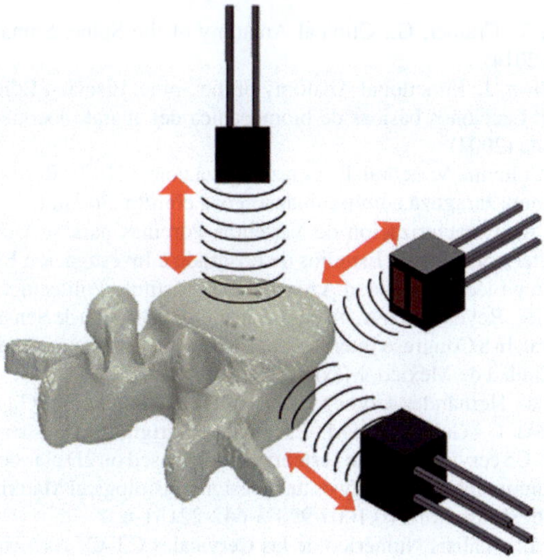

Fig. 15 Positioning of the sensors on the vertebrae

displacement of each of the specimens along each of their faces (superior, frontal, and lateral) and is sufficiently precise to measure the movements of each of the indicated faces.

These data may be used by the physician to observe the correct evolution of the patient after a fusion or corporectomy surgery. In addition, the values reported in this paper indicate the displacement of specimens according to the stability criterion of White and Panjabi, that is, if displacement is greater than 3.5 mm there is instability. This would indicate an unwanted displacement of the vertebra by some accident or sudden movement.

Acknowledgements The authors kindly acknowledge the support given to the National Polytechnic Institute and the National Council for Science and Technology (CONACYT).

References

1. Los problemas de salud. (2017). Retrieved 3 June 2017, from http://www.lasaludes.com/tiposdecirugíadecolumna/
2. Peter, F., Ullrich Jr., F.: Anatomía de la columna vertebral y dolor de espalda. spine-health. (2017). Retrieved 2 May 2017, from https://www.spinehealth.com/espanol/anatomia-de-la-columna-vertebral/anatomia-de-la-columnavertebral-y-dolor-de-espalda
3. Bogduk, N.: Clinical Anatomy of the Lumbar Spine and Sacrum. Church Livingstone, Edinburgh (2005)
4. Peter F., Ullrich Jr., P.: ¿Qué es una hernia de disco, un pinzamiento de un nervio, un disco protuberante?. Spine-healt. (2017). Retrieved 7 May 2017, from https://www.spine-health.com/espanol/hernia-de-disco/que-es-una-hernia-de-discoun-pinzamiento-de-un-nervio-un-disco

5. Cramer, G., Darby, S., Cramer, G.: Clinical Anatomy of the Spine, Spinal Cord, and ANS. Elsevier, St. Louis (2014)
6. Middleditch, A., Oliver, J.: Functional Anatomy of the Spine. Elsevier, Edinburgh (2005)
7. Viladot Voegeli, A.: Lecciones básicas de biomecánica del aparato locomotor, pp. 105–120. MASSON, Barcelona (2004)
8. Biomecánica de la Columna Vertebral Traumazaragoza.com. (2017). Retrieved 15 June 2017, from http://www.traumazaragoza.com/traumazaragoza.com/Inicio.html
9. Fuerte Hernández, A.: Caracterización de Vertebras Porcinas para su Uso en Aplicaciones Biomecánicas (Master). Sección de Estudios de Posgrado e Investigación Escuela Superior de Ingeniería Mecánica y Eléctrica Unidad Azcapotzalco, Instituto Politécnico Nacional (2010)
10. Vazquez-Machorro, R., Reyes-Cruz, A., Atonal-Sánchez, J.: Sistema de Sensado para Implantes de Columna Vertebral. In 8 Congreso Internacional de Ingeniería Electromecánica y de Sistemas (CIIES), pp. 2–5. Ciudad de México, México (2016)
11. Beltrán-Fernández, J., Hernández-Gómez, L., Urriolagoitia-Calderón, G., González-Rebatú, A., Urriolagoitia-Sosa, G., Galán Vera, M., Escalante-Rodríguez, E.: Assessment of the structural integrity of C3–C5 cervical porcine vertebrae model based on 2D classic CAD, 3D scanner and 3D computed tomography. In: Analysis and Design of Biological Materials and Structures, pp. 3–17 (2012). http://dx.doi.org/10.1007/978-3-642-22131-6_1
12. Beltrán Fernández, J.: Análisis Numérico de las Cervicales C3–C7 Asociado al Problema de Latigazo Cervical (Ph.D.). Sección de Estudios de Posgrado e Investigación, Escuela Superior de Ingeniería Mecánica y Eléctrica Unidad Zacatenco, Instituto Politécnico Nacional (2007)
13. Beltrán-Fernández, J., Valdez, N., Rebatu, A., Hernández Gómez, L.: Biomechanical design of a vertebral distractor for fractured bodies useful in kyphoplasty. In: Advanced Structured Materials, pp. 89–122 (2015) http://dx.doi.org/10.1007/978-3-319-19470-7_7
14. Torrente Artero, O.: ARDUINO Curso práctico de formación, pp. 61–77. Alfaomega, México, D.F. (2013)
15. 5Hertz Electrónica. (2017). Retrieved 5 Feb 2017, from http://5hertz.com/tutoriales/?p=571
16. 5Hertz Electrónica. (2017). Retrieved 4 Feb 2017, from http://5hertz.com/tutoriales/?p=228

Solution Behavior Near an Envelope of Characteristics in Planar Flow of a Material Obeying the Double Slip and Rotation Model

Sergei Alexandrov and Alexander Pirumov

Abstract Planar flow of an incompressible rigid plastic material obeying a special case of the double slip and rotation model is studied in the vicinity of an envelope of characteristics. An orthogonal coordinate system whose base vectors are normal and tangent to the envelope is adopted. It is shown that solutions are in general singular. In particular, the magnitude of the shear strain rate in coordinate system chosen approaches infinity near the envelope of characteristics. An asymptotic representation for the stresses and velocities in the vicinity of such envelopes is derived. This representation can be used in numerical codes to overcome a difficulty caused by the singular behavior of exact solutions.

Keywords Double slip and rotation model · Envelope of characteristics
Singularity · Planar flow · Granular material

1 Introduction

Models of pressure dependent plasticity are known to be applicable to several different types of material such as soils, granular materials, concrete and traditional metals. Reviews of such models have been provided in [1, 2]. In the present paper, a special case of the double slip and rotation model [3, 4] is considered under plane strain conditions. This model is also a variant of the double shearing model proposed in [5]. The elastic portion of the strain rate tensor is neglected. Therefore, the model under consideration is rigid plastic. It is also known [3, 4] that the system of equations comprising the constitutive equation and the equilibrium equations is hyperbolic.

S. Alexandrov (✉)
Institute for Problems in Mechanics, 101-1 Prospect Vernadskogo, 119526 Moscow, Russia
e-mail: sergei_alexandrov@spartak.ru

A. Pirumov
Moscow Technological University, 78 Prospect Vernadskogo, 119526 Moscow, Russia
e-mail: alpirumov@mail.ru

© Springer International Publishing AG, part of Springer Nature 2019

A. Öchsner and H. Altenbach (eds.), *Engineering Design Applications*,
Advanced Structured Materials 92, https://doi.org/10.1007/978-3-319-79005-3_24

Therefore, solutions of boundary value problems may include an envelope of characteristics. In the case of several rigid plastic models, envelopes of characteristics are a source of singular behavior of solutions [6–8]. This singularity may cause difficulties with numerical solutions. Therefore, it is of importance to investigate the asymptotic behavior of solutions for the double slip and rotation model in the vicinity of envelopes of characteristics. It is worthy of note that available analytic solutions for this model are singular [9, 10]. In the present paper, a general asymptotic representation of solutions in the vicinity of envelopes of characteristics is derived under plane strain conditions.

An applied aspect of the theory of singular solutions is that this theory can be used for predicting the generation of narrow fine grain layers in the vicinity of frictional interfaces [11]. Such layers are often generated in various processes [12]. On the other hand, an envelope of characteristics usually coincides with friction surfaces if the maximum friction law is adopted. It is worthy of note that models of pressure-dependent plasticity are applicable to traditional metals [13–15]. Therefore, the main result of the present paper can be used in conjunction with the theory proposed in [11].

2 Material Model

For a planar deformation of an incompressible material the constitutive equations of the double slip and rotation model in arbitrary orthogonal coordinates (q, s) are

$$\left(\sigma_{qq} + \sigma_{ss}\right) \sin \varphi + \sqrt{\left(\sigma_{qq} - \sigma_{ss}\right)^2 + 4\sigma_{qs}^2} = 2k \cos \varphi \tag{1}$$

$$\xi_{qq} + \xi_{ss} = 0 \tag{2}$$

$$\sin 2\psi \left(\xi_{qq} - \xi_{ss}\right) - 2 \cos 2\psi \xi_{qs} - 2 \sin \varphi \left(\omega_{qs} + \Omega\right) = 0 \tag{3}$$

where (1) is the Coulomb-Mohr yield criterion, (2) the incompressibility equation and (3) an equation connecting the principal directions of stress and of deformation rate. The quantity k is the cohesion, φ the angle of internal friction, ψ denotes the orientation of the principal stress axis corresponding to the major principal stress σ_1 relative to the q-direction, measured anti-clockwise positive (Fig. 1), $\xi_{qq}, \xi_{ss}, \xi_{qs}$ denote the components of the deformation-rate tensor and ω_{qs} is the only non-zero spin (vorticity) component.

The quantity Ω is the intrinsic spin due to grain rotation. In general it is an unknown variable which is governed by the equation of rotational motion. In a special case of the model considered in the present paper it is assumed that

$$\Omega = 0. \tag{4}$$

It is a reasonable assumption for a number of processes [4]. The system of equations comprising (1)–(3) together with the stress equilibrium equations is hyperbolic

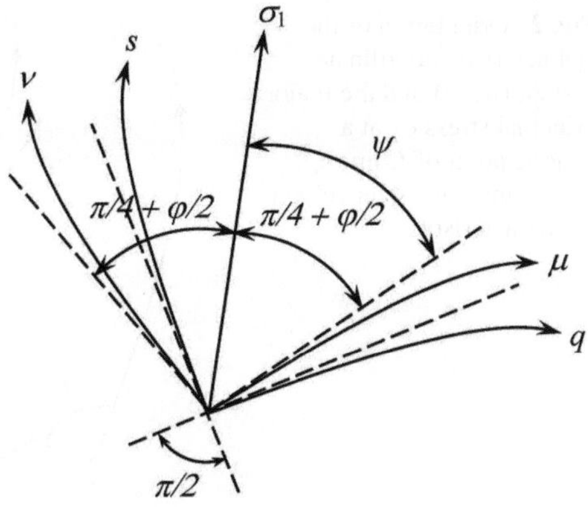

Fig. 1 Local coordinate system $(q,\ s)$, characteristic coordinate system $(\mu,\ \nu)$ and the orientation of the major principal stress σ_1

[4]. The characteristics for the stresses and the velocities coincide, and there are only two distinct characteristic directions at a point. The angle between each of the characteristic directions and the principal stress axis corresponding to the major principal stress σ_1 is equal to $\pi/4 + \varphi/2$. Therefore, the equations for determining the characteristics are (Fig. 1)

$$\frac{H_s ds}{H_q dq} = \tan\left(\psi - \frac{\pi}{4} - \frac{\varphi}{2}\right), \quad \frac{H_s ds}{H_q dq} = \tan\left(\psi + \frac{\pi}{4} + \frac{\varphi}{2}\right) \tag{5}$$

where H_q and H_s are the scale factors for the q- and s-lines, respectively. The first equation in (5) determines a family of $\mu-$ characteristics and the second a family of $\nu-$ characteristics.

3 Solution Behavior Near an Envelope of Characteristics

Let L be a tool surface (line in a generic plane of flow). The tool is regarded as fixed. It is convenient to use a coordinate system whose s-lines are straight and orthogonal to L. With no loss of generality, it is possible to assume that the curve $s = 0$ coincides with L and to put $H_s = 1$. Then, s is the normal distance from the tool surface. Also, the s-lines are directed away from the rigid tool and towards the plastic material. Therefore, $s \geq 0$ in the subsequent analysis. It is assumed that an envelope of characteristics coincides with L. Therefore, it follows from (5) that

$$\psi = \frac{\pi}{4} + \frac{\varphi}{2} \quad \text{or} \quad \psi = -\frac{\pi}{4} - \frac{\varphi}{2} \tag{6}$$

at $s = 0$. The first case is illustrated in Fig. 2. In this case, L is an envelope of $\mu-$ characteristic lines. In the other case, L is an envelope of $\nu-$ characteristic lines.

Since $H_s = 1$, the stress equilibrium equations are [16]

Fig. 2 Orientation of the characteristic coordinate system $(\mu,\ \nu)$ and the major principal stress σ_1 at a generic point of L-line coinciding with an envelope of characteristics

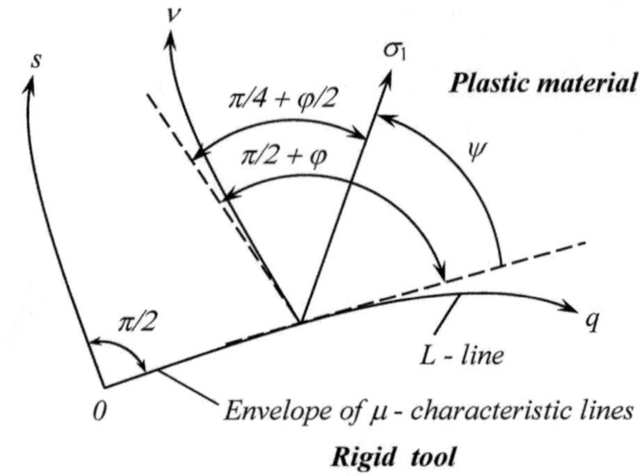

$$\frac{\partial\sigma_{qq}}{\partial q}+H_q\frac{\partial\sigma_{qs}}{\partial s}+2\sigma_{qs}\frac{\partial H_q}{\partial s}=0,\quad H_q\frac{\partial\sigma_{ss}}{\partial s}+\frac{\partial\sigma_{qs}}{\partial q}+\left(\sigma_{ss}-\sigma_{qq}\right)\frac{\partial H_q}{\partial s}=0. \quad (7)$$

Also, the components of the deformation-rate tensor and the spin component are expressed as [16]

$$\xi_{qq}=\frac{1}{H_q}\frac{\partial u_q}{\partial q}+\frac{u_s}{H_q}\frac{\partial H_q}{\partial s},\quad \xi_{ss}=\frac{\partial u_s}{\partial s},\quad 2\xi_{qs}=\frac{\partial u_q}{\partial s}+\frac{1}{H_q}\frac{\partial u_s}{\partial q}-\frac{u_q}{H_q}\frac{\partial H_q}{\partial s},$$
$$2\omega_{qs}=\frac{\partial u_q}{\partial s}-\frac{1}{H_q}\frac{\partial u_s}{\partial q}+\frac{u_q}{H_q}\frac{\partial H_q}{\partial s} \quad (8)$$

where u_q and u_s are the velocity components in the $(q,\ s)$ coordinate system. Since the tool is regarded as fixed, the component u_s should satisfy the boundary condition

$$u_s=0 \quad (9)$$

for $s=0$. Using (4) and (8), Eqs. (2) and (3) can be rewritten in the form

$$\frac{1}{H_q}\frac{\partial u_q}{\partial q}+\frac{u_s}{H_q}\frac{\partial H_q}{\partial s}+\frac{\partial u_s}{\partial s}=0,$$
$$2\sin 2\psi\frac{\partial u_s}{\partial s}+(\cos 2\psi+\sin\varphi)\frac{\partial u_q}{\partial s}+\frac{(\cos 2\psi-\sin\varphi)}{H_q}\left(\frac{\partial u_s}{\partial q}-u_q\frac{\partial H_q}{\partial s}\right)=0. \quad (10)$$

It follows from (6) that $\cos 2\psi+\sin\varphi=0$ at $s=0$ in either case. Therefore, if $\left|\partial u_q/\partial s\right|<\infty$ then the second equation in (10) is a characteristic relation. In this case, the line L is not an envelope of characteristics, which contradicts the original assumption. Therefore, it is necessary to assume that

$$\left|\frac{\partial u_q}{\partial s}\right| \to \infty \tag{11}$$

as $s \to 0$. In this case one term in (10) reduces to the expression $0 \cdot \infty$. The subsequent analysis is based on the following assumptions:

(i) stresses and velocities are bounded everywhere
(ii) derivatives with respect to q are bounded everywhere
(iii) solutions are represented by a power series of s in the vicinity of $s = 0$.

In particular,

$$u_q = U_{q0} + U_{q1}s^{\beta} + o\left(s^{\beta}\right), \quad u_s = U_{s1}s^{\omega} + o\left(s^{\omega}\right), \quad \psi = \psi_0 + \psi_1 s^{\gamma} + o\left(s^{\gamma}\right) \tag{12}$$

as $s \to 0$. Here the boundary condition (9) has been taken into account. Also, U_{q0}, U_{q1}, U_{s1}, and ψ_1 are independent of s, $\beta > 0$, $\omega > 0$, and $\gamma > 0$. It is evident from (6) and (12) that $\psi_0 = \pm\left(\pi/4 + \varphi/2\right)$. In what follows, it is assumed that

$$\psi_0 = \frac{\pi}{4} + \frac{\varphi}{2}. \tag{13}$$

The case $\psi_0 = -\left(\pi/4 + \varphi/2\right)$ can be treated in a similar manner. It follows from (12) and (13) that

$$\begin{aligned}
\sin 2\psi &= \cos\varphi - 2\psi_1 \sin\varphi s^{\gamma} + o\left(s^{\gamma}\right), \\
\cos 2\psi + \sin\varphi &= -2\psi_1 \cos\varphi s^{\gamma} + o\left(s^{\gamma}\right), \\
\cos 2\psi &= -\sin\varphi - 2\psi_1 \cos\varphi s^{\gamma} + o\left(s^{\gamma}\right), \\
\cos 2\psi - \sin\varphi &= -2\sin\varphi - 2\psi_1 \cos\varphi s^{\gamma} + o\left(s^{\gamma}\right)
\end{aligned} \tag{14}$$

as $s \to 0$. Differentiating u_q given in (12) with respect to s leads to

$$\frac{\partial u_q}{\partial s} = \beta U_{q1} s^{(\beta-1)} + o\left[s^{(\beta-1)}\right] \tag{15}$$

as $s \to 0$. Equations (11) and (15) combine to give

$$0 < \beta < 1. \tag{16}$$

It is worthy of note that the range $\beta < 0$ is not permissible since $\left|u_q\right| \to \infty$ as $s \to 0$ in this case as follows from (12). Substituting (12) in the first equation in (10) yields

$$\left[\frac{dU_{q0}}{dq} + o(1)\right] + \left\{H_q U_{s1} \omega s^{(\omega-1)} + o\left[s^{(\omega-1)}\right]\right\}$$

as $s \to 0$. It is seen from this equation that $\omega = 1$ and Eq. (12) for u_s becomes

$$u_s = U_{s1}s + o(s) \tag{17}$$

as $s \to 0$. Substituting Eq. (12) for u_q together with (14) and (17) into the second equation in (10) results in

$$\left[\cos\varphi U_{s1} + \frac{\sin\varphi U_{q0}}{H_q}\frac{\partial H_q}{\partial s} + o(1)\right] - \left\{\psi_1 \cos\varphi U_{q1}\beta s^{(\beta+\gamma-1)} + o\left[s^{(\beta+\gamma-1)}\right]\right\} = 0 \quad (18)$$

as $s \to 0$. It follows from (18) that

$$\beta + \gamma = 1. \quad (19)$$

The stress components can be represented as [5]

$$\sigma_{qq} = -p + t\cos 2\psi, \quad \sigma_{ss} = -p - t\cos 2\psi, \quad \sigma_{qs} = t\sin 2\psi \quad (20)$$

where

$$p = -\frac{\sigma_{qq} + \sigma_{ss}}{2}, \quad t = \frac{1}{2}\sqrt{\left(\sigma_{qq} - \sigma_{ss}\right)^2 + 4\sigma_{qs}^2}. \quad (21)$$

Using (21), the yield criterion (1) can be rewritten as

$$t - p\sin\varphi = k\cos\varphi. \quad (22)$$

Substituting (20) and (21) into (7) yields

$$\begin{aligned} &-\frac{\partial p}{\partial q} + \cos 2\psi\frac{\partial t}{\partial q} - 2t\sin 2\psi\frac{\partial \psi}{\partial q} \\ &+H_q\left(\sin 2\psi\frac{\partial t}{\partial s} + 2t\cos 2\psi\frac{\partial \psi}{\partial s}\right) + 2t\sin 2\psi\frac{\partial H_q}{\partial s} = 0, \\ &H_q\left(\frac{\partial p}{\partial s} + \cos 2\psi\frac{\partial t}{\partial s} - 2t\sin 2\psi\frac{\partial \psi}{\partial s}\right) \\ &-\sin 2\psi\frac{\partial t}{\partial q} - 2t\cos 2\psi\frac{\partial \psi}{\partial q} + 2t\cos 2\psi\frac{\partial H_q}{\partial s} = 0. \end{aligned} \quad (23)$$

By the assumption made, the solution is represented by power series. Therefore,

$$p = p_0 + p_1 s^\alpha + o\left(s^\alpha\right), \quad t = t_0 + t_1 s^\lambda + o\left(s^\lambda\right) \quad (24)$$

as $s \to 0$. Here p_0, p_1, t_0, and t_1 are independent of s. Substituting (24) into (22) shows that

$$t_0 - p_0\sin\varphi = k\cos\varphi, t_1 - p_1\sin\varphi = 0 \quad (25)$$

and

$$\lambda = \alpha. \quad (26)$$

By the assumption made, all derivatives with respect to q are bounded. Therefore, it follows from the first equation in (23) that

$$\left|\sin 2\psi \frac{\partial t}{\partial s} + 2t \cos 2\psi \frac{\partial \psi}{\partial s}\right| < \infty \tag{27}$$

at $s = 0$. Using (12), (14) and (24) it is possible to rewrite (27) as

$$\left|\left(\cos\varphi - 2\psi_1 \sin\varphi s^{\gamma}\right) t_1 \lambda s^{\lambda-1} - 2\left(t_0 + t_1 s^{\lambda}\right)\left(\sin\varphi + 2\psi_1 \cos\varphi s^{\gamma}\right) \psi_1 \gamma s^{\gamma-1}\right| < \infty \tag{28}$$

at $s = 0$. It is seen from (16) and (19) that $\gamma < 1$. Therefore, the term $|2t_0 \sin\varphi\psi_1\gamma|\, s^{\gamma-1}$ approaches infinity as $s \to 0$. In order to cancel this term, it is necessary to assume that

$$\lambda = \gamma \tag{29}$$

and

$$\cos\varphi t_1 - 2t_0 \sin\varphi\psi_1 = 0. \tag{30}$$

Then, the leading term of Eq. (28) becomes

$$\left[2\psi_1 \sin\varphi t_1 - 2\left(2t_0\psi_1 \cos\varphi + t_1 \sin\varphi\right)\psi_1\right] s^{2\gamma-1}. \tag{31}$$

If this term vanishes then the first equation in (23) is the characteristic relation. Then, the friction surface coincides with a characteristic line rather than with an envelope of characteristic lines. Therefore, it is assumed that the coefficient of $s^{2\gamma-1}$ in (31) does not vanish. The term shown in (31) is involved in the first equation in (23). In order to cancel this term, it is necessary to assume that

$$\gamma = \frac{1}{2}. \tag{32}$$

Then, it follows from (19) that

$$\beta = \frac{1}{2}. \tag{33}$$

Since all derivatives with respect to q are bounded, it follows from the second equation in (23) that

$$\left|\frac{\partial p}{\partial s} + \cos 2\psi \frac{\partial t}{\partial s} - 2t \sin 2\psi \frac{\partial \psi}{\partial s}\right| < \infty \tag{34}$$

at $s = 0$. Using (12), (14), (24), (26), (29), and (32) it is possible to rewrite (34) as

$$\left|p_1 - t_1 \sin\varphi - 2t_0 \cos\varphi\psi_1\right| s^{-1/2} < \infty \tag{35}$$

at $s = 0$. Therefore,

$$p_1 - t_1 \sin\varphi - 2t_0 \cos\varphi\psi_1 = 0. \tag{36}$$

Substituting (14) and (24) into (20) and taking into account Eqs. (26), (29) and (32) leads to

$$\begin{aligned}\sigma_{qs} &= t_0 \cos\varphi + (t_1 \cos\varphi - 2t_0\psi_1 \sin\varphi)\sqrt{s} - 2t_1\psi_1 \sin\varphi s + o(s), \\ \sigma_{qq} &= -p_0 - t_0 \sin\varphi - (p_1 + t_1 \sin\varphi + 2t_0\psi_1 \cos\varphi)\sqrt{s} - 2t_1\psi_1 \cos\varphi s + o(s), \\ \sigma_{ss} &= -p_0 + t_0 \sin\varphi - (p_1 - t_1 \sin\varphi - 2t_0\psi_1 \cos\varphi)\sqrt{s} + 2t_1\psi_1 \cos\varphi s + o(s),\end{aligned} \tag{37}$$

as $s \to 0$. Equations (30), (36) and (37) combine to give

$$\begin{aligned}\sigma_{qs} &= t_0 \cos\varphi - 2t_1\psi_1 \sin\varphi s + o(s), \\ \sigma_{qq} &= -p_0 - t_0 \sin\varphi - 2(t_1 \sin\varphi + 2t_0\psi_1 \cos\varphi)\sqrt{s} - 2t_1\psi_1 \cos\varphi s + o(s), \\ \sigma_{ss} &= -p_0 + t_0 \sin\varphi + 2t_1\psi_1 \cos\varphi s + o(s),\end{aligned} \tag{38}$$

as $s \to 0$. Substituting (12), (17) and (33) into (8) yields

$$|\xi_{qs}| = O\left(\frac{1}{\sqrt{s}}\right) \quad \text{and} \quad |\omega_{qs}| = O\left(\frac{1}{\sqrt{s}}\right) \tag{39}$$

as $s \to 0$.

4 Conclusion

An asymptotic analysis of the system of equations comprising Eqs. (1), (2), (3), and (7) in the vicinity of the envelope of characteristics has been performed. It has been shown that the solution is singular. In particular, the distribution of the shear strain rate and spin component follows Eq. (39). The stress components can be found from (38). The aforementioned singularity may cause difficulties with finding a numerical solution using regular finite elements. Probably, the extended finite element method [17] is capable of solving this type of boundary value problems. To this end, the asymptotic representation of solutions derived in the present paper is required.

If the system of equations comprising the constitutive equations and the equilibrium equations is hyperbolic, then the maximum friction law is defined by the condition that an envelope of characteristics coincides with the friction surface [18]. Therefore, the analysis performed can be used to determine the asymptotic behavior of solutions in the vicinity of maximum friction surfaces. It is worthy of note that the asymptotic representation of the shear strain rate shown in (39) is valid for several other rigid plastic models [6–8, 19, 20].

The semi-analytical solutions provided in [9, 10] confirm the asymptotic representation of solutions in the vicinity of maximum friction surfaces derived in the present paper. The semi-analytical solutions provided in [21–23] are for the double shearing model proposed in [5]. In general, the double shearing model differs from the double slip and rotation model. However, it is possible to verify by inspection that the constitutive equations of these two models coincide in the solutions given in [21–23]. Therefore, these solutions are also valid for the double slip and rotation model. An asymptotic analysis of these solutions shows that they satisfy Eq. (39).

Acknowledgements The research described in this paper has been supported by the grant RFBR-17-01-00624. The second author was partly supported by Moscow Technological University [24].

References

1. Cox, G.M., Thamwattana, N., McCue, S.W., Hill, J.M.: Coulomb-Mohr granular materials: quasi-static flows and the highly frictional limit. Appl. Mech. Rev. **61**, 060802 (2008). https://doi.org/10.1115/1.2987874
2. Goddard, J.D.: Continuum modeling of granular media. Appl Mech Rev. **66**(5), 050801 (2014). https://doi.org/10.1115/1.4026242
3. Harris, D., Grekova, E.F.: A hyperbolic well-posed model for the flow of granular materials. J. Eng. Math. **52**, 107–135 (2005)
4. Harris, D.: A hyperbolic augmented elasto-plastic model for pressure-dependent yield. Acta Mech. **225**, 2277–2299 (2014)
5. Spencer, A.J.M.: A theory of the kinematics of ideal soils under plane strain conditions. J. Mech. Phys. Solids **12**, 337–351 (1964)
6. Alexandrov, S., Richmond, O.: Singular plastic flow fields near surfaces of maximum friction stress. Int. J. Non-Linear Mech. **36**, 1–11 (2001)
7. Alexandrov, S., Lyamina, E.: Singular solutions for plane plastic flow of pressure-dependent materials. Dokl. Phys. **47**, 308–311 (2002)
8. Alexandrov, S., Jeng, Y.R.: Singular rigid/plastic solutions in anisotropic plasticity under plane strain conditions. Continuum Mech. Thermodyn. **25**, 685–689 (2013)
9. Alexandrov, S., Harris, D.: Comparison of solution behaviour for three models of pressure-dependent plasticity: a simple analytical example. Int. J. Mech. Sci. **48**, 750–762 (2006)
10. Alexandrov, S., Harris, D.: An exact solution for a model of pressure-dependent plasticity in an un-steady plane strain process. Eur. J. Mech. A. Solids **29**, 966–975 (2010)
11. Alexandrov, S.E., Goldstein, R.V.: On constructing constitutive equations in metal thin layer near friction surfaces in material forming processes. Dokl. Phys. **60**, 39–41 (2015)
12. Griffiths, B.J.: Mechanisms of white layer generation with reference to machining and deformation processes. Trans. ASME J. Trib. **109**, 525–530 (1987)
13. Spitzig, W.A., Sober, R.J., Richmond, O.: The effect of hydrostatic pressure on the deformation behavior of maraging and HY-80 steels and its implications for plasticity theory. Metall. Trans. A **7**, 1703–1710 (1976)
14. Kao, A.S., Kuhn, H.A., Spitzig, W.A., Richmond, O.: Influence of superimposed hydrostatic pressure on bending fracture and formability of a low carbon steel containing globular sulfides. Trans. ASME J. Eng. Mater. Technol. **112**, 26–30 (1990)
15. Wilson, C.D.: A critical reexamination of classical metal plasticity. Trans. ASME J. Appl. Mech. **69**, 63–68 (2002)
16. Malvern, L.E.: Introduction to the Mechanics of a Continuous Medium. Prentice-Hall, Englewood Cliffs (1969)
17. Fries, T.P., Belytschko, T.: The extended/generalized finite element method: an overview of the method and its applications. Int. J. Numer. Meth. Eng. **84**, 253–304 (2010)
18. Alexandrov, S.: The strain rate intensity factor and its applications: a review. Mater. Sci. Forum **623**, 1–20 (2009)

19. Alexandrov, S., Mustafa, Y.: Singular solutions in viscoplasticity under plane strain conditions. Meccanica **48**, 2203–2208 (2013)
20. Alexandrov, S., Mustafa, Y.: Quasi-static axially symmetric viscoplastic flows near very rough walls. Appl. Math. Model. **39**, 4599–4606 (2015)
21. Pemberton, C.S.: Flow of imponderable granular materials in wedge-shaped channels. J. Mech. Phys. Solids **13**, 351–360 (1965)
22. Marshall, E.A.: The compression of a slab of ideal soil between rough plates. Acta Mech. **3**, 82–92 (1967)
23. Spencer, A.J.M.: Compression and shear of a layer of granular material. J. Eng. Math. **52**, 251–264 (2005)
24. Pankov, V.L.: Effectiveness of incentive mechanism and the potential level meeting the needs of an employee. Herald. MSTU MIREA **1**(4), 288 (2015)

Principal Stress Trajectories in Plane Strain and Plane Stress Plasticity

Sergei Alexandrov and Alexander Pirumov

Abstract The objective of the present work is to develop a method of finding principal stress trajectories in pressure-independent and pressure-dependent plasticity under plane strain and plane stress conditions. It is shown that the geometry of the principal line trajectories is defined, in the general case, by a telegraph equation. In obtaining this general result, two special cases are lost: those corresponding to the cases where one and two families of characteristics are straight. In the case of Tresca and Coulomb-Mohr yield criteria, the results presented are independent of any flow rule that may be chosen to calculate the deformation and also independent of whether elastic strains are included.

Keywords Principal stress · Plane strain · Plane stress · Characteristics · Plasticity

1 Introduction

Several systems of equations are considered in the present paper. One of these systems comprises the two equilibrium equations and an arbitrary pressure-independent plane strain yield criterion in the case of rigid perfectly plastic material. This system of equations coincides with the system that comprises the two equilibrium equations and Tresca's yield criterion in the case of elastic perfectly plastic material under plane strain conditions and under plane stress conditions if a face regime occurs. Another system of equations comprises the two equilibrium equations and the Coulomb-Mohr yield criterion under plane strain conditions and under plane stress conditions if a face regime occurs. The corresponding results are independent of any flow rule that may be chosen to calculate the deformation and also independent of whether elastic

S. Alexandrov (✉)
Institute for Problems in Mechanics, 101-1 Prospect Vernadskogo, 119526 Moscow, Russia
e-mail: sergei_alexandrov@spartak.ru

A. Pirumov
Moscow Technological University, 78 Prospect Vernadskogo, 119526 Moscow, Russia
e-mail: alpirumov@mail.ru

© Springer International Publishing AG, part of Springer Nature 2019
A. Öchsner and H. Altenbach (eds.), *Engineering Design Applications*,
Advanced Structured Materials 92, https://doi.org/10.1007/978-3-319-79005-3_25

strains are included. All these systems of equations are hyperbolic. Therefore, it is possible to introduce characteristic coordinate systems. It is shown that in the general case mapping between the principal line coordinate system (i.e. the coordinate curves of this coordinate system coincide with trajectories of the principal stress directions) and the characteristic coordinate system is given as a solution of the equation of telegraphy. In obtaining this general result, two special cases are lost: those corresponding to the cases where one and two families of characteristics are straight. These special cases are not considered. Mapping between a Cartesian coordinate system and the characteristic coordinate system is determined by evaluating ordinary integrals.

2 Properties of Principal Stress Trajectories

In the case of rigid perfectly plastic solids, the system of stress equations under plane strain and plane stress conditions comprises the two equilibrium equations and a yield criterion. This system can be studied independently of the flow rule. Under plane strain conditions the yield criterion of any incompressible isotropic material which complies with the principle of maximum plastic work can be written as

$$\sqrt{\left(\sigma_{qq}-\sigma_{ss}\right)^2+4\sigma_{qs}^2}=2k \tag{1}$$

where σ_{qq}, σ_{ss} and σ_{qs} are the stress components in an arbitrary orthogonal coordinate system (q, s) and k is the shear yield stress, a material constant. Let us introduce a curvilinear orthogonal coordinate system (ξ, η) whose coordinate curves coincide with trajectories of the principal stress directions. The scale factors for the coordinate curves of this coordinate system are denoted by h_ξ and h_η and the principal stresses by σ_ξ and σ_η. It has been shown in [1] that it is possible to choose such a principal lines based coordinate system that

$$h_\xi h_\eta = 1. \tag{2}$$

Any orthogonal net satisfying (2) determines a net of principal stress trajectories giving a solution to the equilibrium equations and (1).

In the case of plane stress conditions, Eq. (1) and consequently (2) are valid for certain face regimes if Tresca's yield criterion is adopted [2].

Several models of pressure-dependent plasticity, such as [3, 4], are based on the Coulomb-Mohr yield criterion. Under plane strain conditions, this yield criterion reads

$$\left(\sigma_{qq}+\sigma_{ss}\right)\sin\varphi+\sqrt{\left(\sigma_{qq}-\sigma_{ss}\right)^2+4\sigma_{qs}^2}=2c\,\cos\varphi \tag{3}$$

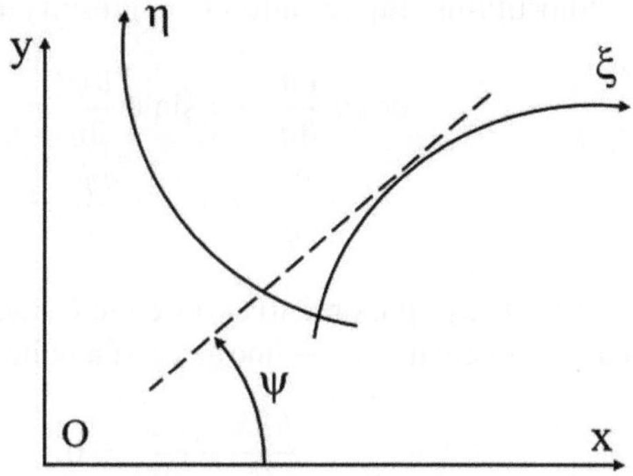

Fig. 1 Principal line based and Cartesian coordinate systems

where the quantity c is the cohesion and φ is the angle of internal friction. It has been shown in [5] that it is possible to choose such a principal lines based coordinate system that

$$h_\eta^b h_\xi = 1 \tag{4}$$

where $b = (t + \sin\varphi) / (t - \sin\varphi)$, $t = 1$ if $\sigma_\xi > \sigma_\eta$ and $t = -1$ if $\sigma_\xi < \sigma_\eta$. Any orthogonal net satisfying (4) determines a net of principal stress trajectories giving a solution to the equilibrium equations and (3). It is evident that Eq. (4) reduces to Eq. (2) at $\varphi = 0$. Equation (4) is also valid under plane stress conditions [2].

3 Method of Constructing Principal Stress Trajectories in Pressure-Independent Plasticity

Introduce a Cartesian coordinate system (x, y). Let ψ be the orientation of the $\xi-$lines relative to the x-axis, measured anticlockwise positive from the x-axis (Fig. 1). It follows from the geometry of this figure that

$$\frac{\partial x}{\partial \xi} = h_\xi \cos\psi, \quad \frac{\partial y}{\partial \xi} = h_\xi \sin\psi, \quad \frac{\partial x}{\partial \eta} = -h_\eta \sin\psi, \quad \frac{\partial y}{\partial \eta} = h_\eta \cos\psi. \tag{5}$$

Consider the yield criterion (1). Substituting Eq. (2) into Eq. (5) yields

$$\frac{\partial x}{\partial \xi} = h\cos\psi, \quad \frac{\partial y}{\partial \xi} = h\sin\psi, \quad \frac{\partial x}{\partial \eta} = -\frac{\sin\psi}{h}, \quad \frac{\partial y}{\partial \eta} = \frac{\cos\psi}{h}. \tag{6}$$

Here $h \equiv h_\xi$. The compatibility equations are

$$\frac{\partial^2 x}{\partial \xi \partial \eta} = \frac{\partial^2 x}{\partial \eta \partial \xi}, \quad \frac{\partial^2 y}{\partial \xi \partial \eta} = \frac{\partial^2 y}{\partial \eta \partial \xi}. \tag{7}$$

Substituting Eq. (6) into Eq. (7) results in

$$\cos\psi\frac{\partial h}{\partial\eta}-h\sin\psi\frac{\partial\psi}{\partial\eta}=\frac{\sin\psi}{h^2}\frac{\partial h}{\partial\xi}-\frac{\cos\psi}{h}\frac{\partial\psi}{\partial\xi},$$
$$\sin\psi\frac{\partial h}{\partial\eta}+h\cos\psi\frac{\partial\psi}{\partial\eta}=-\frac{\cos\psi}{h^2}\frac{\partial h}{\partial\xi}-\frac{\sin\psi}{h}\frac{\partial\psi}{\partial\xi}. \tag{8}$$

It is always possible to choose the Cartesian coordinate system such that its $x-$ axis is tangent to the $\xi-$ line at a given point. In this case $\psi=0$ and Eq. (8) becomes

$$\frac{h\partial h}{\partial\eta}+\frac{\partial\psi}{\partial\xi}=0,\quad h^3\frac{\partial\psi}{\partial\eta}+\frac{\partial h}{\partial\xi}=0. \tag{9}$$

Using a standard technique it is possible to show that this system of equations is hyperbolic. The equations for characteristic curves are

$$\frac{d\xi}{d\eta}=\mp\frac{1}{h^2}. \tag{10}$$

Here and in what follows the upper sign corresponds to $\alpha-$ characteristic curves and the lower sign to $\beta-$ characteristic curves. The characteristic relations along the $\alpha-$ and $\beta-$ curves are

$$d\psi-d\ln h=0\quad\text{on}\quad\alpha-\text{lines},$$
$$d\psi+d\ln h=0\quad\text{on}\quad\beta-\text{lines}. \tag{11}$$

These equations can be immediately integrated to give

$$\psi-\ln h=f_1(\beta)\quad\text{and}\quad\psi+\ln h=f_2(\alpha). \tag{12}$$

Here $f_1(\beta)$ is an arbitrary function of β and $f_2(\alpha)$ is an arbitrary function of α. When both families of characteristics are curved, different choices of the functions $f_1(\beta)$ and $f_2(\alpha)$ involved in (12) merely change the scale of the coordinate curves of the characteristics based coordinate system. Therefore, it is possible to choose with no loss of generality that $f_1(\beta)=2\beta$ and $f_2(\alpha)=2\alpha$. Then, Eq. (12) becomes

$$\psi-\ln h=2\beta\quad\text{and}\quad\psi+\ln h=2\alpha.$$

Solving these equations for ψ and h yields

$$\psi=\alpha+\beta\quad\text{and}\quad h=\exp(\alpha-\beta). \tag{13}$$

Equation (10) can be rewritten as

$$h^2\frac{\partial\xi}{\partial\alpha}+\frac{\partial\eta}{\partial\alpha}=0\quad\text{and}\quad h^2\frac{\partial\xi}{\partial\beta}-\frac{\partial\eta}{\partial\beta}=0.$$

Eliminating h in these equations by means of (13) gives

$$\frac{\partial \xi}{\partial \alpha} + \exp\left[2(\beta - \alpha)\right]\frac{\partial \eta}{\partial \alpha} = 0 \quad \text{and} \quad \frac{\partial \xi}{\partial \beta} - \exp\left[2(\beta - \alpha)\right]\frac{\partial \eta}{\partial \beta} = 0. \tag{14}$$

Introduce the new quantities, ε and ν, as

$$\xi = \varepsilon \exp(\beta - \alpha) \quad \text{and} \quad \eta = \nu \exp(\alpha - \beta). \tag{15}$$

Substituting (15) into (14) results in

$$\frac{\partial \varepsilon}{\partial \alpha} + \frac{\partial \nu}{\partial \alpha} - \varepsilon + \nu = 0 \quad \text{and} \quad \frac{\partial \varepsilon}{\partial \beta} - \frac{\partial \nu}{\partial \beta} + \varepsilon + \nu = 0. \tag{16}$$

It is evident from these equations that

$$\frac{\partial \omega}{\partial \alpha} - \gamma = 0 \quad \text{and} \quad \frac{\partial \gamma}{\partial \beta} + \omega = 0 \tag{17}$$

where

$$\omega = \varepsilon + \nu \quad \text{and} \quad \gamma = \varepsilon - \nu. \tag{18}$$

It is seen from (17) that the quantities ω and γ separately satisfy the equation of telegraphy:

$$\frac{\partial^2 \omega}{\partial \alpha \partial \beta} + \omega = 0 \quad \text{and} \quad \frac{\partial^2 \gamma}{\partial \alpha \partial \beta} + \gamma = 0. \tag{19}$$

In the mathematical theory of plasticity, a number of methods have been developed to solve the equation of telegraphy in conjunction with corresponding boundary conditions [6–8]. Once Eq. (19) has been solved, the dependence of ξ and η on α and β is found by means of (15) and (18).

It is now necessary to determine the dependence of x and y on α and β. Using the chain rule leads to

$$\frac{\partial x}{\partial \alpha} = \frac{\partial x}{\partial \xi}\frac{\partial \xi}{\partial \alpha} + \frac{\partial x}{\partial \eta}\frac{\partial \eta}{\partial \alpha} \quad \text{and} \quad \frac{\partial y}{\partial \alpha} = \frac{\partial y}{\partial \xi}\frac{\partial \xi}{\partial \alpha} + \frac{\partial y}{\partial \eta}\frac{\partial \eta}{\partial \alpha}. \tag{20}$$

Equations (6), (13), (14) and (20) combine to give

$$\frac{\partial x}{\partial \alpha} = \exp(\alpha - \beta)(\cos\psi + \sin\psi)\frac{\partial \xi}{\partial \alpha} \quad \text{and} \quad \frac{\partial y}{\partial \alpha} = \exp(\alpha - \beta)(\sin\psi - \cos\psi)\frac{\partial \xi}{\partial \alpha}. \tag{21}$$

It follows from (15), (17) and (18) that

$$\frac{\partial \xi}{\partial \alpha} = \frac{1}{2}\left(\frac{\partial \gamma}{\partial \alpha} - \omega\right)\exp(\beta - \alpha). \tag{22}$$

Substituting (22) into (21) gives

$$\frac{\partial x}{\partial \alpha} = \frac{1}{2}(\cos\psi + \sin\psi)\left(\frac{\partial \gamma}{\partial \alpha} - \omega\right) \quad \text{and} \quad \frac{\partial y}{\partial \alpha} = \frac{1}{2}(\sin\psi - \cos\psi)\left(\frac{\partial \gamma}{\partial \alpha} - \omega\right). \tag{23}$$

Analogously,

$$\frac{\partial x}{\partial \beta} = \frac{\partial x}{\partial \xi}\frac{\partial \xi}{\partial \beta} + \frac{\partial x}{\partial \eta}\frac{\partial \eta}{\partial \beta} \quad \text{and} \quad \frac{\partial y}{\partial \beta} = \frac{\partial y}{\partial \xi}\frac{\partial \xi}{\partial \beta} + \frac{\partial y}{\partial \eta}\frac{\partial \eta}{\partial \beta}. \tag{24}$$

Using Eqs. (6), (13), (14), (15), (17), and (18) transforms (24) to

$$\frac{\partial x}{\partial \beta} = \frac{1}{2}(\cos\psi - \sin\psi)\left(\frac{\partial \omega}{\partial \beta} + \gamma\right) \quad \text{and} \quad \frac{\partial y}{\partial \beta} = \frac{1}{2}(\sin\psi + \cos\psi)\left(\frac{\partial \omega}{\partial \beta} + \gamma\right). \tag{25}$$

In order to find the mapping into Cartesian space, Eqs. (23) and (25) should be integrated along any convenient path in the $\alpha\beta$ space after the solution to (19) has been found.

4 Method of Constructing Principal Stress Trajectories in Pressure-Dependent Plasticity

Consider the yield criterion (3). Equation (5) is valid. Substituting Eq. (4) into this equation yields

$$\frac{\partial x}{\partial \xi} = h\cos\psi, \quad \frac{\partial y}{\partial \xi} = h\sin\psi, \quad \frac{\partial x}{\partial \eta} = -h^{-1/b}\sin\psi, \quad \frac{\partial y}{\partial \eta} = h^{-1/b}\cos\psi. \tag{26}$$

As before, $h \equiv h_\xi$. Equations (7) and (26) combine to give

$$\begin{aligned} \cos\psi\frac{\partial h}{\partial \eta} - h\sin\psi\frac{\partial \psi}{\partial \eta} &= \frac{\sin\psi}{bh^{(1+1/b)}}\frac{\partial h}{\partial \xi} - \frac{\cos\psi}{h^{1/b}}\frac{\partial \psi}{\partial \xi}, \\ \sin\psi\frac{\partial h}{\partial \eta} + h\cos\psi\frac{\partial \psi}{\partial \eta} &= -\frac{\cos\psi}{bh^{(1+1/b)}}\frac{\partial h}{\partial \xi} - \frac{\sin\psi}{h^{1/b}}\frac{\partial \psi}{\partial \xi}. \end{aligned} \tag{27}$$

It is always possible to choose the Cartesian coordinate system such that its $x-$ axis is tangent to the $\xi-$ line at a given point. In this case $\psi = 0$ and Eq. (27) becomes

$$h^{1/b}\frac{\partial h}{\partial \eta} + \frac{\partial \psi}{\partial \xi} = 0, \quad bh^{(2+1/b)}\frac{\partial \psi}{\partial \eta} + \frac{\partial h}{\partial \xi} = 0. \tag{28}$$

Using a standard technique it is possible to show that this system of equations is hyperbolic. The equations for characteristic curves are

$$\frac{d\xi}{d\eta} = \mp \frac{h^{-(1+1/b)}}{\sqrt{b}}. \tag{29}$$

As before, the upper sign corresponds to $\alpha-$ characteristic curves and the lower sign to $\beta-$ characteristic curves. The characteristic relations along the $\alpha-$ and $\beta-$ curves are given by

$$\begin{aligned} \sqrt{b} d\psi - d\ \ln h = 0 \quad \text{on} \quad \alpha - \text{lines}, \\ \sqrt{b} d\psi + d\ \ln h = 0 \quad \text{on} \quad \beta - \text{lines}. \end{aligned} \tag{30}$$

These equations can be immediately integrated to give

$$\sqrt{b}\psi - \ln h = g_1(\beta) \quad \text{and} \quad \sqrt{b}\psi + \ln h = g_2(\alpha). \tag{31}$$

Here $g_1(\beta)$ is an arbitrary function of β and $g_2(\alpha)$ is an arbitrary function of α. When both families of characteristics are curved, different choices of the functions $g_1(\beta)$ and $g_1(\beta)$ involved in (31) merely change the scale of the coordinate curves of the characteristics based coordinate system. Therefore, it is possible to choose with no loss of generality that $g_1(\beta) = 2\sqrt{b}\beta \cos\varphi$ and $g_2(\alpha) = 2\sqrt{b}\alpha \cos\varphi$. Then, Eq. (31) becomes

$$\sqrt{b}\psi - \ln h = 2\sqrt{b}\beta \cos\varphi \quad \text{and} \quad \sqrt{b}\psi + \ln h = 2\sqrt{b}\alpha \cos\varphi.$$

Solving these equations for ψ and h yields

$$\psi = (\alpha + \beta)\cos\varphi \quad \text{and} \quad h = \exp\left[w\left(\alpha - \beta\right)\right]. \tag{32}$$

where $w = 1 + \sin\varphi$ if $\sigma_\xi > \sigma_\eta$ and $w = 1 - \sin\varphi$ if $\sigma_\xi < \sigma_\eta$. Equation (29) can be rewritten as

$$\frac{\partial \xi}{\partial \alpha} + \frac{h^{-(1+1/b)}}{\sqrt{b}} \frac{\partial \eta}{\partial \alpha} = 0 \quad \text{and} \quad \frac{\partial \xi}{\partial \beta} - \frac{h^{-(1+1/b)}}{\sqrt{b}} \frac{\partial \eta}{\partial \beta} = 0.$$

Eliminating h in these equations by means of (32) gives

$$\sqrt{b}\frac{\partial \xi}{\partial \alpha} + \exp\left[2\left(\beta - \alpha\right)\right] \frac{\partial \eta}{\partial \alpha} = 0 \quad \text{and} \quad \sqrt{b}\frac{\partial \xi}{\partial \beta} - \exp\left[2\left(\beta - \alpha\right)\right] \frac{\partial \eta}{\partial \beta} = 0. \tag{33}$$

Introduce the new quantities, ε and ν, as

$$\xi = \frac{\varepsilon}{\sqrt{b}} \exp\left(\beta - \alpha\right) \quad \text{and} \quad \eta = \nu \exp\left(\alpha - \beta\right). \tag{34}$$

Substituting (34) into (33) results in

$$\frac{\partial \varepsilon}{\partial \alpha} + \frac{\partial \nu}{\partial \alpha} - \varepsilon + \nu = 0 \quad \text{and} \quad \frac{\partial \varepsilon}{\partial \beta} - \frac{\partial \nu}{\partial \beta} + \varepsilon + \nu = 0. \tag{35}$$

Equations (16) and (35) coincide. Therefore, Eqs. (17), (18) and (19) are valid in the case under consideration. Equation (20) is also valid. Then, Eqs. (26), (33), (34) and (20) combine to give

$$\begin{aligned}
\frac{\partial x}{\partial \alpha} &= \frac{\exp[t \sin\varphi(\alpha - \beta)]}{\sqrt{b}} \left(\cos\psi + \sqrt{b}\sin\psi\right)\left(\frac{\partial \varepsilon}{\partial \alpha} - \varepsilon\right),\\
\frac{\partial y}{\partial \alpha} &= \frac{\exp[t \sin\varphi(\alpha - \beta)]}{\sqrt{b}} \left(\sin\psi - \sqrt{b}\cos\psi\right)\left(\frac{\partial \varepsilon}{\partial \alpha} - \varepsilon\right),\\
\frac{\partial x}{\partial \beta} &= \frac{\exp[t \sin\varphi(\alpha - \beta)]}{\sqrt{b}} \left(\cos\psi - \sqrt{b}\sin\psi\right)\left(\frac{\partial \varepsilon}{\partial \beta} + \varepsilon\right),\\
\frac{\partial y}{\partial \beta} &= \frac{\exp[t \sin\varphi(\alpha - \beta)]}{\sqrt{b}} \left(\sin\psi + \sqrt{b}\cos\psi\right)\left(\frac{\partial \varepsilon}{\partial \alpha} + \varepsilon\right).
\end{aligned}$$

Eliminating in these equations ε by means of Eq. (18) and using Eq. (17) results in

$$\begin{aligned}
\frac{\partial x}{\partial \alpha} &= \frac{\exp[t \sin\varphi(\alpha - \beta)]}{2\sqrt{b}} \left(\cos\psi + \sqrt{b}\sin\psi\right)\left(\frac{\partial \gamma}{\partial \alpha} - \omega\right),\\
\frac{\partial y}{\partial \alpha} &= \frac{\exp[t \sin\varphi(\alpha - \beta)]}{2\sqrt{b}} \left(\sin\psi - \sqrt{b}\cos\psi\right)\left(\frac{\partial \gamma}{\partial \alpha} - \omega\right),\\
\frac{\partial x}{\partial \beta} &= \frac{\exp[t \sin\varphi(\alpha - \beta)]}{2\sqrt{b}} \left(\cos\psi - \sqrt{b}\sin\psi\right)\left(\frac{\partial \omega}{\partial \beta} + \gamma\right),\\
\frac{\partial y}{\partial \beta} &= \frac{\exp[t \sin\varphi(\alpha - \beta)]}{2\sqrt{b}} \left(\sin\psi + \sqrt{b}\cos\psi\right)\left(\frac{\partial \omega}{\partial \beta} + \gamma\right).
\end{aligned} \tag{36}$$

In order to find the mapping into Cartesian space, Eq. (36) should be integrated along any convenient path in the $\alpha\beta$ space after the solution to (19) has been found.

5 Conclusions

On the assumption of plane strain and plane stress conditions the system of equations comprising one of widely used yield criteria and the equilibrium equations has been solved in a curvilinear orthogonal coordinate system in which the coordinate curves coincide with trajectories of the principal stress directions. It has been shown that mapping between the principal line based and characteristic coordinate systems are determined by the equation of telegraphy. Then, mapping between the characteristic and Cartesian coordinate systems is determined by evaluating ordinary integrals. Thus, the mapping between the principal line based and Cartesian coordinate systems is given in parametric form with the characteristic coordinates being the parameters.

Acknowledgements The research described in this paper has been supported by RFBR and JSC Russian Railways (Project No. 17-20-03215). The second author was partly supported by Moscow Technological University [9].

References

1. Sadowsky, M.A.: Equiareal pattern of stress trajectories in plane plastic strain. ASME J. Appl. Mech. **63**, A74–A76 (1941)
2. Alexandrov, S.E., Goldstein, R.V.: Trajectories of principal stresses in the plane – stressed state of material obeying the Tresca and Coulomb – Mohr yield conditions. Dokl. Phys. **59**, 460–462 (2014)
3. Spencer, A.J.M.: A theory of the kinematics of ideal soils under plane strain conditions. J. Mech. Phys. Solids **12**, 337–351 (1964)
4. Harris, D.: A hyperbolic augmented elasto-plastic model for pressure-dependent yield. Acta Mech. **225**, 2277–2299 (2014)
5. Alexandrov, S., Harris, D.: Geometry of principal stress trajectories for a Mohr-Coulomb material under plane strain. ZAMM **97**, 473–476 (2017)
6. Hill, R.: The Mathematical Theory of Plasticity. Clarendon Press, Oxford (1950)
7. Chakrabarty, J.: Theory of Plasticity. McGraw-Hill, New York (1987)
8. Druyanov, B.A., Nepershin, R.I.: Problems of Technological Plasticity. Elsevier, Amsterdam (2013)
9. Pankov, V.L.: Effectiveness of incentive mechanism and the potential level meeting the needs of an employee. Herald. MSTU MIREA **1**(4), 288 (2015)

Design and Methodology to Produce Auxiliary Orthopedic Rehabilitation

Alexander Reyes-Cruz, Juan Alfonso Beltrán-Fernández, Luis Héctor Hernández-Gómez, Irving Omar Cazares-Ramírez, Alejandro González-Rebattú y González, Juan Alejandro Flores-Campos and Adolfo López-Lievano

Abstract Nowadays, one billion people live with some form of disability, and face difficulties in the development of activities and integration into society. In future years, this problem will increase. The generation of devices for the rehabilitation of people who suffer some kind of disability is a task that has been studied and attacked for years. The exoskeleton (exo means "outside" in Greek) is an external skeleton that encloses the body of the arthropod as an armor, performing a mechanical function of structural support and protection. Conventional manufacturing processes, present advantages as well as disadvantages, within its disadvantages are the long times for its manufacture and construction, and high cost of manufacturing, since it is not a serial production. The guidelines and actions of the methodology for the manufacture of an exoskeleton of assistance for human gait with problems of acute transverse myelitis.

A. Reyes-Cruz (✉) · J. A. Beltrán-Fernández · L. H. Hernández-Gómez · I. O. Cazares-Ramírez
J. A. Flores-Campos · A. López-Lievano
Instituto Politécnico Nacional-SEPI-ESIME, Unidad Profesional "Adolfo López Mateos", Edificio 5, 3er Piso, Colonia Lindavista, Gustavo A. Madero, 07738 Mexico D.F, Mexico
e-mail: alex_reyes1983@hotmail.com

J. A. Beltrán-Fernández
e-mail: jbeltranf@hotmail.com

L. H. Hernández-Gómez
e-mail: luishector56@hotmail.com

I. O. Cazares-Ramírez
e-mail: cazares0300@hotmail.com

J. A. Flores-Campos
e-mail: jaflores@ipn.mx

A. López-Lievano
e-mail: k_lievano@hotmail.com

A. González-Rebattú y González
Hospital Regional 1ro de Octubre-ISSSTE-Av., Instituto Politécnico Nacional, No. 1669, Lindavista, Gustavo A. Madero, CDMX, 07300 Mexico, Mexico
e-mail: alexrebattu@hotmail.com

© Springer International Publishing AG, part of Springer Nature 2019
A. Öchsner and H. Altenbach (eds.), *Engineering Design Applications*,
Advanced Structured Materials 92, https://doi.org/10.1007/978-3-319-79005-3_26

The design methodology should satisfy the following characteristics:

- Adaptability
- Flexibility
- Reduction of physical prototypes
- Rapidity.

Keywords Orthopedic · Rehabilitation · Exoskeleton · Videogrammetry Human gait

1 Introduction

Today, around one billion people live with some kind of disability, of which, almost 200 million experience considerable difficulties in their functioning [1]. In future years, disability will be an even greater reason of concern because its prevalence is increasing.

An exoskeleton, naturally occurs in some animals, is a result of arthropod adaptation. The exoskeleton (exo means "outside" in Greek) is an external skeleton that encloses the body of the arthropod as an armor, performing a mechanical function of structural support and protection [2].

In engineering, an exoskeleton can be defined as an electromechanical structure, which presents a certain rigidity of support for the body in which is mounted, and has the function of rehabilitating or enhancing the motor function of a human being, depending on its complexity, functions and analysis. Can be classified into different types, however, there are two major classifications, lower limbs and upper limbs [3, 4].

Nowadays, the growing need of people to acquire designs that satisfy different needs that are present, have led to the evolution of conventional manufacturing processes.

Manufacturing can be divided into two stages, the first stage being the discovery of materials and their processes to manufacture goods, and the second stage is the development of systems for manufacturing [5].

Among the most common manufacturing processes are:

- Casting process
- Mechanical metal forming processes
- Chip removal processes
- Processes for forming polymers
- Processing from powders.

All manufacturing processes have some disadvantages, the most common being the porosity present in the parts, and the change of mechanical properties inherent to the process of solidification and cooling (casting) [6, 7], due to the high temperatures of the metal, surface oxidation and poor surface finishes (formed) are produced [8], waste material is generated (detachment), and production time and costs are high [9].

These manufacturing processes used in the production of custom equipment, such as exoskeletons or devices for the rehabilitation of some part of the body of someone, present the disadvantages of high costs when not mass-produced, combined with long manufacturing times, as well as the low aesthetic quality of the designs.

2 Problem

According to the National Institute of Statistic and Geography (INEGI by its initials in Spanish), in the year 2010, the number of people with disabilities is 5,739,270 people, representing 5.1% of the total population, and from this percentage 3,347,849 people have a motor disability, which is defined as the difficulty that a person has to move, walking, moving or climbing stairs due to lack of all or part of their legs; it also includes those who cannot move having their legs, or have restrictions to move, and therefore need help from other persons, wheelchairs or other devices, such as walkers or a prosthetic limb.

Among the problems faced by these people, is the problem of access to equipment for proper therapy, in public or private institutions, or the acquisition of exoskeletons or similar devices, to help support or replace the partial or total functioning of some member of the patient.

Likewise, the design of these equipment presents the disadvantage of not being adequate to the anatomy of the patient, because for the design and for the production of these equipments, only a fraction of the population is phenotypically considered, having as consequence that this equipment is not adequate in many cases with patients who are out of the characteristics of this group.

This paper presents a methodology for the personalized design of robotic assistance systems for the lower extremities in users with motor dysfunction.

3 Methodology

Next, the guidelines and actions of the methodology, for the manufacture of an exoskeleton of assistance for human gait with problems of acute transverse myelitis are presented.

The design methodology should satisfy the following characteristics:

1. Adaptability—Must be able to adapt to different motor needs and degrees of freedom depending on each patient.
2. Flexibility—Generate 3D models, with CAD software which is flexible to adapt in a personalized way to each patient.
3. Reduction of physical prototypes—Analyze the structural stability as well as the kinematics and dynamics of the models with support of computer programs.

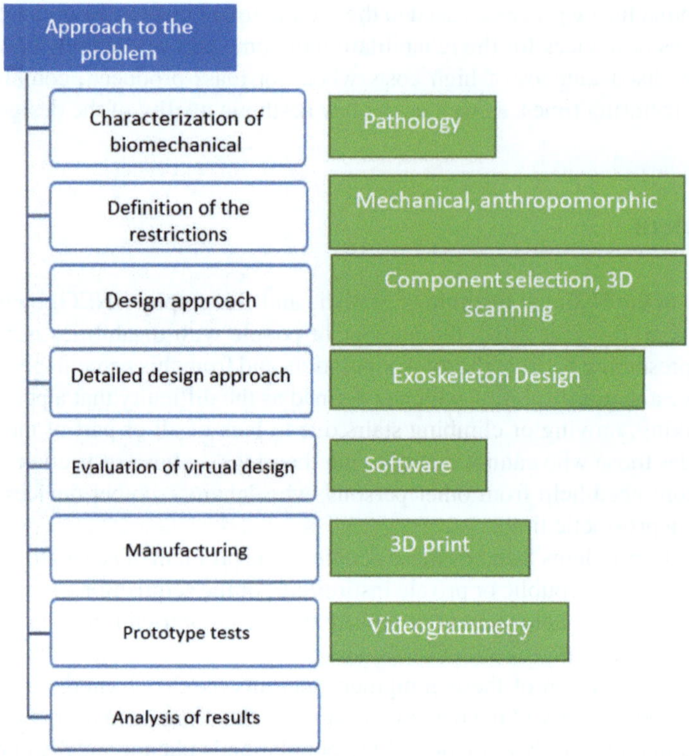

Fig. 1 Stages of the methodology

4. Speed—Production times should be shorter when compared to traditional manufacturing processes (Fig. 1).

4 Case of the Study

In this work, a patient with acute transverse myelitis is presented, with an advance in the pathology considered as total paralysis of lower limbs [10].

A study of the biomechanics of the affected parts was made to define the degrees of freedom of the exoskeleton, as well as the mechanical restrictions that the design will have [11].

With support of videogrammetry, the study of the kinematics and dynamics of the human gait was carried out, to characterize angles speeds, and positions [12].

Data acquisition process using videogrammetry [13]:

Calibration of the camera.

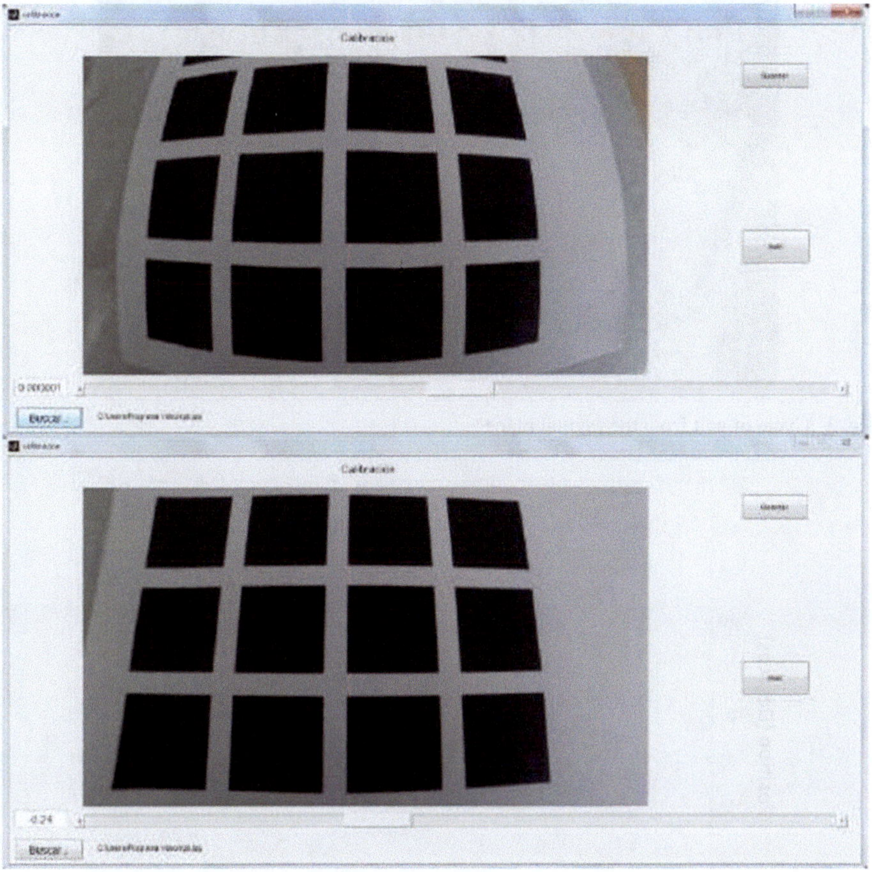

Fig. 2 Calibration and correction of image curvature

It defines the function or the functions that allow to relate the pixels of the captured image, with lengths of the real world (Fig. 2).

1. Digital image preprocessing

This focuses on isolating specific information, where from defined spaces, relevant and irrelevant information is taken. These spaces or thresholds are constraints on the layers of colors. Within this step, are considered processes of expansion and delivery, techniques that allow to increase or decrease the areas of information (Figs. 3 and 4).

The dilatation is defined as:

$$A \oplus B = \left(c \in E^{N} | c = a + b \text{ for some } a \in A \text{ and } b \in b\right) \quad (1)$$

And the delivery as:

Fig. 3 Cycle of gait from the sagittal plane

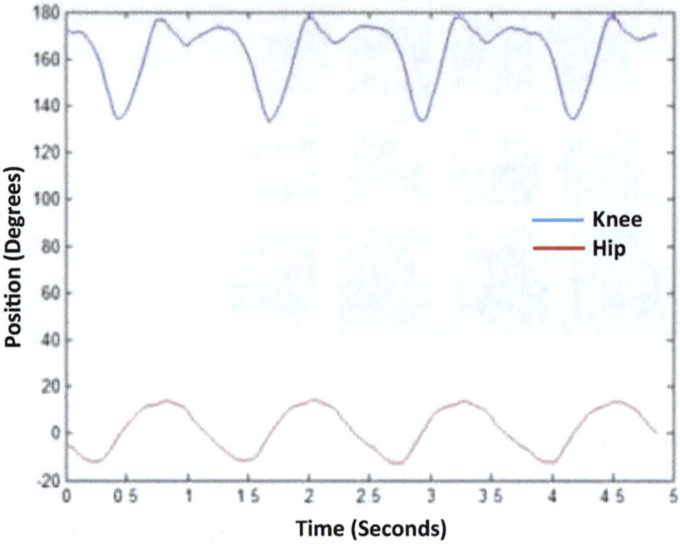

Fig. 4 Parameters of the gait cycle

$$A \ominus B = \left(x \in E^N | B_x \subseteq A\right) \tag{2}$$

Preprocessing

In this process, the markers are located in a plane from centroids, to obtain the desired values.

$$X = \left(\frac{1}{n}\right) \sum_{i=1}^{n} x_i \tag{3}$$

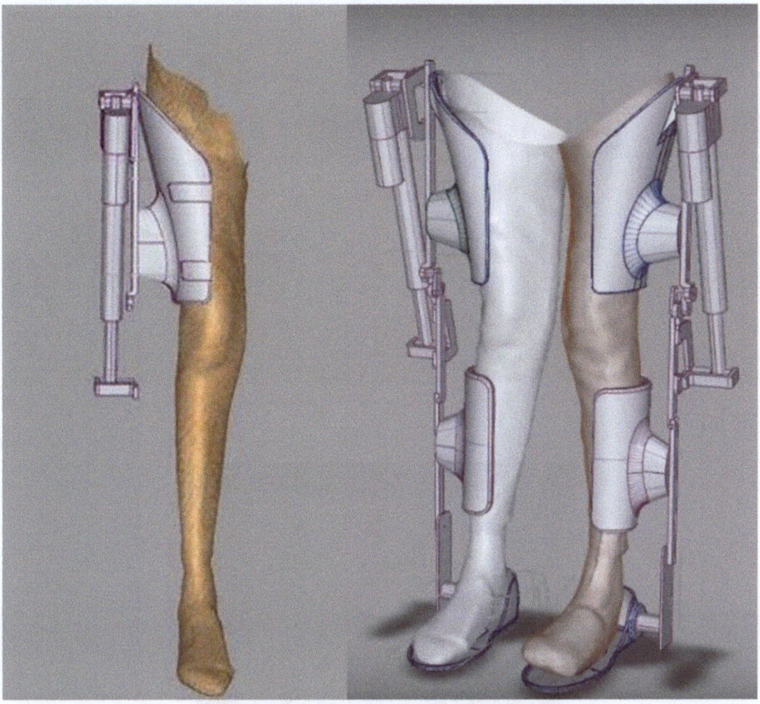

Fig. 5 Conceptual design of the exoskeleton

$$Y = \left(\frac{1}{n}\right)\sum_{i=1}^{n} y_i \tag{2}$$

Once the gait parameters and the degrees of freedom that define the exoskeleton are obtained, a scan of the legs of the patient is performed to make the design, according to their measurements and contours (Fig. 5).

The fabrication of the design is done using 3D printers, made of ABS material reinforced with carbon fiber, which is one of the most used thermoplastics in the manufacture of prototypes of support in orthopedics and prostheses [14–16] (Figs. 6 and 7).

5 Conclusions

The implementation of CAD software for the generation of 3D models, based on scans of a patient, allowed the help of 3D printers, to obtain the parts faster than with conventional manufacturing processes.

In addition, the methodology allows patients to be considered in the 3D modeling process.

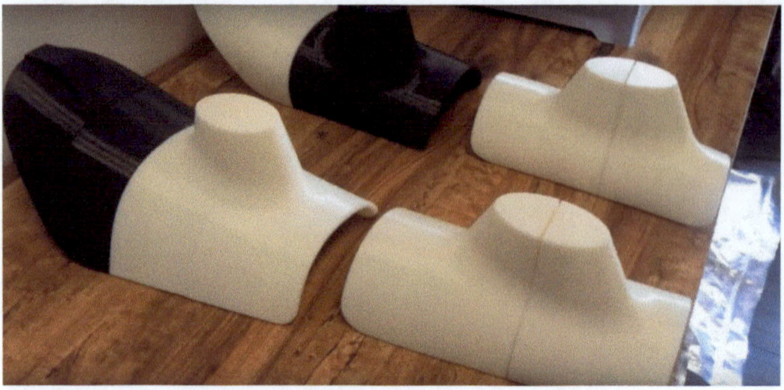

Fig. 6 3D printing

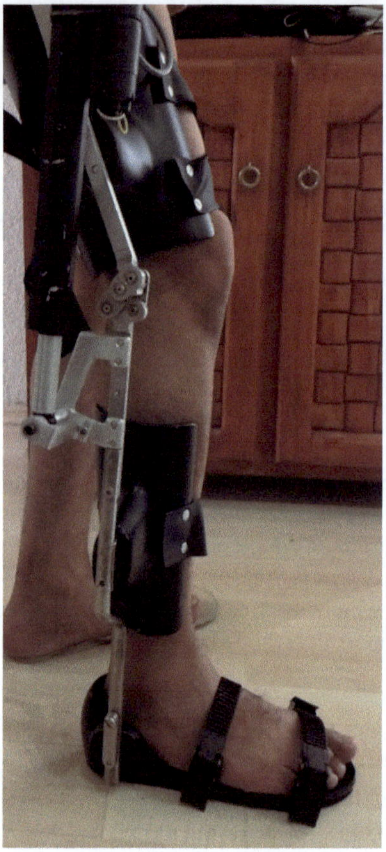

Fig. 7 Assembly with electric actuator

It is important to perform tests for the validation of the kinematics, with the help of Videogrammetry in order to measure the parameters of the supported gait and compare them with a normal gait.

Acknowledgements The authors kindly acknowledge the support given to the National Polytechnic Institute and the National Council for Science and Technology (CONACYT).

References

1. World Health Organization, World Report on Disability, WHO Library Cataloguing-in-Publication Data. www.who.int/disabilities/world_report/2011/report.pdf (2011). Accessed 21 Jan 2017
2. Audesirk, T., Audesirk, G., Byers, B.: Biología, p. 443. Pearson Educación, Mexico (2013)
3. Mena, M., Anabel Sancho Culcay, M., Jazmina, J.: Diseño y construcción de un exoesqueleto de 6 GDL para potenciar las capacidades en personas que han perdido la movilidad parcial del miembro superior derecho (Bachelor). Universidad de las Fuerzas Armadas (2015)
4. Rocon, E., Pons, J.: Exoskeletons in Rehabilitation Robotics. Springer, Berlin (2014)
5. Groover, M.: Fundamentos de manufactura moderna, pp. 3–5. Prentice-Hall Hispanoamericana, Mexico (2007)
6. Waters, T.: Fundamentals of Manufacturing for Engineers, 1st edn. CRC Press, London (2017)
7. Leyensetter, A., Würtemberger, G., Saénz de Magarola, C.: Tecnologia de los oficios metalurgicos. Reverte, Barcelona (1987)
8. Iza Iza, B.: Dimensionamiento y construcción de una roladora manual para laboratorio (Bachelor). Escuela politécnica nacional (2007)
9. Matta-Poveda, J.P.: Diseño de una metodología para la selección de procesos de manufactura usando cartas y bases de datos (Bachelor). Universidad EAN, Facultad de Ingeniería (2012)
10. Oñate Vergara, E., Sota Busselo, I., García-Santiago, J., Gaztañaga Expósito, R., Nogués Pérez, A., Ruiz Benito, M.: Mielitis transversa en inmunocompetentes. An. Pediatr. (2004). https://doi.org/10.1016/s1695-4033(04)78378-2
11. Crowell, H.: Human engineering design guidelines for a powered, full body exoskeleton. Ft. Belvoir, Defense Technical Information Center. http://www.dtic.mil/docs/citations/ADA299539 (1995). Accessed 3 Nov 2016
12. Jiménez-Vázquez, E., Cazares-Ramírez, I., Velazquez-Sanches, A., Urriolagoitia-Sosa, G.: Sistema de análisis de movimiento en los planos sagital y coronal por medio de Videogrametria. In: IV Congreso Nacional de Tecnología Aplicada a Ciencias de la Salud. Tonantzintla, Cholula Puebla (2013)
13. Cazares-Ramírez, I.: Análisis de Marcha Humana con Videogrametría para el Diseño de Exoesqueletos. In: VIII Congreso Internacional de Ingeniería Electromecánica y de Sistemas, pp. 1–4. Ciudad de México (2016)
14. Beltrán-Fernández, J., González, M., Hernández-Gómez, L., González, A., Calderón, G.: Biomechanical prosthesis design of an orbicular cranial cavity. In: Advances in Bio-Mechanical Systems and Materials, pp. 87–94. (2013). http://dx.doi.org/10.1007/978-3-319-00479-2_7
15. Hernández-Gómez, L., Rangel-Elizalde, A., Beltrán-Fernández, J., González-Rebatú, A., Corro-Valdez, N., Urriolagoitia-Calderón, G., Rodríguez-Martínez, R.: An optimization of the manufacturing process of corporeal screws made of synthetic bone. In: Advanced Structured Materials, pp. 43–59. (2015). http://dx.doi.org/10.1007/978-3-319-19470-7_4
16. Moreno Garibaldi, P., Beltrán Fernández, J., Hernández Gómez, L., López Saucedo, F., Corro Valdez, N., López Liévano, A.: Mechanical characterization of new biocompatible composite material focused on maxillofacial applications. DYNA **92**(4), 461–465 (2017). https://doi.org/10.6036/8100

Design and Manufacturing of a Temporomandibular Joint (TMJ) Prosthesis for Mandibular Bone Necrosis Using the Finite Element Method

Juan Alfonso Beltrán-Fernández, Luis Héctor Hernández-Gómez, Juan Luis Cuevas-Andrade, Jesús Eduardo Campa-Zuno, Alfredo de la Peña-Muñoz, José Manuel Guzmán-López, Pablo Moreno-Garibaldi and Adolfo López-Lievano

Abstract In this work, a temporomandibular joint prosthesis system is designed in order to be manufactured and implanted. A computer tomography study of a patient with bone cancer and temporomandibular ankylosis was used. The specific design also considered a local socket to replace the mandibular fossa. The manufacturing of the system used stainless steel (AISI-316L). Mechanical evaluation and numerical testing were done using ANSYS code and PTC CREO Parametric 3.0, while in the biomechanical reconstruction, SCAN IP software was used. The fixation of the prosthesis in the patient is an important requirement and the structural behavior is studied using the finite element method (FEM). In order to design and personalize

J. A. Beltrán-Fernández (✉) · L. H. Hernández-Gómez · J. L. Cuevas-Andrade · J. E. Campa-Zuno
A. de la Peña-Muñoz · J. M. Guzmán-López · P. Moreno-Garibaldi · A. López-Lievano
Instituto Politécnico Nacional, SEPI-ESIME-Unidad Profesional "Adolfo López Mateos", Edificio 5, 3er Piso, Colonia Lindavista, Gustavo A. Madero, 07738 México D.F, Mexico
e-mail: jbeltranf@hotmail.com

L. H. Hernández-Gómez
e-mail: luishector56@hotmail.com

J. L. Cuevas-Andrade
e-mail: jl1z9c88@gmail.com

J. E. Campa-Zuno
e-mail: chuy_campa@hotmail.com

A. de la Peña-Muñoz
e-mail: alfredodlp@outlook.com

J. M. Guzmán-López
e-mail: guzmanlopezjosemanuel@gmail.com

P. Moreno-Garibaldi
e-mail: eagle_gar@hotmail.com

A. López-Lievano
e-mail: k_lievano@hotmail.com

© Springer International Publishing AG, part of Springer Nature 2019
A. Öchsner and H. Altenbach (eds.), *Engineering Design Applications*, Advanced Structured Materials 92, https://doi.org/10.1007/978-3-319-79005-3_27

the implants, quick prototyping printer systems (ABS and PLA) were used, while for the manufacturing process, stainless steel 316L was considered, in order to get low costs. As a result of this research, the chewing process of the patient will be performed.

Keywords Temporomandibular joint disorder (TMJ) · Mandibular cancer Condyllar joint · Ankylosis · Mandibular fossa · Stereolithography · UHMWPE 3D printing models · Additive manufacture

1 Introduction

Mandibular ankylosis is a medical condition that causes mandibular hypomobility. This condition is one of the most severe conditions regarding mandibular hypomobility and it is classified according to the affected area and tissue.

Ankylosis of the temporomandibular joint (TMJ) has many causes. One of the most common causes is trauma. At the same time, this pathology affects oral movements used for eating.

Surgical intervention tries to recover most of the oral aperture and the optimal temporomandibular joint movements. It also tries to reconstruct and reshape it. There are many alternatives for treating temporomandibular joint ankylosis. The treatment may vary depending on each patient.

This pathology is divided into four types:

Type I: The condylar head is still present, but it is deformed.
Type II: There is fusion between the condylar head and the articular surface.
Type III: Osseous blocking between the mandibular branch and the zygomatic bone.
Type IV: TMJ completely deformed.

There are many different surgical procedures for treating ankylosis, as well as different rehabilitation treatments and prosthetics. These prosthetics usually consist of two main components: mandibular component and joint component.

The mandibular component is normally made from a cobalt-chrome alloy and titanium plasma. The joint component is made from polyethylene and titanium screws. The materials used for making these components are selected because of their high biocompatibility, easy accessibility and low prices.

In this case, the main purpose of this work is the realization of a prosthetic device focused on a specific Mexican patient. In Mexico and Latin America, mandibular prostheses do not fit in each patient. It is well known that the implants are fabricated with specific and predetermined sizes that must meet the necessities of the patient in the best way, without considering that this may cause future damage or discomfort. Materials used for fabrication of these prostheses are considerably expensive, which prevents large population sectors from buying one.

1.1 Statement of the Problem

Patients require a prosthetic mandibular device, which is not only accessible, but fits satisfactorily and adapts to their necessities, accelerating the healing process. If the specifications of a temporomandibular joint ankylosis diagnosed patient are known, it is possible to design a personalized prosthetic device that fits every need for the patient. It will also be more accessible due to the low-cost materials used in its fabrication.

1.2 Review of Literature

Nowadays, TMJ ankylosis is one of the most severe mandibular pathologies, regarding oral hypomobility. A few years ago, this illness was surgically treated in Mexico and Latin America. However, different alternatives and methods are available for treating ankylosis.

One of the most common procedures of all times was the surgical intervention. In this procedure, doctors remove the ankylotic blocking, recovering a correct morphology. By doing this, mobility and natural appearance return to the patient.

Surgical intervention consists of different methods and techniques. These are related to the affected area, as well as the severity of the injury. In most of the cases, the ankylotic blocking is removed allowing normal oral movement. These interventions can be done with preauricular or submandibular techniques, depending on the location of the injury.

In past years, specialists observed that performing an osteotomy was a viable procedure, but the recovery and rehabilitation of the patient was slow and uncomfortable. With this in mind, the proposal of using alloplastic and prosthetic materials becomes attractive. These were used for the reconstruction of the TMJ.

These new techniques were attractive because they helped with the rehabilitation of the patient as well as with their physical appearance and oral mobility. Medical experts use these techniques in this time. Materials change with time for these techniques and are compared considering the comfort of the patient and the easy technique of the surgeon. The properties and costs of these materials are the most important in the case of a replacement.

The newest prosthetics consider high resistance and biocompatibility and also are able to adapt to most of the necessities and characteristics of the patients. There are many prosthetic devices available on the market, which differ in sizes, models, materials, shapes and forms. This is done for adapting to most of patients and conditions, because each clinical case differs from one another. One of the main problems of this proposal is the elevated cost of each prosthetic device and the kind of the materials used in its fabrication. The clearest example is titanium alloy, which is a material used in most prostheses and has important specifications and is very expensive.

This material is highly recommended in medical instruments and prosthetic devices because it never reacts with its surroundings. It has a high tolerance to thermic changes and deformation. The main problem of titanium is that is expensive, causing prostheses to be inaccessible to many patients due to their high prices. Also, an important limitation is that, even though prostheses in the market have a wide range of prices, sizes, shapes and forms, it is very hard to fully adapt to the particular necessities of the patients. This can hurt the patient and may produce discomfort.

The most recent and attractive proposal is the implementation of prostheses made specifically for each patient, completely adapted to their necessities and specific conditions. These prostheses are made of stainless steel due to the material's properties, which are similar to the titanium's properties. The advantage of working with stainless steel is that is considerably cheaper than titanium, which will make prostheses accessible to a wider audience [1, 2].

1.3 Clinical Case

This study was based on a 50 years old male patient. The affected zone was a mandibular resection background due to an intraosseous myxoma tumor, which requires a facial reconstruction with predesigned plate and autologous osseous graft fixation.

All the specifications and requirements of the prosthetic device were established in order to be manufactured in stainless steel AISI 316L.

Finally, the procedure for the implantation of the prosthesis needs to be as follows:

- The injured mandibular branch is removed with an osteotomy for further prosthesis implantation.
- Stainless steel hemimandibular prosthesis is implanted with 12 mm screws. The fixation on the patient needs to be validated with the glenoid cavity with mouth aperture and closure movements.
- It is important to mention that these types of prostheses are one of the best options for recovering an aesthetic appearance throughout the years. The reason is that the material of these prostheses is not absorbable.

2 Background

2.1 Mandibular Anatomy

The jaw, also known as the maxilla, is a shaped bone whose concavity is faced backwards. It has two faces (anterior and posterior), as well as two borders (superior and inferior), facilitating its positioning and understanding (Figs. 1 and 2).

The jaw has two mandibular branches which start from posterior extremities towards the superior area, forming a 15° angle, approximately. This angle is known

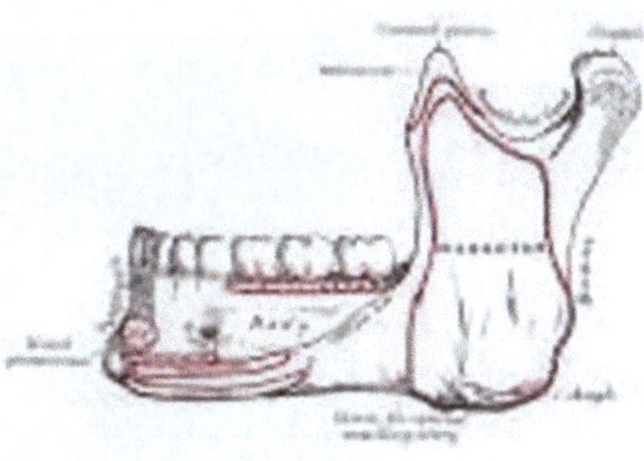

Fig. 1 External surface. Lateral view

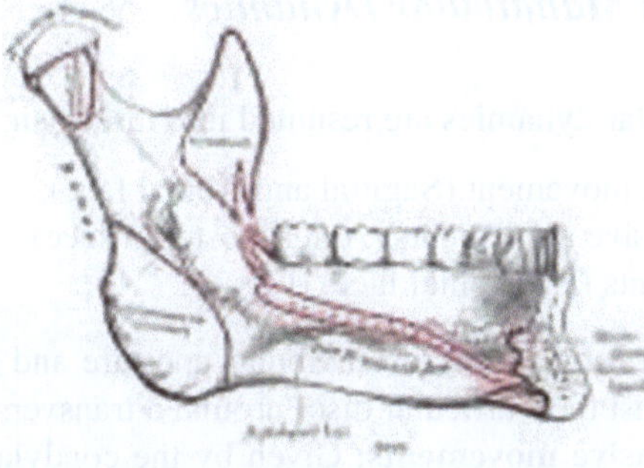

Fig. 2 Interior surface. Lateral view

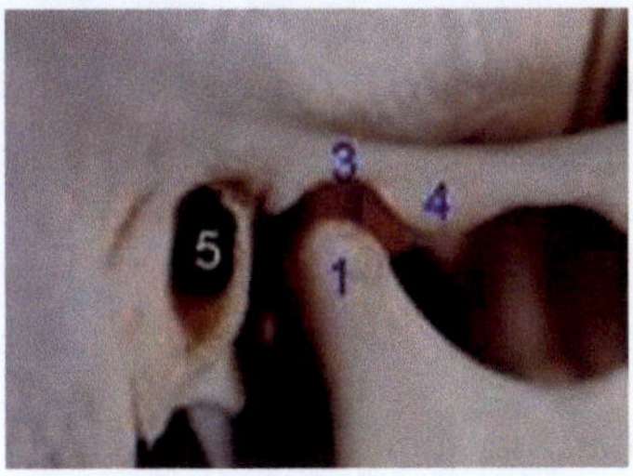

1. Mandibular condyle
2. Articular disc
3. Glenoid cavity
4. Temporal tuber
5. External auditory conduct

Fig. 3 TMJ components

as the mandibular angle or gonion. Each one of these branches present two bone structures in the superior area called coronary apophyses (anterior face) where the temporal muscle inserts, and mandibular condyle (posterior face). Between these two structures, the sigmoid notch is found. The condyle articulates with the mandibular pit of the temporal bone, forming the temporomandibular joint (Fig. 3).

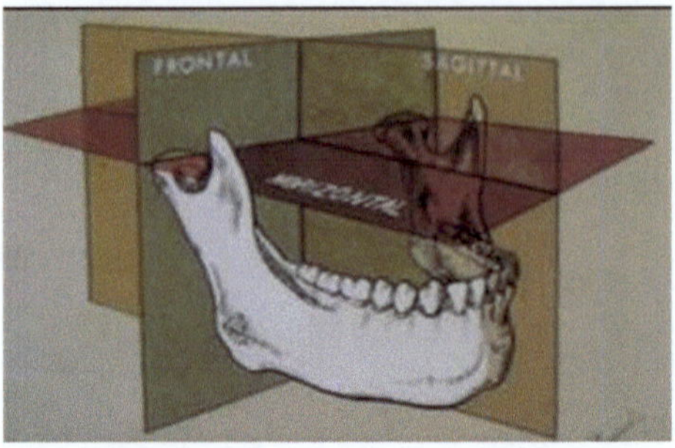

Fig. 4 Mandibular movements faces

2.2 Condylar and Mandibular Dynamics

Condylar and mandibular dynamics are resumed into three pair of movements:

1. Ascent and descent movement (Sagittal and frontal face)
2. Protusive and retrusive movements (Anteroposterior face)
3. Laterality movements (Horizontal face) (Fig. 4).

- **Ascent and descent movements**: Mandibular aperture and closure where both condyles rotate against their articular discs around a transversal axis.
- **Protusive and retrusive movements**: Given by the condylar movement downwards and forwards along the articular eminences. There is only one translation movement.
- **Laterality movements**: They are asymmetric movements. Condyles do not follow the same paths. The side to which the jaw moves is called active side or work side.

2.3 Computerized Tomography

Computerized tomography is known as an imaging technique that obtains an image of the body using X-rays. This procedure is based on emitting a thin beam of X-rays that rotates around the patient. The X-rays create signal that are perceived and processed. Once this is done, transversal images of the body are formed called "slides".

This group of transversal images is more useful than conventional X-ray studies, given that these images can be consequently stocked for creating a 3D model. Having a 3D model presents enormous advantages in the studies of a patient. It is easier to locate precise structures such as tumors or deformations in the organism.

DICOM is a standardized protocol that allows communication between systems and gives the opportunity of having only one image format for storage. This solves any problem that may occur between medical devices.

When an image needs to be analyzed in a hospital or in a health clinic, the DICOM format is always used because the image will not give any useful information of the patient's condition by itself. These images are accompanied by patient data and image acquisition. That is why the use of formats such as JPEG or PNG is not enough. Due to these characteristics, DICOM is recognized in the world of medical sciences as the best for printing, storage and handling of medical images.

2.4 Mandibular Prosthesis

Maxillofacial bone tumors require a reconstruction with the objective of restoring the protective function to provide acceptable aesthetic results. The reconstruction involves a hard tissue replacement implant (HTR) in conjunction with grade 5 titanium, thanks to its high biocompatibility. These implants have demonstrated acceptable integration into hard and soft tissues.

The HTR is made from a polymethylmethacrylate (PMMA) compomer with polyhydroxyhetermethacrylate (PHEMA). These polymer implants can be hinged to titanium structures for added stability and protection.

The total replacement of the temporomandibular joint consists of a glenoid cavity prosthesis. It is based on the actual glenoid cavity, which will articulate with the implanted hemimandibular prosthetic condyle.

3 Biomodelling and CAD Transformation

3.1 Computerized Tomography

In order to make a temporomandibular joint prosthesis for a patient, it is necessary to know its specifications. For this, a computerized tomography is available which shows the damage of each patient. Computer tomography is used to create the model of the prosthesis. To carry out this process, we used the ScanIP program, which allows us to choose a zone of the tomography to be converted to a 3D model using a segmentation algorithm.

Because the tomography was focused in the skull, it was possible to observe the way in which the white color in each section is highlighted, these areas being the ones that need to be selected. The program allows to highlight this color by adjusting both brightness and contrast. It was decided to work with the default brightness, but the contrast was variable for better vision, as shown in Figs. 5 and 6.

The process consists in selecting, cut by cut, the areas of the tomography that will be converted to the third dimension. When selecting a specific pixel of the tomography, the algorithm will look for other related pixels, coloring those pixels of the color of the mask that is being used.

Masks are useful when it is necessary to draw the entire skull, but separate by regions, as in this case in which a mask (blue) was all the skull and upper jaw, and the other mask (red) was the lower jaw, as shown in Fig. 7. This separation was due

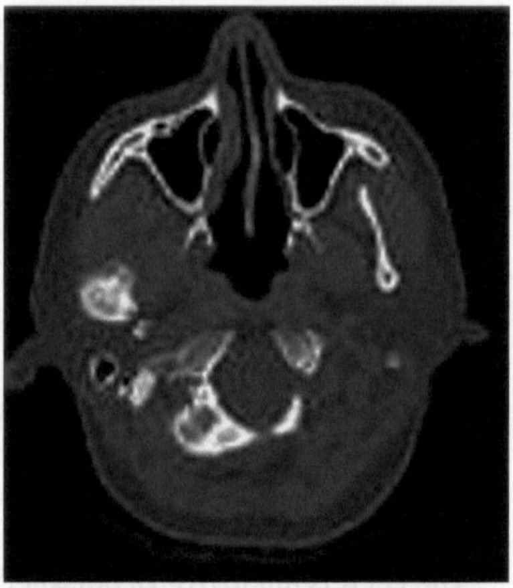

Fig. 5 Tomography slice

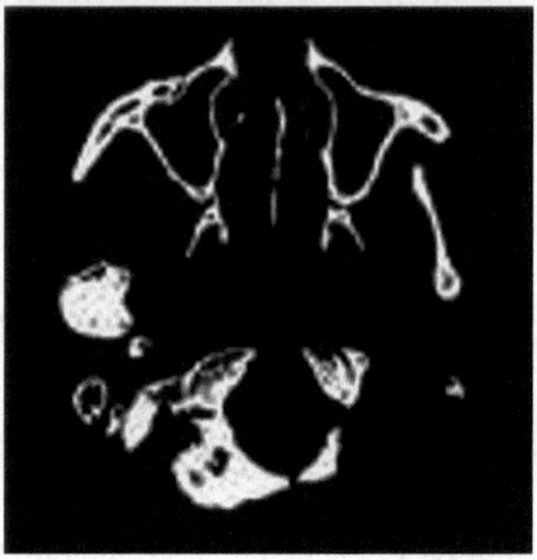

Fig. 6 Tomography slice at different contrast

to the patient's ankylosis being located in the lower jaw, allowing a better analysis of the disease by having modelling the mandible separately.

Another useful tool of the program is the viewing of the tomography in 3 different planes: axial (upper to lower) as in Fig. 8, sagittal (right to left) as in Fig. 9 and coronal A (posterior) as in Fig. 10.

The three different views are useful in difficult areas, for instance: the teeth, as shown in Fig. 8. In general, the axial view was used most of the time; the sagittal view was used for the coloring of the teeth, and the coronal view to omit the vertebrae.

It is important to mention that it had to be very careful in the drawing process, because of when a pixel was selected outside the white area; the program assumed that it was desired to color the non-bone parts, it produces as a consequence an error as shown in Fig. 11.

Once all the slices were painted it was necessary to make a rendering of both masks to visualize how the 3D print would emerge and whether there was any overlapping of the masks or extra floating material that should be removed. As it is only a preview,

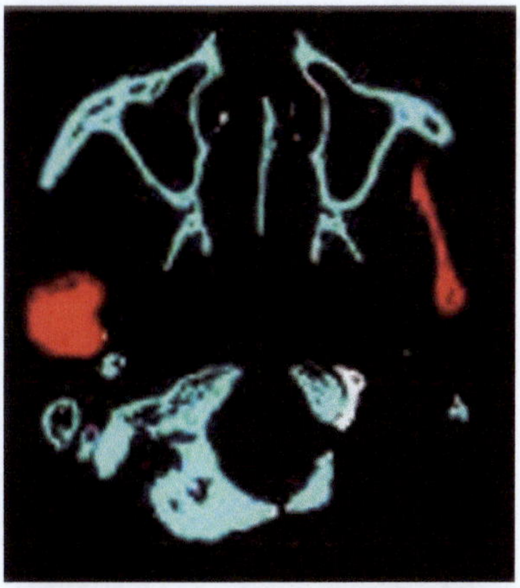

Fig. 7 Cut with both masks selected

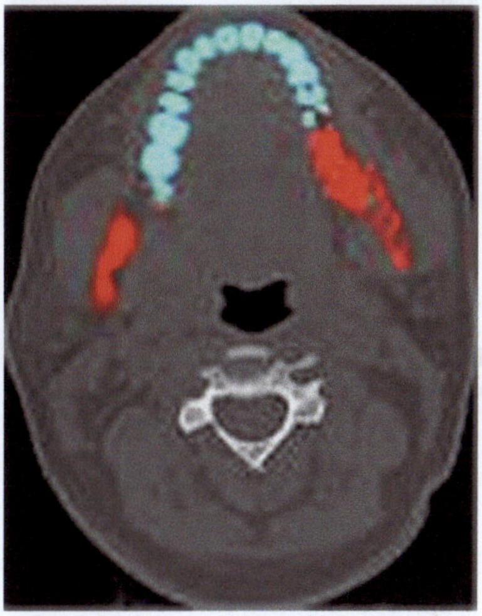

Fig. 8 Slice with axial view

it looks a bit rough and not so well defined; in the next stage, it is possible to see the difference between the preview and the final (Fig. 12).

It is important to emphasize how, in Fig. 13, the increased calcification that the patient suffered in the right condyle is clearly denoted.

3.2 CAD Transformation

Once the biomodelling was finished, the 3D printing model was continued. For this, the first thing that was done was to export from ScanIP both masks in STL format (allowing it to be used in CAD programs) in both ASCII type and binary type, using the "decimate" tool.

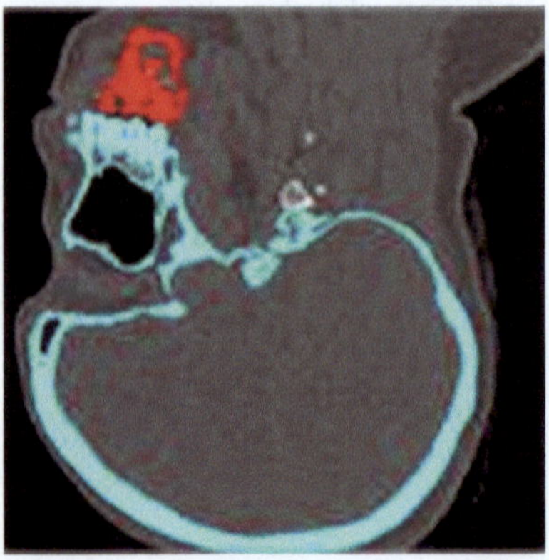

Fig. 9 Slice with sagittal view

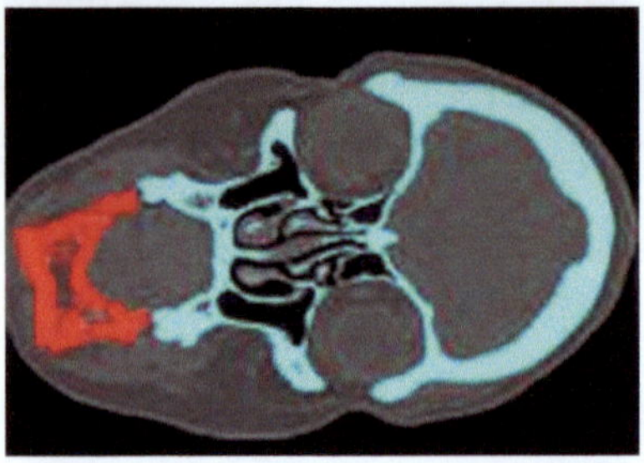

Fig. 10 Slice with coronal view

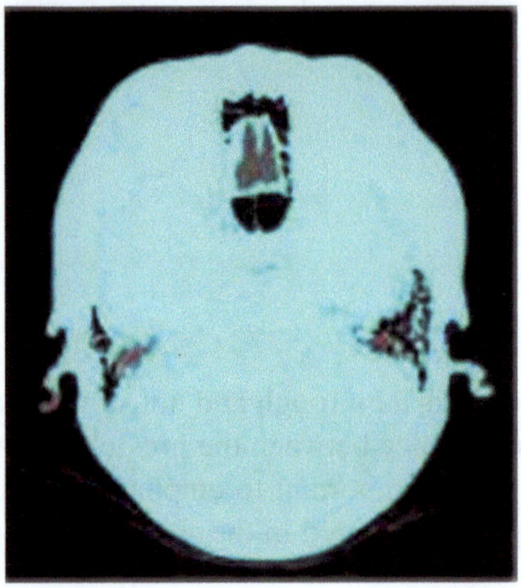

Fig. 11 Error in the selection of the pixel

This tool converts the points cloud of the model into triangles and, depending on how many triangles the program uses, is the quality level with which the model is exported. For this case, 50,000 triangles were used, as shown in Fig. 14.

Each of these exports was uploaded to a program called XYZware, which prepares the model to be sent to the 3D printer. The program allows to use different

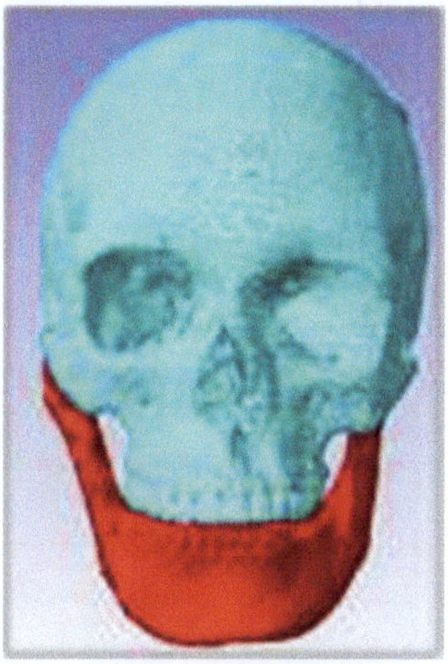

Fig. 12 Frontal view of the render

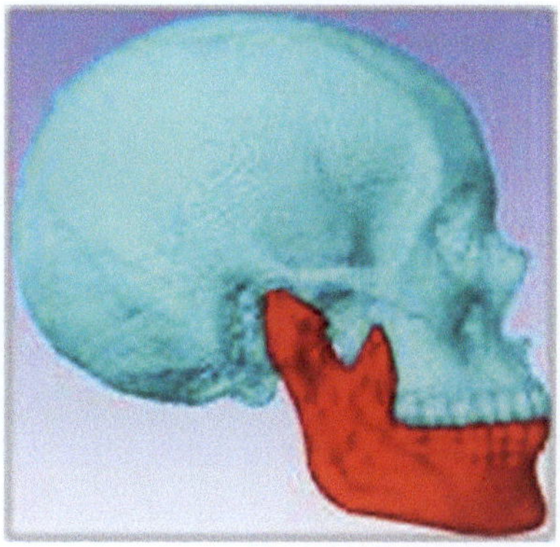

Fig. 13 Lateral view of the render

characteristics to the model to be printed, such as choosing the position, print speed, density, give it a base, supports, among other features.

As mentioned above, the difference in quality between a preview in the ScanIP program and an export with decimate to 50,000 triangles is shown in Figs. 15, 16, 17 and 18.

The skull and upper jaw were the first model to be prepared and it was chosen to include the base and the supports; the base so that when the model is removed from the printer, the chin will not be damaged, and the brackets to give a better finish to the model.

As shown in Fig. 17, once the model is ready to be printed, the supports and the base are shown to improve the 3D printing.

Once the lower jaw was uploaded to the program, some extra material was found on the left side of the jaw, as shown in Fig. 18. It was returned to the tomography and, as shown in Fig. 19, it was found that in some cuts the red mask extended to the upper jaw.

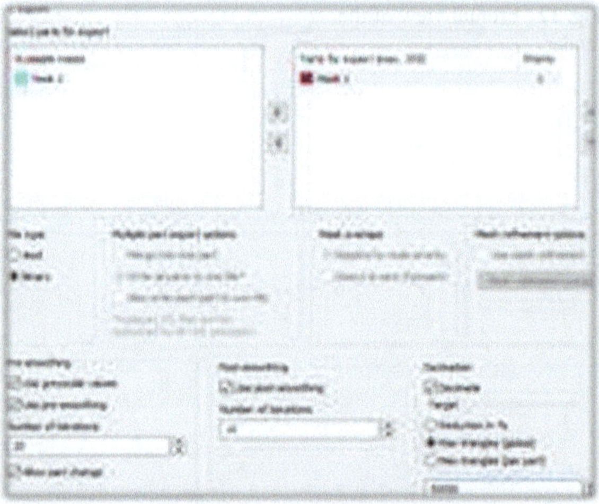

Fig. 14 Exporting menu

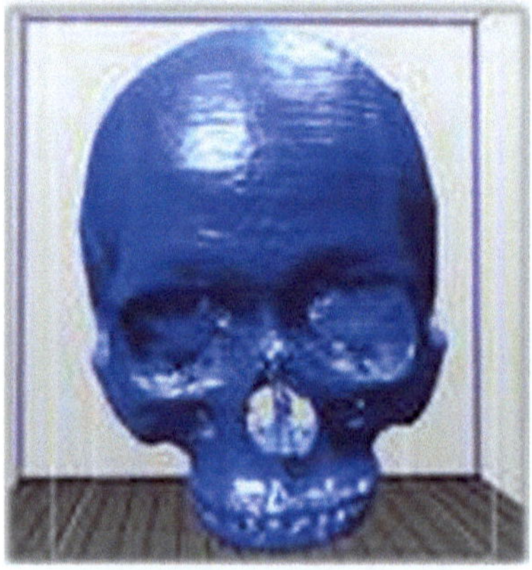

Fig. 15 Frontal view of the skull and superior jaw

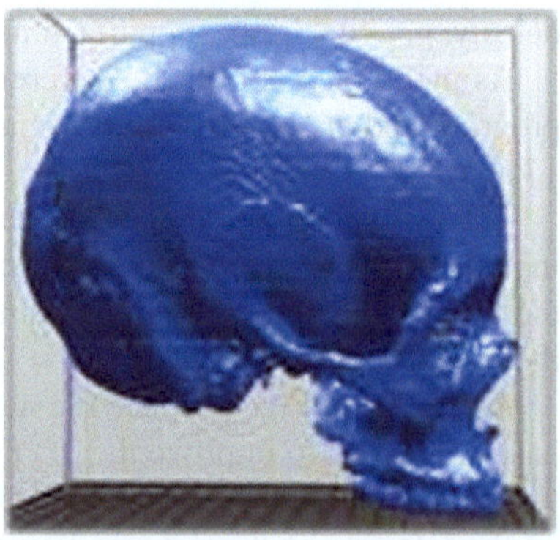

Fig. 16 Lateral view of the skull and superior jaw

Fig. 17 Skull and superior jaw ready to be printed

Fig. 18 Inferior jaw with extra material

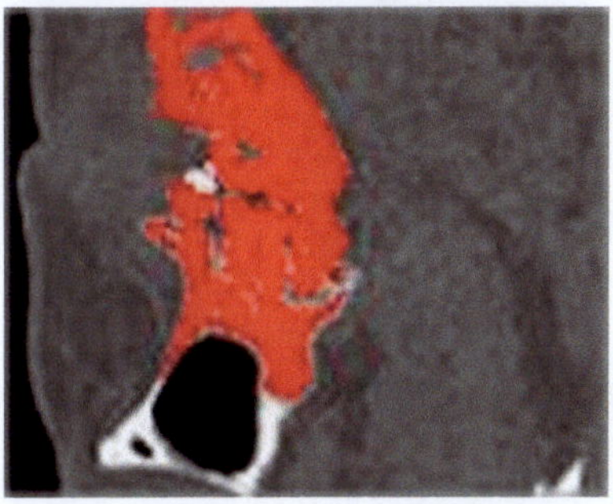

Fig. 19 Cause of the extra material

This was fixed and reloaded to the program, applying the same characteristics as the model before (Figs. 20, 21 and 22).

As shown in Fig. 23, once each model was completed, a screen with important information such as model dimensions, the time it would take to print, and the amount of material that was required to be loaded to the printer, appeared.

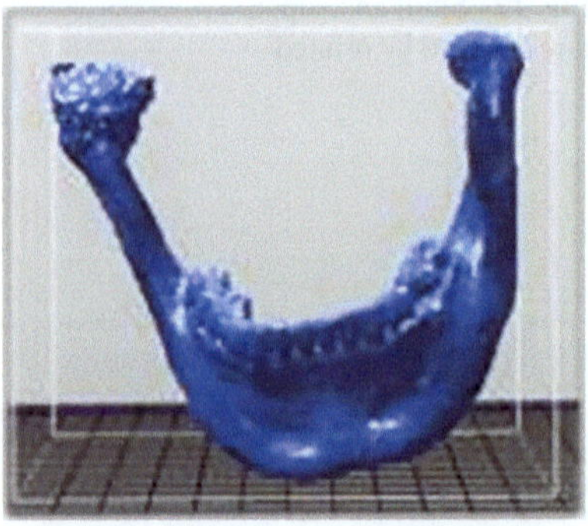

Fig. 20 Frontal view of the inferior jaw

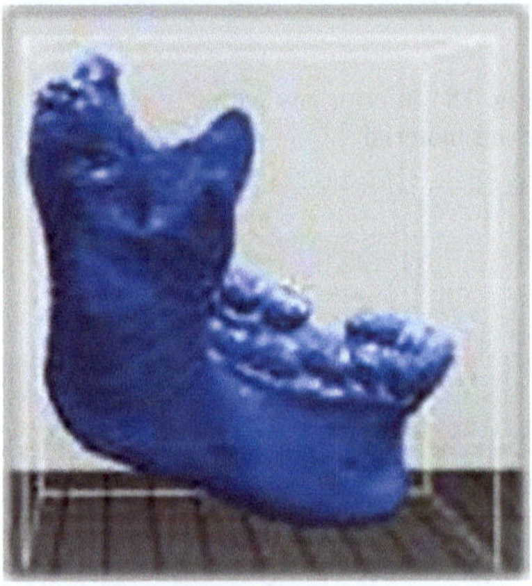

Fig. 21 Lateral view of the inferior jaw

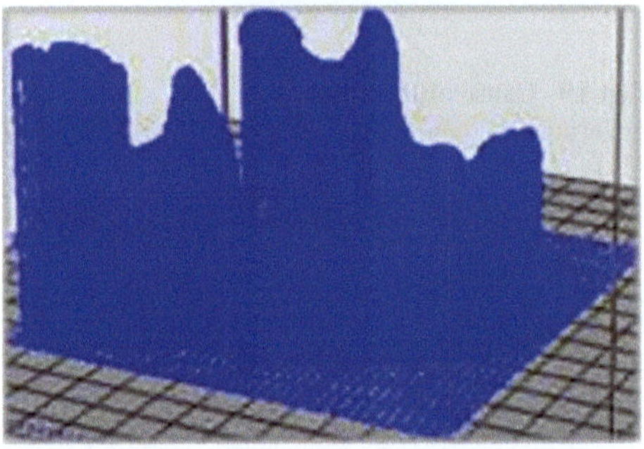

Fig. 22 Inferior jaw ready to be printed

The 3D printer used has an ABS filament limit that can be used by printing, and the skull along with the upper jaw required more than the available limit, so it was decided to cut the top of the skull because it was not necessary for the performance of the prosthesis. After this area was cut, it was necessary to create the 3D printing format in the XYZware software, as shown in Figs. 24 and 25.

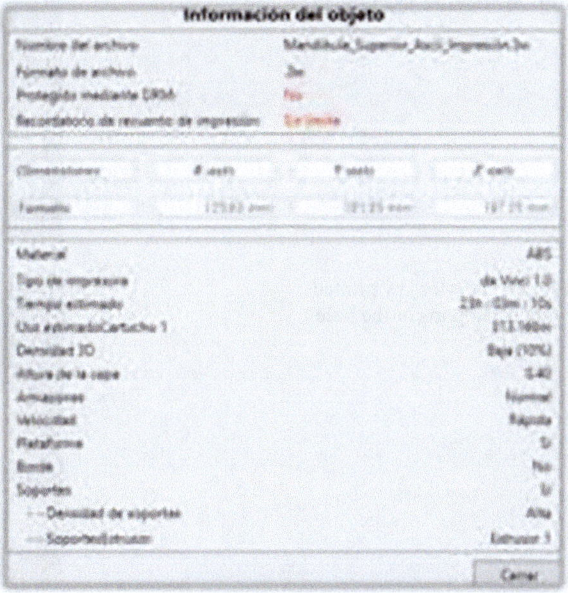

Fig. 23 Printing menu

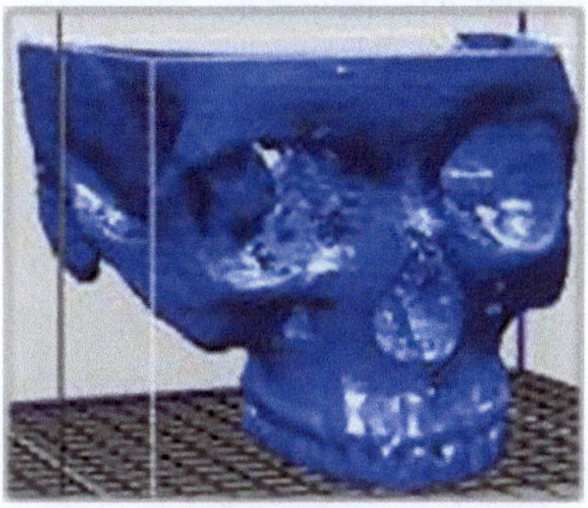

Fig. 24 Frontal view of the cut skull and superior jaw

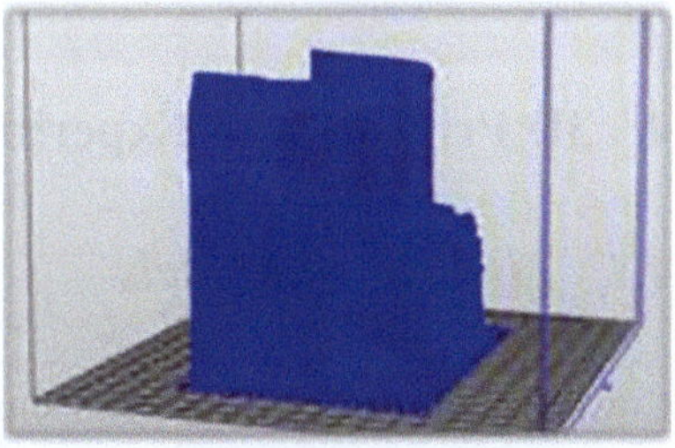

Fig. 25 Cut skull and superior jaw ready to be printed

Fig. 26 Inferior jaw printed

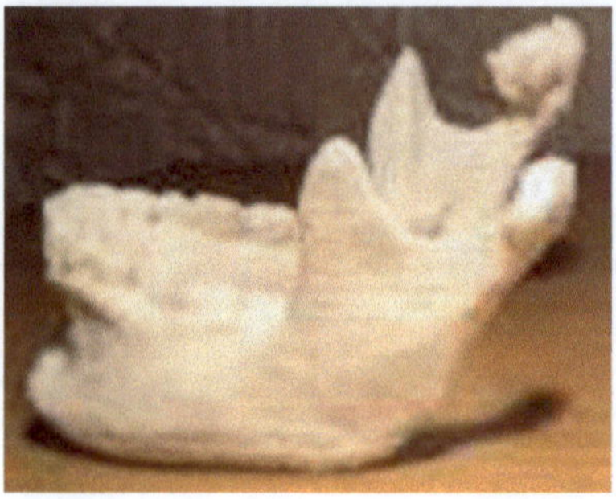

Fig. 27 Inferior jaw printed without supports or the base

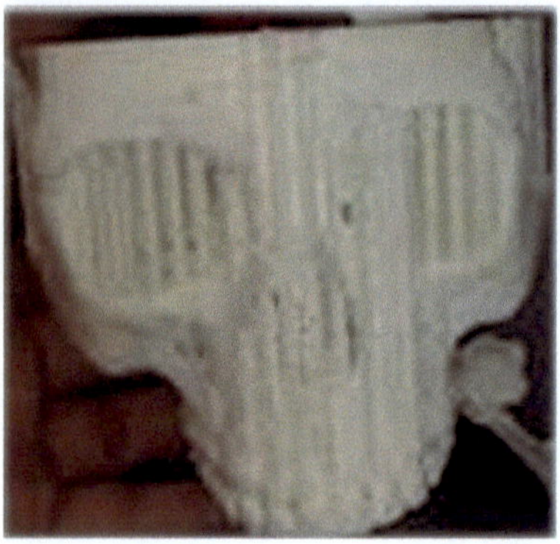

Fig. 28 3D printing of the superior jaw and skull cut with supports and base

4 3D Printing and Experimental Proposal

4.1 3D Printed Models

Considering the cutting process shown in Fig. 25, the final model of the skull and the upper jaw were printed to complete the biomodelling.

Figure 26 showed the lower jaw already printed with the base and the supports, and Fig. 27 showed the lower jaw after having removed both the base and the supports (Fig. 28).

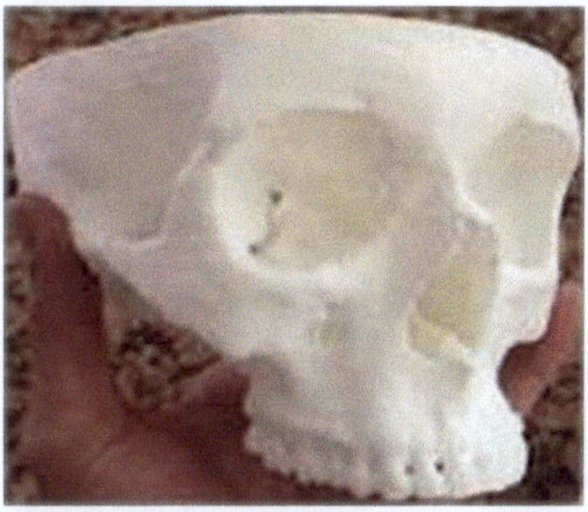

Fig. 29 Cut skull and superior without supports or base

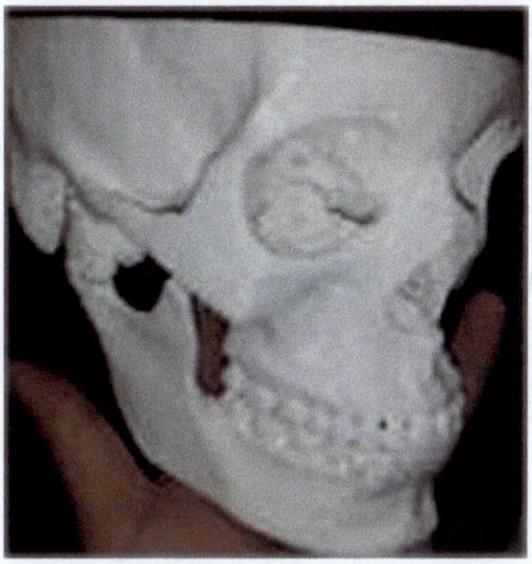

Fig. 30 Prove that both jaws fit jaw

It was necessary to remove all the supports and the base, as shown in Fig. 29. It was verified that both jaws fit correctly, as shown in Fig. 30. It means that the printing and biomodelling process is finished.

4.2 Manual Modelling

Manual modelling consisted of making different proposals for prostheses using malleable and moldable materials. These proposals are made on models of the patient's jaw. These proposals can be made using different materials taken from the 3D printing of the jaw, worked from the tomography provided by the surgeon.

The first step to perform the different proposals of prostheses was, from the 3D printing of the jaw, to make a mold of the same. This mold was subsequently used to make the jaw model with different materials, such as plaster, alginate and acrylic.

A 3D printing mold of the jaw was made. A challenge was that the jaw should not touch the edges of any container so that, when making the mold, the mold wouldn't

Fig. 31 Hanging jaw

Fig. 32 Water and alginate mixture

have any mistake. For this purpose, the mandible was supported by two threads when it was inserted into the container (Fig. 31).

The jaw mold was made with a mixture of alginate and water. 3 parts of water were used for each alginate portion. It had to be mixed vigorously continuously for about 2 min. After this time, the alginate began to harden, so it was necessary to pour the mixture into the container where the object was (Figs. 32 and 33).

Once the mixture was inside the container with the jaw printed, it was necessary to wait between 10 and 15 min for the mixture to dry and the mold passed from a liquid to a slightly more solid state. The mold was removed from the vessel, cuts were made in the mold and the jaw was removed. All this is done with extreme care not to over-break the mold, leaving it unusable (Fig. 34).

When the jaw was out of the mold, it was returned to its original position, placed back into the vessel and sealed with a little more water and alginate mixture.

At this point, they began to mix water and plaster. For this mixture, it was necessary to use 2 portions of plaster per each portion of water used. This mixture takes between 15 and 20 min to become solid, so it is necessary to work with it before this time

Fig. 33 Dry alginate mold

Fig. 34 Alginate mold with printed jaw inside

elapses. Once the mixture was ready, it was poured into the holes in the mold to fill it inside. This was done at the same time as constant light strokes were applied to the vessel to ensure that the mixture settled well inside (Fig. 35).

Finally, after 15 or 20 min for the plaster to solidify, the positive of the mandible was extracted in the same way in which the mandible was removed previously. Some details of the negative were arranged, as they could be small ruptures and excesses of material.

Taking our plaster positive, we attempted to perform different sanding processes so that the condyle was reduced to its original (functional) size, but it was discovered that the plaster is too fragile to support such procedures, so it was decided to change the material and perform another positive with acrylic (Polymethylmethacrylate) (Figs. 36 and 37).

As shown in Fig. 38, the acrylic used is one which includes liquid acrylic and its Nic Tone monomer powder from the MDC Dental brand.

Fig. 35 Plaster mixture being poured into the mold

Fig. 36 Frontal view of the jaw and negative and his negative

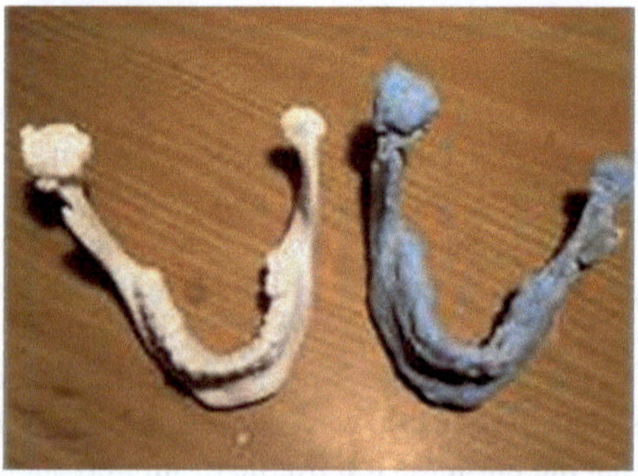

Fig. 37 Superior view of his the jaw

Fig. 38 Liquid acrylic and monomer

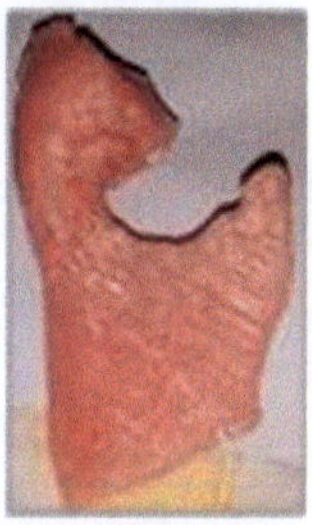

Fig. 39 Acrylic positive of the mandibular branch and the condyle

It was decided to use acrylic because of its high resistance in the molding material in order to be used in the sanding processes. The mechanical properties are shown below:

- High impact resistance
- Rockwell M hardness: 100
- Compressive resistance: 1020 kg/cm^2
- Lightweight
- Density: 1180 kg/m^3.

To perform the acrylic positive, the same procedure was used as the one used for the plaster positive, making the alginate mold and filling it, but this time with acrylic. The main difference is that greater care should be taken in mixing the powdered monomer and the liquid acrylic as it could become corrosive if the correct measurements were not used. Having performed the acrylic positive shown in Fig. 39, it was compared to the printed lower jaw to see if it had been correctly duplicated, as shown in Fig. 40.

Once the problem was fixed, our positive prosthesis was tested by the surgeon to know if he could perform the model of the prosthesis. The medical approval was important in order to define the replacement; however, the mandibular branch was short, and it was necessary to cover everything shown in Fig. 41.

A new mold was made, as shown in Fig. 42, in this case with the specifications that the surgeon advised, in order to produce the positive correctly.

At the end of the positive model, shown in Fig. 43, it was again compared to the printed jaw, as shown in Fig. 44, and it was found that this time the model that would work on the plate was obtained.

Fig. 40 Positive and original comparative

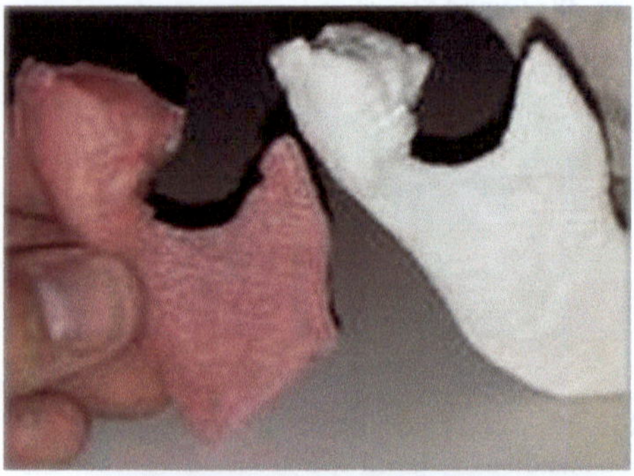

Fig. 41 Area to be duplicated

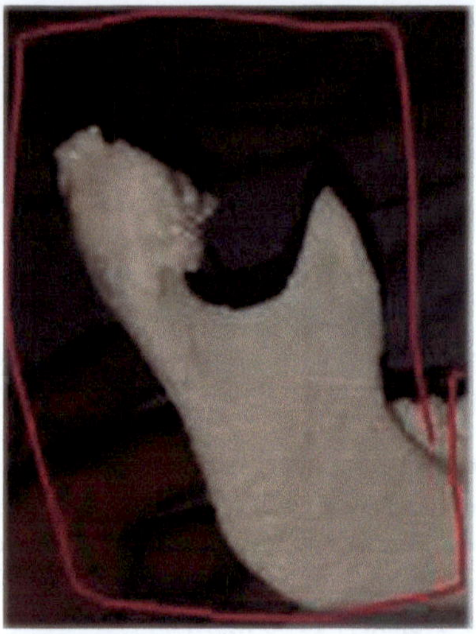

Fig. 42 Alginate mold with the damaged condyle

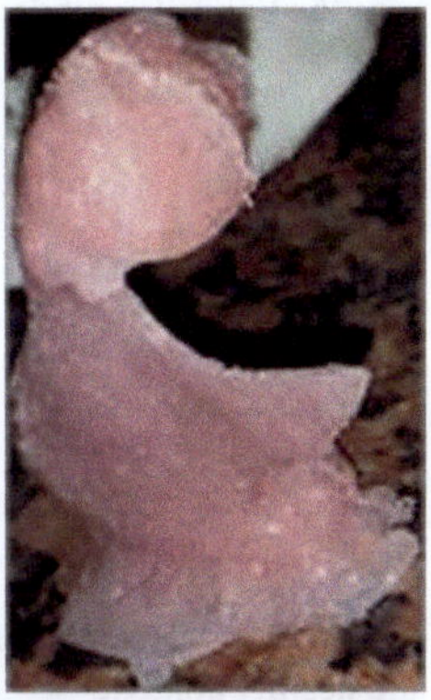

Fig. 43 Correct damaged condyle and mandibular branch model

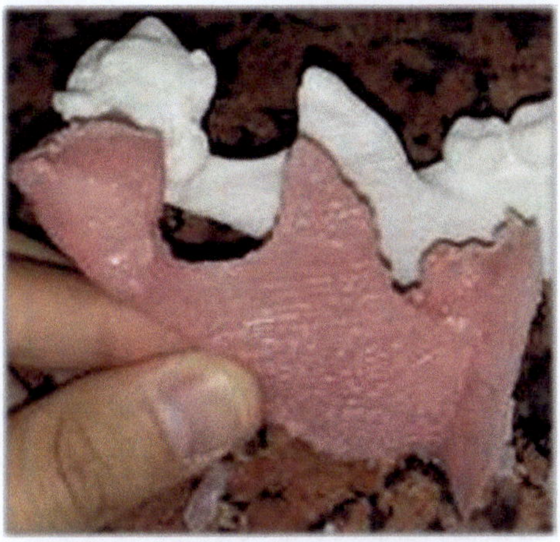

Fig. 44 Positive and correct model comparative

Fig. 45 Dremel sanding the damaged condyle

The sanding procedures were then performed so that the damaged condylar head was reduced to the original (functional) size. For this, a dremel© tool and different types of sandpaper were used, as shown in Fig. 45.

Once the finishing sanding was completed, it was possible to get a similar condyle in comparison to the original size, as shown in Figs. 46, 47 and 48.

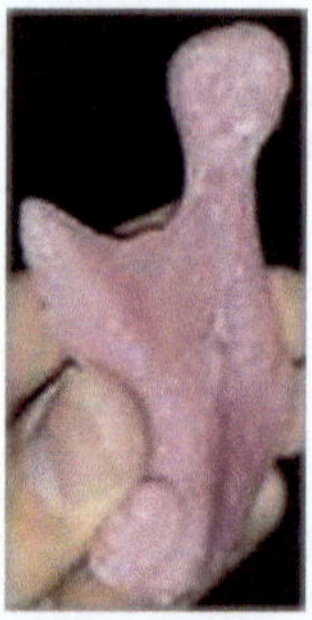

Fig. 46 Sanded condyle

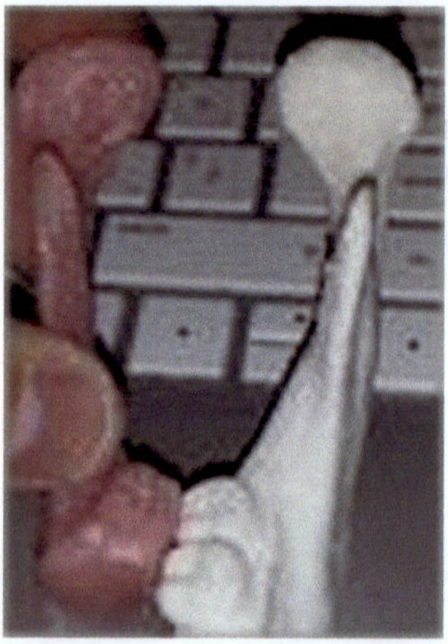

Fig. 47 Damaged condyle and healthy condyle comparative

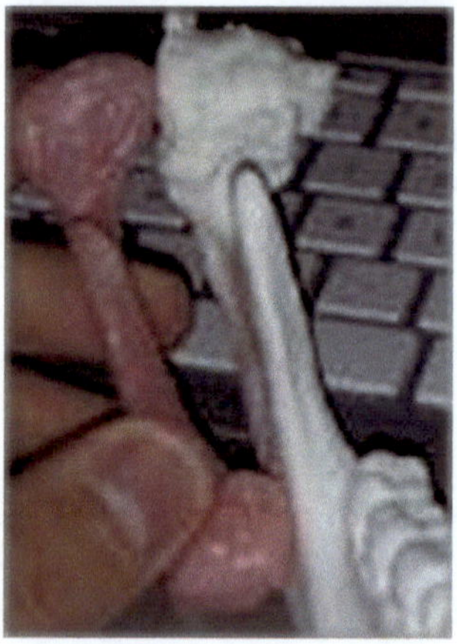

Fig. 48 Damaged condyle and sanded damaged condyle comparative

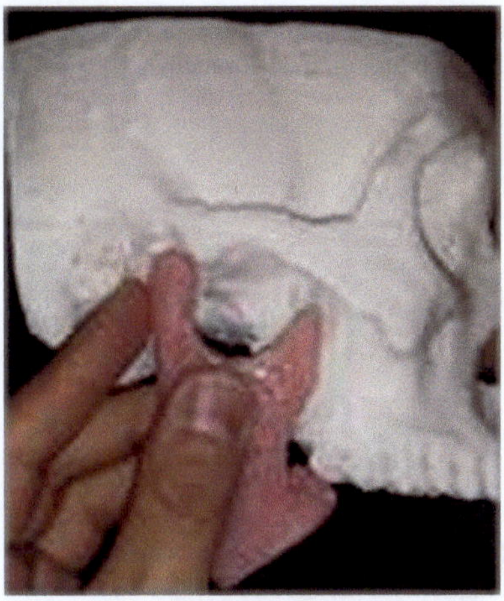

Fig. 49 Proving that the condyle fits in the superior jaw

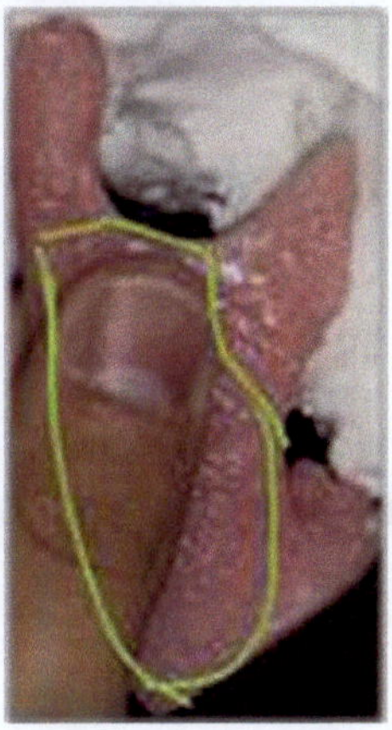

Fig. 50 Suggestion of the work area for the plaque

In order to verify the correct definition, the condyle was placed inside the upper jaw, as shown in Fig. 49.

Taking the model, the next process was divided into two parts: The first was to perform a single positive condyle already sanded and the second was to perform the plaque on the mandibular branch of the model. It was defined a work area where the plaque could be made, which is shown in Figs. 50, 51 and 52).

Using the reference area of work, it was decided to make the proposal of prosthesis in plaster rather than to make it in acrylic model and finish with manual modeling. For the proposal, it was decided to start with the condylar head, followed by the plate to attach the prosthesis to the mandible and ending with the joining of both pieces (Figs. 53, 54, 55 and 56).

When the positive was done in yeast, as shown in Figs. 57 and 58, the final prosthesis in acrylic was fabricated, as shown in Figs. 59 and 60.

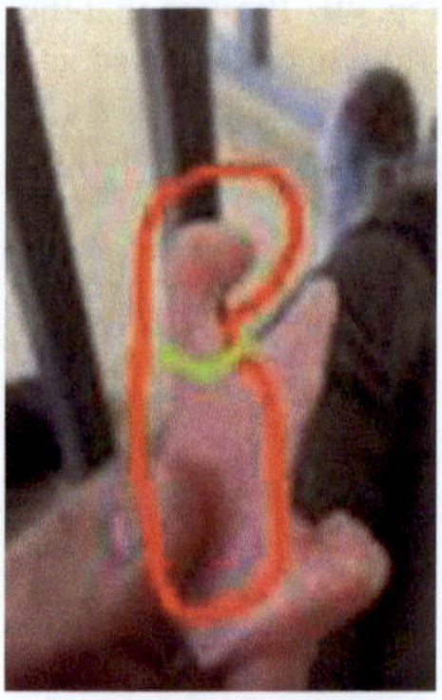

Fig. 51 Suggestion of the cut in the mandibular branch

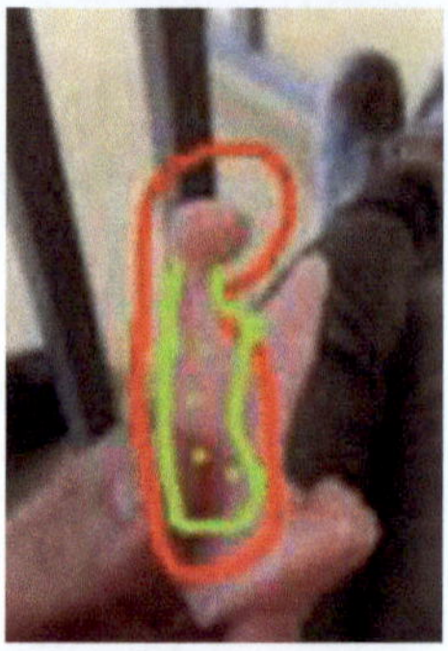

Fig. 52 Suggestion of the plaque with her screws

Fig. 53 Plaster positive of the condylar head

There were some problems with the acrylic plate, because it was necessary to prepare an alginate mold using the proposed plaster prosthesis and create a positive acrylic. It was possible to create the prosthesis (Figs. 61 and 62).

Once the prosthesis had the appropriate shape, the last prosthesis model required an analysis of the places where the holes were to be made in the plate in order to support the model of the prosthesis to our printed jaw. A proposal was sent to the

Fig. 54 Back view

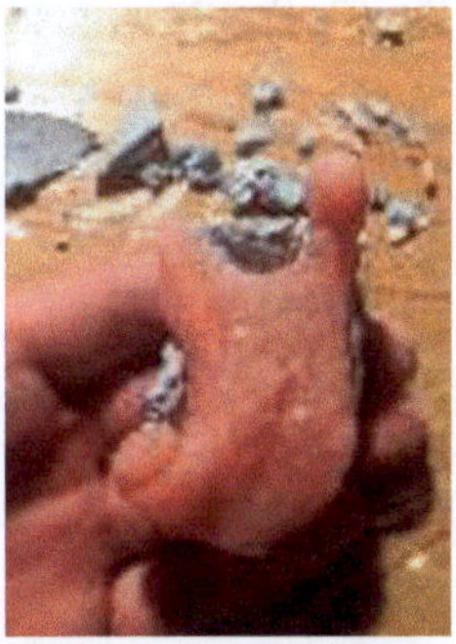

Fig. 55 Frontal view

Fig. 56 Condylar head and plaque

Fig. 57 Prosthesis proposal

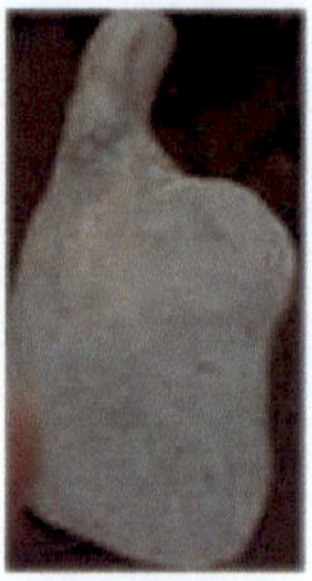
Fig. 58 Prosthesis proposal

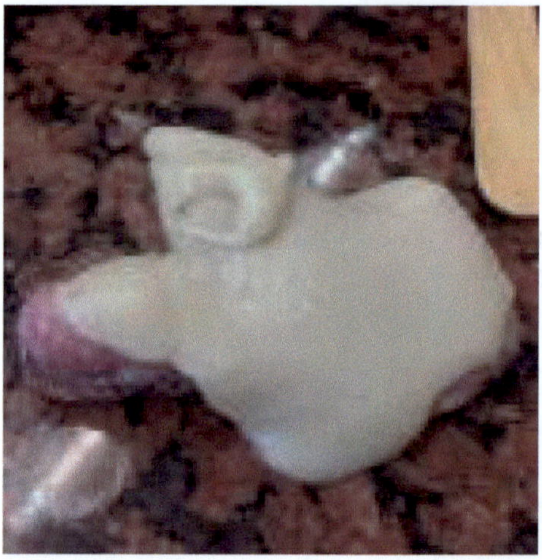
Fig. 59 Acrylic plaque

surgeon (Fig. 63) which he modified (Fig. 64). Finally, the final proposal of the prosthesis model was obtained, which can be seen in Fig. 65.

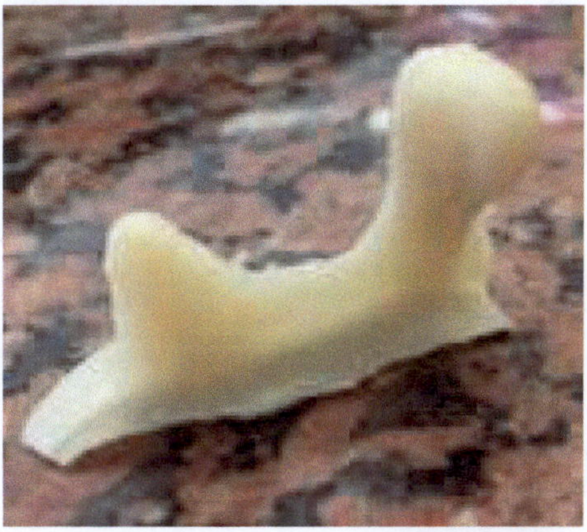

Fig. 60 Acrylic condylar head

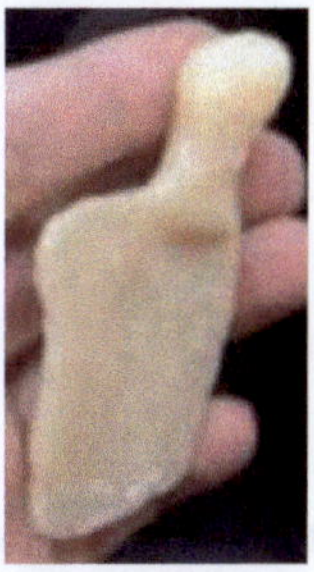

Fig. 61 Acrylic prosthesis proposal

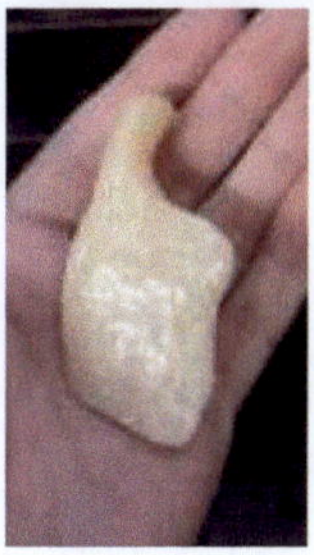

Fig. 62 Acrylic prosthesis proposal

5 Mastication Forces Investigation

In order to know the main forces involved for the numerical simulation, a background about the topic was included.

Fig. 63 Prosthesis proposal

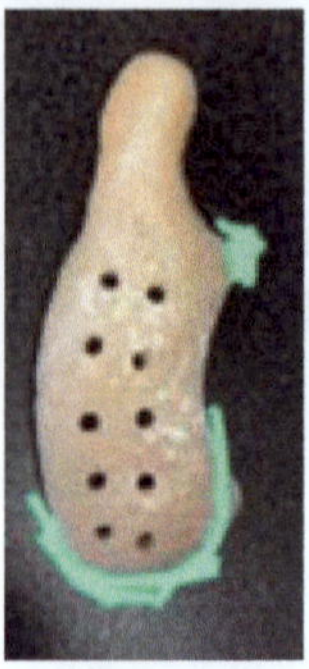

Fig. 64 Surgeon's modifications

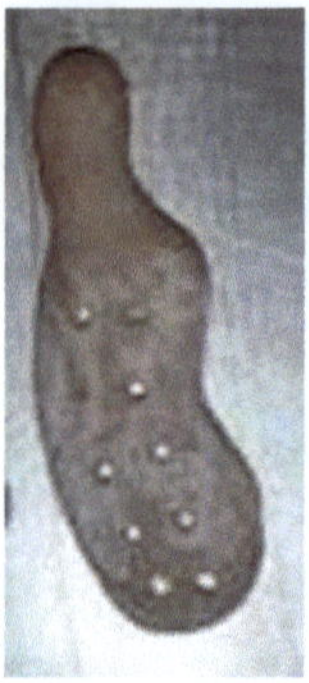

Fig. 65 Final prosthesis proposal

5.1 Mastication Muscles Anatomy

The mastication muscles are a group of muscles that help to perform the movements of the mandible or the TMJ (temporomandibular joint). It is one of the main muscle groups of the head. There is another group that focuses on the muscles of facial expressions.

There are four main muscles that make up the mastication group: the masseter and temporal muscles, shown in Fig. 67, and the medial (internal) pterygoid and lateral (external) pterygoid muscles, shown in Fig. 66.

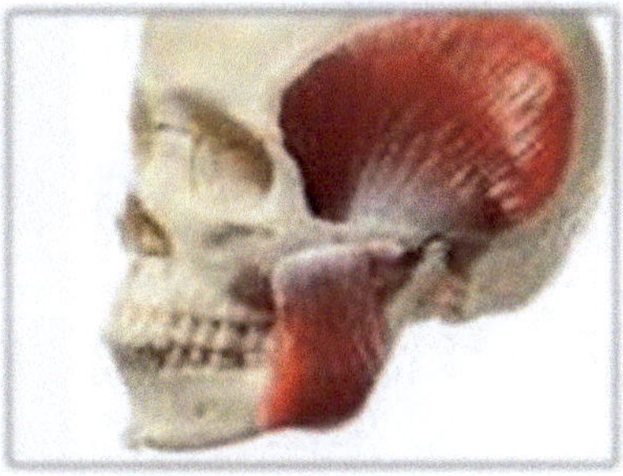

Fig. 66 Internal and external pterygoid muscles

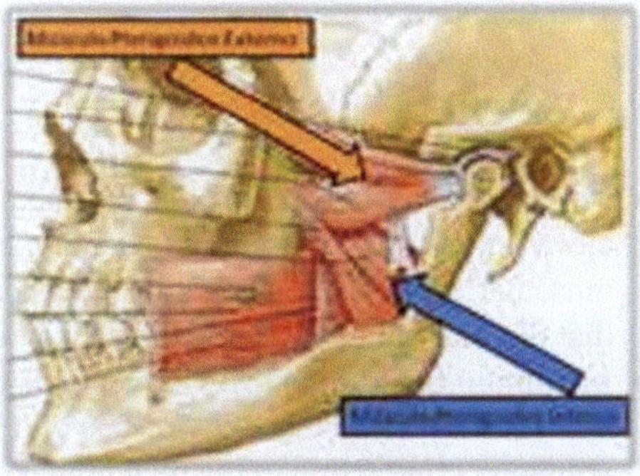

Fig. 67 Temporal and masseter muscles

Also, supra and infrahyoid muscles are involved, but they do not usually have as much impact on the mandible.

- Temporal: It is a strong lever muscle of the mandible located to each side of the head and it is inside the temporary hole above the zygomatic arch and it extends in the form of a fan. When contracted, it raises the jaw and also directs it back.
- Masseter: It is the most superficial of the chewing and palpable muscles when the jaw is closed tightly. It is a large, rectangular muscle. It has two fascicles, one superficial and one deep. When contracted simultaneously, it elevates the jaw [3].
- Internal Pterygoid: It inserts superiorly on the inner side of the pterygoid process. It is a lever muscle of the mandible and due to its position provides small lateral movements.
- External Pterygoid: Extends from the pterygoid process to the neck of the condyle of the mandible. The simultaneous contraction of both external pterygoids produces forward projection movements of the mandible. If contracted in isolation, the jaw executes lateral movements to either side.

There are four main muscles which also help perform the chewing function. These muscles are suprahyoid (digastric, stylohyoid, mylohyoid, genohyoid) and infrahyoids (sternohyoid, omohyoid, sternohyoid, thyrohyoid, buccinator).

The chewing force exerts where it is shown in Fig. 68, but it is divided in 2:

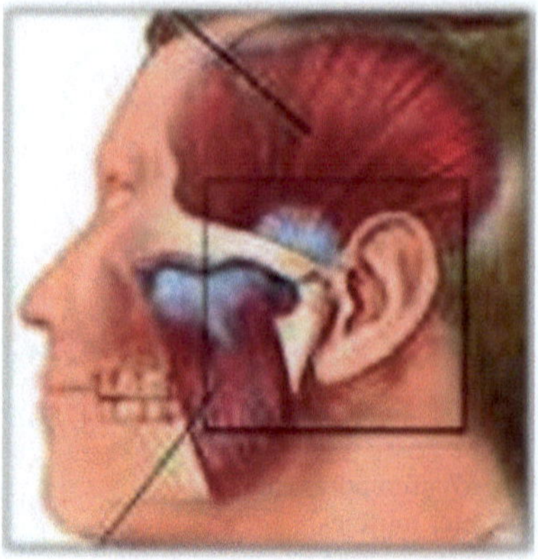

Fig. 68 Mastication forces

- Maximum anatomical FM: It is the maximum contractile power of the mandibular elevating muscles. Based on physiological data, taking into account that 1 cm of muscular cross-section develops between 5 and 12 kg of force, it was calculated that the maximum anatomical masticatory force can be between 210 and 300 kg.
- Maximum functional FM: Corresponds to the force measured between upper and lower jaw. It is calculated by means of a voltage transducer. During the maximum voluntary contraction of the mandibular elevating muscles, values between 60 and 70 kg were recorded. The maximum functional masticatory force represents only a fraction of the maximum anatomical masticatory force.

Different values have been found for the chewing force. Some standard values are: in young adult males, an average value of 727 N (74.15 kg), and for children a value of 425 N (43.35 11 kg) has been reported, while for young women a value of 370 N (37.74 kg); in healthy children with temporal dentition a value of 186.20 N (18.99 kg) and in older adults a value of 181 N (18.46 kg).

The range of values of the normal masticatory process fluctuates between 15 and 25 kg which correspond to 25–35% of the maximum functional masticatory force [4].

6 Model Digitalization

The first attempt was made with a "Kinect" tool used in current video game systems, as shown in Fig. 69. This tool worked with a computer program called "Skanect".

There were different problems when working with both tools. First, the "Kinect" tool had many problems scanning the surroundings, shown in Fig. 70. It failed to detect the model of the prosthesis accurately.

Similarly, the computer program required certain specifications and initial conditions on what was planned to scan, such as area, perimeter, body, and size. Despite trying different options, none managed to function properly.

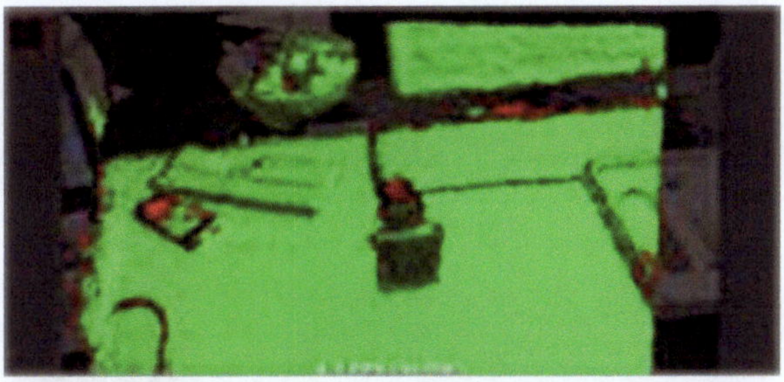

Fig. 69 Kinect tool and Skanect program

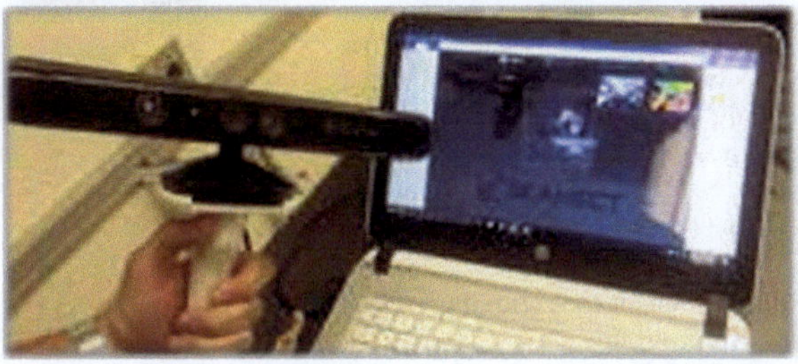

Fig. 70 View of the scanning test using Skanect

After several attempts, and when it came to the conclusion that nothing was being achieved with this method of digitization, it was decided to use a different scan tool, as well as another computer program.

The first scanner used a tool which was called "Sensor" and the computer program it used was called "SENSE". At the beginning of the tests there was a marked improvement in the way of scanning the surroundings, as shown in Figs. 71 and 72, but new problems arose when the piece was digitized. The program could not clearly detect the piece and the computer that was used continued to close the program, which made it impossible to conclude the scanning of the prosthesis in time.

At the end of both tests, the most acceptable result was not close to the required precision. It means that it was required to perform the force simulation in SolidWorks. The scanning results can be seen in Figs. 73 and 74.

An alternative method consisted in taking photographs of the model of the prosthesis from different angles (frontal, shown in Fig. 75, and lateral, shown in Fig. 76).

The pictures were transferred to PTC Creo Parametric program in order to apply CAD modelling using planes and swept commands. The contour and holes are then drawn to this image, and then extruded to both sides, as shown in Fig. 77.

Next, the side picture was placed to complete the digitization of the model. This time the previous operation was repeated to draw the outline of the prosthesis, but also a rectangle was drawn around so that, instead of extruding new material, the surplus material would be removed, as shown in Fig. 78.

Finishing the extrusion, a digitized prosthesis was obtained, as shown in Figs. 79 and 80, concluding the digitalization stage of the model.

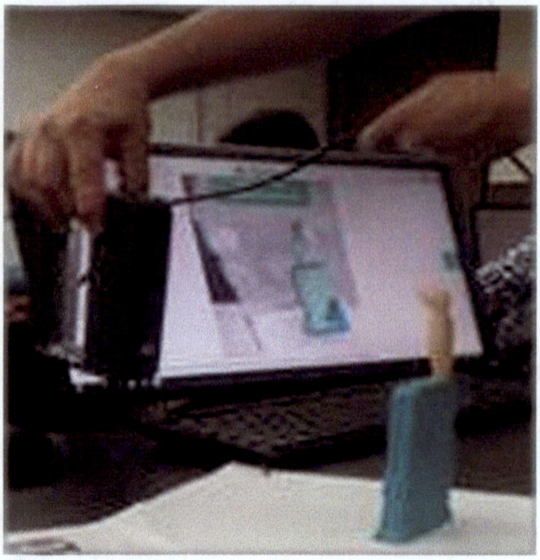

Fig. 71 Test scanning with the program SENSE

Fig. 72 Sensor tool with the SENSE program

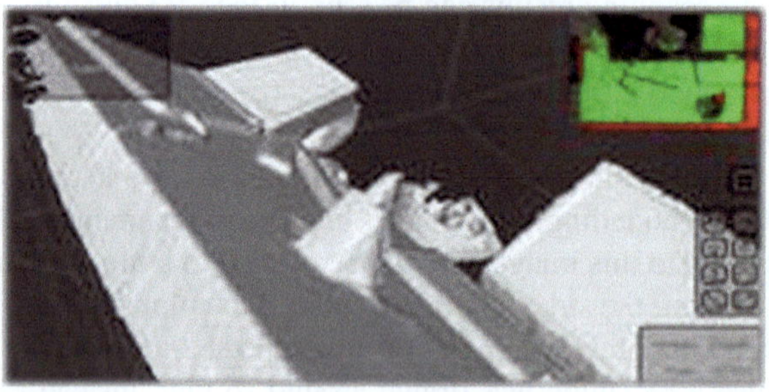

Fig. 73 Scanning the background

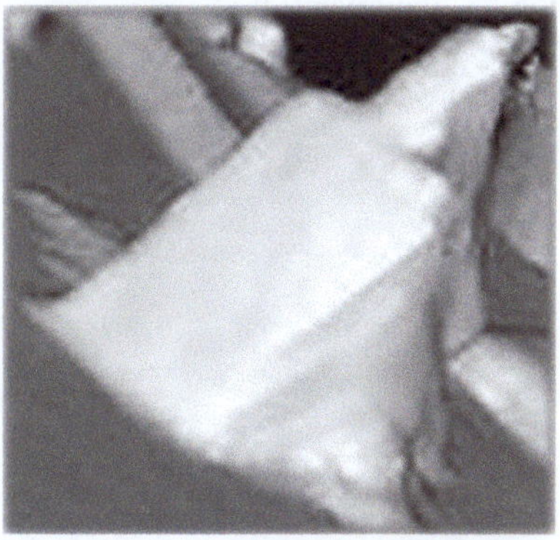

Fig. 74 Prosthesis model scanned

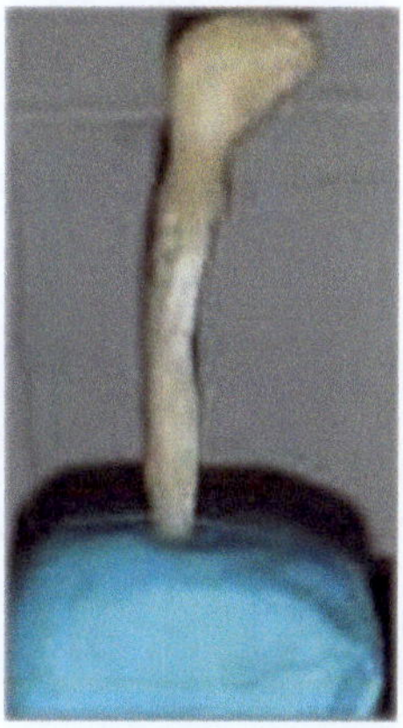

Fig. 75 Frontal photography photography of the model

Fig. 76 Lateral photography of the model

Fig. 77 Extrusion of the front face of the model

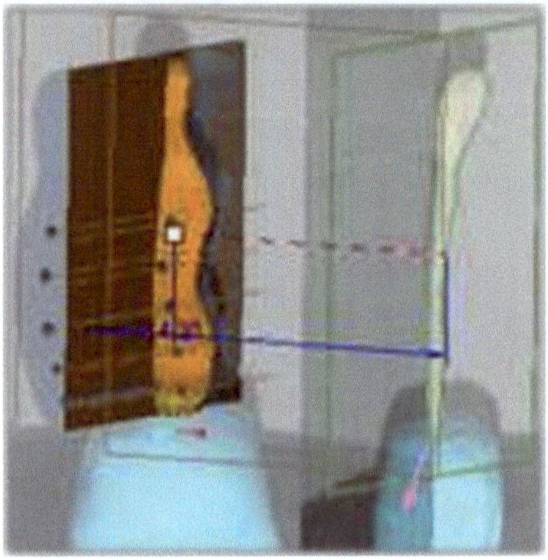

Fig. 78 Removal of material from the side face

7 Numerical Testing

7.1 Finite Element Method

It is a general numerical method for the approximation of solutions of partial differential equations. Converts a problem defined in terms of differential equations into a problem in matrix form that provides the correct result for a finite number of points and then interpolates the solution to the rest of the domain, ultimately resulting in only an approximate solution. The set of points where the solution is exact is called a set of nodes.

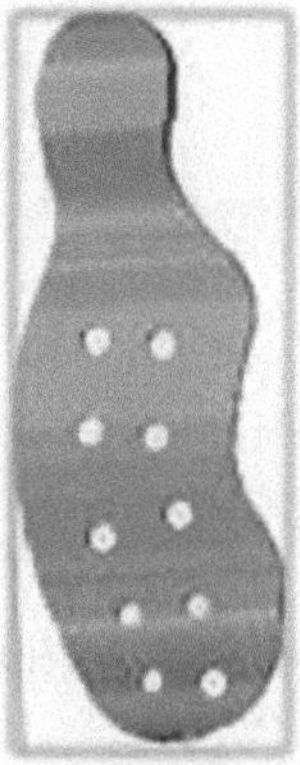

Fig. 79 Front view of the scanned model

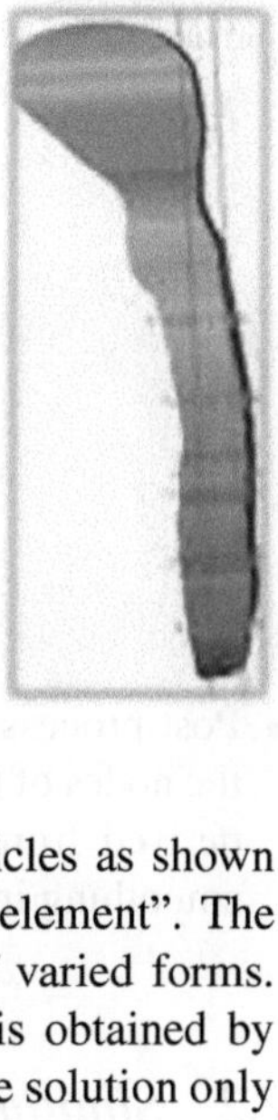

Fig. 80 Side view of the scanned model

This set of nodes, form a network called a mesh formed by reticles as shown in Fig. 81. Each of the reticles contained in said mesh is a "finite element". The set of nodes is obtained by dividing the structure into elements of varied forms. In the rest of points that are not nodes, the approximate solution is obtained by interpolating from the results obtained for the nodes, which makes the solution only approximate.

Discretization of a region is used on which the equations are defined in simple geometric forms called finite elements. The material properties and governing relations in these elements are expressed in terms of the unknown values at the corners of the elements or nodes.

Finite elements are divided into:

- Preprocessing: Definition of geometry, mesh generation, boundary conditions and assignment of properties to materials and other properties.
- Calculation: The result of the preprocess, in a simple non-time dependent problem, allows the generation of a set of N equations and N unknowns, which can be solved with any algorithm for solving systems of linear equations.

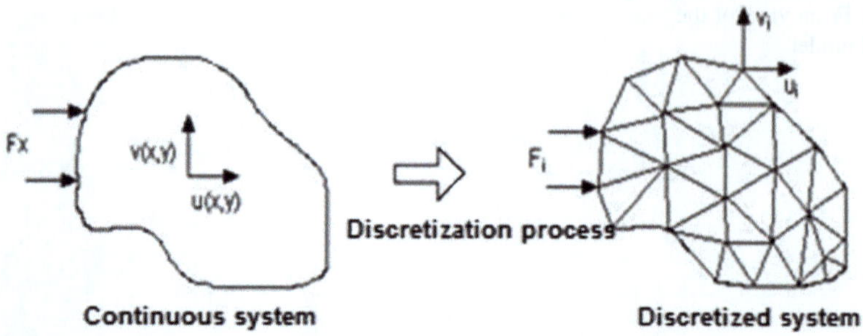

Fig. 81 Discretization by finite element method

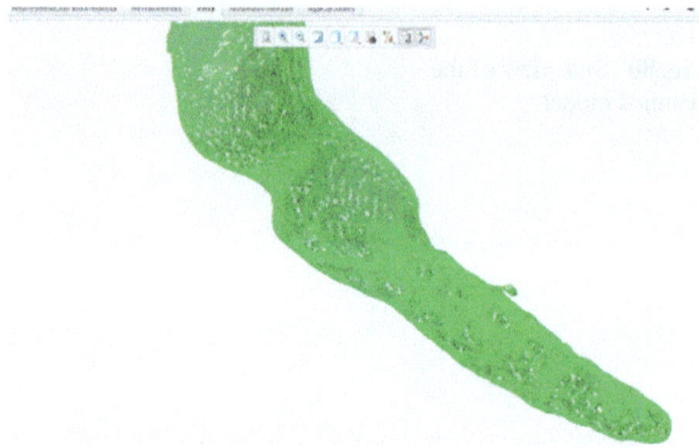

Fig. 82 Model of the prosthesis in the Creo Simulate tool

- Post-processing: The calculation provides values of a certain set of functions in the nodes of the mesh that defines the discretization. In post-processing, the values derived from the values obtained for the nodes are calculated and sometimes smoothing interpolation and even determination of approximation errors. [1]

7.2 *Simulation of Forces in PTC Creo Parametric ©*

Using the model of the prosthesis, the simulation of forces was continued using the analysis of finite elements. Originally, the SolidWorks tool would be used, but our consultant indicated that it would be better to switch to the Creo tool. A performed scanned model was developed and a solid with 841,017 Tethraedrical elements was used for a FEM analysis.

The prosthesis model was uploaded to the Creo Simulate software, as shown in Fig. 82.

In order to perform the test, it was necessary to establish three parameters:

- Strength: 784.8 N (80 kg.) Applied to the top of the condylar head. This number is based on previous research on the masticatory forces of the person. The maximum

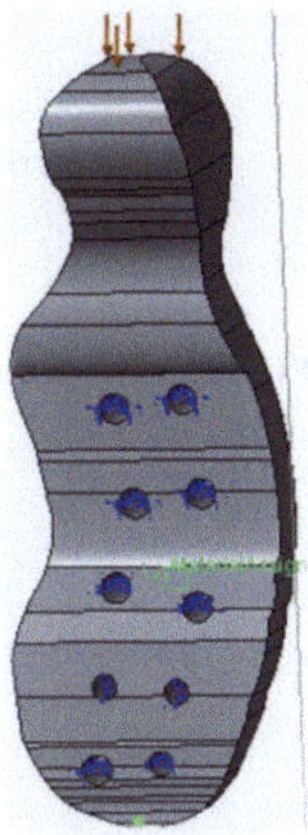

Fig. 83 Model ready to simulate forces

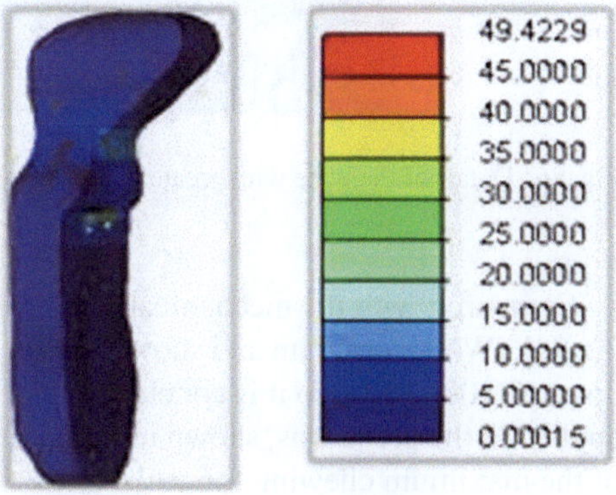

Fig. 84 Stress analysis (MPa)

force is 70 kg. But it was decided to apply an additional 10 kg to ensure the strength of the prosthesis.

- Displacement: applied in the holes of the plate where the screws would be placed because in that area is where the plate would join, with the screws, to the mandibular branch preventing the movement of the plate.
- Material: 316L stainless steel.
  With these parameters, the simulation was run in 4 different ways, as shown in Fig. 83.

Each of the 4 analyses will be explained, in order to compare these results with those obtained in the simulation of the press.

7.2.1 Stress Analysis

Figure 84 shows how the chewing force acts on the prosthesis. The numbers show that the von Mises stress is 49 MPa.

Mechanical properties	Tensile strength: 582 - 591 MPa
	Yield strength: 300 - 331 MPa
	Elongation: 53%
	Hardness: 170 BHN
	Reduction of area: 73%

Fig. 85 Mechanical properties of 316L stainless steel

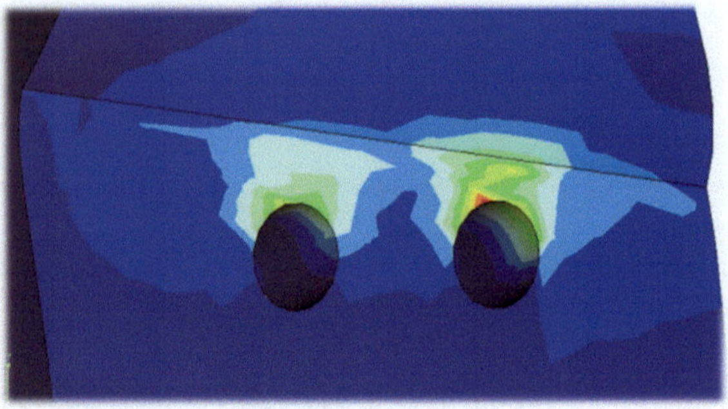

Fig. 86 Detail of the zone with greater stress concentration

Compared with the mechanical properties of 316L stainless steel, obtained from the SolidWorks program and shown in Fig. 85, it is shown how this steel can support up to 170,000 MPa, so it is concluded that, even in the area where the stress is most applied in the prosthesis, shown in Fig. 86, this model will not be affected by the use of the maximum chewing force.

7.2.2 Stress Analysis with Motion

In this analysis is possible to observe displacements of the prosthesis with this load, a movement which is shown represented in Fig. 87. Likewise, this analysis shows that the prosthesis moves adequately, and it is not expected any type of fracture.

7.2.3 Maximum Shear Stress

Figure 88 shows the analysis of maximum shear stress. PTC Creo Simulate was considered for each numerical study.

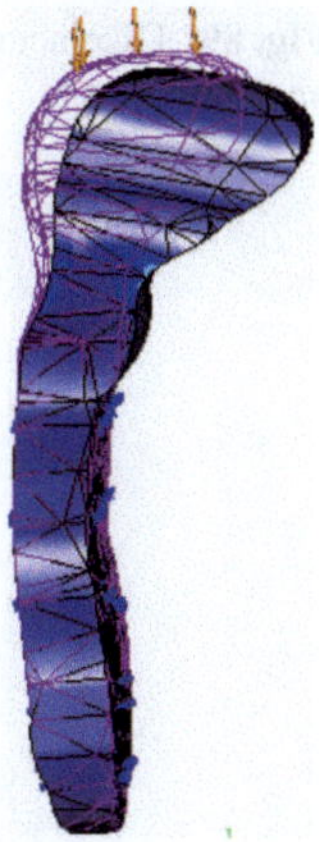

Fig. 87 Stress analysis with motion

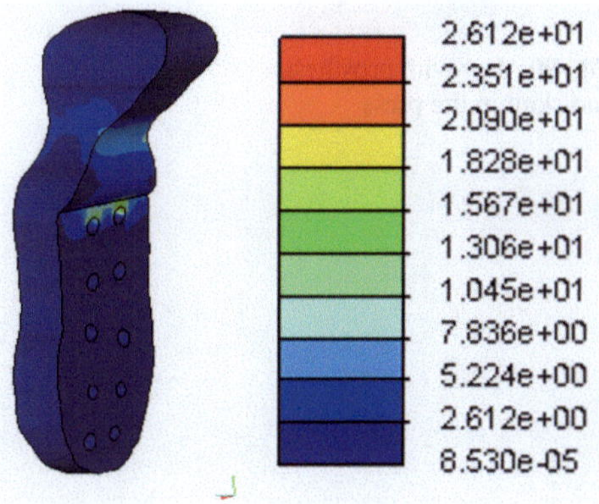

Fig. 88 Maximum shear stress analysis (MPa)

7.2.4 Displacement

In this analysis the point where the displacement is located in the head of the condyle can be seen. It is possible to see in the upper part of the condylar head that 0.011 mm, so it can also be concluded that displacement is not a problem for the prosthesis, as shown in Fig. 89.

7.3 Experimental Testing Using Image Correlation

In order to compare the numerical behavior and the results of the force simulation, it was necessary to perform the fitness calculations by using an experimental press. For this it was necessary to use the model of the prosthesis endorsed by the surgeon and the mandible printed in 3D of the patient.

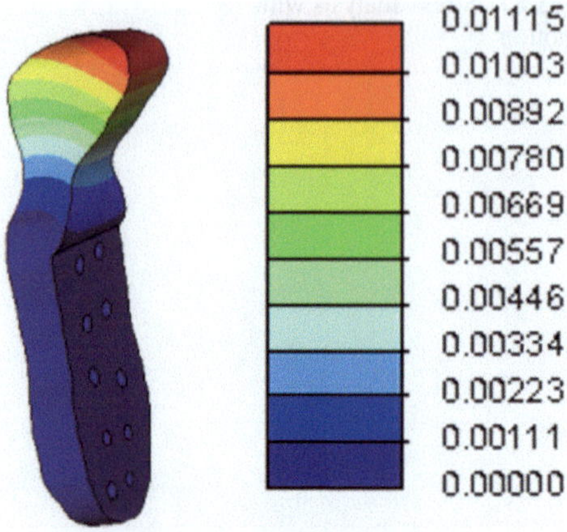

Fig. 89 Displacement analysis (mm)

Fig. 90 Jaw with prosthesis and skull in the press

To do the experiment, the model of the prosthesis was attached to the printed jaw simulating the implantation of the prosthesis in the patient. For this purpose, 5 screws of 2.4 mm in diameter were placed in different holes of the plate of the prosthesis which would be attached to the patient's jaw, as shown in Fig. 90.

The calculation of the forces obtained by the press required to film the jaw while exerting a pressure and the templates upload the file to the computer program called Aramis ©. This program analyzed each frame of the video and it was possible to obtain and analyze the deformation of both, the prosthetic plate and the jaw and skull.

For this method it was necessary to apply a stochastic pattern in the jaw, which allowed the Aramis system to do the recognition, as shown in Fig. 91, and obtain N

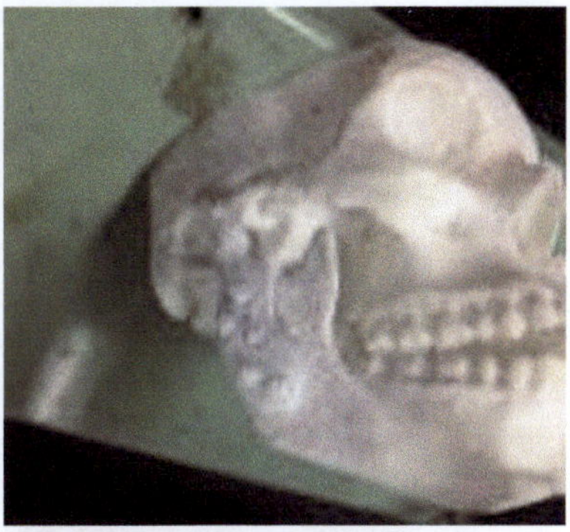

Fig. 91 Jaw, skull and model of mottled prostheses with black color

Fig. 92 Video capture of the force calculation by means of the press

number of points in our prosthesis which would be analyzed by the program Aramis, calculating the deformation between them.

The press simulates for every millimeter that advances. As the chewing force is 80 kg, the conversion was concluded that the press should be 0.0029 mm, an extremely low amount, so that the minimum force was applied to the test.

Once different press tests were made on the jaw and prosthesis, the most accurate video was taken with different angles and approaches, as shown in Fig. 92. Similarly, different photos of the mandible were taken. All these files were transferred to the Aramis © software. Once the files were transferred to the program, it was necessary to create the calculations using the software.

The first step considered to export one of the videos of the press to Aramis ©. The method consists of selecting the area to be analyzed so that the program detects the N numbers of points provided by the mottled with the painting. Once these points were analyzed, Aramis © reported the analysis of each of these tables with the following chart in the video observing the movement of the points and calculating the deformation of the pieces. In order to do this, it was necessary to give initial conditions

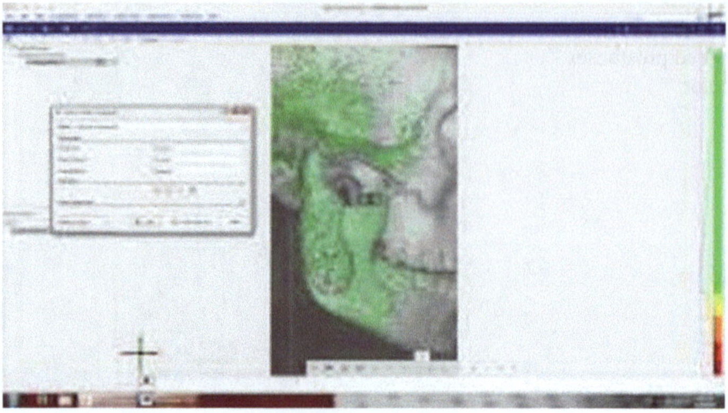

Fig. 93 Selecting the area to be analyzed

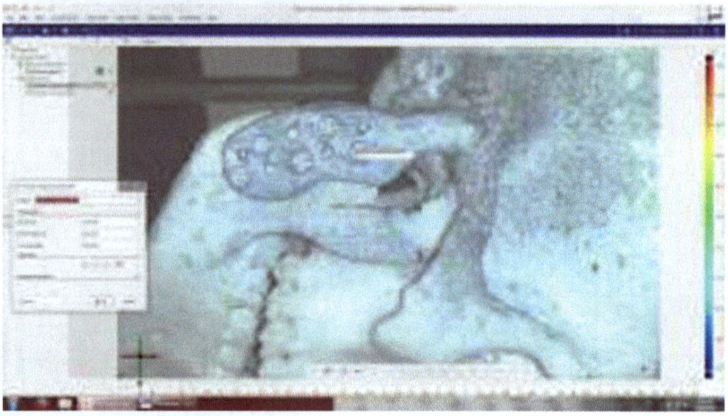

Fig. 94 Detected points in 3D models

to the program, such as approximate number of points, area to be analyzed, original size of the pieces, distances, among others. Once Aramis analyzed all the pictures, a comparison of provided visual results are shown, as in Figs. 93, 94, 95, 96 and 97.

In Fig. 97, it can be observed how the red zone of the condylar head fits with the analysis made in Fig. 89 about the low displacement that occurs in the mastication force.

In addition, in Fig. 96, the time is observed in which the prosthesis is subjected to the highest load. Once the analysis of the results was done, it is concluded that the displacements were lowest.

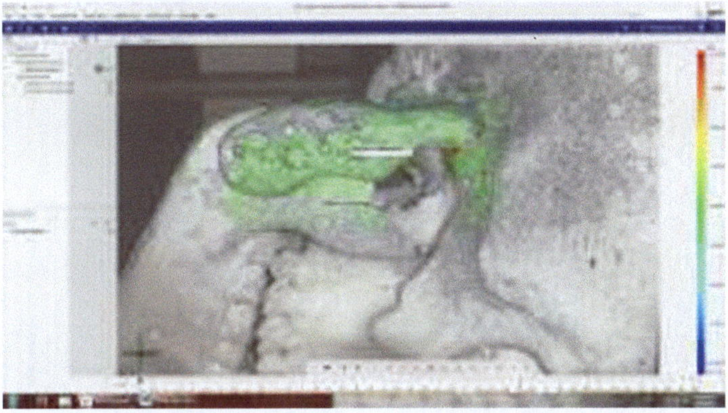

Fig. 95 Comparison image 1

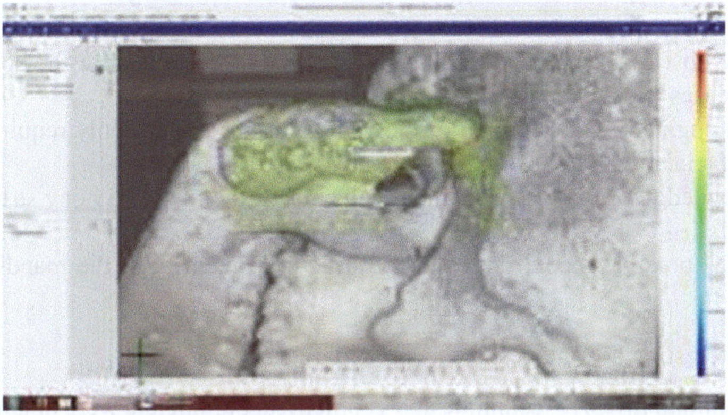

Fig. 96 Comparison image 2

8 Conclusions

- Medical and engineering knowledge has been focused on projects with the finality to reduce the total cost of unadjusted systems for Latin-American patients [5–7].
- Individuals, who are involved in congenital problems or different traumatism treatments, must return to his normal activities, with the best quality of life.
- It was possible to design and simulate a temporomandibular joint prosthesis for a patient with ankylosis in order to recover their mobility and physical appearance.
- Once the simulation of forces could be compared with the simulation of the bite in the press, it was possible get the high resistance in accordance to the designed prosthesis.
- The applied procedure depends on the conditions of the damaged area and the correct alignment of the patient during the tomography session. It is important because of the processing of the models with DICOM files is still a good alternative.

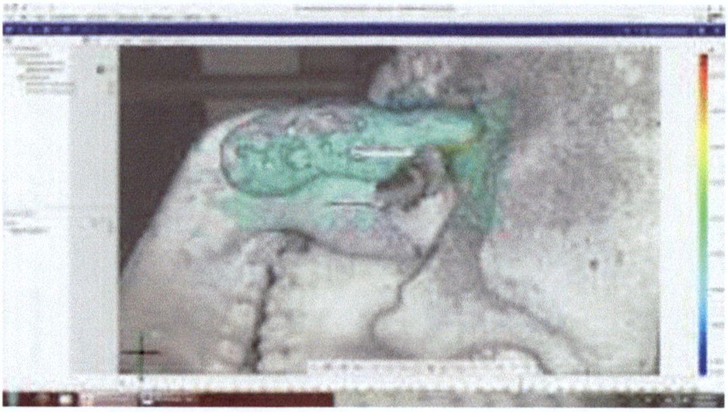

Fig. 97 Comparison image 3

- The low cost of this procedure and its simplicity contributes in the solution of similar cases (Socket and condylar implant).
- Modern engineering evaluation of medical solutions are not only required in the maxillofacial area, but also in the orthopaedic area.
- The knowledge of the biomechanical behavior of the involved bony structures is required.
- The ATM prosthesis contributes in the structural integrity of the mandible zone [8, 9].

Finally, interesting advantages of this procedure can be listed:

- It is a minimal invasive technique and geometrical parameters were obtained from the tomography and adjusted in the 3D quick printing models directly.
- A postsurgical tomography could be studied, in order to verify all the components (socket, screws and ramus).
- Testing of the material as well as mechanical analysis of the design before clinical use is essential for the prosthesis to display optimal results and in order to avoid secondary surgeries.

Acknowledgements The authors acknowledge the support given by the National Council for Science and Technology, CONACYT, Mexico; the National Polytechnic Institute for the means and facilities for the development of this research.

References

1. Aceros y metales Cuautitlán.: Acero inoxidable austenítico 316L Aceros y Metales http://www.acerosymetalescuautitlan.com.mx/catalogo/acero-inodixable/109-acero-inoxidable-austenitico-316l.html (2010). Accessed 18 Sept 2016

2. de Weck, O., Yong, K.I.: Finite element method. Retrieved from http://web.mit.edu/16.810/www/16.810_L4_CAE.pdf
3. Converse, J.M.: Anquilosis condilar mandibular unilateral Acta Odontológica. http://www.actaodontologica.com/ediciones/2008/2/anquilosis_condilar_mandibular_unilateral.asp (2005). Accessed 13 Sept 2016
4. Clínicas Propdental.: Materiales de los implantes dentales Propdental. https://www.propdental.es/implante-dental/por-que-son-de-titanio/ (2013). Accessed 17 Sept 2016
5. Beltrán-Fernández, J., Martínez-Paredes, J., González-Rebattú, M., Hernández-Gómez, L., Ruíz- Muñoz, O.: Customization and numerical simulation of a cranial distractor using computed axial tomography (CAT). In: Properties And Characterization of Modern Materials, pp. 371–398. http://dx.doi.org/10.1007/978-981-10-1602-8_30 (2016)
6. Beltrán-Fernández, J., Garibaldi, P., Saucedo, F., Escalante, E., Hernández-Gómez, L., Calderón, G., Camacho, N.: Design and manufacturing of prosthesis of a jaw for a young patient with articular ankylosis. In: Advanced Structured Materials, pp. 73–87. http://dx.doi.org/10.1007/978-3-319-19470-7_6 (2015)
7. Martínez-Paredes, J., Beltran-Fernández, J.A., Hernandez Gómez, L.H., Gonzales-Rebatú, M., Rangel-Elizalde, A.I.: Reconstruccion de fosa ocular con prótesis de hidroxiapatita y malla de titanio. Journal De Ciencia E Ingeniería **6**(1), 53–60 (2014)
8. Frank, H.: Mandíbula Aula de Anatomía. http://www.auladeanatomia.com/novosite/sistemas/sistema-esqueletico/cranio/esqueleto-da-face/mandibula/ (2000). Accessed 13 Sept 2016
9. Laura, A.: Prótesis total de ATM bilateral ACMCB. http://www.acmcb.es/files/425–5935-DOCUMENT/Aguilera-45-28Gen14.pdf (2009). Accessed 17 Sept 2016

Evaluation of Hot Corrosion Behavior of APS and HVOF Sprayed Thermal Barrier Coatings (TBCs) Exposed to Molten $Na_2SO_4 + V_2O_5$ Salt at 1000 °C

Mustafa Kaplan, Mesut Uyaner, Yasin Ozgurluk, Kadir Mert Doleker and Abdullah Cahit Karaoglanli

Abstract Hot corrosion is a very destructive failure mechanism for thermal barrier coatings (TBCs) during service conditions. In the present study, a CoNiCrAlY is coated using atmospheric plasma spray (APS) and high velocity oxy-fuel (HVOF) techniques which produces TBCs that were exposed to 50% Na_2SO_4 and 50% V_2O_5 molten salts at 1000 °C. The top surface was visually inspected at the end of each four-hour cycle to determine the lifetime of TBCs. The failure criteria required for the termination of the hot corrosion cycles was assumed to be 25% cracking and spallation of top coats. X-Ray Diffraction (XRD) analysis was performed before and after hot corrosion tests in order to observe new phases which may occur in the top coating. After scanning electron microscopy (SEM) images of cross-section samples of hot corrosion tests are completed, samples were taken. As a result, it has been observed that the lifetimes of HVOF bond-coated TBCs are longer.

Keywords Thermal barrier coatings (TBCs) · Atmospheric plasma spray (APS) High velocity oxy-fuel (HVOF) · Hot corrosion · $Na_2so_4 + V_2O_5$ salt

---

M. Kaplan
Department of Metallurgical and Materials Engineering, Graduate School of Natural and Applied Sciences, Selcuk University, 42003 Konya, Turkey
e-mail: mkaplan5442@gmail.com

M. Uyaner
Department of Aeronautical Engineering, Faculty of Aeronautics and Astronautics, Necmettin Erbakan University, 42090 Konya, Turkey
e-mail: muyaner@konya.edu.tr

Y. Ozgurluk (✉) · K. M. Doleker · A. C. Karaoglanli
Department of Metallurgical and Materials Engineering, Bartin University, 74100 Bartin, Turkey
e-mail: ozgurlukyasin@gmail.com

K. M. Doleker
e-mail: mertdoleker@gmail.com

A. C. Karaoglanli
e-mail: cahitkaraoglanli@gmail.com

© Springer International Publishing AG, part of Springer Nature 2019

A. Öchsner and H. Altenbach (eds.), *Engineering Design Applications*, Advanced Structured Materials 92, https://doi.org/10.1007/978-3-319-79005-3_28

1 Introduction

Thermal barrier coatings (TBCs) are widely used to provide thermal insulation to the hot section components of gas turbines and aero engines in order to protect them from thermal degradation and also increase efficiency and durability properties of these components [1–3]. A typical TBC consists of a superalloy substrate that a MCrAlY bond coat is applied onto together with a ceramic top coat of 6–8 wt% yttria partially stabilized zirconia (YSZ). The bond coating provides high temperature oxidation and corrosion resistance, while the top coat offers thermal insulation at high operating temperatures. Many factors influence the lifetime of TBCs during service. Among them, the strength of the bond coat layer against failures such as oxidation and hot corrosion is the critical role of this layer in TBCs. When the TBC is exposed by oxidation with high temperature, oxide formation occurs at the top coat and bond coat interface which is called a thermally grown oxide (TGO) [1, 4–6].

A TGO is formed during the partial deposition of however it usually starts to grow due to the high temperature oxidation between bond coat and ceramic top coat. After that the TGO leads to spallation of the top coat from the bond coat. In TBC systems, spallation arising from TGO is observed as the most common failure. Thus, the growth of TGO leads to degradation and failure of the coating. Furthermore, the thermal expansion mismatch between the bond coat and the ceramic topcoat causes the coating to fail during cooling and heating. The ceramic top coat has a low thermal conductivity and partially low thermal expansion. The metallic bond coats are deposited between the metallic substrate and ceramic top coat for improving adhesion [4, 7, 8].

Another failure mechanism in TBC systems are the one that occur from topcoat phase transformation. Phase transformation and cracking of the top coat has a major impact on the lifetime of the TBCs. At higher operation temperatures, the ceramic top coat undergoes phase transformation from a tetragonal to monoclinic form.

In this study, Atmospheric Plasma Spray (APS) and High Velocity Oxy-Fuel (HVOF) techniques have been used to deposit CoNiCrAlY based metallic bond coats on nickel-based Inconel 718 superalloy substrates. The APS method was only used for deposition of ceramic top coatings with YSZ content. The main purpose of the study is the evaluation of failure mechanisms and lifetime of TBC systems produced by using bond coatings applied by APS and HVOF methods under high temperature and hot corrosion conditions.

2 Experimental Methods

2.1 *Preparation of Substrate Material, Bond and Top Coatings*

The substrates were cut into disks with 1 inch diameters and 4 mm thickness, from a wrought sheet of nickel based superalloy. The substrate surfaces were subject to an ultrasonic cleaning and grit blasting process in order to improve the adherence of the coating. Commercial CoNiCrAlY bond coating (Sulzer-Metco, USA, Amdry 9951, 5–37 μm) and ZrO_2-8 wt% Y_2O_3 top coating (GTV Germany, −45 to +20 μm) powders were deposited on nickel based superalloy substrates. The thickness of the bond and the ceramic top coat are 100 μm and 300 μm, respectively.

2.2 *Hot Corrosion Tests*

In this study, the 99% purified Na_2SO_4 and 98% purified V_2O_5 salt mixtures with a mass ratio of 1:1 was selected as the corrosive salt. The corrosive salt was first spread on the surface of the produced TBC samples uniformly with a salt amount of 10 mg/cm^2. This process was repeated at the beginning of each cycle. After that, prepared specimens are subjected to hot corrosion tests at a temperature of 1000 °C with the periodical cycle of four hours. And then the samples were removed from the furnace and cooled down to room temperature. Hot corrosion tests are continued until approximately 25% degradation of the samples was observed.

2.3 *Characterization*

Microstructures of whole TBCs produced in this study are examined via scanning electron microscopy (SEM, Tescan, MAIA3 XMU, Czech Republic) equipped with energy dispersive spectroscopy (EDS; Oxford Xmax 50, Britain). The content of coats was determined by X-ray diffraction (Rigaku Dmax 2200 PC, Cu Ka radiation, Rigaku, Japan). Surface and cross-sectional microstructural analysis, and phase transformations were carried out to examine corrosive product formation and degradation mechanisms. Some important findings were presented.

3 Results and Discussion

3.1 Characterization of as-Sprayed Coatings

In this study, APS and HVOF techniques were used in the production of TBC systems. While metallic bond coats were produced by the APS and HVOF methods, top coat YSZ was produced only by APS method. As-sprayed surface image of YSZ TBC produced by APS method is shown in Fig. 1. The ceramic top coat, which typically consists of APS-deposited TBCs, has roughness and cracks. As a result of this, the oxygen penetration to YSZ top coat layer is high. The combination of all the conditions leads to the oxidation of the TBCs and the formation of TGO at the bond and top coat interface. At the same time, different corrosive salts and pollutions, which are results of the burning of low-quality fuels, after suffering corrosive penetration through the top coat of TBCs can be cause a damage [5, 8–10].

The microstructure of TBC coatings which is produced using the APS technique contains intense amounts of porous and oxide depending on the production process

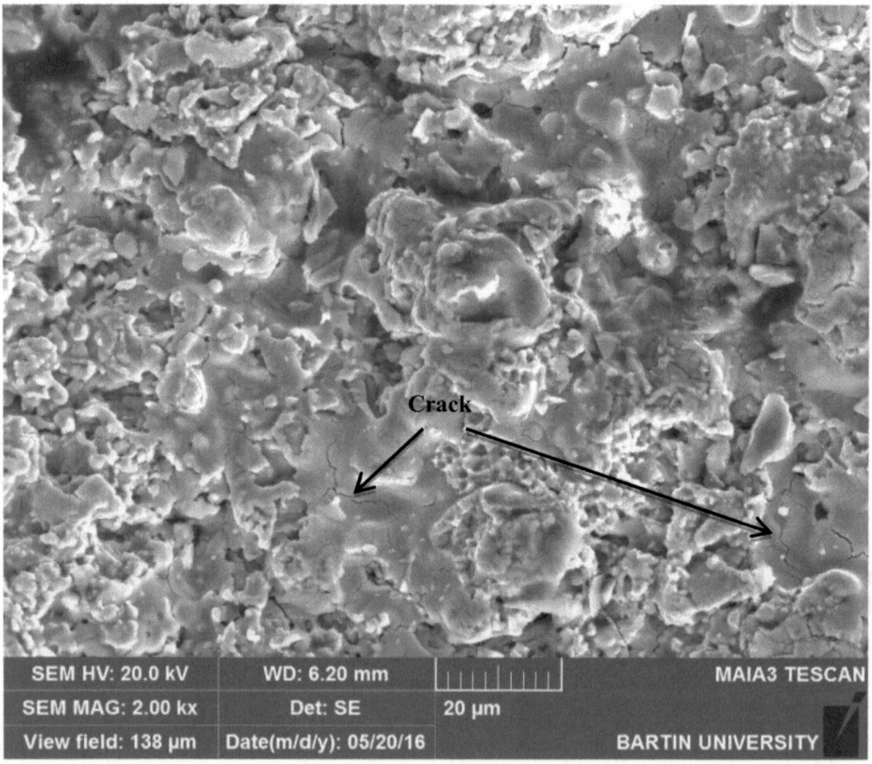

Fig. 1 As-sprayed surface image of YSZ TBC produced by APS method

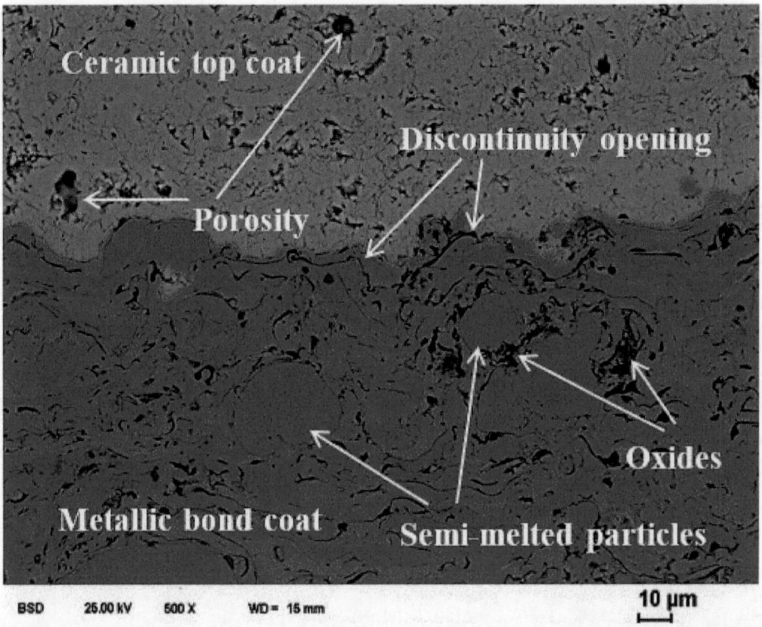

Fig. 2 As-sprayed cross-sectional microstructure of YSZ TBC and CoNiCrAlY bond coat produced with APS method

and spraying condition. This is also a consequence of the fact that the process is carried out open to the atmosphere. The as-sprayed cross-sectional microstructure of YSZ TBCs and CoNiCrAlY bond coatings produced using the APS method is shown in Fig. 2. The metallic bond and ceramic top coat, which was produced by APS methods, had porosities, cracks, discontinuity openings, and oxides. Besides, some melted or semi-melted particles were observed, which is caused by high velocity spraying and the high melting points of the powders. As previously mentioned, the microstructure of the APS coating contains pores and micro-cracks, which is emphasized in the literature on produced by APS technique with YSZ containing coating studies [11–15]. An as-sprayed cross-sectional micrograph of YSZ TBCs produced using by APS method with HVOF CoNiCrAlY bond coat is shown in Fig. 3. When as-sprayed TBCs were evaluated, microstructures of TBCs produced by using APS method have porosity, cracks and discontinuous openings whereas microstructures of TBC produced by the HVOF method have less porosity and content of oxide.

3.2 Hot Corrosion Effect on TBC Samples

It is known that low-quality fuels containing impurities at a high rate (S, V, Na, Ca, K and P) promote hot corrosion depending on the failures in the combustion chambers

Fig. 3 As-sprayed cross-sectional micrograph of YSZ TBC produced by APS method with HVOF CoNiCrAlY bond coat

of gas turbine components. Hot corrosion happens as a result of melted corrosive salts leaking from micro-cracks or porosities on the top coat layer. The top coat has a roughness structure and microcracks (Fig. 1). Thus, it increases the interactions of corrosive salts on the top coat layer. A study by Sreedhar et al. supports this idea [1]. At the same time, repeated melting/solidification cycles of these corrosive salts deposited on the coating at elevated temperatures cause the degradation of the TBC components. In addition, they directly have chemical interaction with yttria (Y_2O_3) which contains t-ZrO_2 phases. This interaction is accelerated by the degradation of TBCs, especially the degradation of the semi-stable structure of the top ceramic (YSZ) coating, the formation of the TGO (α-Al_2O_3) structure, the damage of the metallic bond coating, and the hot corrosion of the superalloy substrate material. Hot corrosion tests were carried out at a temperature of 1000 °C to exceed the melting temperatures of the corrosive salts. Thus, the complete dissolution of the corrosive salts and the interactions with top coat has been achieved [16]. It was found that the corrosive salts primarily react with yttria, the stabilizer of YSZ, which degrades the stable structure of YSZ. Therefore, as the yttria elements are consumed in the structure, the process of t-ZrO_2 to m-ZrO_2 transformation occurs. This phase

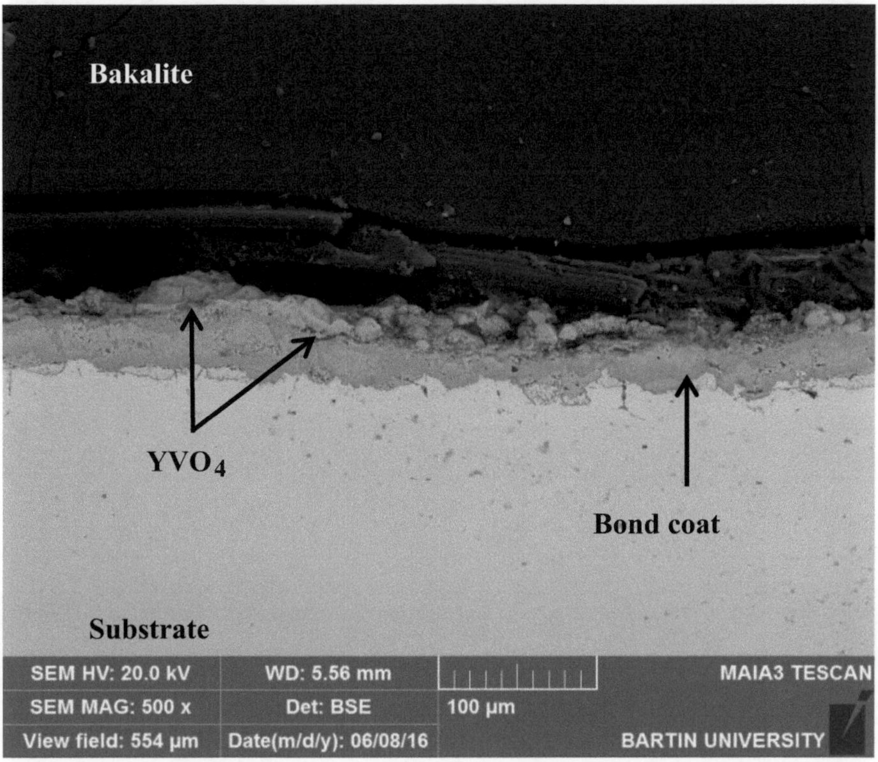

Fig. 4 Cross-sectional micrograph of YSZ TBC and CoNiCrAlY bond coat produced by APS method after the hot corrosion test

transformation causes approximately 3–5% volume expansion of the top coating. Following this, it leads to cracking and spallation of the YSZ coating [9]. In addition, many studies have shown that molten corrosive salts such as Na_2SO_4 and V_2O_5 are completely penetrated by the YSZ top coating at this temperature [5, 17].

A cross-sectional micrograph of the YSZ TBCs and CoNiCrAlY bond coatings produced by the APS method after hot corrosion testing is shown in Fig. 4. As a result of a total of 12 cycles of hot corrosion tests carried out in the 4 h cycle, TBCs failed by spallation from the top coat/bond coat interface. The reason for the spallation of the top coat from the bond coat is the penetration of oxygen through the bond and ceramic coating interface during hot corrosion. Spallation symptoms and cracks were observed on the surface area of coatings from the edge to the center at the end of each cycle. Besides, an YVO_4 corrosion product is observed in Fig. 4. A similar mechanism of failure has also been mentioned in earlier reports. In addition, it has been suggested that the mixed oxides formed in the TGO build up the stress in the structure and accelerate the formation of failure [9, 18–20].

A top surface micrograph of the YSZ and CoNiCrAlY bond coating produced by APS after the hot corrosion test is shown in Fig. 5. After the corrosion test conducted

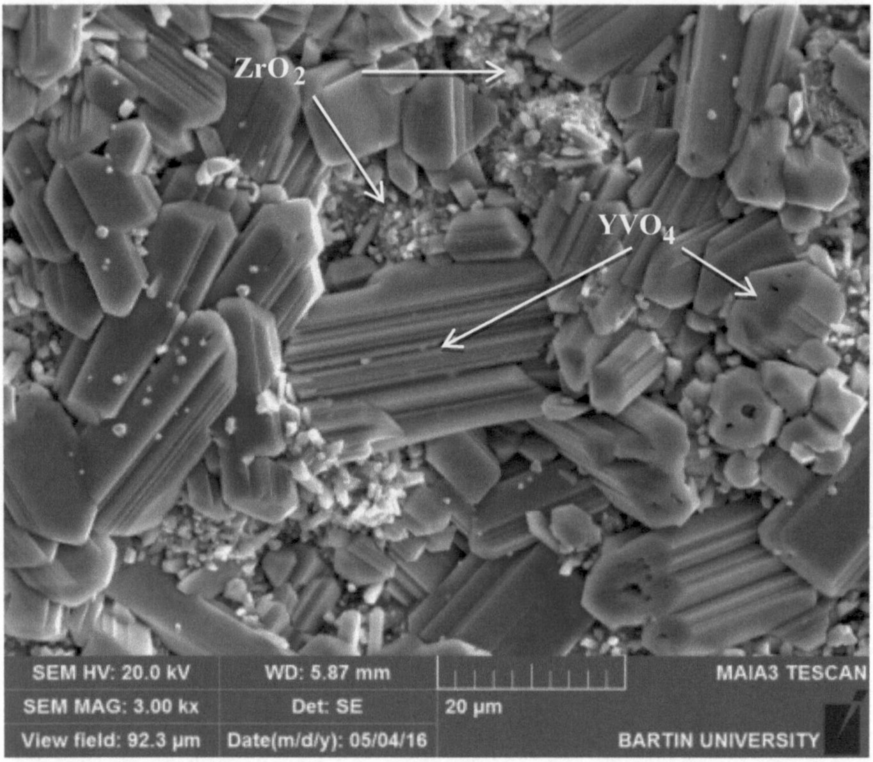

Fig. 5 Top surface micrograph of YSZ TBC and CoNiCrAlY bond coat produced by APS after the hot corrosion test

at 1000 °C for a period of 48 h, the majority of YSZ coating surfaces were coated with corrosion products. Besides, phase transformations occurred with the impact of $Na_2SO_4 + V_2O_5$ molten corrosive salts.

Top surfaces micrographs of the YSZ TBCs and CoNiCrAlY bond coatings produced by APS after the hot corrosion test for (a) 12 h, (b) 24 h, (c) 36 h and (d) 48 h are illustrated in Figs. 5 and 6, YVO_4 rod crystals are observed on the surface of YSZ coatings during the hot corrosion investigation. In the figures ZrO_2 is also shown. It has been observed that the corrosion products (YVO_4) were crystallized to form rod shaped and irregular crystals, especially in the void areas of the coating surface. As the number of hot corrosion cycles increases, corrosion products are more dominant on the coating surface. It was reported that the corrosion product also increases parallel with the increasing number of cycles in the study conducted by Guo et al. [17].

A surface micrograph with EDS analysis of YSZ TBCs and CoNiCrAlY bond coatings produced by the APS method after the hot corrosion test is given in Fig. 7. In the figure, the rod-shaped structures are found by EDS analyses. The rod-shaped structures were formed as YVO_4 after the hot corrosion test. Besides the vanadate

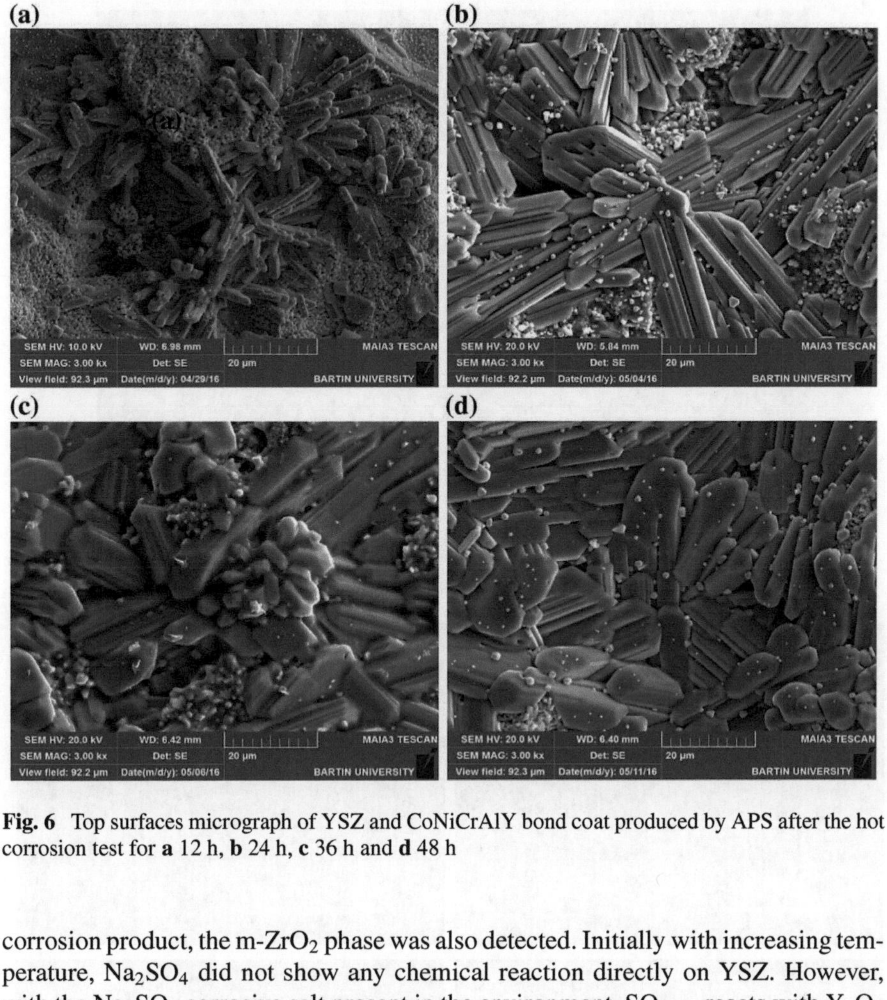

Fig. 6 Top surfaces micrograph of YSZ and CoNiCrAlY bond coat produced by APS after the hot corrosion test for **a** 12 h, **b** 24 h, **c** 36 h and **d** 48 h

corrosion product, the m-ZrO_2 phase was also detected. Initially with increasing temperature, Na_2SO_4 did not show any chemical reaction directly on YSZ. However, with the Na_2SO_4 corrosive salt present in the environment, $SO_{3\,(g)}$ reacts with Y_2O_3 under relative partial pressure to deteriorate the stable structure of YSZ. When Y_2O_3 is absent, a tetragonal to monoclinic phase transformation happens and the stabilization of the tetragonal ZrO_2 phase is lost. The integrity of the structure is destroyed due to the transformation [13, 18, 21, 22].

It is thought that Y_2O_3, which is a stabilizer for the transformation of ZrO_2 to the semi-stable form, occurs after the interaction. As a result, SO_3 and the other melt precipitators accelerated the mechanism of destruction. The reactions occurred during the hot corrosion process are as follows.

$$Na_2SO_4\ (s) \rightarrow Na_2O_{(s)} + SO3_{(g)} \tag{1}$$

$$Na_2O_{(s)} + V_2O_{5(s)} \rightarrow 2NaVO_{3(s)} \tag{2}$$

$$ZrO_2(Y_2O_3)_{(s)} + 2NaVO_{3(s)} \rightarrow m - ZrO_{2(s)} + 2YVO_{4(s)} + Na_2O \tag{3}$$

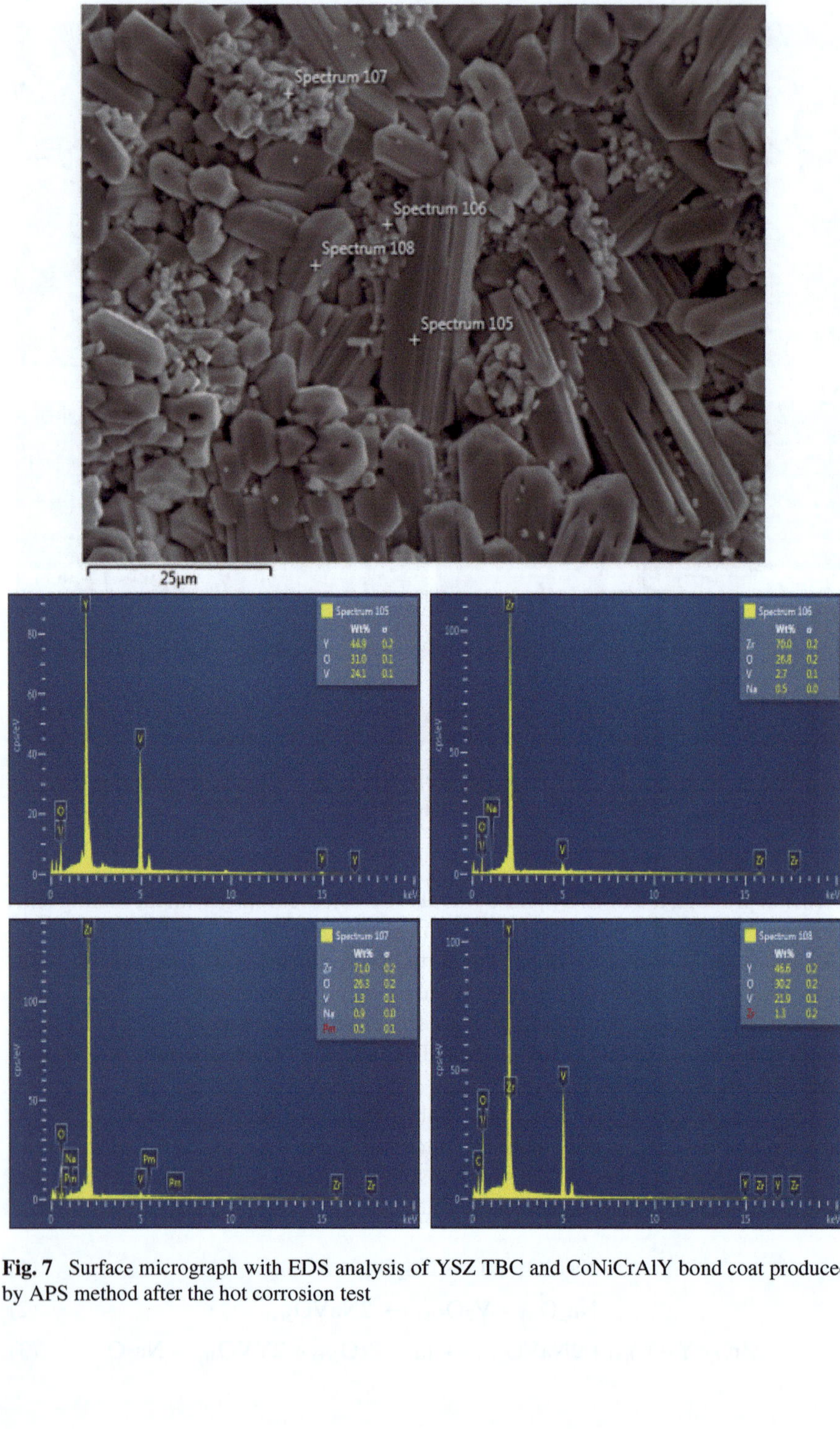

Fig. 7 Surface micrograph with EDS analysis of YSZ TBC and CoNiCrAlY bond coat produced by APS method after the hot corrosion test

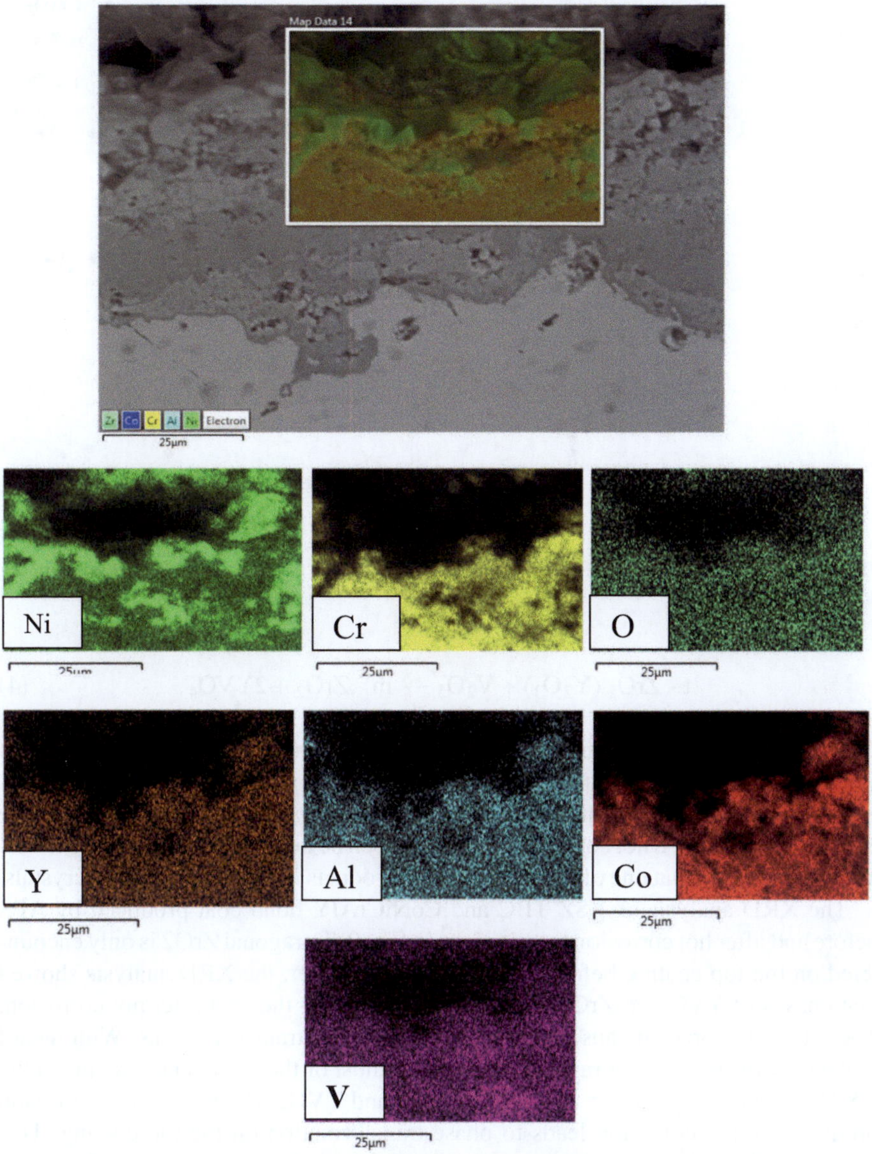

Fig. 8 Cross-sectional micrograph with elemental mapping analysis of YSZ TBC and CoNiCrAlY bond coat produced by APS method after the hot corrosion test

During the process of hot corrosion on YSZ, yttria elements gradually run out. After that, tetragonal zirconia loses semi-stable form and transforms to monoclinic zirconia. The below reaction can also occur [5, 19, 23].

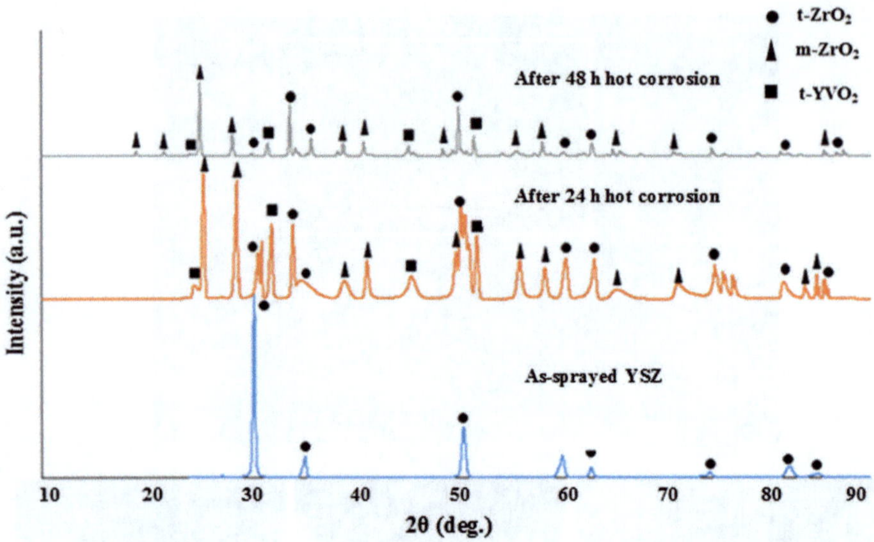

Fig. 9 XRD analysis of YSZ TBC and CoNiCrAlY bond coat produced by APS before and after hot corrosion tests

$$\text{t - ZrO}_2\,(\text{Y}_2\text{O}_3) + \text{V}_2\text{O}_5 \rightarrow \text{m - ZrO}_2 + 2\text{YVO}_4 \tag{4}$$

A cross-sectional micrograph with elemental mapping analysis of YSZ TBCs and CoNiCrAlY bond coatings produced by the APS method after the hot corrosion test is demonstrated in Fig. 8. The elemental mapping analysis after hot corrosion testing can be shown in Fig. 8 hot corrosion products. It can be understood from the elemental mapping analysis that one of the hot corrosion products is YVO_4 rod-shaped crystals.

The XRD analysis of YSZ TBC and CoNiCrAlY bond coat produced by APS before and after hot corrosion tests is given in Fig. 9. Tetragonal ZrO2 is only encountered on the top coating before hot corrosion. However, the XRD analysis showed that phases of YVO_4, m-ZrO_2 and t-ZrO_2, occurred in the coat after hot corrosion. Therefore, the top coat caused deterioration by phase transformations. With regard to the patterns of the as-sprayed TBC samples most of the tetragonal zirconia in the YSZ has transformed to the monoclinic phase and YVO_4 which is the hot corrosion product. So, hot corrosion leads to phase transformation on the top coating. This phase transformation generally disrupts the function of the top coat [4, 9, 24].

A cross-sectional micrograph of TBC produced by APS method with HVOF CoNiCrAlY bond coating after hot corrosion at 1000 °C for 52 h is illustrated in Fig. 10. Primarily, it is observed that TGO formed along the interface of the bond coat and top coat layer which includes mixed oxides under the hot corrosion. Besides, cracks are formed as vertical to the interface TGO/top coat. In addition, the HVOF TBCs exposed to hot corrosion salts exhibited spallation of the top coat at near the bond coat/top coat interface. In similar studies conducted with YSZ top coating in

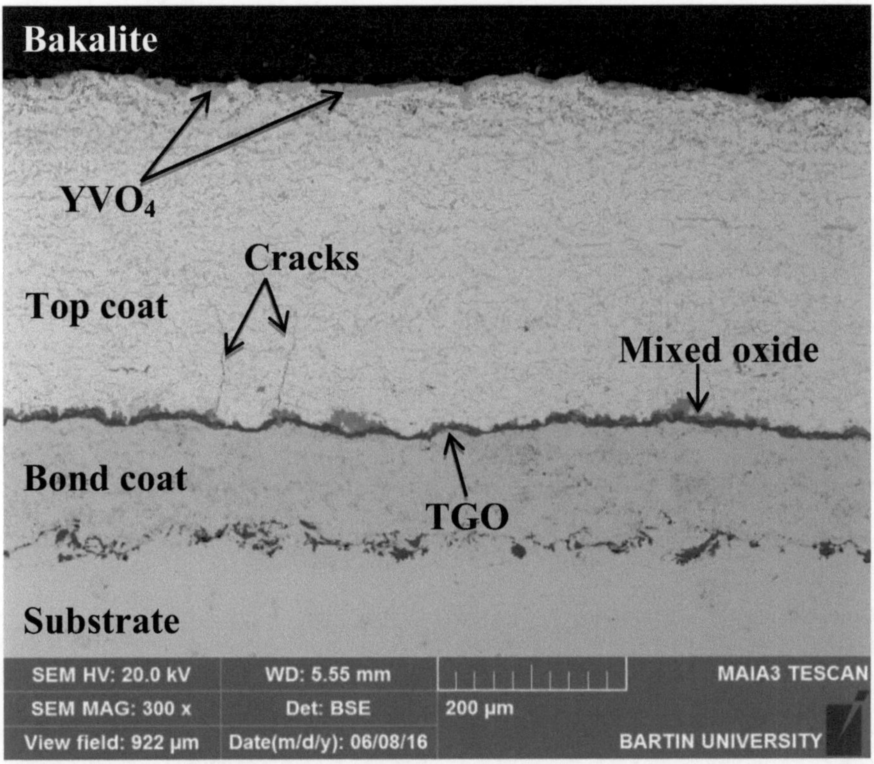

Fig. 10 Cross-sectional micrograph of YSZ TBC produced by APS method with HVOF CoNi-CrAlY bond coat after the hot corrosion test

Na_2SO_4 and V_2O_5V, mixture molten corrosive salts were observed on top coatings of TBC systems. Moreover, it was reported that spallation and phase transformation occurs in the top coatings in the studies [9, 25, 26].

APS TBCs with HVOF sprayed CoNiCrAlY bond coating surface SEM images and EDS analysis after 52 h hot corrosion test are shown in Figs. 11 and 12, respectively. The observation of the surface morphology of the coating showed that the corrosion product, which is YVO4 rod-shaped crystals after the 52 h cyclic hot corrosion test is shown in Fig. 12. These formations have been observed in similar studies in the literature [27, 28]. As shown in Fig. 12, the elements yttria were detected by EDS on surfaces of rod-shaped crystals. So, vanadium may form compounds as vanadate during hot corrosion [27]. Besides, the formation of YVO_4 was confirmed using the EDS analysis in this figure.

Cross-sectional micrographs with elemental mapping analysis of YSZ TBC produced by the APS method with HVOF CoNiCrAlY bond coating after the hot corrosion test is demonstrated in Fig. 13. The corrosion product components can be easily observed by elemental mapping analysis after the hot corrosion test. Besides, oxygen coinciding with the vanadium and yttria elements are shown. In addition to

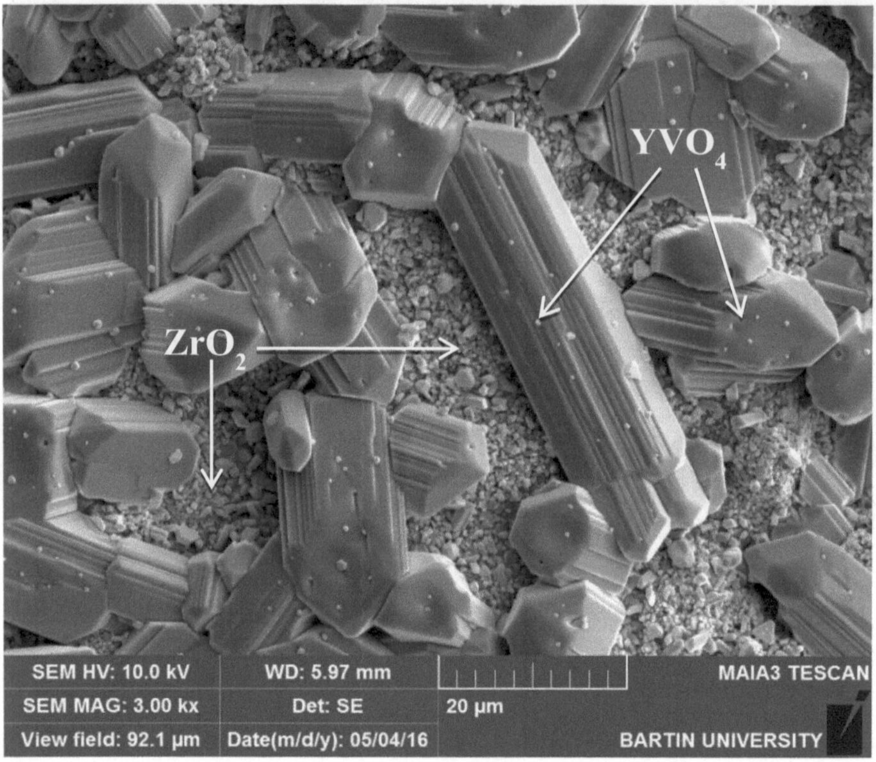

Fig. 11 Surface micrograph of YSZ TBC produced by APS method with HVOF CoNiCrAlY bond coat after the hot corrosion test

the corrosion products, it can be understood that aluminum has been joined to the TGO structure. In addition to EDS and elemental mapping analyses, XRD phase analysis shows an YVO_4 phase, and another hot corrosion product, namely monoclinic ZrO_2, as shown in Figs. 9 and 14. The XRD analysis of the hot corroded YSZ top coating shown in Figs. 9 and 14 confirms the formation of the YVO_4 phase. Figures 9 and 14, the presence of a few tetragonal zirconia phases can be seen, while more number of zirconia monoclinic phases indicating the extensive destabilization of YSZ coating and also TBC system after 52 h hot corrosion testing. Ramachandran et al. [27] reported the presence of the phases.

4 Conclusion

In the present study, the hot corrosion behavior of the two different TBC systems including a CoNiCrAlY bond coat and an YSZ top coat have been compared. The following conclusions are drawn:

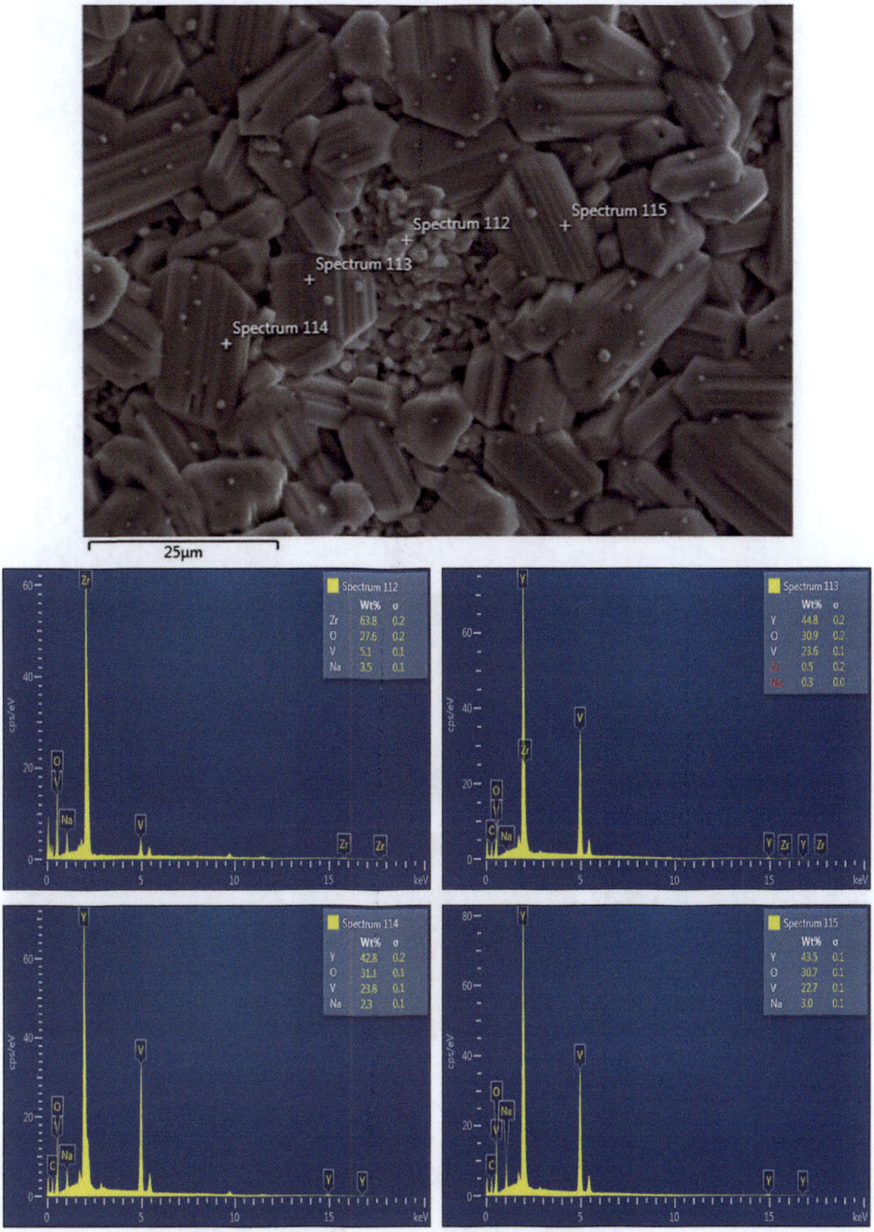

Fig. 12 Surface micrograph with EDS analysis of YSZ TBC produced by APS method with HVOF CoNiCrAlY bond coat after the hot corrosion test

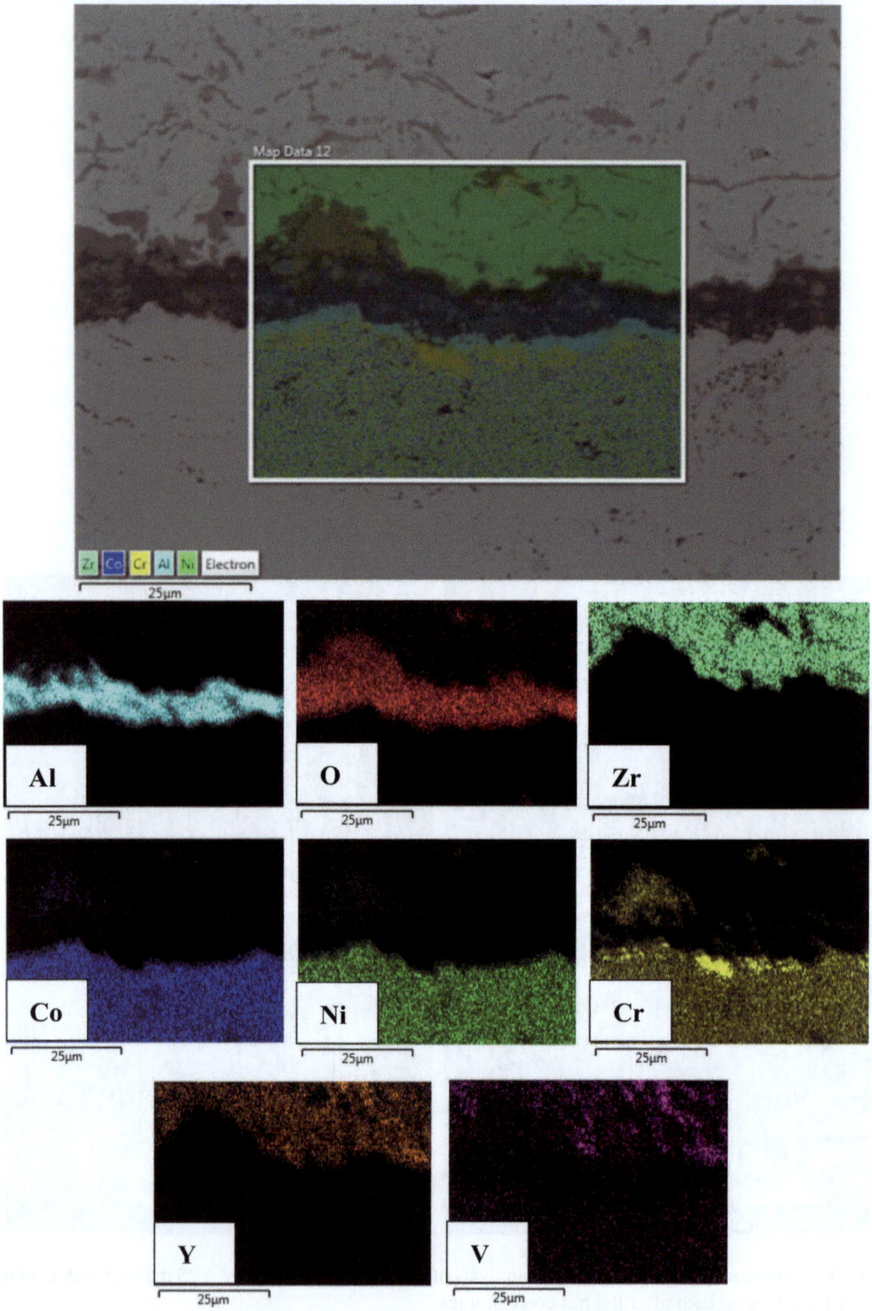

Fig. 13 Cross-sectional micrograph with elemental mapping analysis of YSZ TBC produced by APS method with HVOF CoNiCrAlY bond coat after the hot corrosion test

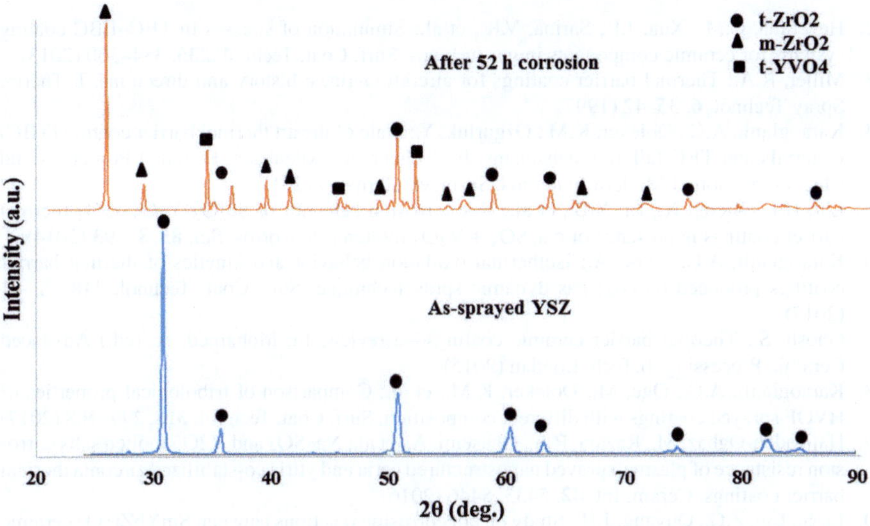

Fig. 14 XRD analysis of HVOF and CoNiCrAlY bond coat produced by APS before and after hot corrosion tests

1. The TBC system produced using the APS technique was found to have high porosity and oxide content whereas the one produced with HVOF technique had low porosity and oxide content.
2. Hot corrosion and failure of the YSZ coating was primarily because of molten corrosive salt containing V_2O_5 and Na_2SO_4.
3. The phases of m-ZrO_2 and YVO_4 crystals promoted crack formation and failure of TBCs. Thus, the molten salt mixture penetrated into the bond coat from cracks and open porosities on the surface of the coating, causing damage. At the same time, the elemental mapping analyses indicate that the molten salts penetrated through cracks.
4. The hot corrosion lifetime of the HVOF–BC system is higher than APS–BC. In addition, the formation rate of APS BC system was faster than that of HVOF BC at interface area with the TGO formation.

Acknowledgements The present study was derived from Ph.D. thesis of Mustafa Kaplan in the University of Selcuk Graduate School of Natural and Applied Sciences, Konya, Turkey. The study was also supported by Selcuk University Scientific Research Projects under the Grant Number of 15201071.

References

1. Sreedhar, G., Raja, V.S.: Hot corrosion of YSZ/Al_2O_3 dispersed NiCrAlY plasma-sprayed coatings in Na_2SO_4-10 wt.% NaCl melt. Corros. Sci. **52**, 2592–2602 (2010)

2. Heverana, C.M., Xua, J.P., Sarina, V.K., et al.: Simulation of stresses in TBC–EBC coating systems for ceramic components in gas turbines. Surf. Coat. Technol. **235**, 354–360 (2013)
3. Miller, R.A.: Thermal barrier coatings for aircraft engines: history and directions. J. Therm. Spray Technol. **6**, 35–42 (1997)
4. Karaoglanli, A.C., Doleker, K.M., Ozgurluk, Y.: State of the art thermal barrier coating (TBC) materials and TBC failure mechanisms. In: Ochner, A., Altenbach, H. (eds.) Properties and Characterization of Modern Materials. Springer, Germany (2017)
5. Liu, H.F., Xiong, X., Li, X.B., et al.: Hot corrosion behavior of Sc_2O_3-Y_2O_3-ZrO_2 thermal barrier coatings in presence of Na_2SO_4 + V_2O_5 molten salt. Corros. Sci. **85**, 87–93 (2014)
6. Karaoglanli, A.C., Turk, A.: Isothermal oxidation behavior and kinetics of thermal barrier coatings produced by cold gas dynamic spray technique. Surf. Coat. Technol. **318**, 72–81 (2017)
7. Ghosh, S.: Thermal barrier ceramic coatings—a review. In: Mohamed, A. (ed.) Advanced Ceramic Processing. InTech, Londan (2015)
8. Karaoglanli, A.C., Oge, M., Doleker, K.M., et al.: Comparison of tribological properties of HVOF sprayed coatings with different composition. Surf. Coat. Technol. **318**, 299–308 (2017)
9. Hajizadeh-Oghaz, M., Razavi, R.S., Ghasemi, A., et al.: Na_2SO_4 and V2O5 molten salts corrosion resistance of plasma-sprayed nanostructured ceria and yttria co-stabilized zirconia thermal barrier coatings. Ceram. Int. **42**, 5433–5446 (2016)
10. Li, S., Liu, Z.G., Ouyang, J.H.: Study on hot corrosion reactions between $SmYbZr_2O_7$ ceramic and vanadium pentoxide at temperatures of 600–1000 °C in air. Mater. Chem. Phys. **130**, 1134–1138 (2011)
11. Liu, L., Xu, H., Xiao, J., et al.: Effect of heat treatment on structure and property evolutions of atmospheric plasma sprayed NiCrBSi coatings. Surf. Coat. Technol. **325**, 548–554 (2017)
12. Aghasibeig, M., Moreau, C., Dolatabadi, A., et al.: Fabrication of nickel electrode coatings by combination of atmospheric and suspension plasma spray processes. Surf. Coat. Technol. **285**, 68–76 (2016)
13. Ajay, A., Raja, V.S., Sivakumar, G., et al.: Hot corrosion behavior of solution precursor and atmospheric plasma sprayed thermal barrier coatings. Corros. Sci. **98**, 271–279 (2015)
14. Tao, C., Wang, L., Cheng, N.L., et al.: Hot corrosion performance of AlO-CrO/NiCoCrAlYTa and AlO/NiCoCrAlYTa coatings deposited by atmospheric plasma spraying. J. Therm. Spray Technol. **25**, 797–805 (2016)
15. Wang, L., Liu, C.G., Zhong, X.H., et al.: Investigation of crack propagation behavior of atmospheric plasma-sprayed thermal barrier coatings under uniaxial tension using the acoustic emission technique. J. Therm. Spray Technol. **24**, 296–308 (2015)
16. Saremi, M., Valefi, Z., Abaeian, N.: Hot corrosion, high temperature oxidation and thermal shock behavior of nano agglomerated YSZ-Alumina composite coatings produced by plasma spray method. Surf. Coat. Technol. **221**, 133–141 (2013)
17. Guo, L., Zhang, C.L., Li, M.Z., et al.: Hot corrosion evaluation of Gd_2O_3-Yb_2O_3 co-doped Y_2O_3 stabilized ZrO_2 thermal barrier oxides exposed to Na_2SO_4 + V_2O_5 molten salt. Ceram. Int. **43**, 2780–2785 (2017)
18. Mohan, P., Yuan, B., Patterson, T., et al.: Degradation of yttria-stabilized zirconia thermal barrier coatings by vanadium pentoxide, phosphorous pentoxide, and sodium sulfate. J. Am. Ceram. Soc. **90**, 3601–3607 (2007)
19. Afrasiabi, A., Saremi, M., Kobayashi, A.: A comparative study on hot corrosion resistance of three types of thermal barrier coatings: YSZ, YSZ + Al_2O_3 and YSZ/Al_2O_3. Mater. Sci. Eng. A **478**, 264–269 (2008)
20. Sidhu, T.S., Prakash, S., Agrawal, R.D.: Hot corrosion performance of a NiCr coated Ni-based alloy. Scripta Mater. **55**, 179–182 (2006)
21. Ahmadi-Pidani, R., Shoja-Razavi, R., Mozafarinia, R., et al.: Evaluation of hot corrosion behavior of plasma sprayed ceria and yttria stabilized zirconia thermal barrier coatings in the presence of Na_2SO_4 + V_2O_5 molten salt. Ceram. Int. **38**, 6613–6620 (2012)
22. Wang, X.Z., Guo, L., Peng, H., et al.: Hot-corrosion behavior of a $La_2Ce_2O_7$/YSZ thermal barrier coating exposed to Na_2SO_4 + V_2O_5 or V_2O_5 salt at 900 °C. Ceram. Int. **41**, 6604–6609 (2015)

23. Li, S., Liu, Z.G., Ouyang, J.H.: Hot corrosion behaviour of $Yb_2Zr_2O_7$ ceramic coated with V_2O_5 at temperatures of 600–800 °C in air. Corros. Sci. **52**, 3568–3572 (2010)
24. Loghman-Estarki, M.R., Razavi, R.S., Edris, H., et al.: Comparison of hot corrosion behavior of nanostructured ScYSZ and YSZ thermal barrier coatings. Ceram. Int. **42**, 7432–7439 (2016)
25. Li, M.Z., Cheng, Y.X., Guo, L., et al.: Preparation of plasma sprayed nanostructured GdPO4 thermal barrier coating and its hot corrosion behavior in molten salts. Ceram. Int. **43**, 7797–7803 (2017)
26. Guo, L., Li, M.Z., Ye, F.X.: Comparison of hot corrosion resistance of $Sm_2Zr_2O_7$ and $(Sm_{0.5}Sc_{0.5})_2Zr_2O_7$ ceramics in Na_2SO_4 + V_2O_5 molten salt. Ceram. Int. **42**, 13849–13854 (2016)
27. Ramachandran, C.S., Balasubramanian, V., Ananthapadmanabhan, P.V.: On the cyclic hot corrosion behaviour of atmospheric plasma sprayed lanthanum zirconate based coatings in contact with a mixture of sodium sulphate and vanadate salts: a comparison with the traditional YSZ duplex and NiCrAlY coated samples. Vacuum **97**, 81–95 (2013)
28. Daroonparvar, M., Yajid, M.A.M., Yusof, N.M., et al.: Investigation of three steps of hot corrosion process in Y_2O_3 stabilized ZrO_2 coatings including nano zones. J Rare Earth **32**, 989–1002 (2014)

Wear Behavior of Severe Shot Peened and Thermally Oxidized Commercially Pure Titanium

Okan Unal, Abdullah Cahit Karaoglanli, Yasin Ozgurluk, Kadir Mert Doleker, Erfan Maleki and Remzi Varol

Abstract In this study, severe shot peening (SSP) has been performed to commercially pure titanium (Grade 2) specimens in order to expose severe plastic deformations to the material surface. S230 shot media with higher air pressure conditions have been selected for SSP. A nanocrystalline layer has been created by means of severe plastic deformation. The nanograined structure has been investigated via XRD, FESEM, and microhardness methods. SSP specimens have been subjected to thermal oxidation treatment in air at 500 and 700 °C for 2 and 16 h. Severe shot peened commercially pure titanium has a thicker oxide layer when compared to as received ones. The results show that both SSP and thermal oxidation improve the wear characteristics.

Keywords Severe shot peening · Thermal oxidation · Wear · Nanograined layer

---

O. Unal
Mechanical Engineering Department, Karabuk University, Demir-Celik Kampusu, 78050 Karabuk, Turkey
e-mail: unalokan78@gmail.com

A. C. Karaoglanli · Y. Ozgurluk (✉) · K. M. Doleker
Metallurgical and Materials Engineering Department, Bartin University, 74100 Kutlubeyyazicilar, Turkey
e-mail: ozgurlukyasin@gmail.com

A. C. Karaoglanli
e-mail: cahitkaraoglanli@gmail.com

K. M. Doleker
e-mail: mertdoleker@gmail.com

E. Maleki
Mechanical Engineering Department, Turkey Sharif University of Technology International Campus, Kish Island, Iran
e-mail: maleky.erfan@gmail.com

R. Varol
Mechanical Engineering Department, Suleyman Demirel University, 32200 Isparta, Turkey
e-mail: remzivarol@sdu.edu.tr

© Springer International Publishing AG, part of Springer Nature 2019
A. Öchsner and H. Altenbach (eds.), *Engineering Design Applications*,
Advanced Structured Materials 92, https://doi.org/10.1007/978-3-319-79005-3_29

1 Introduction

Pure titanium and its alloys have been used for orthopedic implants and prostheses due to their high biocompatibility and corrosion properties. Furthermore, they have the ability to be selected for high temperature applications such as aeronautical and aircraft appliances due to their higher specific strength. However, the alloy has poor high cycle fatigue and tribological properties [1]. In order to increase the resistance of failures, surface severe plastic deformation methods such as shot peening [2–4], SMAT [5] and ultrasonic shot peening [6] have been performed in order to enhance fatigue, corrosion and stress corrosion resistance by creating a nanocrystalline layer on and just below the surface. Besides, surface coating applications are performed to the alloys to prevent high temperature oxidation.

Mechanical surface treatments, and also surface severe plastic deformation methods, are exposed to the materials to create a high strain and dense defect surface on metallic materials for two decades. Ultra-fine grained or nanocrystalline layer formation are of great importance, currently due to its superior mechanical and physical properties [7]. The nanocrystalline layer has been produced by transforming coarse grain to ultra-fine subgrains by means of severe plastic deformation [8]. In recent years, shot peening has been applied to obtain ultra-fine grained layers on the surface. Shot peening parameters are increased to expose high plastic deformation as Almen intensity increases. Shot peening treatment with higher Almen intensities is called "severe shot peening" [9, 10].

Ti and alloys have been manufactured with oxide films on the surface at room temperature, which provides some physical and biological improvements. Titanium alloys first show a passivation layer and the oxidation diffuses throughout the core of the metal. The oxide film is related to the chemical composition of the material, treatment duration and temperature. The film is amorphous and generally constructed of three layers, namely TiO, Ti_2O_3, and the third layer TiO_2. Increasing temperature and duration causes thicker oxide formation [11–13]. In this study, the severe shot peening process before oxidation has been investigated on the oxidation and wear behavior of commercially pure titanium (Grade 2).

2 Experimental Methods

2.1 Preparation of Substrate Material, Bond and Top Coatings

Commercially pure titanium (Grade 2) has been exposed to severe shot peening before thermal oxidation. The specimens were machined to 30–20–10 mm. Then the specimens were heat treated at 810 °C for 2 h and cooled in furnace to room temperature. The annealed CP titanium specimens were shot peened at 27 A Almen intensity (482 kPa—55 s) in a Peenmatic 2000S shot peening machine. The oxidation

Fig. 1 Free ball microabrasion test equipment

behavior of the shot peened specimens was investigated using a high temperature furnace (PLF 130/12, Protherm). The specimens were oxidized in a normal electrically heated furnace in the ambient atmosphere at temperatures of 500 and 700 °C for 2 and 16 h.

The specimens were cut from cross-sections and grounded through 120–1200 grade emery papers then mechanically polished with 6, 3, 1 and 0.25 μm diamond paste. The specimens were etched by the Kroll's reagent. Surface characterization was performed via optical (Nikon Eclipse MA100) and field emission scanning electron microscopy (Carl Zeiss Gemini Sigma).

The peened and oxidized specimens were subjected to the wear (abrasion) process (Fig. 1) using the free ball microabrasion test [7]. The 800 mesh SiC particles were used as an abrasive. SiC particles including 25% SiC mixed with distilled water were used as abrasive solution.

3 Results and Discussion

3.1 Microstructure Analysis

Optical microscope studies reveal in Fig. 2 that SSP exposes severe plastic deformation to the surface and forces the surface microstructure (coarse grains) to plastic deformation. SSP increases the deformed layer up to 50–60 μm. The oriented

(a) Optical view of shot peened pure Ti

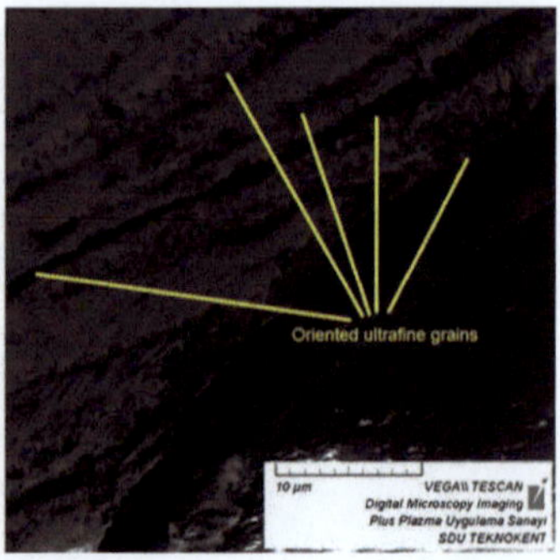

(b) FEGSEM view of shot peened pure Ti

Fig. 2 **a** Optical view of shot peened pure Ti **b** FEGSEM view of shot peened pure Ti

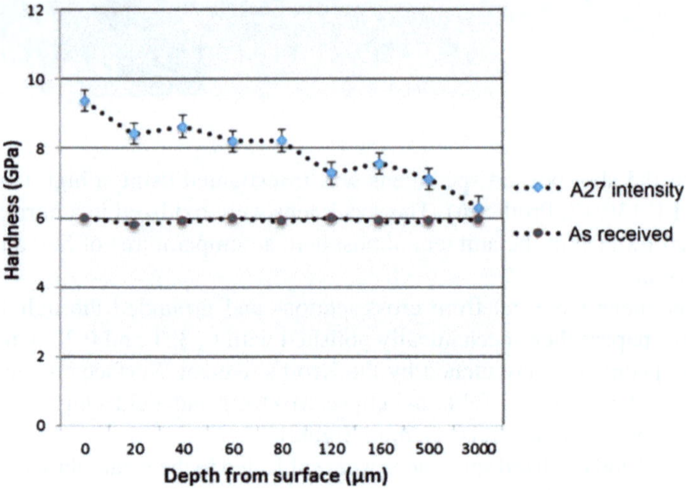

Fig. 3 Ultra microhardness change of shot peened and as received specimens

ultrafine grains can be observed on the right side of the figure. SSP leads to high strain deformation and creates oriented grains [14].

In previous studies, different surface severe plastic deformation methods have been exposed to Ti and Ti–6Al–4V such as fast multiple rotation rolling [15, 16] and ultrasonic nanocrystalline surface modification [6]. These treatments are more effective by means of ultrafine grain occurrence and the layer thickness. However, the SMAT [17] treatment performs a similar effect on the deformed layer thickness when compared to SSP. The shot peened and as-received specimens' ultra microhardness change by nanoindentation is shown in Fig. 3.

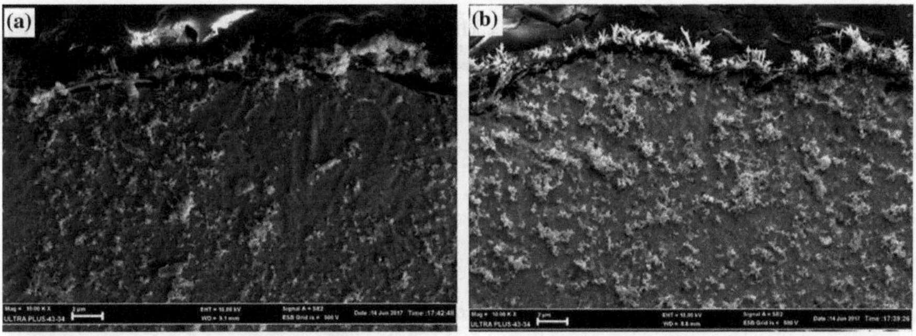

Fig. 4 FESEM images of **a** as received thermally oxidized pure Ti **b** severe shot peened thermally oxidized pure Ti

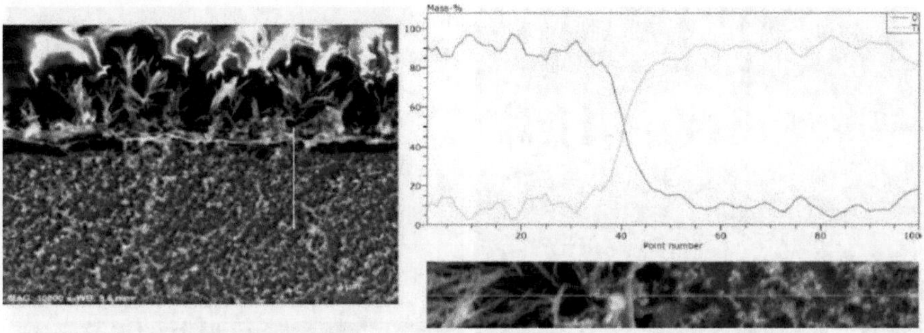

Fig. 5 As-received thermally oxidized at 500 °C pure Ti EDAX analysis

The surface treatment keeps the effect approximately up to 500 μm interior for CP titanium 2. Thermal oxidation tests with 500 °C for 2 h (Fig. 4); the shot peened surface show a dense oxidized layer with compared to as received one. However, thermal oxidation temperature and duration is limited so the structure cannot be effectively determined. It can be hard to state the chemical composition of the C, N and O elements by using EDAX analysis. However, the existence of them can be investigated (Fig. 5). The EDAX analysis is shown of the as received oxidized surface at 500 °C.

The shot peened surface is shown as a dense oxidized layer when compared to the as received one that was oxidized at 500 °C for 16 h (Fig. 6). Pre-shot peening contributes to the diffusion of the oxygen to go deeper. The shot peened surface on the right triggers the exact titanium oxide layer formation when compared to as received one oxidized at 700 °C-2 h. Pre-shot peening contributes to the dislocation of the dense layer; dislocated and ultrafine grained layers have a much higher capacity to diffuse of interstitial alloys. The temperature (700 °C for 16 h) and the duration are adequate for the diffusion of oxygen for both as received and shot peened surfaces (Fig. 7). Shot peening induces homogenous, denser and thicker diffusion layer when

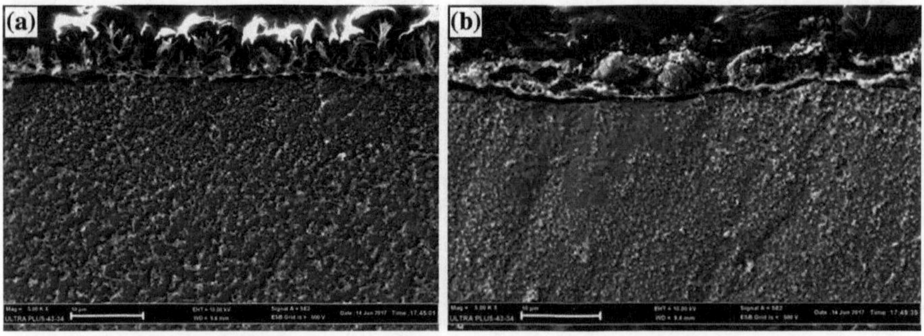

Fig. 6 At 500 °C 16 h **a** as received thermally oxidized pure Ti **b** severe shot peened thermally oxidized pure Ti

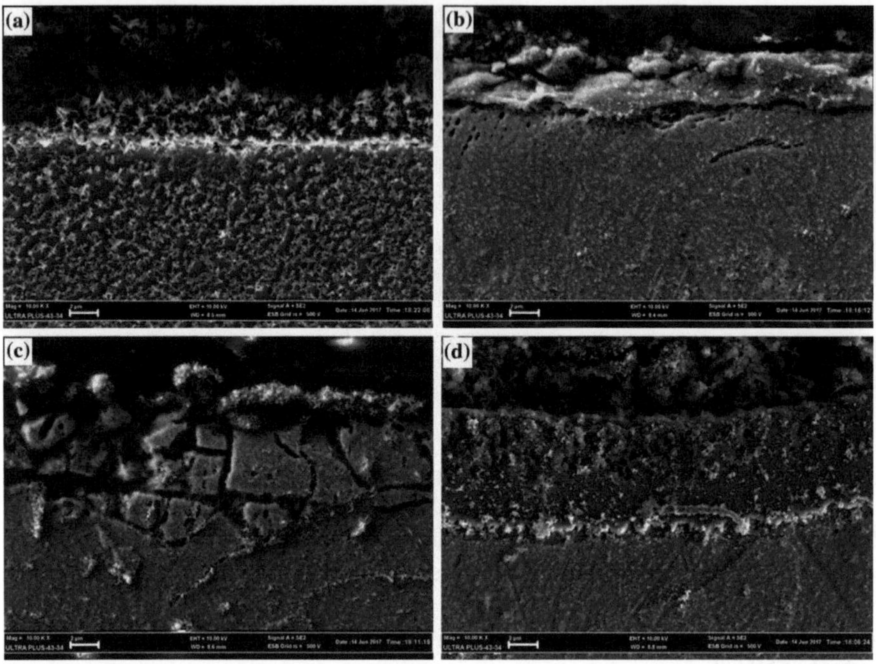

Fig. 7 FESEM observation of oxidized Ti at 700 °C-2 h **a** as received **b** severe shot peened at 700 °C-16 h **c** as received **d** severe shot peened

compared to as received ones. Kanjer et al. shows laser shot peening and air blast shot peening produces the required oxygen diffusion kinetic for under 700 °C [18].

The analysis shows the Ti and O elements throughout the oxide layer (Fig. 8). The transition to the interior results the oxygen diminish besides titanium increase. However, shot peening leads to a very homogenous layer both titanium and oxygen content throughout the layer is closer to each other. The average measurements of the

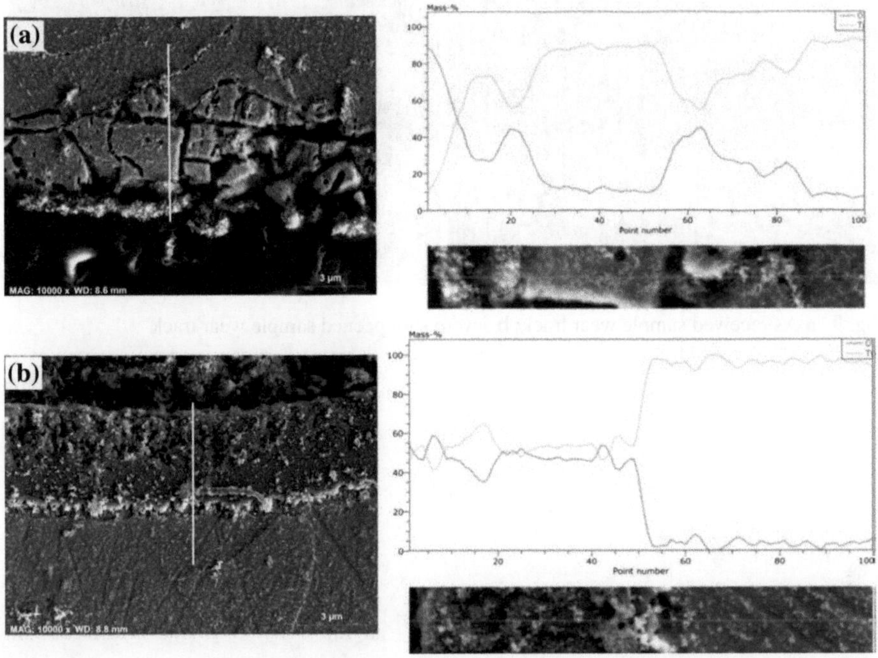

Fig. 8 FESEM EDAX analysis at 700 °C-16 h thermally oxidized pure Ti **a** as received **b** severe shot peened

Table 1 The average measurements of the oxide layers on the surface of pure Ti

The conditions	As received (µm)	Pre shot peening (µm)
500 °C-2 h	0–1	0–1
500 °C-16 h	0–1	1–2
700 °C-2 h	1–2	2–3
700 °C-16 h	4–5	5–6

oxide layers on the surface of both shot peened and as received samples proves that thermal oxidation temperature and duration increases the oxide layer thickness. In addition, pre-shot peening enhances the thickness at the same conditions (Table 1).

3.2 Wear Tests and Analysis

The specimens whose surfaces were peened and also thermally oxidized were subjected to a wear (abrasion) process using the free ball microabrasion test. The 800 mesh SiC particles were used as an abrasive. SiC particles including 25%SiC mixed with distilled water were used as abrasive solution. The volume loss values which

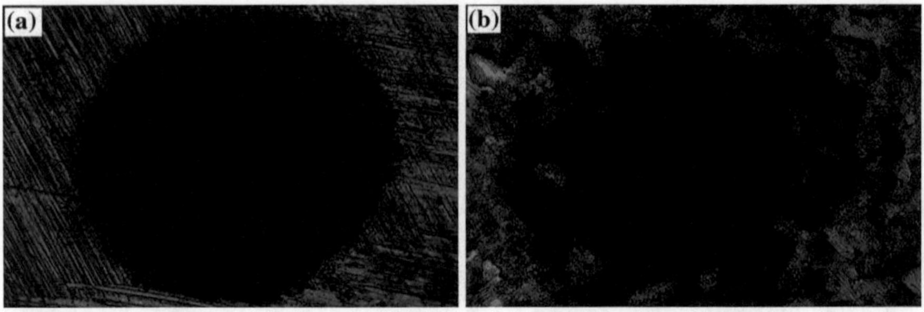

Fig. 9 **a** As received sample wear track; **b** severe shot peened sample wear track

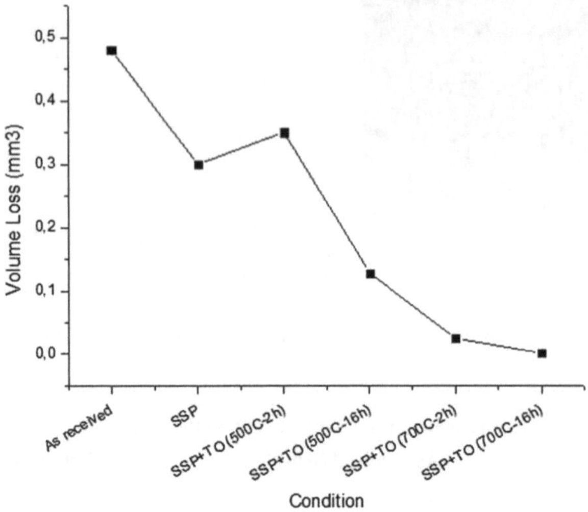

Fig. 10 Volume loss versus condition graph after wear micro abrasion test

resulted in the specimens with peened and non-peened surfaces as a result of the 5 min microabrasion test process applied at 160 rev/min.

In order to detect the influence of severe shot peened and also severe shot peened and thermal oxidation on the wear performance of the pure Ti, as received and severe shot peened specimen were exposed to microabrasion wear tests. Figure 9 shows the wear tracks both as received and peened samples. From the Fig. 9, both surface hardness and also roughness increase the ball encounters much more tackles against wearing of the surface.

According to the wear volume losses, SSP treatment influences the wear resistance positively. However, thermal oxidation after SSP at 500 °C for 2 h reduces the wear resistance capability. Higher durations and temperatures conditions improve the wear characteristics. This could be due to the increase of titanium oxide rich diffusion layer (Fig. 10).

4 Conclusion

From this study, the followings can be concluded;

- A fine grained structure was created in the surface layer of the low carbon steel specimen after a severe shot peening process.
- It is observed that the finer grains are located in the regions close to the surface and that as we move towards the center, the grains become coarse.
- Shot peening triggers the formation of titanium oxide layer and also increases the layer thickness due to the beneficial effect of mechanically deformed unbalanced grain structures.
- It is observed that the wear (abrasion) durability of the specimens improved with shot peening. However, only thermal oxidation conditions at higher temperatures and durations can positively influence the wear durability.

References

1. Pi, Y., Faure, J., Agoda-Tandjawa, G., Andreazza, C., Potiron, S., Levesque, A., et al.: Microstructural characterization of Ti–6Al–4V alloy subjected to the duplex SMAT/plasma nitriding. Microsc. Res. Tech. **76**, 897–903 (2013)
2. Tsuji, N., Tanaka, S., Takasugi, T.: Effects of combined plasma-carburizing and shot-peening on fatigue and wear properties of Ti–6Al–4V alloy. Surf. Coat. Technol. **203**, 1400–1405 (2009)
3. Maleki, E.: Modeling of Severe shot peening effects to obtain nanocrystalline surface on cast iron using artificial neural network. Mater. Today Proc. **3**, 2197–2206 (2016)
4. Maleki, E., Farrahi, G.H., Sherafatnia, K.: Application of artificial neural network to predict the effects of severe shot peening on properties of low carbon steel. In: Öchsner, A., Altenbach, H. (eds.) Machining, Joining and Modifications of Advanced Materials, pp. 45–60. Springer, Singapore (2016)
5. Huang, R., Han, Y.: The effect of SMAT-induced grain refinement and dislocations on the corrosion behavior of Ti–25Nb–3Mo–3Zr–2Sn alloy. Mater. Sci. Eng. C **33**, 2353–2359 (2013)
6. Amanov, A., Cho, I.-S., Kim, D.-E., Pyun, Y.-S.: Fretting wear and friction reduction of CP titanium and Ti–6Al–4V alloy by ultrasonic nanocrystalline surface modification. Surf. Coat. Technol. **207**, 135–142 (2012)
7. Unal, O., Varol, R., Erdogan, A., Gok, M.S.: Wear behaviour of low carbon steel after severe shot peening. Mater. Res. Innov. **17**, 519–523 (2013)
8. Ivanisenko, Y., Lojkowski, W., Valiev, R.Z., Fecht, H.J.: The mechanism of formation of nanostructure and dissolution of cementite in a pearlitic steel during high pressure torsion. Appl. Surf. Sci. **51**, 5555–5570 (2003)
9. Unal, O., Varol, R.: Almen intensity effect on microstructure and mechanical properties of low carbon steel subjected to severe shot peening. Appl. Surf. Sci. 40–47 (2014)
10. Unal, O.: Optimization of shot peening parameters by response surface methodology. Surf. Coat. Technol. **305**, 99–109 (2016)
11. Mitelea, I., Dimian, E., Bordeaşu, I., Crăciunescu, C.: Ultrasonic cavitation erosion of gas nitrided Ti–6Al–4V alloys. Ultrason. Sonochem. **1**, 1544–1548 (2014)
12. Wen, M., Wen, C., Hodgson, P., Li, Y.: Improvement of the biomedical properties of titanium using SMAT and thermal oxidation. Colloid Surf. B. **116**, 658–665 (2014)
13. Mège-Revil, A., Steyer, P., Cardinal, S., Thollet, G., Esnouf, C., Jacquot, P., et al.: Correlation between thermal fatigue and thermomechanical properties during the oxidation of multilayered

TiSiN nanocomposite coatings synthesized by a hybrid physical/chemical vapour deposition process. Thin Solid Films **518**, 5932–5937 (2010)
14. Unal, O., Varol, R.: Surface nanostructuring of AISI 1017 by severe shot peening. Res. Rep. Metals. **1**, 1–4 (2017)
15. Li, Y., Sun, K., Liu, P., Liu, Y., Chui, P.: Surface nanocrystallization induced by fast multiple rotation rolling on Ti–6Al–4V and its effect on microstructure and properties. Vacuum **101**, 102–106 (2014)
16. Unal, O., Cahit Karaoglanli, A., Varol, R., Kobayashi, A.: Microstructure evolution and mechanical behavior of severe shot peened commercially pure titanium. Vacuum **110**, 202–206 (2014)
17. Anand Kumar, S., Ganesh Sundara Raman, S., Sankara Narayanan, T.S.N., Gnanamoorthy, R.: Influence of counterbody material on fretting wear behaviour of surface mechanical attrition treated Ti–6Al–4V. Tribol. Int. **57**, 107–114 (2013)
18. Kanjer, A., Optasanu, V., Lavisse, L., de Lucas, Marco, MdC, Dejardin S., François, M., et al.: Influence of mechanical surface treatment on high-temperature oxidation of pure titanium. Oxid. Met. **88**, 383–395 (2017)

Characterization of Scaffold Structures for the Development of Prostheses and Biocompatible Materials

Luis Héctor Hernández-Gómez, Juan Alfonso Beltrán-Fernández, Marcelino Ramírez-Jarquín, Itzel Bantle-Chávez, Carolina Alvarado-Moreno, Alejandro González-Rebattú y González, Mauricio González-Rebattú y González, Juan Alejandro Flores-Campos, Pablo Moreno-Garibaldi, Nefi Pava-Chipol and Salatiel Pérez-Montejo

Abstract In the present work, the importance and consideration of the use of scaffold structures for biomaterials is highlighted in order to be applied in prosthetic

---

L. H. Hernández-Gómez · J. A. Beltrán-Fernández (✉) · M. Ramírez-Jarquín · I. Bantle-Chávez · C. Alvarado-Moreno · J. A. Flores-Campos · P. Moreno-Garibaldi · N. Pava-Chipol · S. Pérez-Montejo
Unidad Profesional Adolfo López Mateos "Zacatenco" Col. Lindavista, Escuela Superior de Ingeniería Mecánica y Eléctrica, Instituto Politécnico Nacional, Sección de Estudios de Posgrado e Investigación Edificio 5, 2do Piso, C.P. 07738 Ciudad de México, Mexico
e-mail: jbeltranf@hotmail.com

L. H. Hernández-Gómez
e-mail: luishector56@hotmail.com

M. Ramírez-Jarquín
e-mail: onilcram@outlook.com

I. Bantle-Chávez
e-mail: itzi.bantle@gmail.com

C. Alvarado-Moreno
e-mail: aly-oroshiwa@hotmail.com

J. A. Flores-Campos
e-mail: jaflores@ipn.mx

P. Moreno-Garibaldi
e-mail: pmg170588@gmail.com

N. Pava-Chipol
e-mail: sowbran@gmail.com

S. Pérez-Montejo
e-mail: perez_ms01@hotmail.com

A. González-Rebattú y González · M. González-Rebattú y González
Hospital Regional 1 de Octubre – ISSSTE, Av. Instituto Politécnico Nacional, #1669, Lindavista, 07300 Gustavo A. Madero, CDMX, Mexico
e-mail: alexrebattu@hotmail.com

M. González-Rebattú y González
e-mail: mauriciorebattu@yahoo.com.mx

© Springer International Publishing AG, part of Springer Nature 2019
A. Öchsner and H. Altenbach (eds.), *Engineering Design Applications*, Advanced Structured Materials 92, https://doi.org/10.1007/978-3-319-79005-3_30

cases where it is possible to consider a three dimensional porous structure, maintaining the mechanical properties that are necessary for the right functioning of the models under preoperative and postoperative loads. The requirements for biodegradable and bioactive properties of the models are indicated by surgical specialists, with the purpose of considering both the shape and type of the scaffold and its influence in the building of diverse mechanical supports. A specific maxillofacial application case of models, a mandibular implant, whose numerical-experimental properties and behaviors based on bone, polymers and metal alloy structures used in the manufacturing of models, were used to compare the variation of the scaffold structure types. Furthermore, models printed in 3D using an acrylonitrile butadiene styrene (ABS) polymer with different fiber orientations and type of scaffolds will be considered as an initial part of the case studies.

Keywords Scaffold structures · Prostheses · 3D printing
Biocompatible materials · Mandible

1 Introduction

The characterization of scaffold structures in the tissue engineering field considering its architecture and porosity percentage, is based on computational tridimensional reconstructions and generations of tridimensional meshes for finite-element models.

Tissue engineering (TE) requires specific structures (scaffolds) that are capable of providing support and which are also built in materials capable of allowing cell migration, cell adhesion and also possess certain mechanical properties and a high level of biodegradability.

The tridimensional printing technique (3DP) can be used for the fabrication of scaffolds made of biomaterials, which are employed in the tissue engineering field.

Amongst the ceramics, as an example, materials such as hydroxyapatite and coralline are commonly employed since both are also inorganic components of the bone structure of the skeleton of vertebrates. Specifically for scaffolds built with hydroxyapatite the final pieces need to have a high porosity percentage (40–80%).

The loss of organs and tissues as the consequence of a specific illness or due to an accident is a problem for human's health of big relevance. Alone in the United States 500,000 procedures which required osseous implants were performed in the 2000. The consultancy enterprise Kalorama estimates that in the prostheses market there was an expenditure of 2600 billion dollars in 2010, while in 1980 the prostheses market expenditure for those processes was of 256 billion dollar, showing an increment ten times greater between both years. [1].

Tissue engineering was adjudicated to this science in the spring of 1987 in a meeting at the National Sciences Foundation, although several of the techniques commonly employed in this discipline were used several decades before [2].

However, in 1993 Langer and Vacanti defined tissue engineering as "a multidisciplinary field in which several principles of the engineering and life sciences are applied in order to develop biological replacements in order to restore, maintain or enhance the function of a tissue or organ" [3].

The final purpose of the scaffolds in the tissue engineering field is to provide a structure of support to the cells in order to stimulate their proliferation and differentiation, thus, for their design and development several biological and physical factors need to be taken into consideration, such as their structural properties. The architecture of the scaffolds has to (a) provide the cells enough space to fulfill their function of generating new tissue, (b) allow the access of fluids for the transportation of cells, nutrients and oxygen, together with the elimination of toxins and (c) withstand mechanical loads which are necessary for the cell's stimulation [4, 5].

2 Statement of the Problem

In recent years, years, the ageing population has been steadily increasing, which leads to the constant demand of a better quality of life. However, bone injuries are one of problems, which usually need treatment. In many cases they are incapacitated because of the consequences by being the consequence of irreparable tissue loss due to traumatic injuries, cancer, autoimmune or genetic diseases.

In accordance with the statistics of the National Institute of Statistics and Geography, there were 664,000 individuals who required a prosthesis in Mexico.

The cost of a prosthesis is high. For example, prosthesis for an upper limb can cost 20,000 American dollars.

Taking these antecedents into account, it is necessary to optimize the design of such prostheses in order to maintain their resistance properties and avoid stress shielding. One alternative is the use of scaffolding structures because of their property of keeping the functionality and stability in the assembly of the bone area, allowing for bone regeneration. In order to obtain the best results, an evaluation of the scaffold structures is required. Besides, alternative materials can be used in conjunction with manufacturing techniques.

3 Theoretical Concepts

Amongst the most important structural properties, are the porosity level, the interconnectivity between the pores and the specific superficial area, together with the distribution of the pores' size [6].

3.1 Porosity and Interconnectivity Between the Pores

Scaffolds need to possess an open pore geometry with a surface and microstructure with a high porosity percentage to allow the growth and reorganization of the cells in an in vitro environment and also to provide the necessary space for the neovascularization of the surrounding tissues in vivo. The highly porous microstructure with porous interconnected meshes is essential to assure the uniform spatial distribution of the cells, their survival, proliferation and migration in vitro [7]. The porosity of the scaffold (superior to 90%) and the level of interconnection of the meshes have a direct influence on the diffusion of nutrients and physiological gases, allowing the elimination of toxins, metabolic waste molecules and sub products left by the cells that penetrate the scaffold's structure [8, 9].

In the case of bigger scaffolds, the high porosity and interconnectivity of the pores allows the use of bioreactors so that hydrodynamic micro environments can be created with minimum diffusion limitations, imitating the natural conditions of the interstitial liquid in vivo in order to obtain bigger and better organized cell colonies.

3.2 Size of the Pore

The diversity of the tissue's architecture requires a different micro environment for its regeneration that includes the use of scaffolds with an optimum size of the pore. For the regeneration of osseous tissue in vitro, researchers of [10, 11] indicate the need for pores of a size of 200–400 μm, while Yoshikawa et al. [12] successfully employed a scaffold with a pore size of 500 μm. Scaffolds with pores with a size of 20 mm and 125 mm are frequently used in order to regenerate the skin of adult mammals [13] and a size of 150 mm is used to regenerate liver tissue [14]. When the size of the pores is too small, the consequence will be the occlusion of the pores [15], and also the cell penetration will be obstructed and also the generation of a cell cluster inside the scaffolds.

3.3 Mechanical Properties

Because of acting like staffs used to guide the tissue regeneration, the scaffolds have to present a high mechanical resistance during the cell cultivation in vitro in vitro in order to keep the required spaces for the cell's growth and the development of the cell cluster [16].

During the regeneration of the tissues that have to resist the load (e.g. cartilage, bone), additional elements related to the mechanical properties of the scaffolds have to be considered. [9]. To allow the early development of the treated area, the degrad-

Table 1 Biomaterials applications

Material	Modulus of elasticity (GPa)	Tensile strength (MPa)	Rupture strain %	Density (g/cm^3)	Fracture toughness (MPa m 1/2)
Hydroxyapatite	40–115	147–294 (compression)		3.16	1.5–2.0
Compact bone	13–19	120–150		1.86–2.02	2–12
Cartilage	0.01–0.1	10–30	80		
Collagen	1.0	50–100	10		
Achilles tendon	0.43	11	5–8		

able shell has to keep enough mechanical resistance in order to control any tension or physiological load imposed during the engineering construction.

The degradation of the scaffolds has to be properly adjusted so that it will retain enough structural integrity until the newly regenerated tissue has replaced the support function of the scaffolds. Table 1 shows some of the principal properties used in the present study case [9].

3.4 *Biomaterials*

According to the Conference of the European's Society Consensus of Biomaterials the concept will designate a "material designed to act in an interfacial way with biological systems with the purpose of evaluate, treat, increase o replace a certain tissue, organ or function of the body" [17].

Biomaterials can have a natural or synthetic origin or, even a combination of both. Amongst the classification of the biomaterials the prefix "bio" is used to emphasize its biocompatibility. Several types of biomaterials are known, some metals, ceramics, polymers and composites or hybrids [18].

Biomaterials have different applications, to mention a few: joint replacement (hip, knee), plaques and bone cement, repairing of osseous defects, artificial tendons and ligaments, dental implants (artificial heart, skin substitutes, artificial kidney, heart-lung machine), as well as the organs of the senses such as cochlear replacements, intraocular lenses, contact lenses and corneal attachment [19].

Table 2 shows a wide range of applications of the mentioned biomaterials.

A general way to illustrate the advantages and principal functions of the scaffolds is shown in Fig. 1 [20, 21].

The scaffolds have been investigated as an alternative approach to bone graft substitutes, which are made with metal, ceramic and polymer. Also they are used for the replacements of lost or damaged bone tissues.

Table 2 Some applications of certain biomaterials

Applications	Materials
Joint replacement (hip, knee)	Titanium, Ti–Al–V alloys, stainless steel, polyethylene
Bone plaques	Stainless steel, cobalt-chromium alloys
Repair of bone defects	Hydroxyapatite (HAP)
Artificial ligaments and tendons	Teflon, Dacron
Dental implants	Titanium, aluminum oxide, calcium phosphate

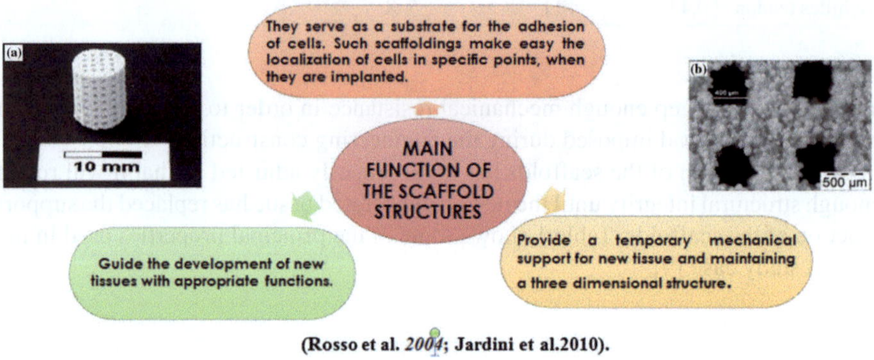

Fig. 1 Main functions of the scaffolded structures

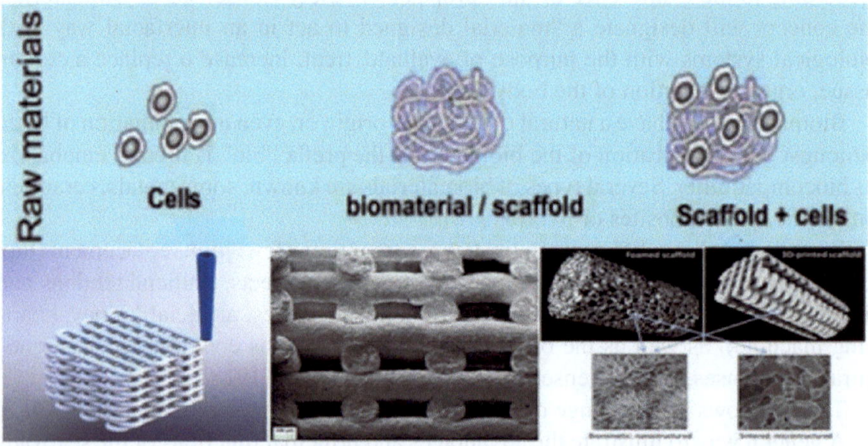

Fig. 2 Kind of strategies of the tissue engineering

Figure 2 illustrates three strategies used in the tissue engineering's field for the purposes previously mentioned:

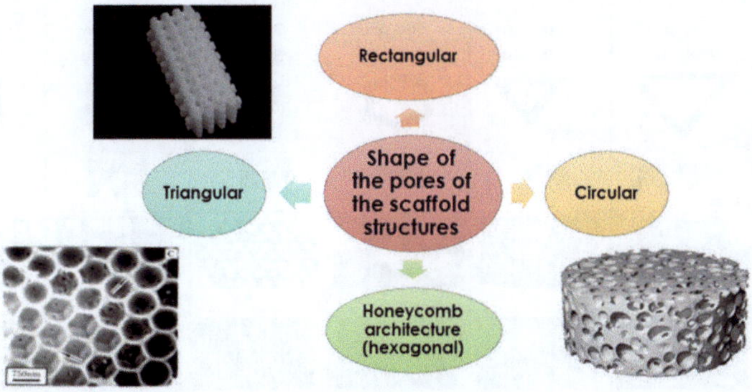

(Binil Starly, Lauren Shor, 2005)

Fig. 3 Shape of scaffolds

(a) Isolated cells
(b) Scaffolds
(c) Cells with scaffolds.

The main achievements of the tissue engineering and their main goal is the repair and formation of new tissue through the use of scaffolds (support) on which cells can grow, as shown in Fig. 3.

There are two different pore sizes (500 and 1000 μm) and the main consideration, when lattice structures are used, is to obtain the Minimum Feature Size, so that the smallest minimum unit cell size, the minimum wall thickness, as well as a good resolution of the lattice can be obtained. It is important to keep in mind if the required geometry can be additively manufactured and if the weight reduction benefits the strength of the arrangement so that it is similar to the bulk material as shown in Fig. 4 [22–24].

4 Materials and Methods

The methodology followed for the development of the case study with the mentioned variations is shown in Scheme 1.

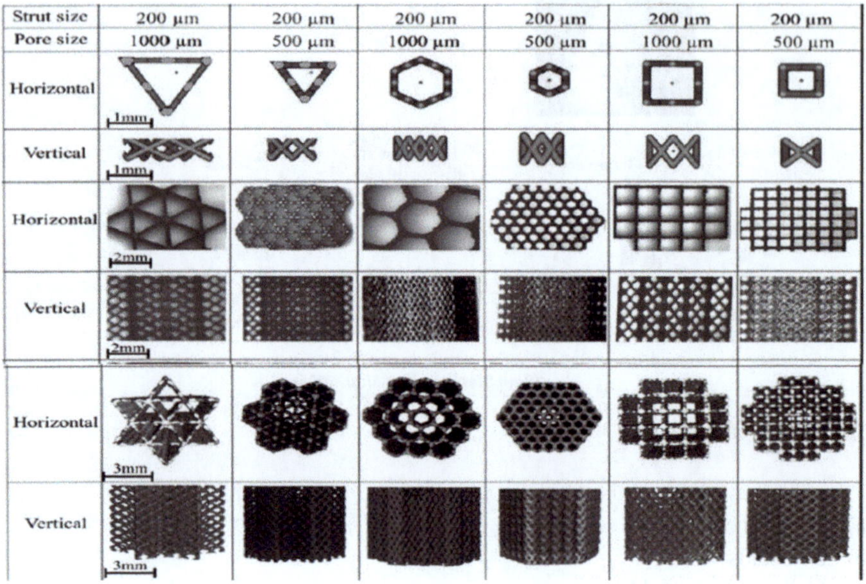

Fig. 4 Pore size of the scaffolds in accordance with the direction and geometry of the meshes

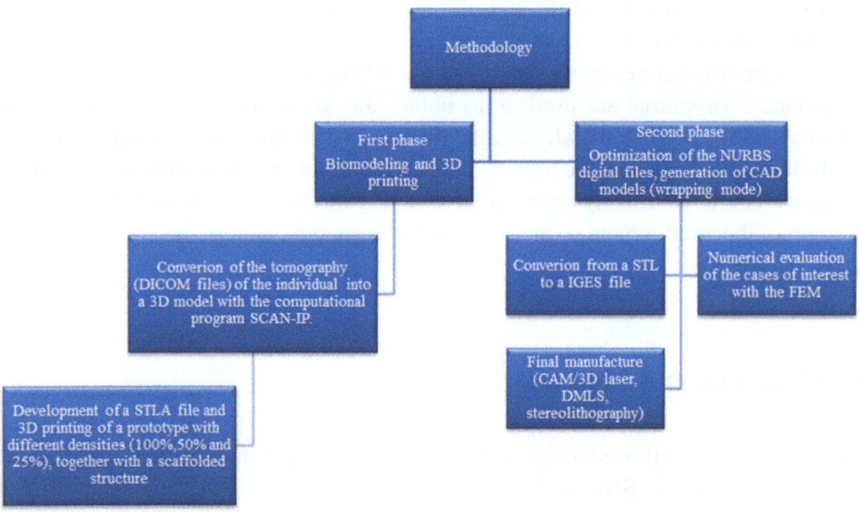

Scheme 1 Methodology description

4.1 Prosthesis of a Hemimandible

4.1.1 Description

The case is related to a biomechanical prosthesis of a hemimandible for a patient with an illness known as osteonecrosis, presenting a severe affectation of the trabecular and cortical structure of the mandible (63–80%).

As the osteonecrosis progresses, small fractures take place and a personalized implant for the Mexican patient is required. Some features were asked to be considered:

- Reduction of the total weight.
- It has to be produced with 3D metal printing (Titanium alloy or Stainless Steel 316L, medical grade).

For the development of the mandible prosthesis three cases were considered:

- Case 1—100% density with 1,016,587 triangles.
- Case 2—20% density, infill strut and 1 mm thickness (1 mm wall thickness = 1 mm), 988,982 triangles.
- Case 3—10% density, infill strut and 0.5 mm thick.
- ***For the 3 cases the characteristics of the strut were x0—Honeycomb*** (Figs. 5 and 6).

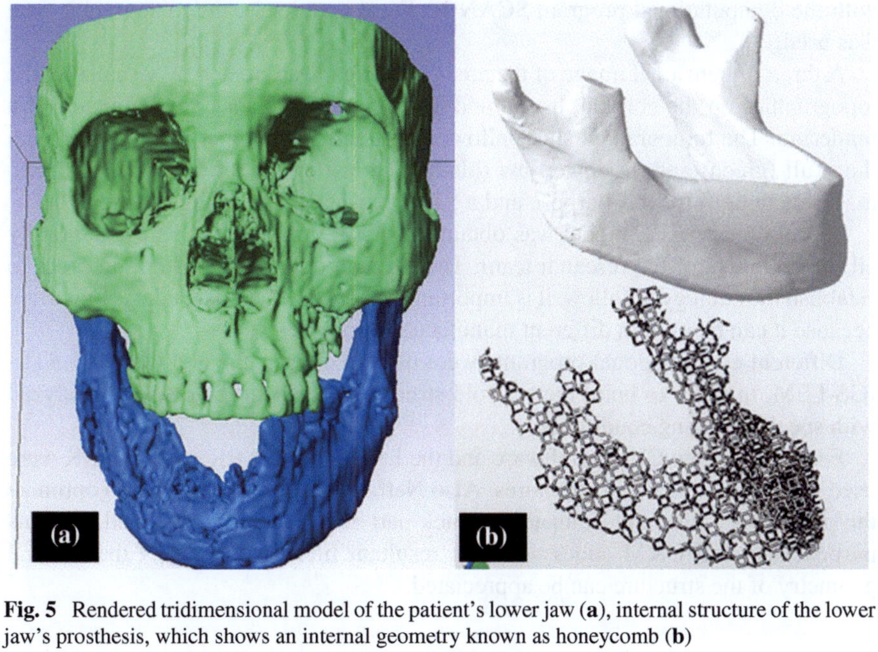

Fig. 5 Rendered tridimensional model of the patient's lower jaw (**a**), internal structure of the lower jaw's prosthesis, which shows an internal geometry known as honeycomb (**b**)

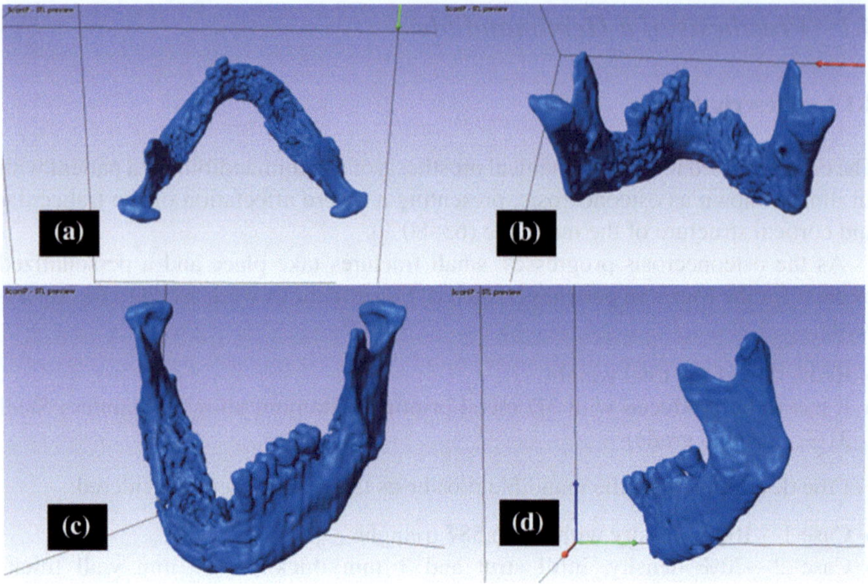

Fig. 6 Evaluation of the patient's tomography in the coronal (**a**), posterior (**b**), angular (**c**) and lateral plane (**d**), showing the inferior mask

For this case the three-dimensional image of the affected bone area was modeled with the computational program SCAN-IP. For this purpose, the tomographic data was used.

A three-dimensional image of the areas of interest was created by visualizing the topographies on the coronal, lateral and anteroposterior planes and by means of a rendering. The tomographic slices allowed to identify the upper jaw together with the skull (green) and the lower jaw (blue). The last one is the affected area. Two masks were used for this purpose and a STL file was obtained (Figs. 7 and 8).

A preliminary ABS model was obtained with 3D printing for its evaluation by all the members of the research team. The surgeon defined the criteria in order to establish the strategy to follow. It is important to mention that this model is important because it can be used in different manufacture processes.

Different computational programs were used to convert the archives OBJ-STL-IGS-FEM, in order to build the scaffold structures. Such structures were analyzed with specific loading conditions.

For instance, the CURA Software and the LATTICE SPARK AUTODESK were used to create the scaffold structures. Also Netfabb Autodesk was used to optimize the preliminary files and evaluate the thick part and the supports created. For this purpose Obj and G&M codes were the resultant files. In the Fig. 9 the internal geometry of the structure can be appreciated.

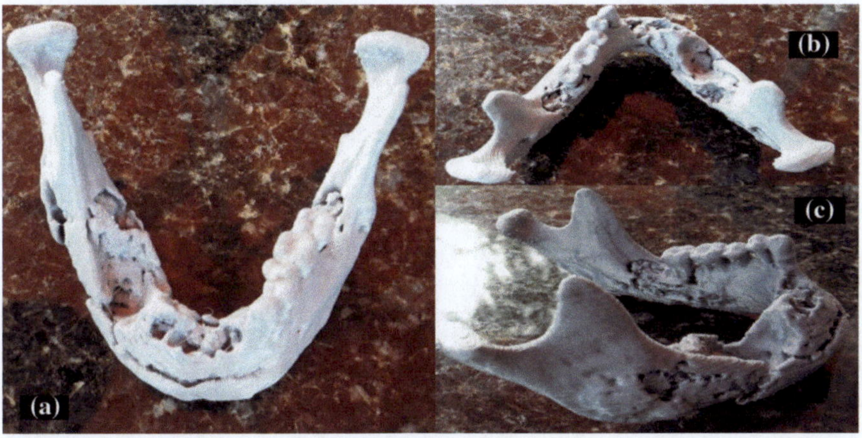

Fig. 7 Lower jaw affected by osteonecrosis, printed in ABS plastic, appreciated from above (**a**), from behind (**b**) and laterally (**c**)

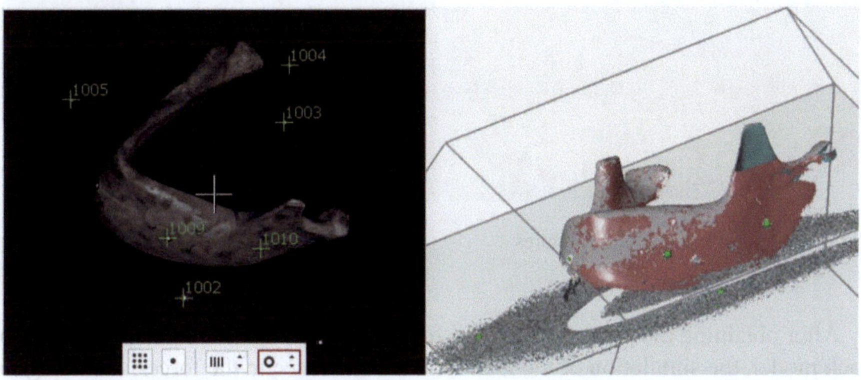

Fig. 8 3D printed model and scanning process for the mandibular prosthesis. Interface's change of coloration that indicates with a cherry color the entrance of new information and with a grey color the information previously captured

5 Numerical Analysis

For the numerical analysis of the models, different .IGS files were exported to the Ansys program. The thickness for each case was evaluated and a numerical analysis was carried out with a FEM code in order to determine its structural integrity in accordance with the preliminary description.

- **Case 1—100 % density**.
- **Case 2—20 % density, infill strut and 1 mm thickness**.
- **Case 3—10 % density, infill strut and 0.5 mm thickness** (Fig. 10).

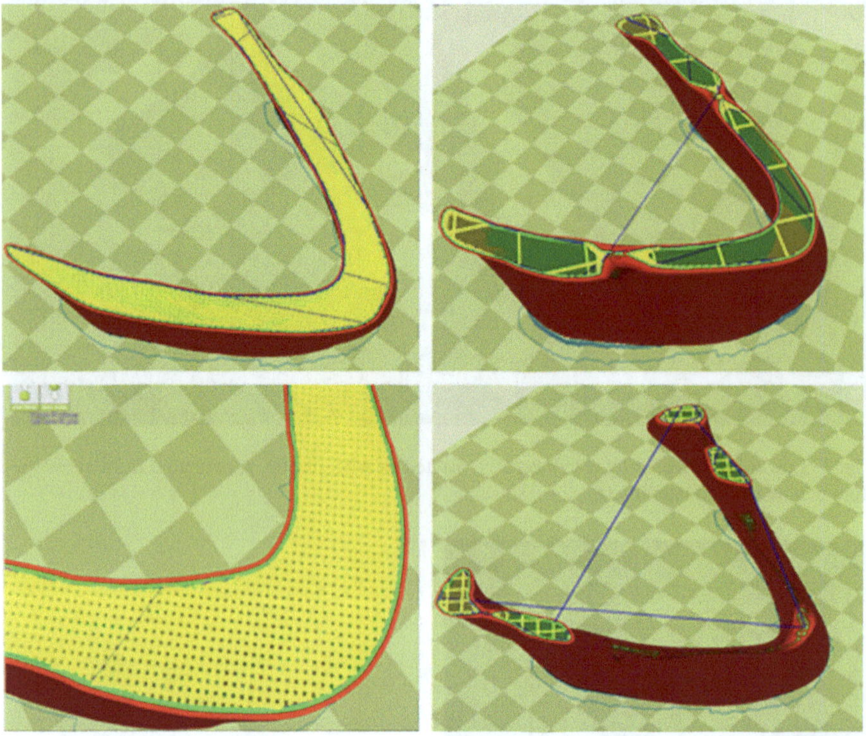

Fig. 9 Details of the scaffold structures

After obtaining the pieces in their .stl format and adding the internal scaffolds to each model, the simulation of the loads and the reference to the forces according to the established parameters in the endorsed articles, making reference of the strength of the muscles, according to the material assigned to the models during the simulation process, is conducted.

Likewise, anatomical data must be taken into consideration for the design and development of the prosthesis. The bite force produced during mastication is generated by multiple muscles that work simultaneously and in conjunction on the lower jaw's surface. In order to determine the influence of each muscle during the production of a certain force, an anatomical model is used for the analysis of the strain distribution [3]. A coordinate system with its origin set on the projection of the center of the mandible's coronoid process in relation with the symmetric plane is used in order to establish the direction and degree of the force vectors. The x, y, z axes are projected along the intercondylar axis. The muscles that act over the jaw, the masseter, temporal, internal pterygoid, external pterygoid and digastric, create the forces Fm, Ft, Fi, Fe, Fo, respectively. Balancing the forces and force moments on the lower jaw, three equations are obtained for the force's magnitude of the five muscles, altogether with the magnitude and direction of the articulation's force reaction.

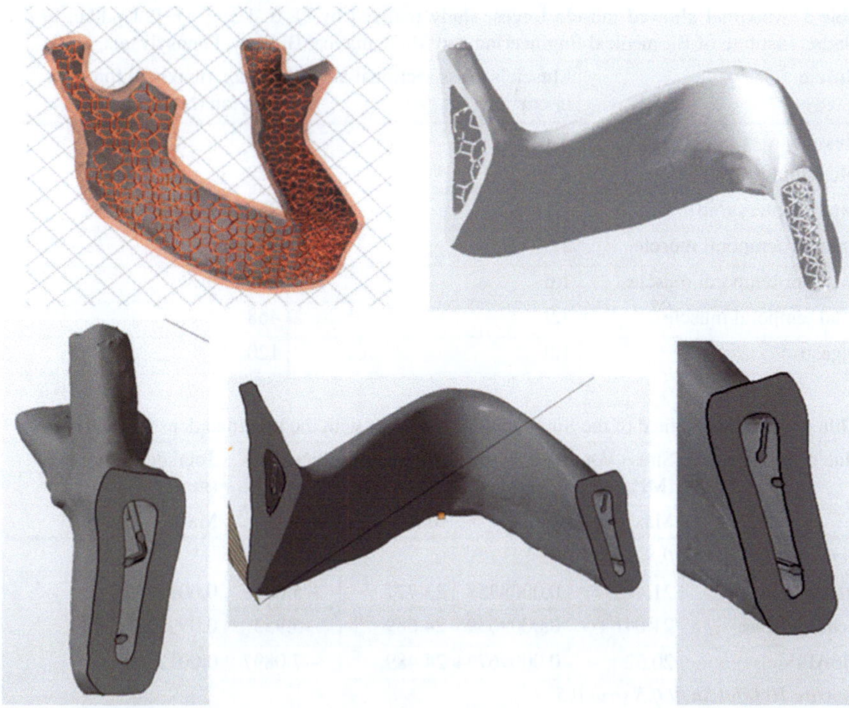

Fig. 10 Thickness and type of strut of the scaffold structures

The following table shows the values of the force's magnitude according to the transversal cross area obtained from the calculation of the equations. The muscle's forces will be used for the force's simulation in the computational programm Creo Simulate on the articulated model in order to make a numerical analysis of an accurate anatomical model.

Taking into account the maximum force values that will be used as a reference (shown in Table 3), the process of simulation of the loads and the assignation of the materials to the models through the ANSYS program is performed. Three cases are considered for the process and in each one the lower jaw was set to the collocation of internal scaffolds, creating a hollow piece with a stability given by the scaffolded structures known as honeycomb, under the following specifications: model 1, density 5, Shell 0.5, Strut 0.5; model 2, density 10, Shell 0.5, Strut 0.5 and model 3, density 20, Shell 0.5, Strut 0.5 with three materials endorsed by the FDA (Healthy bone, stainless steel 316l, titanium alloy Ti6Al4V), together with the parameters of a healthy bone in order to make a more accurate comparison (Fig. 11).

Hereunder, the results obtained for each one of the models with the proposed materials and the complete mandible, particularly for the "Displacement", "Stress Von Mises" and "Maximum Shear Stress" tests, are shown in Fig. 12. Furthermore, the numerical results are organized in the following table (see Table 4). The models

Table 3 Maximal allowed muscle forces, study realized by Q. S. Kang, D. P. Updike, Eric P. Salathe, Institute of Biomedical Engineering and Mathematical Biology, Pennsylvania, USA

Muscle	Muscle's cross sectional area in cm^2	F max. (maximal force) in Newton (N)
Masseter	3.4	408
Internal pterygoid	1.9	228
External pterygoid	2.1	252
Anterior temporal muscle	2.6	312
Posterior temporal muscle	1.6	192
Total temporal muscle	3.9	468
Digastric	1.0	120

Table 4 Results obtained of the Simulation of the loads with the specified densities and loads

Material	Stress Von Mises (MPa)		Maximum principal stress (MPa)		Total deformation (mm)	
	Max	Min	Max	Min	Max	Min
Density 5 Infill Shell 0.5 strut 0.5						
Stainless Steel	21.445	0.0008358	23.777	−5.0741	0.0012435	0
Healthy Bone	21.041	0.00090395	24.089	−5.953	0.099222	0
TI6AI4V	20.52	0.00096734	24.489	−7.0897	0.0012764	0
Density 10 Infill Shell 0.5 strut 0.5						
Stainless Steel	22.504	6.82E−10	20.995	−8.6213	0.0014646	0
Healthy Bone	22.1	5.93E−10	21.239	−9.604	0.013175	0
TI6AI4V	22.566	6.73E−10	21.545	−10.873	0.0029003	0
Density 20 Infill Shell 0.5 strut 0.5						
Stainless Steel	26.755	0	35.313	−10.597	0.0016111	0
Healthy Bone	26.303	0	35.963	−11.791	0.12877	0
TI6AI4V	25.707	0	36.704	−13.351	0.0028335	0

are solid structures with an internal scaffolded geometry with a honeycomb pattern. The objective of these simulations is to prove the level of optimization of every single material of the proposed model and also if the internal structure is capable of showing the necessary resistance of the natural loads created by the muscles and other anatomical structures (Fig. 13).

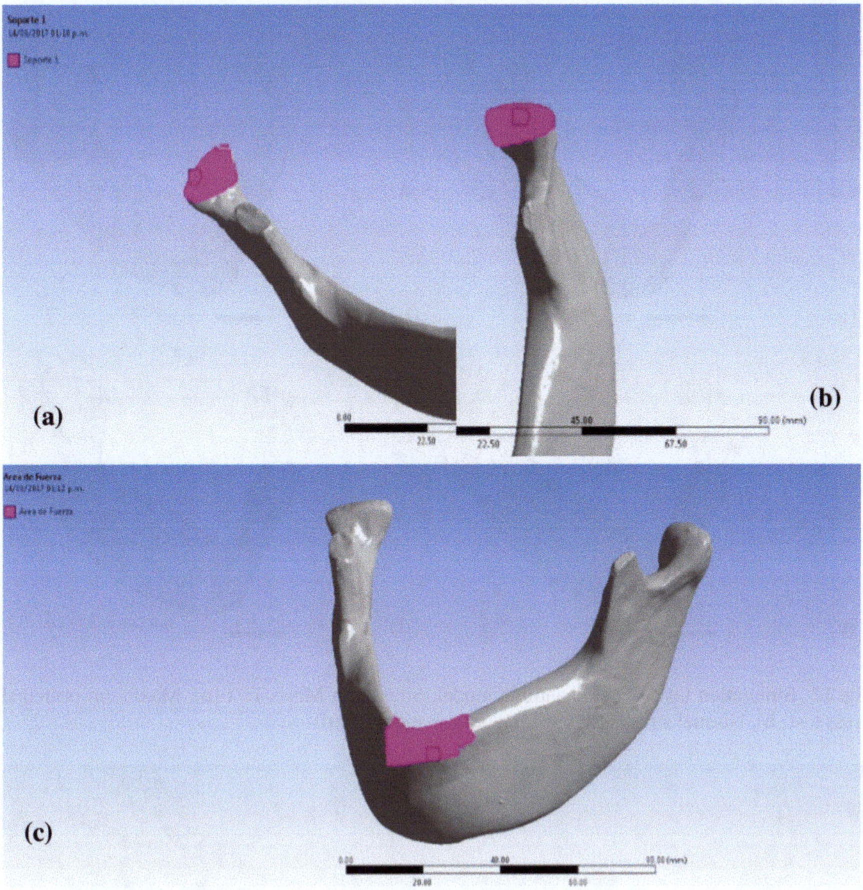

Fig. 11 Placement of the vectors for force recognition over the lower jaw (**a**, **b**), located on both rami of the mandible and on the border (**c**)

6 Results

In order to select the right material, the maximum shear stress principle is applied. For instance, the evaluation of the obtained results concerning the von Mises stress test are compared against the mechanical tensile strength of each material. Likewise, maximum shear stress test indicates the maximum shear stress localized in a determined area and, as a consequence, it is taken as a base for the material's evaluation. This means that the material is resistant enough to withstand the muscular forces.

To analyze these values, the mechanical tensile strength of each material must be taken into account and also be compared against the medium values obtained from the von Mises Stress test. First the values of the test must be inferior to the limits of the mechanical tensile strength, because all the values found in a range of

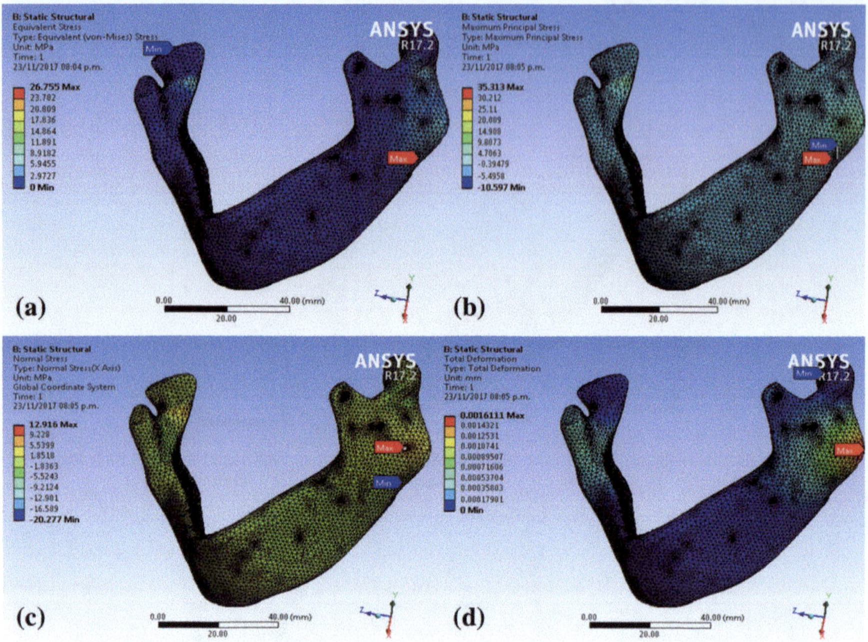

Fig. 12 Simulation tests of the mandible model, Stress von Mises Test (**a**), Maximum principal stress test (**b**), Normal stress test (**c**), Total deformation test (**d**)

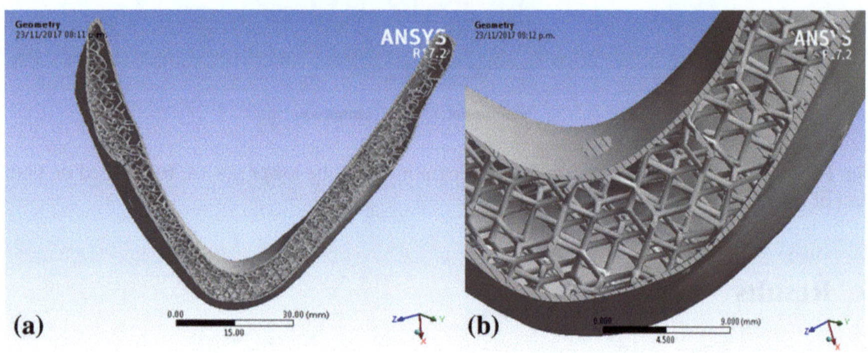

Fig. 13 Strut of the scaffold localized inside the original jaw model

values inferior to the limit are indicators of an optimal material's resistance. It also states that the material must not lose its original form by deformation and should be considered as an option of which the prosthesis should be built. For the evaluation of the second test the medium values of the mechanical tensile strength material need to be compared with the Maximum Shear Stress test. If the values of the tests are inferior to the medium values of the mechanical tensile strength, the material will be capable of resisting the loads (Tables 5, 6, 7 and 8).

Table 5 Comparative values of the mechanical tensile strength of the first limit (superior limit), second limit (inferior limit) for stainless steel 316L, healthy bone and Ti6Al4V

Material	Satinless steel (MPa)	Healthy Bone	TI6AL4V (MPa)
Mechanical Resistance to Traction limit 1	490	–	800
Mechanical Resistance to Traction limit 2	520	167 MPa	1450
MRT Medium value for limit 1	245	–	400
MRT Medium value for limit 1	260	83.5 MPa	725

7 Discussion of the Results

All the obtained results indicate that the internal structure of honeycomb scaffolds are able to resist the loads to which the models were submitted, not being the density the key factor, since when comparing the values of the mechanical tensile strength with the results obtained through the tests, a value of a 95% inferior to the limits is obtained when the models are built in stainless steel 316L. If the titanium alloy is assigned, the percentage is a 97–98% inferior to the limits of the mechanical tensile strength, making this material the best for its construction. Likewise, it can be appreciated that when the material has a higher density level, the percentage that is inferior to the mechanical tensile strength's limit is higher too, showing also a higher stability of the models.

Similar research has demonstrated that the 3D printing systems will contribute to optimize and personalize the new implant in benefit of the patients, however high level computational resources are necessaries to determine its structural integrity [25–28].

8 Conclusions

- Promising results have been obtained. A scaffold structure can improve the design of prosthesis.
- The reduced quantity of 3D printed material does not affect in a radical way the mechanical properties of the prosthesis and the models have proven to conserve their stability.
- The pattern of compressive stress-strain behavior depends on the porosity of the scaffolds.
- The compressive stiffness, yield strength and yielding strain are dependent of the density percentage in each model.

Table 6 Results in percentage of the first model, with a density of 5, Shell 0.5, Strut 0.5 after comparing the values of both mechanical tensile strength limits with the results of the von Mises Stress Test and the medium values of the mechanical tensile strength with the results of the Maximum Shear Stress test

Stainless steel	
490 MPa	***Stress von Mises***
	21.445
Percentage between "Stress von Mises vs. Limit 1"	4.376530612
100%—Percentage of the case	95.62346939
520 MPa	**Stress von Mises**
	21.445
Percentage between "Stress von Mises vs. Limit 2"	4.124038462
100%—Percentage of the case	95.87596154
245 MPa	***Maximum Shear Stress***
	23.777
Maximum Shear Stress versus Limit 1	9.704897959
100%—Percentage of the case	90.29510204
260 MPa	***Maximum Shear Stress***
	23.777
Maximum Shear Stress versus Limit 2	9.145
100%—Percentage of the case	90.855
Healthy bone	
135 MPa	***Stress Von Mises***
	21.041
Percentage between "Stress von Mises vs. Limit 1"	15.58592593
100%—Percentage of the case	84.41407407
167 MPa	***Stress von Mises***
	21.041
Percentage between "Stress von Mises vs. Limit 2"	12.5994012
100%—Percentage of the case	87.4005988
67.5 MPa	***Maximum Shear Stress***
	24.089
Maximum Shear Stress versus Limit 1	35.68740741
100%—Percentage of the case	64.31259259
83.5 MPa	***Maximum Shear Stress***
	24.089
Maximum Shear Stress versus Limit 2	28.8491018
100%—Percentage of the case	71.1508982

(continued)

Table 6 (continued)

TI6AI4V	
800 MPa	***Stress von Mises***
	20.52
Percentage between "Stress von Mises vs. Limit 1"	2.565
100%—Percentage of the case	97.435
1450 MPa	***Stress von Mises***
	20.52
Percentage between "Stress von Mises vs. Limit 2"	1.415172414
100%—Percentage of the case	98.58482759
400 MPa	***Maximum Shear Stress***
	24.489
Maximum Shear Stress versus Limit 1	6.12225
100%—Percentage of the case	93.87775
725 MPa	***Maximum Shear Stress***
	24.489
Maximum Shear Stress versus Limit 2	3.377793103
100%—Percentage of the case	96.6222069

Table 7 Results in percentage of the first model, with a density of 10, Shell 0.5, Strut 0.5 after comparing the values of both mechanical tensile strength limits with the results of the von Mises Stress Test and the medium values of the mechanical tensile strength with the results of the Maximum Shear Stress test

Stainless steel	
490 MPa	***Stress von Mises***
	22.504
Percentage between "Stress von Mises vs. Limit 1"	4.592653061
100%—Percentage of the case	95.40734694
520 MPa	***Stress von Mises***
	22.504
Percentage between "Stress von Mises vs. Limit 2"	4.327692308
100%—Percentage of the case	95.67230769
245 MPa	***Maximum Shear Stress***
	20.995
Maximum Shear Stress versus Limit 1	8.569387755
100%—Percentage of the case	91.43061224
260 MPa	***Maximum Shear Stress***

(continued)

Table 7 (continued)

	20.995
Maximum Shear Stress versus Limit 2	8.075
100%—Percentage of the case	91.925
Healthy bone	
135 MPa	***Stress von Mises***
	22.1
Percentage between "Stress von Mises vs. Limit 1"	16.37037037
100%—Percentage of the case	83.62962963
167 MPa	***Stress von Mises***
	22.1
Percentage between "Stress von Mises vs. Limit 2"	13.23353293
100%—Percentage of the case	86.76646707
67.5 MPa	***Maximum Shear Stress***
	21.239
Maximum Shear Stress versus Limit 1	31.46518519
100%—Percentage of the case	68.53481481
83.5 MPa	***Maximum Shear Stress***
	21.239
Maximum Shear Stress versus Limit 2	25.43592814
100%—Percentage of the case	74.56407186
TI6AI4V	
800 MPa	***Stress von Mises***
	22.566
Percentage between "Stress von Mises vs. Limit 1"	2.82075
100%—Percentage of the case	97.17925
1450 MPa	***Stress von Mises***
	22.566
Percentage between "Stress von Mises vs. Limit 2"	1.556275862
100%—Percentage of the case	98.44372414
400 MPa	***Maximum Shear Stress***
	21.545
Maximum Shear Stress versus Limit 1	5.38625
100%—Percentage of the case	94.61375
725 MPa	***Maximum Shear Stress***
	21.545
Maximum Shear Stress versus Limit 2	2.971724138
100%—Percentage of the case	97.02827586

Table 8 Results in percentage of the first model, with a density of 20, Shell 0.5, Strut 0.5 after comparing the values of both mechanical tensile strength limits with the results of the von Mises stress test and the medium values of the mechanical tensile strength with the results of the maximum shear stress test

Stainless steel	
490 MPa	***Stress von Mises***
	26.755
Percentage between "Stress von Mises vs. Limit 1"	5.460204082
100%—Percentage of the case	94.53979592
520 MPa	***Stress von Mises***
	26.755
Percentage between "Stress von Mises vs. Limit 2"	5.145192308
100%—Percentage of the case	94.85480769
245 MPa	***Maximum Shear Stress***
	35.313
Maximum Shear Stress vs. Limit 1	14.41346939
100%—Percentage of the case	85.58653061
260 MPa	***Maximum Shear Stress***
	35.313
Maximum Shear Stress versus Limit 2	13.58192308
100%—Percentage of the case	86.41807692
Healthy bone	
135 MPa	***Stress von Mises***
	26.303
Percentage between "Stress von Mises vs. Limit 1"	19.4837037
100%—Percentage of the case	80.5162963
167 MPa	***Stress von Mises***
	26.303
Percentage between "Stress von Mises vs. Limit 2"	15.7502994
100%—Percentage of the case	84.2497006
67.5 MPa	***Maximum Shear Stress***
	35.963
Maximum Shear Stress versus Limit 1	53.27851852
100%—Percentage of the case	46.72148148
83.5 MPa	***Maximum Shear Stress***
	35.963
Maximum Shear Stress versus Limit 2	43.06946108
100%—Percentage of the case	56.93053892
TI6AI4V	
800 MPa	***Stress von Mises***
	25.707

(continued)

Table 8 (continued)

Percentage between "Stress von Mises vs. Limit 1"	3.213375
100%—Percentage of the case	96.786625
1450 MPa	***Stress von Mises***
	25.707
Percentage between "Stress von Mises vs. Limit 2"	1.772896552
100%—Percentage of the case	98.22710345
400 MPa	***Maximum Shear Stress***
	36.794
Maximum Shear Stress versus Limit 1	9.1985
100%—Percentage of the case	90.8015
725 MPa	***Maximum Shear Stress***
	36.794
Maximum Shear Stress versus Limit 2	5.075034483
100%—Percentage of the case	94.92496552

- Different materials can be used in order to build a scaffold structure and the fibers can be oriented in order to get the best structural results.
- DLP (Digital Light Processing)/SLA (Stereolithography)/DMLS (Direct Metal Laser Sintering) 3D printer systems are other alternatives for the production of scaffold models.
- The next step of this research is the validation of the results with experimental analysis.
- Complex FEM analyses were expected in order to get difficult meshing process.

Acknowledgements The authors kindly acknowledge the support given to the National Polytechnic Institute and the National Council for Science and Technology (CONACYT).

References

1. Centers for Medicare and Medicaid Services.: Office of the Actuary, National Health Statistics Group, National Health Care Expenditures Data, 08 Aug 2013 (2012)
2. Alejandro, C.R., Hotza, D., Salmoria, G.V.: Fabricación de Andamios de Hidroxiapatita Por Impresión Tridimensional. Rev. Latinam. Metal. Mat. **34**(2), 262–274 (2014)
3. Langer, R., Vacanti, J.P.: Tissue engineering. Science **260**, 920–926 (1993)
4. Lanza, R., Langer, R., Vacanti, J., Freed, L.E., Vunjak-Novakovic, G.: Principles of Tissue Engineering, 2nd edn. Academic Press. Part II: In Vitro Control of Tissue Development. Tissue Engineering Bioreactors (2000)
5. Ratner, B.D., Hoffman, A.S., Schoen, F.J., Lemons, J.E.: Biomaterials Science: An Introduction to Materials in Medicine, 2nd edn. Academic Press (2004)
6. Van Cleynenbreugel, T., Schrooten, J., Van Oosterwyck, H., Van Vander Floten, J.: Micro-CT-based screening of biomechanical and structural properties of bone tissue engineering scaffolds. Med Bio. Eng. Comput. **44**, 517–525 (2006)

7. LeGeros, R.Z., LeGeros, J.P.: Calcium phosphate biomaterials: preparation, properties, and biodegradation. In: Wise, D.L., Trantolo, D.J., Altobelli, D.E., Yaszemski, M.J., Gresser, J.D., Schwartz, E.R. (eds.) Encyclopedia Handbook of Biomaterials and Bioengineering Part A: Materials, vol. 2, pp. 1429–1463. Marcel Dekker, New York (1995)
8. Vacanti, J.P., Morse, M.A., Saltzman, W.M., Domb, A.J., Peter-Atayde, A., Langer, R.: Selective cell transplantation using bioabsorbable artificial polymers as matrices. J. Pediatr. Surg. **23**(1), 3–9 (1988)
9. Mikos, A.G., Sarakinos, G., Lyman, M.D., Ingber, D.E., Vacanti, J.P., Langer, R.: Prevascularization of porous biodegradable polymers. Biotechnol. Bioeng. **42**, 716–723 (1993)
10. Robinson, B., Hollinger, J.O., Szachowicz, E., Calvarial, Brekke J.: Bone repair with porous d, l-polylactide. Otolaryngol. Head Neck Surg. **112**(6), 707–713 (1995)
11. Boyan, B.D., Hummert, T.W., Dean, D.D., Schwartz, Z.: Role of material surfaces in regulating bone and cartilage cell response. Biomaterials **17**(2), 137–146 (1996)
12. Yoshikawa, T., Ohgushi, H., Tamai, S.: Intermediate bone forming capability of prefabricated osteogenic hydroxyapatite. J. Biomed. Mater. Res. **32**(3), 481–492 (1996)
13. Yannas, I.V., Lee, E., Orgill, D.P., Skrabut, E.M., Murphy, G.F.: Synthesis and characterization of a model extracellular matrix that induces partial regeneration of adult mammalian skin. Proc. Natl. Acad. Sci. U. S. A. **86**, 933 (1989)
14. Kim, S.S., Utsunomiya, H., Koski, J.A., Wu, B.M., Cima, M.J., Sohn, J., Mukai, K., Griffith, L.G., Vacanti, J.P.: Survival and function of hepatocytes on a novel three-dimensional synthetic biodegradable polymer scaffold with an intrinsic network of channels. Ann. Surg. **228**(1), 8–13 (1998)
15. Rout, P.G.J., Tarrant, S.F., Frame, J.W., Davies, J.E.: Interaction between primary bone cell cultures and biomaterials. Part 3: a comparison of dense and macroporous hydroxyapatite. In: Pizzoferratto, A., Ravaglioli, P.G., Lee, A.J.C. (eds.) Bioceramics and Clinical Applications, pp. 591–596. Elsevier, Amsterdam (1988)
16. Brekke, J.H.: A rationale delivery of osteoconductive proteins (a review). Tissue Eng. **2**, 97 (1996)
17. Havelin, L.I.: The Norwegian arthroplasty register, 1st edn. Brit. Editorial Soc. Bone Joint Surg. 88–95 (1999)
18. Mateus Gaviria, E.: Biometales, Mercado Potencial. Metal Actual **26**(1), 22 (2013)
19. Ratner, B.: Biomaterials Science, 1st edn. Elsevier/Academic Press, Amsterdam (2013)
20. Osteonecrosis.: Clinic DAM Madrid. Spain. https://www.clinicadam.com/salud/5/007260.html. Accessed 21 Jan 2017
21. Fernández López, R.G., Arellano Flores, Á.M., Serrano, S.N.V.: Mandibular osteonecrosis linked to bisphosphonates—a clinical study. http://www.scielo.org.mx/scielo.php?script=sci_arttext&pid=S1870199X2013000100008 (2013). Accessed 24 Jan 2017
22. Kang, Q.S., Updike, D.P., Salathe E.P.: Theoretical Prediction of Muscle Forces on the Mandible During Bite. Institute of Biomedical Engineering and Mathematical Biology, Pennsylvania, USA (1990). Accessed 24 Jan 2017
23. Castellano Navarro, J.M., Navarro Navarro, R., Chiriño González, Rodríguez Álvarez, J.P.: Mandibular fractures. Canarias Med. Surg. **4**(11) (2006)
24. Sierra-Martínez, E., Cienfuegos, R.M.: Treatment of the mandibular angle with the AO system. Plast. Surg. MG **14**(3), 126–131 (2004)
25. Beltrán-Fernández, J., Hernández-Gómez, L., Urriolagoitia-Calderón, G., González-Rebatú, A., Urriolagoitia-Sosa, G., Galán Vera, M., Escalante-Rodríguez, E.: Assessment of the Structural Integrity of C3–C5 Cervical Porcine Vertebrae Model Based on 2D Classic CAD, 3D Scanner and 3D Computed Tomography. Anal. Des. Biol. Mater. Struct. 3–17 (2012). http://dx.doi.org/10.1007/978-3-642-22131-6_1
26. Beltran and cols: Modelling of a cervical plate and human cervical section C3-C5 under compression loading conditions using the finite element method. Appl. Mech. Mater. **13–14**, 49–56 (2008)

27. Moreno-Garibaldi, P., Beltrán-Fernández, J.A., Hernández-Gómez, L.H., López-Saucedo, F., Corro-Valdez, N., López-Liévano, A.: Caracterización mecánica de material compuesto enfocado hacia aplicaciones maxilofaciales. DYNA-Ingeniería e Industria **94**(4) (2017)
28. Beltrán-Fernández, J., Valdez, N., Rebatu, A., Hernández Gómez, L.: Biomechanical Design of a Vertebral Distractor for Fractured Bodies Useful in Kyphoplasty. Adv. Structured Mater. 89–122 (2015)

Evaluation of the Structural Integrity of a Boiling Water Reactor Skirt Under Stationary and Transient Loading Conditions

Yunuén López Grijalba, Luis Héctor Hernández Gómez, Pablo Ruiz López, Guillermo Manuel Urriolagoitia Calderón, Gilberto Soto Mendoza, Alejandra Armenta Molina and Juan Alfonso Beltrán Fernández

Abstract In this paper, the stress analysis of the support structure (skirt) of a Boiling Water Reactor (BWR-5) vessel is presented. Three transient loading conditions, which take place during the life of the plant, were evaluated, in order to determine the structural integrity under fatigue loading. The concept of Cumulative Usage Fatigue (CUF) was used. The guidelines of the subsection NB and subsection NF of Section III of the ASME Code were taken into account in the evaluation in conjunction with the Finite Element Analyses (FEA). In addition, the environmental fatigue penalty factors (F_{en}) were also considered during the evaluation. The environmental conditions affected the integrity of the skirt. Besides, a perturbation in the stresses

Y. L. Grijalba · L. H. H. Gómez (✉) · G. M. U. Calderón · G. S. Mendoza · A. A. Molina · J. A. B. Fernández
Sección de Estudios de Posgrado e Investigación, Instituto Politécnico Nacional, ESIME Unidad Zacatenco, Edificio 5, 3er Piso, Unidad Profesional Adolfo López Mateos. Col. Lindavista, 07738 Mexico City, Mexico
e-mail: luishector56@hotmail.com

Y. L. Grijalba
e-mail: yunuen.grijalba@gmail.com

G. M. U. Calderón
e-mail: urrio332@hotmail.com

G. S. Mendoza
e-mail: gilberto789456@hotmail.com

A. A. Molina
e-mail: alejandra_armenta_m@esimez.mx

J. A. B. Fernández
e-mail: jbeltranf@hotmail.com

P. R. López
Comisión Nacional de Seguridad Nuclear y Salvaguardias, Dr. José Ma. Barragán no. 779, Colonia Narvarte, Del. Benito Juárez, C. P., 03020 Mexico City, Mexico
e-mail: pruiz@cnsns.gob.mx

© Springer International Publishing AG, part of Springer Nature 2019

A. Öchsner and H. Altenbach (eds.), *Engineering Design Applications*, Advanced Structured Materials 92, https://doi.org/10.1007/978-3-319-79005-3_31

was observed due to the dynamic forces generated by the jet pumps, which increased the CUF in the skirt.

Keywords Nuclear power plant (NPP) · Cumulative usage fatigue (CUF) Environmental factor (F_{en}) · ASME section III · Submodeling FEM

Abbreviations

BWR	Boiling Water Reactor
CUF	Cumulative Usage Factor
FEA	Finite Element Analysis
NPP	Nuclear Power Plant
FEM	Finite Element Method
PWR	Pressure Water Reactor
PVR	Pressure Vessel Reactor
AEF	Assisted Environmental Fatigue
ppm	Parts per million
DO	Dissolved Oxygen
CS	Carbon Steel
LAS	Low Alloy Steel

1 Introduction

Recent studies of environmental fatigue have shown that there is a significant reduction in fatigue life due to environmentally assisted fatigue. Fatigue tests are usually performed at a constant temperature and in a controlled atmosphere. In fact, the conditions of operations of nuclear plants involve variations of temperature and parameters of operation, among other factors. The safety aspects concerning the management of the ageing programs for nuclear power plants (NPP) require the implementation of effective programs for the detection and mitigation of the effects of degradation of the systems, components, and structures that are important for safety, in order to ensure their integrity and functionality throughout their life time.

Boiling Water Reactors (BWRs) incorporate unique structural features that are different to Pressure Water Reactors (PWRs) in the response to the expected conditions during a severe accident. These structural differences and the operating procedures have mainly focused on the accidents that may occur in the period preceding the removal of the waste fuel from the core in the lower chamber of the Pressure Reactor Vessel (PVR) [1].

Accordingly, studies have been carried out with respect to this zone of the reactor. The reactor vessel contains fluids at high temperatures which results in a heat flow through the skirt [2]. The heat flow and the resultant thermal stresses give rise to "the

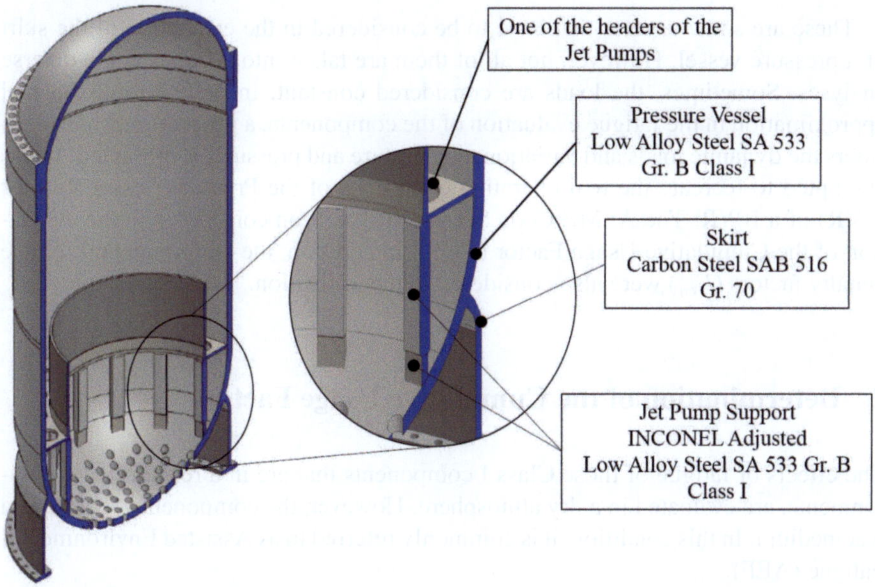

Fig. 1 Junction of the housing, head and skirt of a BWR 5

hot skirt problem" which is generalized for determining the stress distribution near the joint of the shell, lower cap and skirt (see Fig. 1) of a vertical pressure vessel. It is given by the combination of any of the following situations: (1) Temperature change takes place axially, along any of the elements shown in Fig. 1; (2) Temperature varies linearly through the thickness of an element; (3) There is internal pressure at the shell and the head of the vessel, but not in the skirt; (4) There is an additional axial load, due to the reaction of the jet pumps; (5) External moments and loads can be applied to the union of the three elements during extraordinary or emergency conditions.

All these loads have been modelled in accordance with the geometry of the skirt. Besides, materials are considered to be elastic. As the thermal stresses are reduced, the material has the stiffness necessary to withstand dynamic loads [3].

There are other aspects which have been analyzed in the open literature. This is the case of Wilkie [4]. The case, in which explosions could take place close to the skirt due to the concentration of hydrogen (H_2), was analyzed. It could cause great structural damage of the reactor support system. Other studies have focused on determining if there will be a collapse in the bottom head of the vessel if it is torispherical [5].

Debris may be deposited at the bottom of the vessel, making the bottom thicker. Such debris can reach the melting point. The lower head would not be affected until all debris is removed from this area.

The results of Hodge and coworkers [6] revealed that there are many penetration faults that can occur due to the weakening of the welded parts of the control rods at the bottom of the vessel or by tubes welded to the wall of the reactor vessel.

These are some aspects that have to be considered in the evaluation of the skirt of a pressure vessel. However, not all of them are taken into account in the diverse analyses. Sometimes, the loads are considered constant. In order to make a real approximation in the fatigue evaluation of the components, a procedure, which considers the dynamic loads and variation temperature and pressure, is presented. It was attempted to recreate the real operating conditions of the Pressure Vessel Reactor (PVR) of a BWR. The ASME Code Section III has been considered in the evaluation of the Cumulative Usage Factor (CUF). In addition, the environmental fatigue penalty factors (F_{en}) were also considered in the evaluation.

2 Determination of the Cumulative Usage Factor

The effects of fatigue of those Class I components that are in a reactor water environments, are evaluated in a dry atmosphere. However, the components operate in a wet medium. In this condition, it is commonly referred to as Assisted Environmental Fatigue (AEF).

The evaluation and selection of Class I components is performed according to NUREG/CR-6260 [7]. The CUF evaluation of Class I components is straightforward and may require a relatively simple procedure. With this information, environmental fatigue factors (F_{en}) are applied according to the existing CUF values [8].

According to the ASME Code [9], the CUF is calculated as the summation of incremental fatigue contributions derived from the fluctuation of thermal and mechanical stresses in a metal component. This value must be lower than 1.0 to demonstrate an acceptable design behavior. As is specified in the ASME Code, the fatigue analysis procedure consists of the following steps: (1) The number of stress cycles (n_i), that will take place on each one of the transient loading conditions on the life of the component of interest, is determined; (2) The stress amplitude of the cycles of each transient loading conditions is calculated; (3) The previous mentioned stress intensity S_{alt} of amplitudes is calculated. With this data, the (4) The fatigue life (N_i) of each transient loading condition is established with the fatigue curve of the ASME Code; and (5) The Cumulative Usage Factors (CUF) are evaluated for each transient loading conditions. The Cumulative Usage Factor is determined by:

$$CUF = \sum_{i=1}^{M} \frac{n_i}{N_i} \tag{1}$$

where:

n_i number of stress cycles of a transient loading condition i
N_i number allowable cycles of a transient loading condition i
M total number of transients

Table 1 Parameter used in the calculation of F_{en}

For carbon and low-alloy steel NUREG/CR-6583 [12]:	
S	0.015 weight percent or higher, Sulfur content
DO	0.5 ppm or higher, Level of Dissolved Oxygen in the primary fluid
T	350 °C or higher, temperature of the metal in contact with the primary fluid
$\acute{\epsilon}$	0.001%/sec or lower, strain rate

3 Fatigue Environmental Factor F_{en} Formulations

In accordance with [10], the Environmental Cumulative Usage Factor, CUF_{en}, can be calculated in accordance with guidance from either NUREG/CR-5704 [11] (for austenitic stainless steels), NUREG/CR-6583 [12] (for carbon and low alloy steels), or NUREG/CR-6909 [13] (for all materials).

Regarding the evaluation of the structural integrity of a joint made of low alloy steel, the parameters shown in Table 1 will be used. In this case, the worst case value will be considered due to the uncertainty and lack of data.

According to the new rules established by NUREG/CR-6909 [13] the F_{en} should be calculated by the following equations:

For carbon steel (CS)

$$F_{en} = exp\left(0.632 - 0.101S^*T^*O^*\dot{\varepsilon}^*\right) \quad (2)$$

For Low Alloy Steel (LAS):

$$F_{en} = exp\left(0.702 - 0.101S^*T^*O^*\dot{\varepsilon}^*\right) \quad (3)$$

Where S^*, T^*, O^* and $\acute{\epsilon}^*$ are the transformed sulfur content, service temperature, dissolved oxygen (DO), and strain rate, respectively, which are defined as follows:

$S*$ = 0.001 for S $\leq$ 0.001 wt%
=S for S $\leq$ 0.015 wt%
=0.015 for S > 0.015 wt%
$T*$ =0 for T < 150 °C
=(T-150) for 150 $\leq$ T $\leq$ 350 °C
$O*$ =0 for dissolved oxygen, DO $\leq$ 0.04 parts per million (ppm)
=ln(DO/0.04) for 0.04 ppm < DO < 0.5 ppm
=ln(12.5) for DO > 0.5 ppm
$\acute{\epsilon}*$ =0 for strain rate, $\acute{\epsilon}$ > 1%/s
=ln $\dot{\varepsilon}$ for 0.001 $\leq \acute{\epsilon} \leq$ 1%/s
=ln(0.001) for $\acute{\epsilon}$ < 0.001%/s

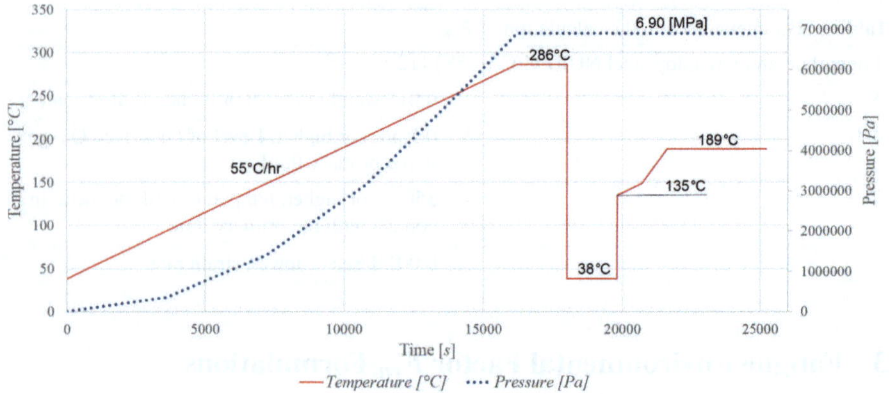

Fig. 2 Behavior of pressure and temperature of transient loading condition 1

4 Analyzed Cases

Three transient loading conditions were considered in the evaluation of the joint between the skirt and the vessel (Fig. 1). The first transient loading condition is referred to as the start-up and turbine roll. It takes place when the reactor is put into operation. It was postulated that a BWR has operated for 27 years. Besides, it is assumed that the reactor is stopped every 18 months, because it is refueled. At the beginning of this transient loading condition, there is a temperature ramp, which varies 55 °C/h [14]. It is illustrated in the Fig. 2.

For the case of transient loading condition 2, the phase of reduction to 0% power and hot stand was considered. In other words, the reactor is stopped and cooled, and the associated variations in temperature and pressure are shown in Fig. 3. These conditions take place during every refueling outage. The number of cycles is the same as in the case of the transient loading condition 1.

The third transient loading condition is referred when a complete plant shutdown occurs due to a loss of the feedwater systems, before the reactor is depressurized to a point where the shutdown cooling system can be placed into operation (see Fig. 4). This is a critical condition, which has a low probability to take place during the whole life of the plant. In accordance with the data registered, these events have occurred in some BWR plants around the world on an average of 8 times during the performance of quarterly surveillance test [15]. This condition is uncommon. However, it was considered in this paper, in order to have a severe combinations of transient loading conditions.

For comparison, a static analysis was carried out. The hydrostatic test was evaluated. The internal pressure is 8.95 MPa at 38 °C.

The thermal and structural analyses were made using the Finite Element Method (FEM). ANSYS® code was used to solve. The area of interest was the union of a BWR Pressure Vessel with its skirt. The materials of these elements are shown in

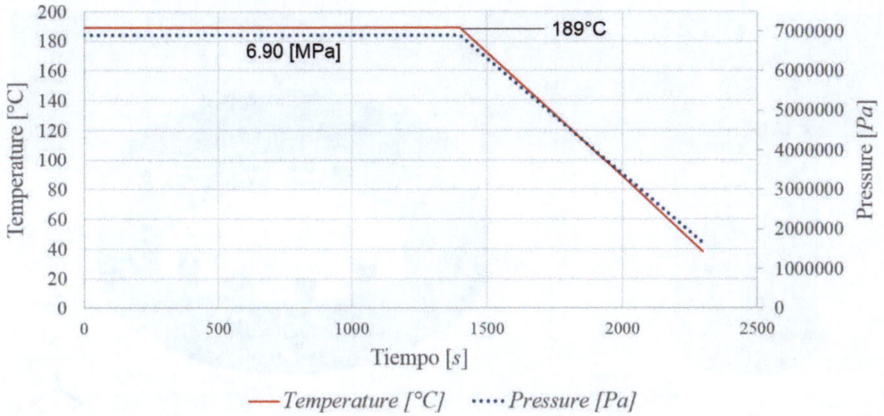

Fig. 3 Behavior of pressure and temperature of transient loading condition 2

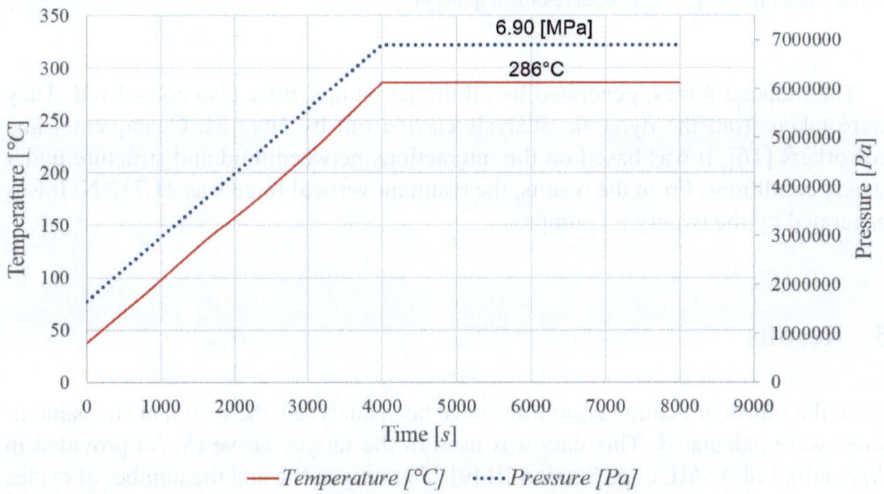

Fig. 4 Behavior of pressure and temperature of transient loading condition 3

Fig. 1. The model required a large number of elements. Therefore, the solutions were obtained by employing the technique of submodeling. A mesh sensitivity study was conducted for the submodel in order to improve the convergence of the peak stresses of each transient. Figure 5 illustrates the finite element models and summarizes the sensitivity analysis. Best results are obtained when the evaluation of the elements is close to 1. A great number of these elements are in this range. It made a great improvement of quality of the mesh. The whole model has 197,682 nodes and 100,496 elements, meanwhile the submodel was meshed with a total of 41,447,542 nodes and 29,087,795 elements.

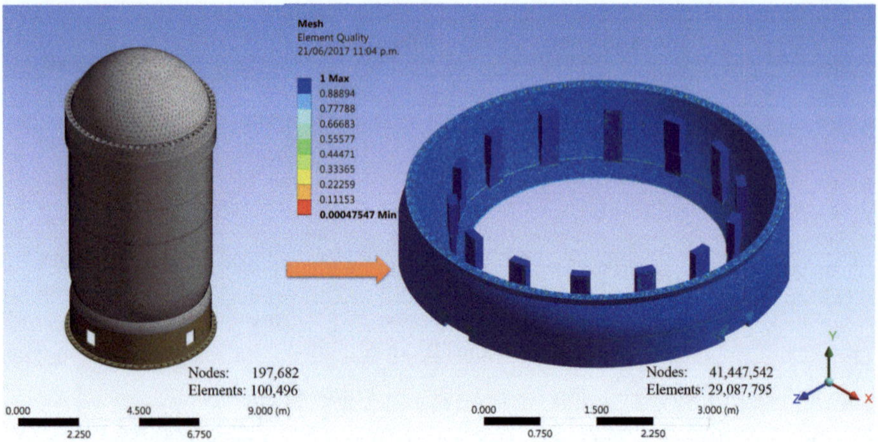

Fig. 5 Mesh model and submodel (element quality)

The induced forces, generated by all the jet pumps, were also considered. They were taken from the dynamic analysis carried out by Noel M. Cuahquentzi and coworkers [16]. It was based on the interactions between fluid and structure under steady conditions. From the results, the resultant vertical force was 21,712 N. It was generated by the twenty jet pumps.

5 Results

Once the transient loading conditions have been analyzed, the resultant stress intensities were calculated. This data was used in the fatigue curve (S, N) provided in Appendix I of ASME code Section III [9]. The fatigue life and the number of cycles of each transient were used to determine the Cumulative Usage Factor. The results are shown in Table 2. The total Cumulative Usage Factor of the three transients is below 1.

A comparison of the three transient analyses against the static analysis was performed. The range of the transient Von Mises stresses are lower than those of the hydrostatic test (Fig. 6). Besides, the transient stresses have a periodic variation along the circumference in which the vessel is supported. It can be considered that this variation is due to the reaction of the jet pumps.

Transient loading condition 3 has a series of higher stresses than those obtained from transient loading condition 1 and 2. However, the CUF obtained (see Table 2) is greater for the first two transients. It is considered that their recurrence increment their influence on the results and increase the fatigue damage. It should be remembered that transient loading conditions 1 and 2 take place each time the reactor is refueled, whereas transient loading condition 3 may or not be present in combination with

Table 2 Cumulative Usage Factor obtained

Fatigue Curves used of ASME Code Section III [9]		
Material evaluated in the section Low Alloy Steel SA 533 Gr.B Class I		
UTS (Ultimate Tensile Strength) used	680.21 MPa [98.65612 ksi]	
Cumulative Usage Factor obtained in each transient	1	0.2571429
	2	0.00225
	3	0.0074074
	CUF_{Total}	0.2668003

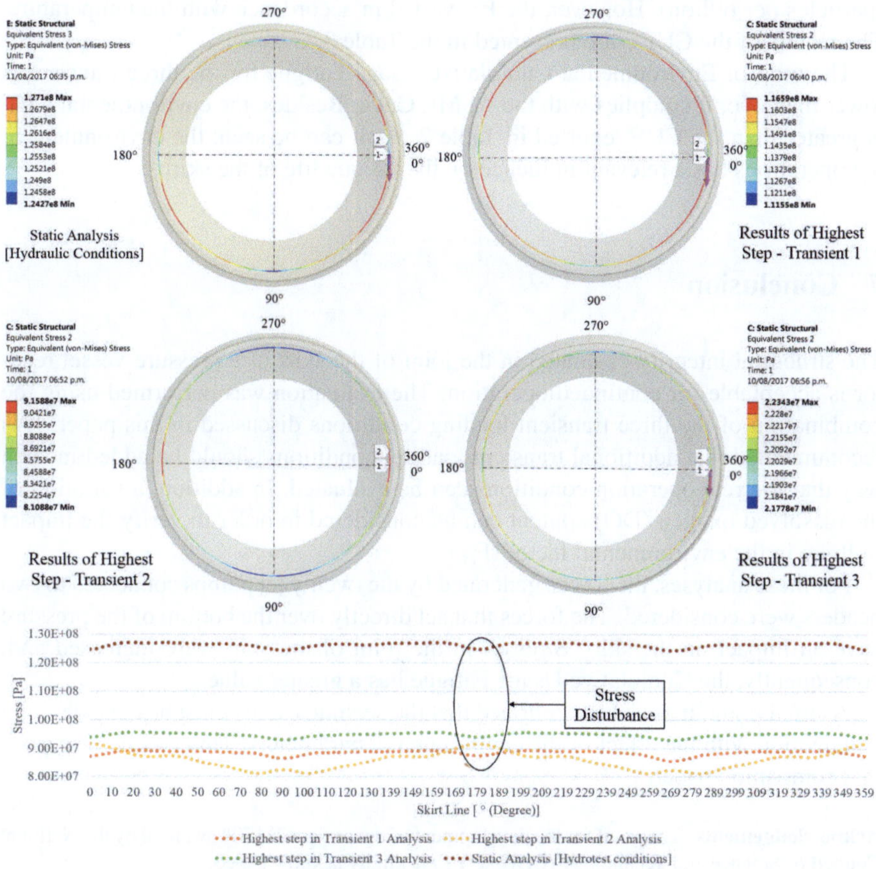

Fig. 6 Comparison of results obtained by FEM in the four cases of study

the other two. Regarding the refueling outage, transient loading condition 1 shows higher variation in pressure and temperature. Therefore, a higher CUF is generated.

Table 3 Environmental Cumulative Usage Factor obtained

Transient Pair	1	0.520098651
	2	0.004540564
	3	0.014959477
	CUF_{en}	0.539596

6 F_{en} Calculations

F_{en} calculations were performed for the three transient cases. According to the values provided in Table 1, the DO concentration for the three transient was set to 5 ppb (particles per billion). However, the F_{en} varied in accordance with the temperature. The results of the CUF_{en} are presented in the Table 3.

The total of Environmental Cumulative Usage Fatigue for the three transient is lower than one. It complies with the ASME Code. Besides, the environmental CUF is greater than the CUF reported in Table 2. As it can be seen, the environment of the operations has a relevant influence on the fatigue life of the skirt.

7 Conclusion

The structural integrity evaluated in the joint of the skirt of a pressure vessel reactor is acceptable for continued operation. The evaluation was performed using the combination of the three transient loading conditions discussed in this paper. It is recommended that additional transient loading conditions should be added in such way that the real operation conditions can be evaluated. In addition, a variation of the dissolved oxygen (DO) content can be considered in order to verify the impact of these in the environmental factor (F_{en}).

For these analyses, the forces generated by the twenty jet pumps connected to two headers were considered. The forces that act directly over the bottom of the pressure have an impact on the skirt. Stresses in the joint of the skirt were increased and, consequently, the Cumulative Usage Fatigue has a greater value.

As final point, it must be mentioned that the seismic events must be considered in conjunction with the effect of the environmental parameters. This will be discussed in a companion paper.

Acknowledgements The grant for the development of the project 211704 awarded by the National Council of Science and Technology (CONACYT) is kindly acknowledged.

References

1. NUREG/CR-2940, Greene, S.R.: Realistic simulation of severe accidents in BWRs computer modeling requirements. ORNL/TM8517, April 1984
2. McManus, J.P.: Thermal stress analysis of pressure vessels with cylindrical skirt supports. Thesis and Dissertation Collection, Calhoun: The NPS Institutional Archive, Dudley Knox Library, Monterey, California, USA, 93943, 1964
3. Cruz, J.R.B., Mattar, M.N., de Miranda, C.A.J., Bezerra, L.M.: Aspects of design and stress classification of a PWR support structure. Structural Mechanics Division, COPES/IPEN-CNEN/SP, Sao Paulo, Brazil
4. Wilkie, D.: Reactor vessel support skirt concerns. Crowd Sourced Information & Analysis Without Focus On Profit, SimplyInf.Org, consulted in November 2016
5. Mkrtchyan, L., Schau, H., Wolf, W., Holzer, W., Wernicke, R., Trieglaff, R., Stress analyses for reactor pressure vessels by the example of a product line '69 Boiling Water Reactor. Kerntechnik, Agosto 2011. https://doi.org/10.3139/124.110173
6. Hodge, S.A., Ott, L.J.: BWR reactor vessel head failure modes. Boiling Water Reactor Severe Accident Technology (BWRSAT) Program, Oak Ridge National Laboratory, Oak Ridge, Tennessee, 10 de Mayo 1989
7. NUREG/CR-6260 INEL-95/0045, Ware, A.G., Morton, D.K., Nitzel, M.E.: Application of NUREG/CR-5999 Interim fatigue curves to selected nuclear power plants components. Idaho Technologies Company, February 1995
8. EPRI 1024995, Chu, S.: Environmentally assisted fatigue screening. Process and technical basis for identifying EAF limiting locations, Final report. Palo Alto, California, August 2012
9. ASME Boiler and Pressure Vessel Code, Section III, Division 1—Subsection NB, Class I Components: Rules for construction of nuclear facility components. American Society of Mechanical Engineers, 2010
10. NUREG-1801, Revision 2: Generic Aging Lessons Learn (GALL) Report. U.S. Nuclear Regulatory Commission, December 2010
11. NUREG/CR-5704 (ANL-98/31): Effects of LWR coolant environments on fatigue design curves of austenitic stainless steels, 1999
12. NUREG/CR-6583 (ANL-97/18): Effects of LWR coolant environments on fatigue design curves of carbon and low-alloys steels, 1998
13. NUREG/CR-6909 (ANL-06/08): Effects of LWR coolant environments on the fatigue life of reactor materials, Final report, February 2007
14. Ranganath, S., Carter, B.: Simplified methods for including environmental effects in ASME code fatigue analysis for BWR2, XGEN engineering, Electric Power Research Institute, U.S., NRC Office, Washington, DC, March 2010
15. Poloski, J.P., Grant, G.M., Gentillon, C.D., Galyean, W.J., Roesener, W.S.: Reactor core isolation cooling system reliability, 1987–1993, NUREG/CR-5500, Idahi National Engineering Laboratory, Nuclear Risk Management Technologies Department, Lockheed Martin Idaho Technologies Company, Idaho Falls, Idaho 83415, June, 1997
16. Cuahquetzi, N.M., Gómez, L.H.H., López, P.R., Calderón, G.U., Fernández, J.A.B., Sosa, G.U., Cuamatzi, E.F., Ramírez, A.O., Ángeles, B.R.: "Evaluation of the structural integrity of the jet pumps of a Boiling Water Reactor under hydrodynamic loading", Defect and diffusion forum, vol. 348, pp. 261–270, Trans Tech Publications, Switzerland (2014)

Effective Thermal Conductivity of Fiber Reinforced Composites Under Orientation Clustering

Zia Javanbakht, Wayne Hall and Andreas Öchsner

Abstract A parametric finite element analysis was carried out to investigate the sensitivity of the effective thermal conductivity of fibers to orientation clustering. Randomly-positioned fibers with von Mises orientation distributions were used in different considerations and volume fractions to generate the dispersion in a partitioned representative volume element. It was found that increasing the fiber volume fraction increases the thermal conductivity; this improvement is significant specially when a preferred orientation is detected with a cluster-free state. Further reinforcement of the composite is made possible by increasing the maximum principal value of the orientation tensor provided that the principal direction is set accordingly. Furthermore, clustering index does not seems to be affected by volume fraction when an equal distribution is present in partitions.

1 Introduction

Heat dissipation can be problematic in composites due to their low thermal conductivity and anisotropic nature [19]. There are several parameters which affect the thermal conductivity of fiber reinforced composites among which is the orientation of the fibers; it plays a key role in defining the internal structure of the composite [4, 26]. Additionally, increasing the fiber volume fraction will initiate additional interactions between the fibers specially in the concentrated regions. The effect of

Z. Javanbakht (✉) · W. Hall
Griffith School of Engineering and Built Environment, Griffith University, Gold Coast, QLD 4214, Australia
e-mail: zia.javanbakht@gmail.com

W. Hall
e-mail: w.hall@griffith.edu.au

A. Öchsner
Faculty of Mechanical Engineering, Esslingen University of Applied Sciences, Kanalstrasse 33, 73728 Esslingen, Germany
e-mail: andreas.oechsner@gmail.com

© Springer International Publishing AG, part of Springer Nature 2019
A. Öchsner and H. Altenbach (eds.), *Engineering Design Applications*,
Advanced Structured Materials 92, https://doi.org/10.1007/978-3-319-79005-3_32

these two parameters couples together in non-dilute regions of, for instance, molding flows during injection molding processes, i.e., volume fractions more than 1% are more susceptible to fiber clustering [36]. Further complication is introduced when the effect of interfacial thermal barrier resistance is considered. Namely, not merely the volume fraction but also the specific area of the inclusions is also influential [10].

Due the aforementioned complications, among many others, evaluation of the thermal conductivity of various fibers of composites has been an attraction to researchers. Effective thermal conductivity of fibers in the transverse direction [3, 18, 27, 34], in-plane [4, 40], and longitudinal direction [24, 43] has been investigated in the literature. Furthermore, three dimensional pseudo-random distribution of fibers is considered in [8, 14, 15, 22, 41]. The anisotropic improvement of heat conduction is also investigated in several publications, see [9, 39]. Many of the analytical approaches tend to provide wide bounds, specially for high contrast heterogeneous media, rather than specific values [21]. In addition, high contrast fiber reinforced composites show more sensitivity to clustering [23]. Therefore in this context, computational methods are favored despite the fact that the results are more model-dependent. Such side effects should be smoothed by selecting appropriate representative volume elements (RVEs) and explicit consideration of fiber orientation.

The orientation of fibers in a composite can be described statistically in a continuous manner using orientation distribution functions [30]. The discretized alternative is using orientation tensors [1, 37] which are evaluated for various regions of the medium to detect orientation clustering of fibers [25, 26, 38, 42]. Many high order orientation tensors are available in the literature [6] but the second order orientation tensor is more frequent due to its simplicity. Although a second order orientation tensor is not deemed accurate enough to represent the micro-structure of a composite [29], it is adequate for calculations pertaining to thermal conductivity [36].

In the current study, a mean field approach is adopted to correlate the effect of clustered fibers to the effective thermal conductivity. Computational analyses were carried out using the finite element (FE) method. Most of the numerical researches have used auxiliary programming languages, such as Mathematica [8], to carry out the generation of pseudo-random fiber distribution. This requires a transfer of data between programs and some manual handling of the operations. Herein, an integrated Python script is used along with the MSC Marc commercial finite element package in a manner that the connection is automatically established and maintained. Furthermore, all supplementary calculations and post-processing tasks were handled by means of the same facility, for detailed information see [12]. A list of the used symbols in the current work is also provided for the convenience of the reader, see Table 1.

2 Eigenvalue Analysis of Spherical Distributions

The direction of every short fiber can be indicated by a unit vector ($\mathbf{p} = \mathbf{p}(\theta, \phi)$), see Fig. 1. The locus of the end points of such vectors is the surface of a unit sphere

Table 1 List of symbols

γ	Uniform mesh density
θ_{p}	Rotation of the principal direction transformation
ϕ_{f}	Volume fraction of the fiber
λ_i	i-th eigenvalue
μ	Mean value of the von Mises distribution
κ	Concentration parameter of the von Mises distribution
ψ	Probability distribution function
ΔT	Prescribed temperature gradient
Δz	Length over which temperature gradient is formed, i.e., the distance between the boundary edges
a	Element edge length
k_{eff}	Effective thermal conductivity of the composite
k_{f}	Thermal conductivity of the fiber
k_{m}	Thermal conductivity of the matrix
A_0	Cross-sectional area perpendicular to direction of the flux
$\dot{Q}$	Total heat flux magnitude
$\mathbf{e}_i$	Orthonormal unit basis
$\mathbf{e}_i^{\mathrm{p}}$	Orthonormal unit basis of the principal direction
$\mathbf{p}$	Fiber direction vector
$\mathbf{A}$	Second-order orientation tensor
$\mathbf{A}^*$	Second-order orientation tensor (spectral form)
$\mathbf{Q}$	Second-order transformation tensor

in 3D space—a spherical distribution. Considering fibers as the reinforcements of the composite makes their direction insignificant, i.e., both $\mathbf{p}$ and $-\mathbf{p}$ will have the same reinforcing effect. Therefore, the locus of the end points will reduce to a unit hemisphere in 3D or a unit half circle in 2D. Namely, in a 2D Euclidean coordinate system, one degree of freedom is enough to specify the orientation of a fiber, i.e., $\mathbf{p} = \mathbf{p}(\phi)$.

Using the summation convention, the representative planar unit vector of a fiber is denoted as:

$$\mathbf{p} = p_i \mathbf{e}_i, \quad i = 1, 2, \tag{1}$$

where $\mathbf{e_i}$ is the i-th basis unit vector of the coordinate system, and p_i is the i-th component of the unit vector $\mathbf{p}$.

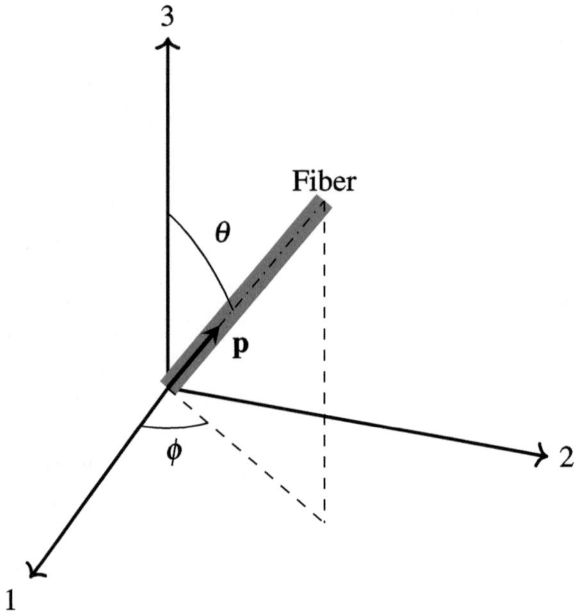

Fig. 1 Representation of the orientation of a general fiber in the 3D space using a unit vector **p**

The 2nd-order planar orientation tensor (**A**) can be defined as [1]:

$$A_{ij} = \int_0^{2\pi} \psi_\phi(\phi) p_i p_j d\phi, \tag{2}$$

where ψ_ϕ is the planar probability distribution function; p_i and p_j are the i-th and j-th components of the unit vector **p**, respectively. Equation (2) can be rewritten for a discrete case consisting of $\mathrm{N_f}$ fibers [36]:

$$A_{ij} \approx \frac{1}{N_\mathrm{f}} \sum_{k=1}^{N_\mathrm{f}} p_i^k p_j^k. \tag{3}$$

The average orientation tensor of the entire sample $(\overline{\overline{A_{ij}}})$ is obtained if N_f is the total number of fibers in the sample.

The orientation tensor is symmetric (A_{ij}=A_{ji}) and normalized to a unit length ($A_{ii} = 1$). Therefore, only two independent components exist for a planar orientation tensor, e.g., A_{11}, and A_{12}. Namely, A_{11} quantifies the proportion of alignment along the 1-axis and A_{12} indicates the deviation of the coordinate axes from the principal directions of the orientation tensor [16].

The spectral form (**A***) of the orientation tensor (**A**) can be acquired [5]:

$$\mathbf{A}^* = \lambda_i \mathbf{e}_i^\mathrm{p} \otimes \mathbf{e}_i^\mathrm{p}, \tag{4}$$

whereλ_i are the real eigenvalues and $\mathbf{e}_i^{\mathrm{p}}$ are the normed eigenvectors forming the basis of the orthonormal coordinate system. A unique rotation $\mathbf{Q}$ is used to transform the orientation tensor, by an angle equal to θ_{p}, to the principal directions to obtain:

$$\mathbf{A}^* = \begin{bmatrix} \lambda_1 & 0 \\ 0 & \lambda_2 \end{bmatrix}. \tag{5}$$

An ellipse can be used to graphically represent the principal directions and values, i.e., the axes of the ellipse are directed along the principal directions and scaled according to the principal values. The major axis points to the preferred direction of the majority of the local fibers. One extreme case is to have all fibers aligned in one direction which results in a stretched ellipse with a zero minor axis, i.e., a line along the major axis. The other extreme case is a circle which indicates no preferred orientation of the fibers [1, 2].

Analysis of the orientation of a total N_{f} fibers in a region can be done by partitioning the RVE into N_{p} partitions:

$$N_{\mathrm{f}} = \sum_{l=1}^{N_{\mathrm{p}}} N_{\mathrm{f}l}, \tag{6}$$

where $N_{\mathrm{f}l}$ is the total number of fibers in the l-th partition. The orientation state of the l-th partition ($\overline{A_{ij}^l}$) is:

$$\overline{A_{ij}^l} = \sum_{k=1}^{N_{\mathrm{f}l}} A_{ij}^{kl}, \tag{7}$$

and the orientation state of the whole sample is:

$$\overline{\overline{A_{ij}}} = \sum_{k=1}^{N_{\mathrm{f}l}} A_{ij}^{k}. \tag{8}$$

From the mixing-analogy method, the clustering index [CI] is defined as [36]:

$$[\mathrm{CI}]_{ij} = \frac{N_{\mathrm{p}} - 1}{N_{\mathrm{f}} - 1} \frac{[\sigma_{\mathrm{bp}}^2]_{ij}}{[\sigma_{\mathrm{t}}^2]_{ij}}, \tag{9}$$

where $[\sigma_{\mathrm{bp}}^2]_{ij}$ is the variance between the partitions, and $[\sigma_{\mathrm{t}}^2]_{ij}$ is the total variance. Finally, the type-1 clustering index can be expressed in terms of the orientation state as follows:

$$[\mathrm{CI}]_{ij} = 1 - \frac{\sum_{l=1}^{N_p} \sum_{k=1}^{N_{fl}} (A_{ij}^{kl} - \overline{A_{ij}^{l}})^2}{\sum_{l=1}^{N_p} \sum_{k=1}^{N_{fl}} (A_{ij}^{kl} - \overline{\overline{A_{ij}^{l}}})^2}, \tag{10}$$

where the nominator indicates the sum of the variances within each partition and the denominator is the variance of the whole RVE.

3 Methodology

In the current study, the FE method [13, 31–33] was used to estimate the effective thermal conductivity of randomly oriented short fibers with clustering. A Python script was used to create the parametric FE model, run the job submissions, and post-process the resulting data. This procedure was embedded in the MSC.Marc (version 2017.0) commercial FE package, and thus had the advantage of automatizing the whole process [12, 14, 15].

Dimensionless values were used throughout the investigation, and thus normalized results will be discussed. All the FE models were generated based on a 2D prototype RVE (20×20) which consisted of 4-node bilinear isoparametric planar heat transfer elements (type 39) for the matrix phase and straight 2-node link elements (type 36) for the fibers. The contrast of the properties was chosen to be 600, i.e., thermal conductivities of 1, and 600 for the matrix and fibers, respectively. An aspect ratio of 40 was acquired for the fibers, i.e., a fiber length of 5 and a diameter of 0.125, to stay within the range of 20–60 aspect ratios of the short fibers [20].

Considering the effectiveness of boundary conditions [17], the recommended prescribed temperature boundary condition was used [11]. A temperature gradient of 100 was imposed to set up a steady-state flux across the RVE, see Fig. 2. After steady state heat transfer analyses, the reaction flux was collected from the right-hand boundary nodes and summed up to obtain the total heat flux [35]. Then, the effective thermal conductivity of the RVE was calculated using Fourier's law [7]:

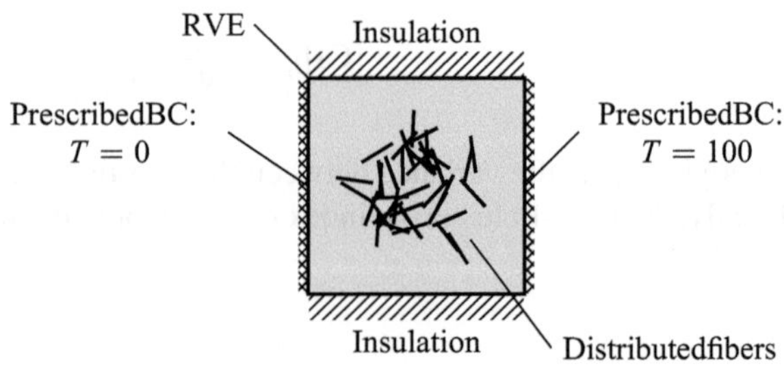

Fig. 2 Schematic illustration of the prototype RVE

$$k_{\text{eff}} = \frac{\dot{Q}}{A_0} \cdot \frac{\Delta z}{\Delta T}, \quad (11)$$

where k_{eff} is the effective thermal conductivity of the RVE, $\dot{Q}$ is the sum of the generated reaction fluxes on the right-hand boundary, A_0 is the assumed cross-sectional area (with a thickness of unit length) perpendicular to direction of the flux, Δz is the distance between the edges of the sample, and ΔT is the prescribed temperature gradient.

Note that mesh sensitivity analyses were carried out to remove the mesh dependency of the values of effective thermal conductivity. It was carried over a range of uniform mesh densities (γ) from 1 to 35 [18]:

$$\gamma = \frac{1}{a}, \quad (12)$$

where a was the edge length of the used square elements. The results showed that the mesh sensitivity was diminished for γ over 20 and thus, it was selected as the prototype mesh density for all the simulations.

The following assumptions were made for creating all the models:

- The thermal conductivity is independent of the temperature.
- The transverse conductivity of the fibers is negligible.
- A perfect bonding exists between the fibers and matrix, i.e., the thermal barrier resistance is disregarded.

The RVE was divided into four equal partitions which incorporated fibers with equal volume fractions. The location of the fibers was selected randomly and their orientation followed a von Mises distribution [28]. The mean value of the distribution was kept to $\mu = 0$ so the fibers were oriented along the direction of the heat flow, see Fig. 3. However, in order to investigate the effect of orientation distribution, the concentration parameter (κ) was modified from 0 to 50 in the two leftmost partitions to consider the random and highly oriented cases, respectively.

Finally, a statistical analysis was carried out for each sample by dividing the RVE to four equal partitions. The orientation tensors of the fibers were obtained along with their principal values and directions in each partition and the whole RVE. Finally, the clustering index was calculated for the RVE and correlated with the values of the effective thermal conductivity.

4 Results and Discussion

The main focus of the current study was on randomly-oriented fibers, and thus the selection of a pseudo-random algorithm will highly affect the result of the study. Therefore, a von Mises distribution was considered for the orientations to obtain a more controllable approach. In order to challenge the validity of the clustering

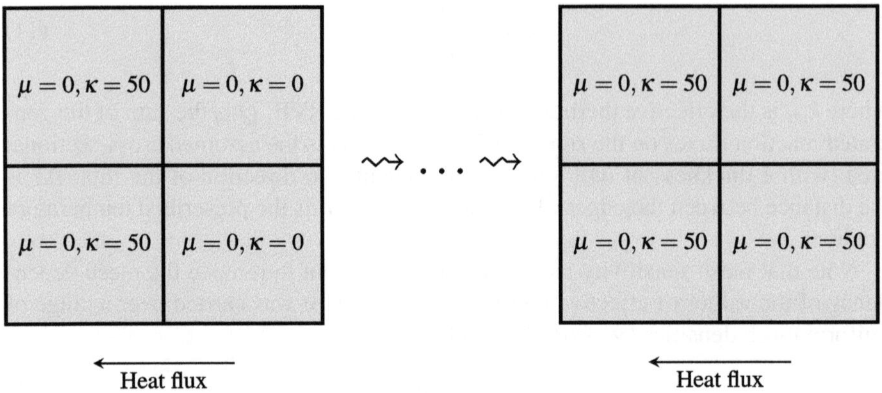

Fig. 3 Two extreme cases of fiber distribution in the RVE—the fiber volume fraction is kept the same in each partition

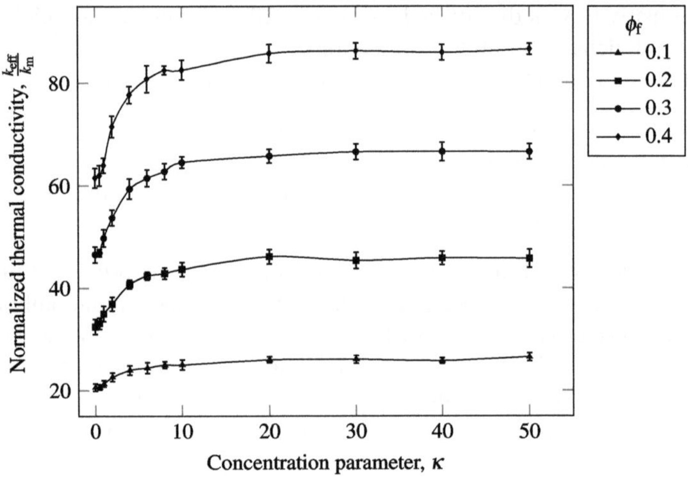

Fig. 4 Normalized effective thermal conductivity for different concentration parameters and volume fractions

index, two rightmost partitions undergone a transition from a random orientation state ($\kappa = 0$) to a more concentrated state ($\kappa = 50$). This parameter change was applied to volume fractions above the non-dilute limit, i.e., volume fractions from 10–40%.

The effective thermal conductivity was also calculated in each case and normalized by means of the thermal conductivity of the matrix.

Increasing the fiber volume fraction increases the effective thermal conductivity of the composite, see Fig. 4. Increasing the concentration parameter of the von Mises distribution locates more fibers about the heat flow direction, and thus the thermal conductivity increases. Orienting fibers has a strong effect on improving the thermal

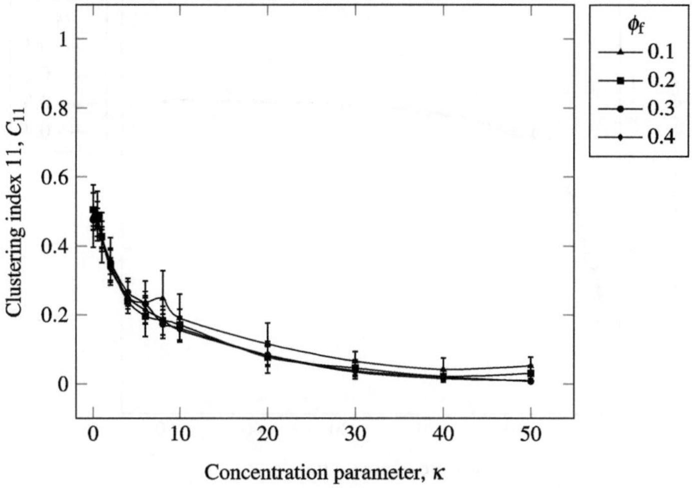

Fig. 5 Clustering index 11 for different concentrations and volume fractions

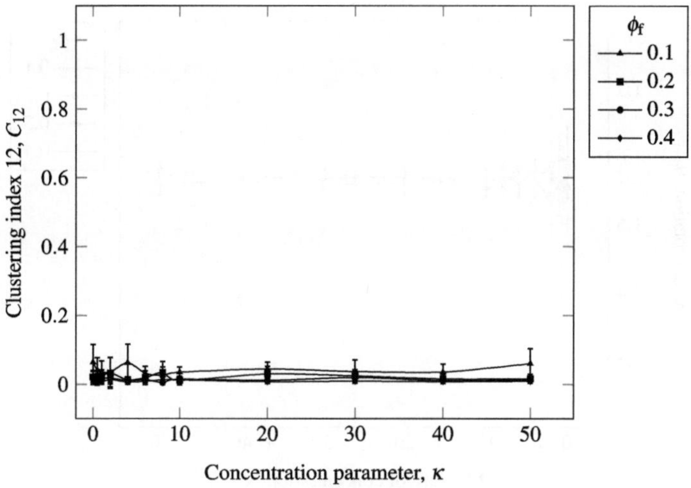

Fig. 6 Clustering index 12 for different concentrations and volume fractions

conductivity since a higher conductivity is acquired in an oriented state compared to a random distribution with a higher volume fraction of fiber reinforcements. Nevertheless, the effect of orientation concentration seems to diminish for concentration parameters more than $\kappa = 20$. This is accompanied by a decrease in the standard deviation which is highly noted in lower fiber volume fractions.

Detection of clustering was investigated using the clustering index obtained from the mixing analogy [36]. In the case of clustering, it is assumed that a preferred orientation exists along which fibers locally co-orient themselves. The clustering index

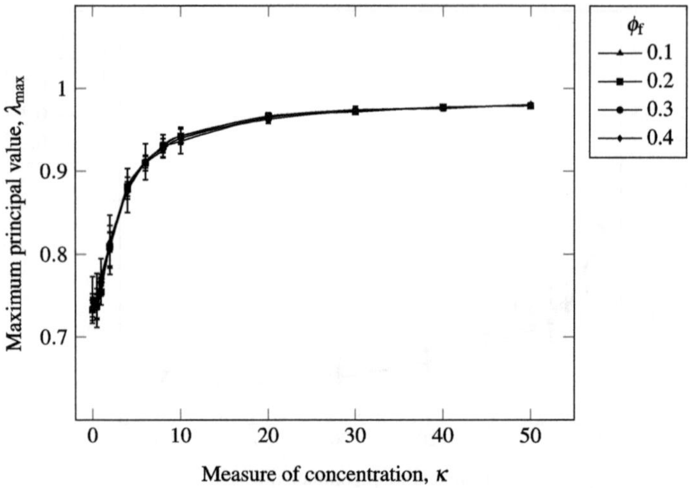

Fig. 7 Maximum principal value of the orientation tensor for different concentrations and volume fractions

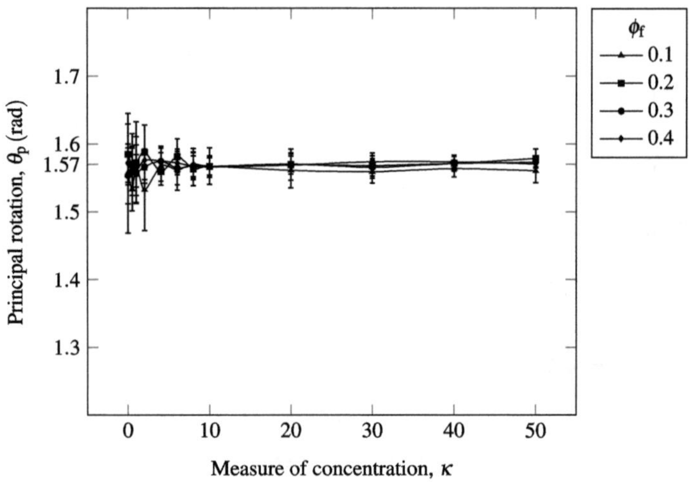

Fig. 8 Rotation to principal direction with respect to the 1-direction

varies between 0 and 1 where the former indicates a cluster-free state and the latter indicates a clustered state. In the orientation tensor, the A_{11} component indicates the strength of fiber alignment along the direction of the heat flux whereas the A_{12} component quantifies the dissimilarity in the principal directions of orientation in the partitions of the RVE. In Figs. 5 and 6, the clustering index for the components 11 and 12 of the orientation tensor is illustrated, respectively. A clustered state is detected for A_{11} component in the low values of the concentration parameter whereas no clustering is found for the A_{12} component. In addition, a variation is observed for the low concentration parameter region, i.e., κ less than 20. By increasing the concentration parameter, the clustering state for A_{11} weakens and becomes negligible for concentration parameters more than 30. Note that no interesting change is observed for the other index, i.e., it indicates a more or less similar principal direction in all cases, see Fig. 8. More importantly, no remarkable change is detected by increasing the volume fraction. Namely, the orientation tensor is not affected in the case of equal distribution of fiber volume. The maximum principal value of the orientation tensor approaches 1 by increasing the concentration parameter, see Fig. 7. Namely, orientation of the fibers is enforced along the flow direction by increasing the greatest eigenvalue provided that the principal direction remains the same.

5 Conclusion

It can be concluded that increasing the fiber volume fraction increases the effective thermal conductivity of the composite. This is highly affected by the orientation state of the fibers, i.e., a clustered fiber distribution may inhibit the heat flux. Moreover, orienting the fibers along the heat flow facilitates heat transfer—provided that clustering is minimum. This is true specially when the principal direction of the fibers remains along the heat flow and the clustering index does not change along this direction. Finally, by increasing the maximum principal value of the orientation tensor, alignment of the fibers can be further enforced along the flow direction which results in an efficient use of fibers for improving thermal conductivity. This approach can be used to characterize the composite and determine its degree of efficiency in terms of fiber engagement both in its thermal and mechanical aspects.

References

1. Advani, S.G., Tucker, C.L.: The use of tensors to describe and predict fiber orientation in short fiber composites. J. Rheol. **31**(8), 751–784 (1987)
2. Advani, S.G., Tucker, C.L.: Closure approximations for three-dimensional structure tensors. J. Rheol. **34**(3), 367–386 (1990)
3. Agrawal, A., Satapathy, A.: Mathematical model for evaluating effective thermal conductivity of polymer composites with hybrid fillers. Int. J. Therm. Sci. **89**, 203–209 (2015)
4. Behzad, T., Sain, M.: Measurement and prediction of thermal conductivity for hemp fiber reinforced composites. Polym. Eng. Sci. **47**(7), 977–983 (2007)
5. Bertram, A., Glüge, R.: Solid mechanics. Springer International Publishing, Cham (2015)
6. Du Chung, H., Kwon, T.H.: Invariant-based optimal fitting closure approximation for the numerical prediction of flow-induced fiber orientation. J. Rheol. **46**(1), 169–194 (2002)
7. Fiedler, T., Hosseini, S., Belova, I.V., Murch, G.E., Öchsner, A.: A refined finite element analysis on the thermal conductivity of perforated hollow sphere structures. Comput. Mater. Sci. **47**(2), 314–319 (2009)
8. Gomez-Munoz, J.L., Bravo-Castillero, J.: Calculation of effective conductivity of 2D and 3D composite materials with anisotropic constituents and different inclusion shapes in mathematica. Comput. Phys. Commun. **179**(4), 275–287 (2008)

9. Gori, F., Corasaniti, S., Ciparisse, J.F.: Theoretical prediction of the anisotropic effective thermal conductivity of composite materials. In: Proceedings of the ASME international mechanical engineering congress and exposition—2012, p. 91. ASME, New York (2013)
10. Hasselman, D., Johnson, L.F.: Effective thermal conductivity of composites with interfacial thermal barrier resistance. J. Compos. Mater. **21**(6), 508–515 (1987)
11. Islam, M.R., Pramila, A.: Thermal conductivity of fiber reinforced composites by the fem. J. Compos. Mater. **33**(18), 1699–1715 (1999)
12. Javanbakht, Z., Öchsner, A.: Advanced finite element simulation with MSC Marc: Application of user subroutines. Springer, Cham (2017)
13. Javanbakht, Z., Öchsner, A.: Computational statics revision course. Springer, Cham (2017)
14. Javanbakht, Z., Hall, W., Öchsner, A.: Automatized estimation of the effective thermal conductivity of carbon fiber reinforced composite materials. Defect Diffus. Forum **370**, 177–183 (2016)
15. Javanbakht, Z., Hall, W., Öchsner, A.: Finite element evaluation of effective thermal conductivity of short carbon nano tubes: a comparative study. Defect Diffus. Forum **372**, 208–214 (2016)
16. Javanbakht, Z., Hall, W., Öchsner, A.: The effect of partitioning on the clustering index of randomly-oriented fiber composites: a parametric study. Defect Diffus. Forum **380**, 232–241 (2017)
17. Javanbakht, Z., Hall, W., Öchsner, A.: The effect of substrate bonding on characterization of thin elastic layers: a finite element study. Materialwiss Werkst **48**(5), 456–462 (2017)
18. Javanbakht, Z., Hall, W., Öchsner, A.: Computational evaluation of transverse thermal conductivity of natural fiber composites. In: Improved Performance of Materials, pp. 197–206. Springer, Cham (2018)
19. Kalaprasad, G., Pradeep, P., Mathew, G., Pavithran, C., Thomas, S.: Thermal conductivity and thermal diffusivity analyses of low-density polyethylene composites reinforced with sisal, glass and intimately mixed sisal/glass fibres. Key Eng. Mater. **60**(16), 2967–2977 (2000)
20. Kalpakjian, S., Schmid, S.R.: Manufacturing engineering and technology, 6th edn. Prentice Hall, New York (2010)
21. Kanit, T., Forest, S., Galliet, I., Mounoury, Y., Jeulin, D.: Determination of the size of the representative volume element for random composites: statistical and numerical approach. Int. J. Solids Struct. **40**(13–14), 3647–3679 (2003)
22. Kari, S., Berger, H., Gabbert, U.: Numerical evaluation of effective material properties of randomly distributed short cylindrical fibre composites. Comput. Mater. Sci. **39**(1), 198–204 (2007)
23. Kataoka, Y., Taya, M.: Analysis of mechanical behavior of a short fiber composite using micromechanics based model: effects of fiber clustering on composite stiffness and crack initiation. JSME Int. J. A-Solid M. **43**(1), 46–52 (2000)
24. Korab, J., Stefanik, P., Kavecky, S., Sebo, P., Korb, G.: Thermal conductivity of unidirectional copper matrix carbon fibre composites. Compos. Part A-Appl. S. **33**(4), 577–581 (2002)
25. Krause, M., Hausherr, J.M., Burgeth, B., Herrmann, C., Krenkel, W.: Determination of the fibre orientation in composites using the structure tensor and local x-ray transform. J. Mater. Sci. **45**(4), 888–896 (2010)
26. Lee, K.S., Lee, S.W., Chung, K., Kang, T.J., Youn, J.R.: Measurement and numerical simulation of three-dimensional fiber orientation states in injection- molded short-fiber-reinforced plastics. J. Appl. Polym. Sci. **88**(2), 500–509 (2003)
27. Liu, K., Takagi, H., Osugi, R., Yang, Z.: Effect of lumen size on the effective transverse thermal conductivity of unidirectional natural fiber composites. Key Eng. Mater. **72**(5), 633–639 (2012)
28. Mardia, K.V., Zemroch, P.J.: Algorithm as 86: the von Mises distribution function. Appl Stat-J Roy St C **24**(2), 268 (1975)
29. Muller, V., Bohlke, T.: Prediction of effective elastic properties of fiber reinforced composites using fiber orientation tensors. Key Eng. Mater. **130**, 36–45 (2016)
30. Nomura, S., Kawai, H., Kimura, I., Kagiyama, M.: General description of orientation factors in terms of expansion of orientation distribution function in a series of spherical harmonics. J. Polym. Sci. A-2 Polym. Phys. **8**(3), 383–400 (1970)

31. Öchsner, A.: Computational statics and dynamics: an introduction based on the finite element method. Springer, Singapore (2016)
32. Öchsner, A.: A project-based introduction to computational statics. Springer International Publishing, Cham (2018)
33. Öchsner, A., Merkel, M.: One-dimensional finite elements: an introduction to the FE method. Springer, Berlin (2013)
34. Qian, L., Pang, X., Zhou, J., Yang, J., Lin, S., Hui, D.: Theoretical model and finite element simulation on the effective thermal conductivity of particulate composite materials. Compos. Part B-Eng. **116**, 291–297 (2016)
35. Rakovsky, S.K.: Analytical tools and industrial applications for chemical processes and polymeric materials. Apple Academic Press, Point Pleasant (2014)
36. Ranganathan, S., Advani, S.G.: Characterization of orientation clustering in short-fiber composites. J. Polym. Sci. B Polym. Phys. **28**(13), 2651–2672 (1990)
37. Scheidegger, A.E.: On the statistics of the orientation of bedding planes, grain axes, and similar sedimentological data. In: Thomas Nolan, B. (ed) Geological Survey Professional Paper 525, Geological Survey Research, vol. 525, pp. 164–167, United States Government Printing Office, Washington, D.C. (1965)
38. Sliseris, J., Andra, H., Kabel, M., Wirjadi, O., Dix, B., Plinke, B.: Estimation of fiber orientation and fiber bundles of mdf. Mater. Struct. **49**(10), 4003–4012 (2016)
39. Tian, T., Cole, K.D.: Anisotropic thermal conductivity measurement of carbon-fiber/epoxy composite materials. Int J Heat Mass Tran **55**(23–24), 6530–6537 (2012)
40. Villiere, M., Lecointe, D., Sobotka, V., Boyard, N., Delaunay, D.: Experimental determination and modeling of thermal conductivity tensor of carbon/epoxy composite. Compos. Part A-Appl. S **46**, 60–68 (2013)
41. Wang, M., Kang, Q., Pan, N.: Thermal conductivity enhancement of carbon fiber composites. Appl. Therm. Eng. **29**(2–3), 418–421 (2009)
42. Yang, C., Huang, H.X., Li, K.: Investigation of fiber orientation states in injection-compression molded short-fiber-reinforced thermoplastics. Polym. Compos. **31**(11), 1899–1908 (2010)
43. Yang, R., Chen, G., Dresselhaus, M.S.: Thermal conductivity of simple and tubular nanowire composites in the longitudinal direction. Phys. Rev. B **72**(12), 111 (2005)

MIX
Papier aus verantwortungsvollen Quellen
Paper from responsible sources
FSC® C105338
www.fsc.org

If you have any concerns about our products, you can contact us on
ProductSafety@springernature.com

In case Publisher is established outside the EU, the EU authorized representative is:
Springer Nature Customer Service Center GmbH
Europaplatz 3, 69115 Heidelberg, Germany

Printed by Libri Plureos GmbH
in Hamburg, Germany